Annual Reports

on the Progress of Chemistry

Volume 71, 1974

SECTION A

Physical and
Inorganic Chemistry

The Chemical Society

Burlington House, London W1V 0BN

ISBN: 0 85186 070 2
ISSN: 0069–3022

Contributors

R. P. Bell, M.A., F R.S., *University of Stirling*
B. Cleaver, B.A., Ph.D., D.I.C., *University of Southampton*
R. J. Cross, B.Sc., Ph.D., *University of Glasgow*
G. W. Fraser, M.Sc., Ph.D., *University of Strathclyde*
G. V. Jagannathan, M.Sc., *University of St. Andrews*
A. Maccoll, B.Sc., Ph.D., F.R.I.C., *University College London*
S. M. Nelson, M.Sc., Ph.D., D.Phil., *Queen's University of Belfast*
D. W. A. Sharp, M.A., Ph.D., A.R.I.C., F.R.S.E., *University of Glasgow*
W. V. Steele, B.Sc., Ph.D., *University of Stirling*
B. Stevens, M.A., D.Phil., *University of South Florida, U.S.A.*
C. Thomson, B.Sc., Ph.D., *University of St. Andrews*
I. C. Walker, B.Sc., Ph.D., *University of Stirling*
M. G. H. Wallbridge, B.Sc., Ph.D., *University of Warwick*
G. R. Woolley, B.Sc., Ph.D., *University of St. Andrews*
P. A. H. Wyatt, B.Sc., Ph.D., F.R.I.C., F.R.S.E., *University of St. Andrews*

Set in Times on Monophoto Filmsetter and printed offset by
J. W. Arrowsmith Ltd., Bristol, England

Made in Great Britain

Preface

The measures taken as a result of the recommendations arising from the Meeting of Senior Reporters during 1974 intended to reduce the length of each Section (mentioned in the Preface to the 1973 Reports) have been markedly successful, *viz.* Section A: 1973, 568 pp; 1974, 386 pp

    Section B: 1973, 760 pp; 1974, 600 pp.

However, the conclusions of a Working Party (Convenor, Professor M. R. Truter) formed under the auspices of the Tertiary Publications Committee to consider future policy, were that a further decrease in size was necessary for *Annual Reports* to remain a viable publication. Restrictions should be placed on the number of references to be cited, which would cause Reporters to evaluate the *key* papers in their field. It was further recommended that rather than attempting to cover *all* areas of organic or inorganic chemistry in each year's Reports, certain topics of current interest should be selected. A critical over-view of the recent progress (*e.g.* over the past 2—3 years) which reviewed no more than *ca.* 100 key references should be given, in line with the policy already adopted for the Physical Chemistry Section.

If implemented, these recommendations would mark a radical change in the content of *Annual Reports*, and it was agreed that it would be advisable to make a preliminary survey of the needs of readers; it was hoped to carry this out during 1975.

For Section A, Professors R. P. Bell, F.R.S., and D. W. A. Sharp, F.R.S.E., have once again undertaken the duties of Senior Reporter for the Physical Chemistry and Inorganic Chemistry parts respectively. Professor Sharp has now completed his term of office and the Society wishes to express its especial appreciation of his efforts for the past *five* years, an unprecedented spell of duty in modern times. We welcome as new Senior Reporters for Section B, Dr. M. F. Ansell and Professor P. G. Sammes, and thank them for their success in persuading Reporters to adhere closely to their commissioned chapter lengths, which should result in a considerable easing of the economic picture for these 1974 Reports.

Contents

vii

PART I
PHYSICAL CHEMISTRY

1 Introduction

By R. P. BELL
Department of Chemistry, University of Stirling, Stirling FK9 4LA

Part I continues the practice of recent years in presenting a number of reports, not necessarily confined to the most recent literature, which deal with topics of particular interest to the Reporter, but which attempt to give a critical assessment which will be useful to the general reader. As an experiment authors have been asked to limit themselves to about 100 key references in each article: apart from the economy in space and expense, it is hoped that this will render the text less intimidating to the non-specialist. In this respect an exception has been made in the article on *ab initio* quantum calculations (Chapter 2), since it is thought that the average reader will be more interested in the variety of species for which meaningful calculations can be made than in the detailed methods employed.

This year's reports place a good deal of emphasis on simple species, many of them quite unfamiliar to the chemist, on excited states of molecules, and on the application of physical techniques which have not yet come into general use. The success of *ab initio* quantum calculations in predicting the properties of simple molecules (Chapter 2) gives confidence in their use for predicting the energy and geometry of unstable or unknown species. For reactions involving three atoms appreciable progress has been made in the much more difficult task of computing the energy surfaces which form the basis of the theoretical treatment of collision and scattering phenomena, reviewed in last year's *Annual Reports*. Chapter 4 deals with a particular type of collision process, between simple molecules and slow electrons: although this is a difficult experimental technique, improvements in resolving power have yielded much information about energy states which are not revealed by optical spectroscopy, and about short-lived negative ions formed by electron attachment. The behaviour of more complex molecules when excited by radiation, was reviewed last year under the title *Luminescence Spectroscopy*. This year Chapter 3 deals with a particular aspect of this field, the formation of molecular complexes by excited species, which is more widespread than was at one time supposed.

In the 1973 *Annual Reports* five chapters in Section A described various aspects of chemical kinetics. This year considerably less space is devoted to this field in view of the *Specialist Periodical Reports* on Reaction Kinetics, the first volume of which appeared in early 1975. The kinetic effects of heavy-atom isotopic substitution are reviewed in Chapter 5, and the effects of high pressures on reaction rates in Chapter 8. Advances in experimental methods have rendered both of these techniques more readily accessible, and they could probably be applied more widely to studies of mechanism than is the case at present. In particular, volumes of activation are conceptually more directly related to model mechanisms than are other quantities commonly used for this purpose, such as entropies of activation, isotope and solvent effects, or Hammett parameters.

Chapter 8 also describes the effects of high pressures on equilibrium phenomena, and the two preceding chapters are concerned with other thermodynamic topics. Chapter 6 reviews the present state of knowledge about the thermochemistry of organometallic compounds, especially simple metallic alkyls: it is interesting to note that the crucial factor in obtaining reliable enthalpy values often lies in the difficulty of defining exactly the chemical and physical nature of the initial and final states, which are often critically dependent on exact calorimetric conditions. Chapter 7 reports on the thermodynamics of a wide variety of simple inorganic reactions at high temperatures. These often involve vapour phases of variable composition, and the use of mass spectrometry has done much to resolve the ambiguities inherent in earlier work. This chapter provides an excellent example of the way in which studies initiated because of their practical importance in fields such as metallurgy and ceramics can yield results of prime theoretical interest.

2 Ab initio Calculations on Small Molecules, and Potential Energy Surfaces

By C. THOMSON

Department of Chemistry, University of St. Andrews, St. Andrews, KY16 9ST

Glossary of Abbreviations

CGTO	contracted gaussian-type orbitals
DZ (+ P)	double zeta (+ polarization)
FSGO	floating spherical gaussian orbitals
GTO	gaussian type orbitals
GVB	generalized valence bond
H–F	Hartree–Fock
HFPD	Hartree–Fock with proper dissociation
ICSCF	internally consistent self-consistent field
IEPA	independent electron pair approximation
MCSCF	multi-configuration self-consistent field
OVC	optimized valence configuration
RHF	restricted Hartree–Fock
SCF	self-consistent field
S(T)O	Slater (type) orbital
UHF	unrestricted Hartree–Fock

1 Introduction

Recent developments in quantum chemistry with particular emphasis on the theoretical methods which are currently in use were reported by Duke in 1971,[1] but there has not been a detailed report of recent work on small molecules for several years, although some work was reported in 1970.[2] The present review is restricted to a discussion of *ab initio* calculations on small molecules, and the *ab initio* investigation of potential energy (P.E.) surfaces involving such species. It should be emphasized that this review deals primarily with investigations which seem to the author to be particularly interesting to the non-specialist in theoretical chemistry. A more detailed review of the recent work in theoretical chemistry is to be found in two recent Specialist Periodical Reports.[3,4]

[1] B. J. Duke, *Ann. Reports* (*A*) 1971, **68**, 3.
[2] E. Steiner, *Ann. Reports* (*A*), 1970, **67**, 5.
[3] 'Theoretical Chemistry', ed. R. N. Dixon, (Specialist Periodical Reports), The Chemical Society, London, 1974, Vol. 1.
[4] 'Theoretical Chemistry', ed. R. N. Dixon and C. Thomson, (Specialist Periodical Reports), The Chemical Society, London, 1975, Vol. 2.

Results of calculations are quoted in atomic units, which are used by the majority of theoretical chemists and also in the Specialist Periodical Reports.[3,4] In these units, energy values (E) are in hartree (1 hartree = 27.210 eV = 4.36 × 10^{18} J), and distances are in bohr (1 bohr = 0.529 Å = 5.2918 × 10^{-11} m).

A number of recent books and review articles have dealt with various aspects of ab initio calculations and the books by Schaefer[5] and Cook[6] provide excellent introductions for the non-specialist. A comprehensive bibliography by Richards et al.[7] has recently been updated. Reviews of recent work on diatomic species are provided by Wahl[8] and Goodisman.[9] Browne and Matsen[10] have reviewed recent work on three- and four-electron molecules, and the proceedings of two recent conferences which have appeared in book form[11,12] deal with many different aspects of ab initio work. Other general reviews are cited in Ref. 4.

Section 2 deals with ab initio investigations of small molecules containing up to four atoms, and Section 3 with ab initio investigations of P.E. surfaces. Calculations on large organic molecules were reviewed last year by Clark[13] and are also reviewed in ref. 4 by Duke.

The ab initio calculations discussed here involve the solution of the non-relativistic Schrödinger equation, assuming the Born–Oppenheimer approximation. Calculations on a few molecules removing this approximation have been reviewed elsewhere.[4] The various methods used to obtain approximate solutions to Schrödinger's equation have been discussed by Duke[1] and no radically new methods have appeared since that report. The total energy, the orbital energy, and the wavefunction and properties computed from it are the main molecular quantities discussed here.

2 Ab initio Calculations on Diatomic Molecules

Investigations on diatomic species AH, A_2, and AB, have involved two main areas, firstly, the testing of new computational methods, and secondly, the extension of calculations on several familiar molecules to their excited states.

H_2.—There have been several investigations on H_2, and one of particular interest is the direct calculation of the Brueckner orbitals.[14] A single-determinant wavefunction constructed from these orbitals has maximum overlap with the exact

5   H. F. Schaefer, tert., 'Electronic Structure of Atoms and Molecules', Addison Wesley, Boston, 1972.
6   D. B. Cook, 'Ab-Initio Valence Calculations in Chemistry', Wiley, London, 1974.
7   W. G. Richards, T. E. H. Walker, and R. Hinkley, 'Bibliography of ab-initio molecular wave functions', O.U.P., 1971, Supplement 1974.
8   A. C. Wahl, in 'Theoretical Chemistry', ed. W. Byers Brown, MTP International Review of Science, Physical Chemistry, Series 1, 1972, Vol. 1.
9   J. Goodisman, 'Diatomic Interaction Potential Theory, Vol. 1', Academic Press, New York, 1972.
10 J. C. Browne and F. A. Matsen, Adv. Chem. Phys., 1973, 23, 161.
11 'Energy, Structure and Reactivity', ed. D. W. Smith and W. B. McRae, John Wiley, New York, 1973.
12 'Computational methods for large molecules and localized states in solids', ed. F. Herman, A. D. McLean, and R. K. Nesbet, Plenum Press, New York, 1973.
13 D. T. Clark, Ann. Reports (B), 1972, 69, 40.
14 V. Staemmler and M. Jungen, Theor. Chim. Acta, 1972, 24, 152.

wavefunctions. Such a function is particularly suitable for the study of correlation effects in molecules. For the H_2 ground state, this wavefunction gave more reliable values for one-electron properties than the SCF wavefunction, and also gave reliable P.E. curves.

Recent work on H_2 using Boys' transcorrelated wavefunction method[15,16] has been reported by Handy.[17] In this method, a single-configuration Slater determinant Φ is multiplied by a specific correlation factor C of the form

$$C = \prod_{i>j}^{n} f(\mathbf{r}_i, \mathbf{r}_j) \tag{1}$$

for n electrons, and since f can depend on r_{ij}, Boys showed that, using the operator $C^{-1}HC$, very accurate wavefunctions can be obtained without the n^4 integral problem which occurs in conventional methods. The early calculations on He, Be, Ne, and LiH[15,16] gave the most accurate wavefunctions to date for these species, but the integrals were evaluated numerically. Handy[17] has recently shown how these can be evaluated analytically in terms of a gaussian basis set. Even with a small basis set, the correlation energy (E_c) obtained was impressive; for LiH 92% of E_c was obtained; for H_2O 70%. The extension of this method to larger molecules may not be easy, but if possible it promises reliable values of E_c with less computational effort than, for example, conventional configuration interaction (CI) calculations. One problem with this method is the difficulty, compared with conventional variational methods, of evaluating expectation values, a point emphasized by Armour.[18,19]

Although nuclear spin–spin coupling constants J_{AB} are readily obtained from n.m.r. spectra, their evaluation theoretically has proved difficult because of the necessity of including the excited states which are involved in the equation for J_{AB}. A recent calculation[20] of J_{HD} in HD has shown that if the zeroth order wavefunction is correlated (*via* a large CI calculation) and all singly- or doubly-excited triplets included in the first-order correction, the computed value of J_{HD} of 43.48 Hz is in good agreement with the experimental value of 42.94 $\pm$ 0.1 Hz. It should be possible to extend the method to larger systems and further applications should be most interesting. A variety of excited states of H_2 have been studied using various methods and these are discussed in ref. 4. Among these,[21] the use of the generalized valence bond (GVB) method by Goddard and co-workers (which has been reviewed recently[22]), is of particular interest. We return to this method later. The thermodynamic properties of H_2 have been computed by Kosloff *et al.* from *ab initio* wavefunctions.[23]

[15] S. F. Boys and N. C. Handy, *Proc. Roy. Soc.*, 1969, **A309**, 209; **A310**, 43, 63.
[16] S. F. Boys and N. C. Handy, *Proc. Roy. Soc.*, 1969, **A311**, 309.
[17] N. C. Handy, *Mol. Phys.*, 1972, **23**, 1.
[18] E. A. G. Armour, *Mol. Phys.*, 1973, **25**, 993.
[19] E. A. G. Armour, *Mol. Phys.*, 1973, **26**, 1093.
[20] J. Kowalewski, B. Roos, P. Siegbahn, and R. Vestin, *Chem. Phys.*, 1974, **3**, 70.
[21] D. L. Huestis and W. A. Goddard, tert., *Chem. Phys. Letters*, 1972, **16**, 157.
[22] W. A. Goddard, tert., T. H. Dunning, jun., W. J. Hunt, and P. J. Hay, *Accounts Chem. Res.*, 1973, **6**, 383.
[23] R. Kosloff, R. D. Levine, and R. B. Bernstein, *Mol. Phys.*, 1974, **27**, 981.

Diatomic Hydrides.—Recent calculations have focused attention on various excited states with the wavefunctions being evaluated in extensive CI calculations. Docken and Hinze[24,25] in two important papers have examined the P.E. curves for five valence excited states of LiH by the MCSCF method. Using a large STO basis set, the computed dissociation energy of the ground state was only 0.003 hartree less than experiment. Computed P.E. curves and one-electron properties are in reasonable agreement with experimental data where these are available.

LiH has also been extensively studied by the GVB method developed by Goddard and co-workers.[22] This method employs a single-determinant wavefunction with singly occupied orbitals, *i.e.* different spatial orbitals for different spins, and can correctly describe molecular dissociation. A number of variants have been described[26] which differ primarily in the spin-coupling schemes. Very interesting discussions of the bonding in various small molecules using this method have been described and reviewed,[22,26] and applications to both ground and excited states of LiH reported. Its relation to other methods has been discussed.[26d] The original VB method has also been applied recently to several first-row hydrides by Gallup and co-workers.[27] The IEPA method is based on Sinanŏglu's many-electron theory[28] and has been used with some success recently to study electron correlation in a variety of first-row hydrides, including LiH.[29,30]

Calculations beyond H–F on LiH and other hydrides are neither easy nor inexpensive and are thus not readily applicable to larger molecules. Therefore attempts to obtain the correlation energy semi-empirically are of considerable interest, and two recent papers by Lie and Clementi are of particular importance.[31,32] The authors compute E_c in terms of a functional of the Hartree–Fock density

$$E_c = \int 0.02096(1.2 + \rho_m^{\frac{1}{3}})^{-1}\rho_m^{\frac{4}{3}}\,dv + \int 0.02096 \ln(1 + 2.39\rho_m^{\frac{1}{3}})\rho_m\,dv \qquad (2)$$

where

$$\rho_m = \sum_i \bar{n}_i\rho; \qquad \bar{n}_i = n_i \exp[-0.5(2 - n_i)^2] \qquad (3)$$

n_i is the orbital occupation number. This form ensures that the computed and experimental atomic correlation energies agree for first-row atoms and is

[24] K. K. Docken and J. Hinze, *J. Chem. Phys.*, 1972, **57**, 4928.
[25] K. K. Docken and J. Hinze, *J. Chem. Phys.*, 1972, **57**, 4936.
[26] (a) W. A. Goddard, tert., *Phys. Rev.*, 1967, **157**, 81; (b) W. E. Palke and W. A. Goddard, tert., *J. Chem. Phys.*, 1969, **50**, 4524; (c) W. A. Goddard, tert. and R. C. Ladner, *J. Amer. Chem. Soc.*, 1971, **93**, 6750; (d) W. J. Hunt, P. J. Hay, and W. A. Goddard, tert., *J. Chem. Phys.*, 1972, **57**, 738; (e) R. C. Ladner and W.A. Goddard, tert., *J. Chem. Phys.*, 1969, **51**, 1073.
[27] G. A. Gallup, *Internat. J. Quantum Chem.*, 1972, **6**, 899; G. A. Gallup, *Adv. Quantum Chem.*, 1973, **7**, 113; J. M. Norbeck and G. A. Gallup, *Internat. J. Quantum Chem.*, 1973, 7S, 161.
[28] O. Sinanŏglu, *Adv. Chem. Phys.*, 1964, **6**, 315; 1969, **14**, 237.
[29] R. Ahlrichs and W. Kutzelnigg, *J. Chem. Phys.*, 1968, **48**, 1819.
[30] M. Jungen and R. Ahlrichs, *Theor. Chim. Acta*, 1970, **17**, 339.
[31] G. C. Lie and E. Clementi, *J. Chem. Phys.*, 1974, **60**, 1275.
[32] G. C. Lie and E. Clementi, *J. Chem. Phys.*, 1974, **60**, 1288.

subsequently used to calculate E_c in hydrides from H–F values of ρ, or from MCSCF calculations, in which enough Slater determinants were included to ensure the correct dissociation behaviour for the wavefunction. The latter wavefunctions are referred to as HFPD (Hartree–Fock with proper dissociation). The computed value of the binding energy was much improved, together with the variation along the series, and spectroscopic constants were also significantly better. Extensions to large molecules should be much easier than direct calculations of E_c, since H–F calculations on large molecules are becoming increasingly reliable.[4] Calculations on other hydrides using conventional CI have also been reported, of particular note being a very extensive study of the P.E. curves of BeH ($A^2\Pi$ and $X^2\Sigma^+$) by Bagus *et al.*[33] Agreement with experiment was excellent, but it was pointed out that with less than complete CI there can be spurious maxima in the curves. BH has been extensively studied by Csizmadia's group,[34,35] both the ground and five excited singlet states, using average natural orbitals and extensive CI. Results were in good agreement with experiment. (Table 1 compares the

Table 1 *Calculated energies and spectroscopic constants of* BH/a.u.

Method	E	D_e	r_e	$\omega_e(\times 10^2)$	Ref.
H–F	-25.1315	0.102	2.268	1.140	*a*
VB	-25.1454	0.110	2.527	1.258	*b*
VB	-25.1453	0.107	2.536	—	*c*
SO-GVB	-25.1664	0.121	2.360	—	*d*
Bethe–Goldstone	-25.1723	0.169	2.331	1.139	*e*
CI	-25.1797	0.120	2.412	0.984	*b*
Separated pair	-25.2054	0.142	2.343	1.324	*f*
CI	-25.2150	0.131	2.357	1.259	*g*
Experimental	-25.289	0.131	2.336	1.078	*h*

(*a*) P. E. Cade and W. M. Huo, *J. Chem. Phys.*, 1967, **47**, 614. (*b*) J. C. Browne and E. M. Greenwalt, *Chem. Phys. Letters*, 1970, **7**, 363. (*c*) J. F. Harrison and L. C. Suen, *J. Mol. Spectroscopy*, 1969, **29**, 432. (*d*) R. J. Blint, W. A. Goddard, tert., R. C. Ladner, and W. E. Palke, *Chem. Phys. Letters*, 1970, **5**, 302. (*e*) G. A. Van der Velde and W. C. Nieuwpoort, *Chem. Phys. Letters*, 1972, **13**, 409. (*f*) E. L. Mehler, K. Ruedenberg, and D. M. Silver, *J. Chem. Phys.*, 1970, **52**, 1181. (*g*) S. A. Houlden and I. G. Csizmadia, *Theor. Chim. Acta*, 1974, **35**, 173. (*h*) J. W. C. Johns, F. A. Grimm, and R. F. Porter, *J. Mol. Spectroscopy*, 1967, **22**, 435.

results of recent work on BH.) Mehler has also used a new method similar to IEPA to study correlation effects in BH and BH$^+$.[36] Goddard and co-workers have discussed in considerable detail GVB calculations on LiH[26] and BH[37] and

[33] P. S. Bagus, C. M. Moser, P. Goethals, and G. Verhaegen, *J. Chem. Phys.*, 1973, **58**, 1886.
[34] S. A. Houlden and I. G. Csizmadia, *Theor. Chim. Acta*, 1973, **30**, 209.
[35] S. A. Houlden and I. G. Csizmadia, *Theor. Chim. Acta*, 1974, **35**, 173.
[36] E. L. Mehler, *Theor. Chim. Acta*, 1974, **35**, 17.
[37] C. F. Melius and W. A. Goddard, tert., *J. Chem. Phys.*, 1972, **56**, 3348; R. J. Blint and W. A. Goddard, tert., *J. Chem. Phys.*, 1972, **57**, 5296; W. J. Hunt, P. J. Hay, and W. A. Goddard, tert., *J. Chem. Phys.*, 1972, **57**, 738; W. A. Goddard, tert and R. J. Blint, *Chem. Phys. Letters*, 1972, **14**, 616; R. J. Blint and W. A. Goddard, tert., *Chem. Phys.*, 1974, **3**, 297.

their various excited states, with particular emphasis on the electron distribution. Rydberg states have been studied by Mulliken.[38] Schaefer's first-order wavefunction method[5,39] has also been applied to this species.[40] The important molecules CH^{41} and CH^{+} [42] have been studied using large basis sets and extensive CI. The GVB orbital description of the CH molecule is illuminating.[43] In contrast to most workers, Tantardini and Simonetta[44] have reported minimal basis conventional VB calculations, with surprisingly good results. The FSGO method has been applied to a variety of diatomic species.[45]

The species NH, NH^{+}, and OH have all been studied in detail, especially NH, for which various CI wavefunctions were reported,[46] and also MCSCF calculations,[47] the latter giving impressive agreement with the experimental values of the spectroscopic constants. The P.E. curves of the OH radical were investigated by Bondybey et al.[48] with first-order wavefunctions, whilst MCSCF calculations were reported by Karo et al.[49] Goddard and co-workers have also investigated OH by the GVB procedure.[50]

The problem of the accurate calculation of dipole moments has been thoroughly reviewed by Green, who investigated OH and OD[51] among other diatomics, and concluded that an accuracy of ± 0.06 D is currently feasible. OH^{-} has been studied by the IEPA method by Lishka.[52] Both the theoretical P.E. curves[48] and the theoretical dipole moment function have been computed for HF, the latter from MCSCF calculations.[53] A comparison of minimal basis MO and perfect pairing wavefunctions for HF has appeared[54] and also a careful eight-configuration MCSCF study.[55]

Various calculations on hydrides containing elements below the first row have appeared. SiH has been studied using CI by Wirsam,[56] NeH and NeH^{+} by Bondybey et al.,[48] and TiH by Scott and Richards[57] who obtained approximate SCF wavefunctions and predict a $^{4}\Phi$ ground state. A very detailed study of MnH

[38] R. S. Mulliken, *Internat. J. Quantum Chem.*, 1971, **3**, 83.
[39] H. F. Schaefer, tert., *J. Chem. Phys.*, 1971, **54**, 2207.
[40] P. K. Pearson, C. F. Bender, and H. F. Schaefer, tert., *J. Chem. Phys.*, 1971, **55**, 4235.
[41] G. C. Lie, J. Hinze, and B. Liu, *J. Chem. Phys.*, 1972, **57**, 625 (this paper contains extensive references to earlier work); 1973, **59**, 1872, 1887.
[42] S. Green, P. S. Bagus, B. Liu, A. D. McLean, and M. Yoshimine, *Phys. Rev.*, 1972, A**5**, 1614; M. Yoshimine, S. Green, and P. Thadeus, *Astrophys. J.*, 1973. **183**, 899.
[43] P. J. Hay, W. J. Hunt, and W. A. Goddard, tert., *J. Amer. Chem. Soc.*, 1972, **94**, 8293.
[44] G. F. Tantardini and M. Simonetta, *Chem. Phys. Letters*, 1972, **14**, 170.
[45] P. H. Blustin and J. W. Linnett, *J.C.S. Faraday II*, 1974, **70**, 327, 826.
[46] S. V. O'Neil and H. F. Schaefer, tert., *J. Chem. Phys.*, 1971, **55**, 394.
[47] W. J. Stevens, *J. Chem. Phys.*, 1973, **58**, 1264; G. Das and A. C. Wahl, *J. Chem. Phys.*, 1972, **56**, 1769; G. Das, A. C. Wahl, and W. J. Stevens, *J. Chem. Phys.*, 1974, **61**, 433.
[48] V. Bondybey, P. K. Pearson, and H. F. Schaefer, tert., *J. Chem. Phys.*, 1972, **72**, 1123.
[49] A. M. Karo, M. Krauss, and A. C. Wahl, *Internat. J. Quantum Chem.*, 1973, **7S**, 143.
[50] S. L. Guberman and W. A. Goddard, tert., *J. Chem. Phys.*, 1970, **53**, 1803.
[51] S. Green, *J. Chem. Phys.*, 1973, **58**, 4327; S. Green, *Adv. Chem. Phys.*, 1974, **25**, 179.
[52] H. Lishka, *Theor. Chim. Acta*, 1973, **31**, 39.
[53] G. C. Lie, *J. Chem. Phys.*, 1974, **60**, 2991.
[54] R. E. Bruce, K. A. R. Mitchell, and M. L. Williams, *Chem. Phys. Letters*, 1973, **23**, 504.
[55] M. Krauss and D. Neumann, *Mol. Phys.*, 1974, **27**, 917.
[56] B. Wirsam, *Chem. Phys. Letters*, 1971, **10**, 180.
[57] P. R. Scott and W. G. Richards, *J. Phys. (B)*, 1974, **7**, 500.

by Bagus *et al.* shows the utility of SCF calculations for these species.[58] Bauschlicher and Schaefer[59] have examined the basis orbitals to be used in such calculations and concluded that the use of fully contracted gaussian orbitals (CGTO) for inner shells is permissible. Extensive CI calculations on HCl with different basis sets have been reported by Petke and Whitten.[60] However, even a 206 configuration wavefunction only recovered *ca.* 4% of E_c. The inclusion of Cl $3d$ functions improved E and also the molecular properties.

Homonuclear Diatomic Species.—These molecules continue to be studied at various levels of approximation. Possibly the most significant new paper is the extension by Clementi and Lie of their semi-empirical calculations of E_c referred to above to these molecules.[32] The *same* density functional was used, and HFPD wavefunctions were evaluated. followed by E_c. In this case, however, rather more care has to be taken in the choice of the reference function if the same degree of accuracy is to be achieved as for AH. However, the success in these papers for this approach suggests that extensions to larger molecules will be fruitful.

Goddard and co-workers[61] have studied Li_2 and Li_2^+ using GVB wavefunctions, and a near H–F calculation on Li_2^+ has also been reported.[62] C_2 and C_2^- have been studied by Barsuhn using a GTO lobe basis set and CI.[63] Predictions as to hitherto unobserved states of the latter were made.

Mulliken has reported an interesting investigation of the correlation diagram of N_2 using optimized SCF wavefunctions over a large range of internuclear distances.[64] There are several interesting features of these diagrams which are not expected from purely qualitative discussions. A number of other more specialized calculations on N_2 are discussed in ref. 4. The most extensive calculation yet published on N_2 is that by Langhoff and Davidson[65] who in an extensive CI calculation obtained *ca.* 63% of E_c, carrying out the calculation in terms of both ICSCF[66] and canonical orbitals.

Schaefer has studied O_2 using first-order wavefunctions with 128 configurations.[67] The computed dissociation energy was 4.27 eV (experiment = 5.21 eV). The GVB method, both with and without CI, has also been applied to O_2.[68] A recent calculation of the Rydberg states[69] has used a modified Hamiltonian[70] which improves the virtual orbitals so that they are better approximations to excited-state orbitals, with results in good agreement with experiment. Various

[58] P. S. Bagus and H. F. Schaefer, tert., *J. Chem. Phys.*, 1973, **58**, 1844.

[59] C. W. Baushlicher, jun., and H. F. Schaefer, tert., *Chem. Phys. Letters*, 1974, **24**, 412.

[60] J. D. Petke and J. L. Whitten, *J. Chem. Phys.*, 1972, **56**, 830.

[61] W. A. Goddard, tert., *J. Chem. Phys.*, 1968, **48**, 1008, 5337.

[62] G. A. Henderson, W. T. Zemke, and A. C. Wahl, *J. Chem. Phys.*, 1973, **58**, 2654.

[63] J. Barsuhn, *Z. Naturforsch.*, 1972, **27a**, 1031; *J. Phys. (B)*, 1974, **7**, 155.

[64] R. S. Mulliken, *Chem. Phys. Letters*, 1972, **14**, 137.

[65] S. R. Langhoff and E. R. Davidson, *Internat. J. Quantum Chem.*, 1974, **8**, 61.

[66] E. R. Davidson, *J. Chem. Phys.*, 1972, **57**, 1999.

[67] H. F. Schaefer, tert., *J. Chem. Phys.*, 1971, **54**, 2207.

[68] W. J. Hunt, P. J. Hay, and W. A. Goddard, tert., *J. Chem. Phys.*, 1972, **57**, 538.

[69] D. C. Cartwright, W. J. Hunt, W. Williams, S. Trajmar, and W. A. Goddard, tert., *Phys. Rev.*, 1973, **8A**, 2436.

[70] W. J. Hunt and W. A. Goddard, tert., *Chem. Phys. Letters*, 1969, **6**, 414.

excited states of O_2 have been studied by Morukuma,[71] and a detailed study of the fine structure has appeared,[72] using CI wavefunctions and a variety of basis sets. About two-thirds of the observed splitting is ascribed to spin–orbit effects.

Photoelectron spectral measurements have prompted near H–F calculations on the $1s$ hole states of $O_2{}^+$.[73] Agreement with experiment was found only if the restriction of the MO symmetry to g or u was lifted, corresponding to the singly occupied $1s$ orbital being localized on one of the two O atoms. The electron affinity of O_2 has been estimated from OVC calculations on $O_2{}^-$, with the computed value close to experiment.[74]

The MCSCF (OVC) method has been much used for the study of F_2, and recent work has been reviewed.[8,75] The ground and excited states have been studied including CI by both VB and MO methods.[76]

Studies on molecules containing second-row atoms of this type have been few in number, but several calculations on Ne_2 have been published.[77] Na_2 has been studied at the MCSCF level,[78] and also the P.E. curve of Cl_2, using a VB–CI wavefunction.[79] The calculated D_e was, however, only *ca.* 29 % of the experimental value and very little improved over the H–F value.

An interesting and detailed comparison of the bonding in P_2 and N_2 has been reported by Mulliken and Liu, who found near H–F wavefunctions and examined the influence of $3d$ orbitals on the bonding.[80]

Heteronuclear Diatomic Species.—There have been several calculations on diatomics containing rare-gas atoms. Details of calculations of interaction energies for relatively unstable species can be found in ref. 4. A particularly interesting recent calculation is a near H–F study of KrF and KrF$^+$.[81] Only the latter gave a non-repulsive P.E. curve, with $D_e = 0.07$ hartree and $R_e = 3.3$ bohr. More recently the P.E. curves for XeF ($^2\Sigma^+$ and $^2\Pi$) were computed,[82] and only a weak Van der Waals interaction was predicted. This result is not consistent with interpretations of several experimental studies.

[71] K. Morukuma and H. Kohnishi, *J. Chem. Phys.*, 1971, **55**, 402.

[72] S. R. Langhoff, *J. Chem. Phys.*, 1974, **61**, 1708.

[73] P. S. Bagus and H. F. Schaefer, tert., *J. Chem. Phys.*, 1972, **56**, 224.

[74] W. T. Zemke, G. Das, and A. C. Wahl, *Chem. Phys. Letters*, 1972, **14**, 310; M. Krauss, D. Neumann, A. C. Wahl, G. Das, and W. Zemke, *Phys. Rev.*, 1973, **7A**, 69.

[75] G. Das and A. C. Wahl, *J. Chem. Phys.*, 1972, **56**, 1769.

[76] E. Kasseckert, *Z. Naturforsch.*, 1973, **28a**, 704.

[77] A. Conway and J. N. Murrell, *Mol. Phys.*, 1974, **27**, 873; G. Starkschall and R. G. Gordon, *J. Chem. Phys.*, 1971, **56**, 2801; E. Kochanski, *Chem. Phys. Letters*, 1974, **25**, 380; W. J. Stevens, A. C. Wahl, M. A. Gardner, and A. M. Karo, *J. Chem. Phys.*, 1974, **60**, 2195.

[78] A. M. Karo, M. Krauss, and A. C. Wahl, *Internat. J. Quantum Chem.*, 1973, **7S**, 143.

[79] T. G. Heil, S. V. O'Neil, and H. F. Schaefer, tert., *Chem. Phys., Letters*, 1970, **5**, 253.

[80] R. S. Mulliken and B. Liu, *J. Amer. Chem. Soc.*, 1971, **93**, 6738.

[81] B. Liu and H. F. Schaefer, tert., *J. Chem. Phys.*, 1971, **55**, 2369; P. S. Bagus, B. Liu, and H. F. Schaefer, tert., *J. Amer. Chem. Soc.*, 1972, **94**, 6635.

[82] D. H. Liskow, H. F. Schaefer, tert., P. S. Bagus, and B. Liu, *J. Amer. Chem. Soc.*, 1973, **95**, 4056.

There have been several calculations on the alkali oxides. An extensive study of LiO,[83] both SCF and CI, gave results in good agreement with experiment. A later paper[84] reported results on AlO and reviewed the computation of spectroscopic band intensities. Several excited states of AlO and AlO^+ were also studied by Schamps[85] and by Wahl and co-workers,[86] who also investigated NaO and its ions.[87]

Experimental uncertainty as to the ground-state configuration of alkaline-earth oxides has resulted in several calculations on these species. Several papers on various states of BeO have appeared, the ground state being $X^1\Sigma^+$.[88] H–F calculations on MgO predict a $^3\Pi$ ground state; however, since the order of the $^3\Pi$ and $^1\Sigma^+$ states of BeO is interchanged when CI is included, the MgO ground-state question is still not completely resolved.[89,90]

Most *ab initio* calculations neglect relativistic effects, but for high atomic numbers relativistic energies are large. A minimal basis calculation on the 90-electron PbO,[91] and comparison of the valence-shell electronic structure with that of CO, have been carried out. The errors in D_e and in spectroscopic constants at this level are comparable, hence relativistic effects may be neglected without sizeable errors in chemically interesting properties. A near H–F calculation of the low-lying states of FeO has appeared,[92] and also the results of a limited CI treatment on various states. The lowest $^5\Sigma^+$ state was subjected to extensive CI but it was concluded that this state is not the ground state.

Recent calculations on CO have focused attention on the excited states, but a very large CI study of the ground state obtained *ca.* 70% of E_c.[93] MCSCF calculations have been reported of the quadruple moments of CO, N_2, and NO^+, with reasonable agreement with experiment.[94] It was emphasized that correlation must be included in order to obtain reliable values. The accurate calculation of dipole moments for CO and CS has been reviewed by Green[95] who describes his extensive studies on this subject.[96]

The P.E. curves for 72 excited states of SiO have been reported by Heil and Schaefer,[97] who used a full CI and a minimal basis set. Such calculations have been reviewed by Schaefer,[5] and they seem to give reliable results for this type

[83] M. Yoshimine, *J. Chem. Phys.*, 1972, **57**, 1108.
[84] M. Yoshimine, A. D. McLean, and B. Liu, *J. Chem. Phys.*, 1973, **58**, 4412.
[85] J. Schamps, *Chem. Phys.*, 1973, **2**, 352.
[86] G. Das, T. Janis, and A. C. Wahl, *J. Chem. Phys.*, 1974, **61**, 1274.
[87] P. A. G. O'Hare and A. C. Wahl, *J. Chem. Phys.*, 1972, **56**, 4516.
[88] H. F. Schaefer, tert., *J. Chem. Phys.*, 1971, **55**, 176; S. V. O'Neil, P. K. Pearson, and H. F. Schaefer, tert., *Chem. Phys., Letters*, 1971, **10**, 404; P. K. Pearson, S. V. O'Neil, and H. F. Schaefer, tert., *J. Chem. Phys.*, 1972, **56**, 3938.
[89] J. Schamps and H. Lefebvre-Brion, *J. Chem. Phys.*, 1972, **56**, 573.
[90] N. J. Stagg and W. G. Richards, *Mol. Phys.*, 1974, **27**, 787.
[91] G. M. Schwenzer, D. H. Liskow, H. F. Schaefer, tert., P. S. Bagus, B. Liu, A. D. McLean, and M. Yoshimine, *J. Chem. Phys.*, 1973, **58**, 3181.
[92] P. S. Bagus and H. J. T. Preston, *J. Chem. Phys.*, 1974, **59**, 2986.
[93] A. K. Q. Siu and E. R. Davidson, *Internat. J. Quantum Chem.*, 1970, **4**, 223.
[94] F. P. Billingsley and M. Krauss, *J. Chem. Phys.*, 1974, **60**, 2767.
[95] S. Green, *Adv. Chem. Phys.*, 1974, **25**, 179.
[96] S. Green, *J. Chem. Phys.*, 1970, **52**, 3100; 1971, **54**, 827; 1972, **56**, 729; 1972, **57**, 2830.
[97] T. G. Heil and H. F. Schaefer, tert., *J. Chem. Phys.*, 1972, **56**, 958.

of problem. The predictions as to the order of the states should be useful for interpreting spectroscopic data.

Several papers have dealt with NO, NO^+, and NO^- (see ref. 4 for details). The most recent CI study by Thulstrup et al.[98] has compared theoretical values with a variety of experimental results. Green has computed CI wavefunctions and spin densities for the $^2\Sigma^+$ state.[99]

Mulliken and Liu[80] have obtained SCF wavefunctions for PO, and more extensive CI calculations of the P.E. curves for a variety of excited states have been reported by Tseng and Grein,[100] and also by Roche and Lefebvre-Brion.[101] Calculations on ClO and FO and their ions at the near H–F[102,103] level gave results in good agreement with the relatively sparse experimental data.

Diatomic metal halide molecules have been studied by various experimental techniques and there have been several theoretical studies of these. One interesting paper dealt with the study of the effect of the basis set on the charge distribution for LiF.[104] It is again noted that minimal basis sets tend to give unrealistic charge distributions.

In recent years some of the computational problems which have hampered VB calculations have been solved and among recent work with this method we note in particular an investigation of TiF^{3+}.[105] However, a pseudo-potential approach to the core-electrons was adopted.

CF and its ions have been studied in detail, with the dipole moment predicted in the sense C^-F^+.[106] It is thought unlikely that the sign is incorrect but MCSCF calculations are needed. Calculations on several excited states were reported with a different basis set,[107] and these authors also obtained the same sign for the dipole moment of the ground state. SiF and its ions have also been studied by Wahl et al.,[106,108] and also NF and PF and their ions.[109] Beyond H–F calculations (minimal basis — full CI) on NF and NF^+ by Anderson et al. gave significantly different results from the SCF calculation.[110] Other fluorides studied by Wahl et al. were SF, SeF, and AsF.[111–113] A large number of states of CN have been studied by O'Neil and Schaefer[114] and also by Das et al.[86]

98 P. W. Thulstrup, E. W. Thulstrup, A. Anderson, and Y. Öhrn, J. Chem. Phys., 1974, **60**, 3975.
99 S. Green, Chem. Phys. Letters, 1972, **13**, 552; 1973, **23**, 115.
100 T. J. Tseng and F. Grein, J. Chem. Phys., 1973, **59**, 6563.
101 A. L. Roche and H. Lefebvre-Brion, J. Chem. Phys., 1973, **59**, 1914.
102 P. A. G. O'Hare and A. C. Wahl, J. Chem. Phys., 1970, **53**, 2469.
103 P. A. G. O'Hare and A. C. Wahl, J. Chem. Phys., 1971, **54**, 3770.
104 J. E. Williams and A. Streitwieser, jun., Chem. Phys. Letters, 1974, **25**, 507.
105 P. J. Carrington and P. G. Walton, Mol. Phys., 1973, **26**, 705.
106 P. A. G. O'Hare and A. C. Wahl, J. Chem. Phys., 1971, **55**, 666.
107 J. A. Hall and W. G. Richards, Mol. Phys., 1972, **23**, 331.
108 P. A. G. O'Hare, J. Chem. Phys., 1973. **59**, 3842.
109 P. A. G. O'Hare and A. C. Wahl, J. Chem. Phys., 1971, **54**, 4563.
110 A. Anderson and Y. Öhrm, J. Mol. Spectroscopy, 1973, **45**, 358.
111 P. A. G. O'Hare and A. C. Wahl, J. Chem. Phys., 1970, **53**, 2834.
112 P. A. G. O'Hare, J. Chem. Phys., 1974, **60**, 4084.
113 P. A. G. O'Hare, A. Batana, and A. C. Wahl, J. Chem. Phys., 1973, **59**, 6495.
114 T. G. O'Neil and H. F. Schaefer, tert., J. Chem. Phys., 1971, **54**, 2573.

Agreement between the two sets of calculations was good, although many more configurations are needed in the case of the conventional CI calculation. NS and BC are other diatomic open-shell species which have been studied recently.[115,116]

A number of calculations on interhalogens such as ClF have been discussed elsewhere,[4] the most extensive work being a DZ + P study of many of these species, including a DZ basis set calculation on IBr.[117]

3 *Ab initio* Calculations on Triatomic Molecules

The past two years have seen a large increase in the number of *ab initio* calculations on triatomic molecules, particularly potential energy surface calculations (see below). As in Section 2, we will discuss only a few representative examples of such calculations.

Hydrides AH₂.—Calculations on linear BeH_2, including correlation by the IEPA method, have been reported by Ahlrichs and Kutzelnigg.[118] There have also been several VB calculations[119,120] and a comparison of both MO and VB calculations.[121]

The most extensive study to date is the large CI calculation of Hosteny and Hagstrom[122] who obtained *ca.* 55% of E_c, using an STO basis set. LiH_2 has been investigated by Goddard *et al.* using the GVB method.

Several authors have investigated the stability of species containing rare-gas atoms. KrF_2 has been extensively studied,[123,124] the most accurate wavefunction being obtained by Bagus *et al.*[124] from a CI calculation. Only if CI is included is the molecule predicted to be found, but the computed D_e is only *ca.* 1/3 of the experimental value. Electron correlation effects were shown to be very important. XeF_2 has also been studied at the SCF level.[125]

The dihalide ions HX_2^- (X = halogen) have been studied extensively experimentally, and Almlöf[125] has performed near H–F calculations on HF_2^-, which is predicted to be symmetrical. This species was also studied by Noble and Kortzeborn.[126] Both Janoschek[127] and Thomson *et al.*[128] have investigated $ClHCl^-$, the latter paper showing that this species should be bound, but the

[115] P. A. G. O'Hare, *J. Chem. Phys.*, 1971, **54**, 4124.
[116] J. E. Kouba and Y. Öhrn, *J. Chem. Phys.*, 1970, **53**, 3923.
[117] P. A. Straub and A. D. McLean, *Theor. Chim. Acta*, 1972, **32**, 227.
[118] R. Ahlrichs and W. Kutzelnigg, *Theor. Chim. Acta*, 1968, **10**, 377.
[119] R. G. A. R. Maclagan and G. W. Schnuelle, *J. Chem. Phys.*, 1971, **55**, 5431.
[120] G. A. Gallup and J. M. Norbeck, *Chem. Phys. Letters*, 1973, **21**, 495.
[121] K. A. R. Mitchell and T. Thirunamachandran, *Mol. Phys.*, 1972, **23**, 947.
[122] R. P. Hosteny and S. A. Hagstrom, *J. Chem. Phys.*, 1973, **58**, 4396.
[123] G. A. D. Collins, D. W. J. Cruickshank, and A. Breeze, *Chem. Comm.*, 1970, **884**; *J.C.S. Faraday II*, 1974, **70**, 393.
[124] P. S. Bagus, B. Liu, and H. F. Schaefer, tert., *J. Amer. Chem. Soc.*, 1972, **94**, 6635.
[125] J. Almlöf, *Chem. Phys. Letters* 1972, **17**, 49.
[126] P. N. Noble and R. N. Kortzeborn, *J. Chem. Phys.*, 1970, **52**, 5375.
[127] R. Janoschek, *Theor. Chim. Acta*, 1973, **29**, 57.
[128] C. Thomson, D. A. Clark, and T. Waddington, to be published.

radical HCl_2 is probably not bound. An SCF study of OHO^{3-} has also been reported.[129]

An interesting recent study of HCN and HNC, including extensive CI, was carried out in order to see if HNC could be responsible for a radio frequency galactic emission.[130] Computation of the known geometry of HCN enabled the authors to predict confidently the geometry and B_0 of HNC. The authors conclude that the emitter could be HNC.

An extensive study of a large number of excited states of HCN has given useful criteria for describing these excited states in MO terms and examination of 12 low-lying states showed that the experimental assignment of the $\tilde{B}^1 A''$ state is probably incorrect.[131]

The isoelectronic molecules HBS[132] and HCP[133,134] have been the subject of near H–F calculations. A large number of one-electron properties[134] were computed and compared with recent experimental data. HCC[135] and HBO[136] are two other examples of unstable species recently investigated by SCF calculations. It should be emphasized that accurate SCF calculations on unstable intermediates can yield reliable information on geometry and one-electron properties which is not readily accessible experimentally.

CO_2 has been studied many times in the past,[4,5] but only recently have extensive calculations on the excited states appeared. Using a DZ + P GTO basis set, open-shell SCF calculations on 13 excited states were reported by Winter et al.[137] and used in interpreting experimental spectra. The molecules Li_2O and Al_2O are both predicted to be linear in SCF calculations,[138] a result in disagreement with semi-empirical work. Li_2O has been the subject of separated-electron pair type calculations.[139]

OCC and TiCO are both unstable species and Thomson and Wishart have studied the former,[140] predicting the equilibrium geometry in near H–F calculations. Goddard and Mortola[141] obtained GVB wavefunctions for TiCO and $TiCO^+$, and gave a useful discussion of carbonyl bonding in the light of their calculations.

[129] H. Blum, R. Frey, H. S. H. Gunthard, and T.-K. Ha. *Chem. Phys.*, 1973, **2**, 262.
[130] P. K. Pearson, G. L. Blackman, H. F. Schaefer, tert., B. Rees, and U. Wählgren, *Astrophys. J.*, 1973, **184**, L19.
[131] G. M. Schwenzer, S. V. O'Neil, H. F. Schaefer, tert., C. P. Baskin, and C. F. Bender, *J. Chem. Phys.*, 1974, **60**, 2787.
[132] C. Thomson, *Theor. Chim. Acta*, 1974, **35**, 237.
[133] J. B. Robert, H. Marsmann, I. Absar, and J. R. Van Wazer, *J. Amer. Chem. Soc.*, 1971, **93**, 3320.
[134] C. Thomson, *Chem. Phys. Letters*, 1974, **25**, 59.
[135] J. Barsuhn, *Astrophys. Letters*, 1972, **12**, 169.
[136] C. Thomson and B. J. Wishart, *Theor. Chim. Acta*, 1974, **35**, 267.
[137] N. W. Winter, C. F. Bender, and W. A. Goddard, tert., *Chem. Phys. Letters*, 1973, **20**, 489.
[138] E. L. Wagner, *Theor. Chim. Acta*, 1974, **32**, 296.
[139] T. K. Liu and D. D. Ebbing, *Internat. J. Quantum Chem.*, 1972, **6**, 297.
[140] C. Thomson and B. J. Wishart, *Theor. Chim. Acta*, 1973, **31**, 347.
[141] A. P. Mortola and W. A. Goddard, tert., *J. Amer. Chem. Soc.*, 1974, **96**, 1.

A very extensive set of calculations on various linear molecules containing B, N, and C has been reported by Thomson[142] which show that BNC, BCC, and BCB are predicted to be more stable than their isomers. The results on BCC are consistent with the e.s.r. description as a σ-radical.

A similar study of NCN, NNC, CNC, and CCN[143] gave results in reasonable agreement with experimental data. One important conclusion from these calculations, carried out with various STO basis sets, was that discussions of the bonding and charge distribution using the Mulliken population analyses can lead to quite wrong conclusions unless basis sets of at least DZ + P quality are used for the calculations. The spin–spin interaction in NCN, CNN, and OCC has been computed by Williams,[144] with results which were not in particularly good agreement with experiment.

There have been many interesting studies of non-linear species. H_3^+ has been extensively studied and most recently the excited states have been investigated using a large CI.[145] Handy has suggested a new form for the correlation factor and studied H_3^+ with this method.[146]

Although Walsh's rules predict BeH_2^+ to be linear, both SCF and VB–CI calculations[147] show the ground state to be an electrostatically bound complex with $^2A_1$ symmetry and $\angle HBeH = 20°$. BH_2 has been thoroughly studied by Bender and Schaefer,[148] and by Goddard et al.,[149] and most recently by Staemmler and Jungen[150] who in IEPA calculations computed a large number of properties for both the $^2A_1$ and $^2B_1$ states. Jungen has also studied BH_2^+.[151] Walsh's rules have been reviewed by Buenker et al.[152] and further studied by Stenkamp and Davidson.[153]

The power of *ab initio* calculations is well illustrated by several calculations on CH_2, where theory firmly predicts a bond angle of ca. 134°, and this has led to a revision of the earlier experimental value.[154] Details of earlier work are to be found in the recent papers by Bender et al.[155] There have also been several VB

[142] C. Thomson, *J. Chem. Phys.*, 1973, **58**, 216.
[143] C. Thomson, *J. Chem. Phys.*, 1973, **58**, 841.
[144] G. R. Williams, *Chem. Phys. Letters*, 1974, **25**, 602.
[145] L. J. Schaad and W. V. Hicks, *J. Chem. Phys.*, 1974, **61**, 1934.
[146] N. C. Handy, *Mol. Phys.*, 1973, **26**, 169.
[147] R. D. Poshusta, D. W. Klint, and A. Liberles, *J. Chem. Phys.*, 1971, **55**, 252.
[148] C. F. Bender and H. F. Schaefer, tert., *J. Mol. Spectroscopy*, 1971, **37**, 423.
[149] W. A. Goddard, tert. and R. J. Blint, *Chem. Phys. Letters*, 1972, **14**, 616.
[150] V. Staemmler and M. Jungen, *Chem. Phys. Letters*, 1972, **16**, 187.
[151] M. Jungen, *Chem. Phys. Letters* 1970, **5**, 241.
[152] R. J. Buenker and S. D. Peyerimhoff, *Chem. Rev.*, 1974, **74**, 127.
[153] L. Z. Stenkamp and E. R. Davidson, *Theor. Chim. Acta*, 1973, **30**, 283.
[154] W. A. Lathan, W. Hehre, and J. A. Pople, *J. Amer. Chem. Soc.*, 1971, **93**, 808; W. A. Lathan, W. Hehre, L. A. Curtiss, and J. A. Pople, *J. Amer. Chem. Soc.*, 1971, **93**, 6377; J. del Bene, *Chem. Phys. Letters*, 1971, **9**, 68; S. V. O'Neil, H. F. Schaefer, tert., and C. F. Bender, *J. Chem. Phys.*, 1971, **55**, 162.
[155] D. R. McLaughlin, C. F. Bender, and H. F. Schaefer, tert., *Theor. Chim. Acta*, 1972, **25**, 362; C. F. Bender, H. F. Schaefer, tert., D. R. Franceschetti, and L. C. Allen, *J. Amer. Chem. Soc.*, 1972, **94**, 6888; J. F. Harrison and L. C. Allen, *J. Amer. Chem. Soc.*, 1969, **91**, 807; J. F. Harrison, *J. Amer. Chem. Soc.*, 1971, **93**, 4112; G. F. Tantardini, M. Raimondi, and M. Simonetta, *Internat. J. Quantum Chem.*, 1973, **7**, 893; P. J. Hay, W. J. Hunt, and W. A. Goddard, tert., *Chem. Phys. Letters*, 1972, **13**, 30.

calculations on this molecule, and also GVB calculations.[155] Calculations on the excited states[154—156] and in particular the magnitude of $\Delta E(^1A_1 - {}^3B_1)$ have shown that inclusion of electron correlation is of crucial importance. The most extensive calculations give $\Delta E = 0.014 \pm 0.004$ hartree. A large number of molecular properties were computed by Staemmler[157] and the most extensive CI calculation was used to obtain the spin–spin splitting parameters.[158]

The most recent calculations on NH_2 including CI *via* first-order wavefunctions gave useful information on various states of this molecule.[159] $NH_2{}^+$ has also been carefully studied recently.[156,160] Of particular note is the predicted $\Delta E(^1A_1 - {}^3B_1)$, which is about twice as large as in CH_2.

Brown and Williams[161,162] have used the unrestricted Hartree–Fock (UHF) method to study a large number of triatomic radicals, including NH_2. The $^2B_2$ state was predicted to have $\angle HNH = 26.6°$, and Thomson and Brotchie[163] in near H–F calculations confirm this. However, this state is unbound and seems to correspond to $H_2 + N^*$. Some analogous states of BF_2 are bound (see below). Brown *et al.*[161] have also computed spin-dependent properties from their wavefunctions. Minimal basis sets were used throughout.

The water molecule is of paramount importance, and a very near H–F calculation investigated a large number of ground-state properties.[164] An alternative method has been used by Thomsen and Swanstrøm.[165] VB calculations have also been reported and most recently electron correlation has been included in some detail: for details we refer the reader to ref. 4. The most accurate wavefunction currently available is due to Meyer[166] and the results were discussed by Schaefer.[5] About 85% of E_c was obtained. A large number of papers have dealt with the excited states, and ref. 167 cites other recent work. H_2O^+ and H_2S^+ have also been studied recently.[168]

H_2F^+ has only recently been observed and the optimum geometry has been computed by Diercksen *et al.*[169] Lishka[1] and Leibovici[170] have also studied

156 S. Y. Chu, A. K. Q. Siu, and E. F. Hayes, *J. Amer. Chem. Soc.*, 1972, **94**, 2969.
157 V. Staemmler, *Theor. Chim. Acta*, 1973, **31**, 49.
158 J. F. Harrison, *J. Chem. Phys.*, 1971, **54**, 5413; S. R. Langhoff and E. R. Davidson, *Internat. J. Quantum Chem.*, 1973, **7**, 759; J. F. Harrison and R. C. Liedtke, *J. Chem. Phys.*, 1973, **58**, 3106.
159 C. F. Bender and H. F. Schaefer, tert., *J. Chem. Phys.*, 1971, **55**, 4798.
160 J. F. Harrison and C. W. Eakers, *J. Amer. Chem. Soc.*, 1973, **95**, 3467.
161 R. D. Brown and G. R. Williams, *Chem. Phys.*, 1974, **3**, 19.
162 R. D. Brown and G. R. Williams, *Mol. Phys.*, 1973, **25**, 673.
163 C. Thomson and D. A. Brotchie, *Mol. Phys.*, 1974, **28**, 301.
164 T. H. Dunning, jun., R. M. Pitzer, and S. Aung, *J. Chem. Phys.*, 1972, **57**, 5044.
165 K. Thomsen and P. Swanstrøm, *Mol. Phys.*, 1973, **26**, 735, 751.
166 W. Meyer, *Internat. J. Quantum Chem.*, 1971, **55**, 341.
167 W. A. Goddard, tert. and W. J. Hunt, *Chem. Phys. Letters*, 1974, **24**, 464.
168 H. Sakai, S. Yanabe, T. Yanabe, K. Fukui, and N. Kato, *Chem. Phys. Letters*, 1974, **25**, 541.
169 G. H. F. Diercksen, W. von Niessen, and W. P. Kraemer, *Theor. Chim. Acta*, 1973, **31**, 205.
170 C. Leibovici, *Internat. J. Quantum Chem.*, 1974, **8**, 193.

this species. There have also been recent calculations on H_2S[171] and H_2Si,[172] the latter calculations including the excited states.

Hydrides HAB.—The HO_2 radical has for many years been postulated as an important intermediate, but only recently has its geometry been definitely established *via* accurate calculations,[173] which show $R(O—H) = 1.84$ bohr, $R(O—O) = 2.70$ bohr, $\angle HOO = 104.6°$. Gole and Hayes also studied the excited $^2A''$ and $^2A'$ states.[174]

HCO has been investigated recently using near H–F wavefunctions[175] and also by the UHF method.[176] Thomson and Brotchie[175] computed isotropic hyperfine coupling constants and Botschwina[176] force constants for this molecule. HCO^+ has also been studied using extensive CI (6343 configurations);[177] however, although this molecule has been suggested as the species responsible for an observed astrophysical emission line, this has also been attributed to HCC.[135] Clearly more work is needed on this problem.

HNF and HBF are related to NH_2 and NF_2 and BH_2 and BF_2 and have been studied by Brotchie and Thomson[178] and by Brown,[161] using different methods. Some low-lying excited states were also investigated. Calculations on HNO and on HON using small GTO basis sets enabled Peslak *et al.* to compute electron density maps.[179] CHF, CHF^+, and NHF^+ are related molecules studied by Harrison.[160] HCF[179,180] and $HOCl$[181] have been investigated with rather small basis sets.

AB$_2$ Molecules.—This type of molecule, with many more electrons, presents a greater computational problem and most studies have been of SCF wavefunctions, relatively few, however, being close to the H–F limit.

Ozone, O_3, is described rather poorly at the SCF level and extensive CI calculations have been reported by Heaton *et al.*[182] and by Goddard (GVB–CI).[183] Various other excited-state calculations have been reported, and more extensive discussion is to be found elsewhere (ref. 4).

[171] S. Rothenberg, R. H. Young, and H. F. Schaefer, tert., *J. Amer. Chem. Soc.*, 1970, **92**, 3243; B. Roos, and P. Siegbahn, *Theor. Chim. Acta*, 1971, **21**, 368.

[172] B. Wirsam, *Chem. Phys. Letters*, 1972, **14**, 214.

[173] D. H. Liskow, H. F. Schaefer, tert., and C. F. Bender, *J. Amer. Chem. Soc.*, 1971, **93**, 6734.

[174] J. L. Gole and E. F. Hayes, *J. Chem. Phys.*, 1972, **57**, 360.

[175] C. Thomson and D. A. Brotchie, *Internat. J. Quantum Chem.*, 1974, **8S**, 277.

[176] P. Botschwina, *Chem. Phys. Letters*, 1974, **29**, 98.

[177] U. Wåhlgren, B. Liu, P. K. Pearson, and H. F. Schaefer, tert., *Nature Phys. Sci.*, 1973, **246**, 4.

[178] D. A. Brotchie and C. Thomson, *Chem. Phys. Letters*, 1973, **22**, 338.

[179] J. Peslak, jun., D. S. Klett, and C. W. David, *J. Amer. Chem. Soc.*, 1971, **93**, 5001.

[180] H. Kim and J. R. Sabin, *Chem. Phys. Letters*, 1973, **20**, 215; T.-K. Ha, *J. Mol. Structure*, 1973, **18**, 486; D. P. Chong, F. G. Herring, and D. McWilliams, *Chem. Phys. Letters*, 1974, **25**, 568.

[181] G. L. Bendazzoli, D. G. Lister, and P. Palmieri, *J.C.S. Faraday II*, 1973, **69**, 791.

[182] M. H. Heaton, A. Pipano, and J. J. Kaufman, *Internat. J. Quantum Chem.*, 1972, **6S**, 181.

[183] P. J. Hay and W. A. Goddard, tert., *Chem. Phys. Letters*, 1972, **14**, 46; P. J. Hay, T. H. Dunning, jun., and W. A. Goddard, tert., *Chem. Phys. Letters*, 1973, **23**, 457. W. R. Wadt and W. A. Goddard, tert., *J. Amer. Chem. Soc.*, 1974, **96**, 1689.

LiO$_2$ has an isosceles triangle ground state ($^2A_2$) but a similar shape ($^2B_2$) is close in energy.[184] A mixed basis set method has been tested on LiO$_2$.[185]

The excited states of NO$_2$ have been studied by Hay,[186] and the NO$_2$$^+$ ground state in SCF calculations by Cremashi *et al.*[187]

McCain and Palke investigated a large number of triatomic radicals and their isotropic and anisotropic hyperfine coupling constants in a minimal basis STO–SCF study.[188]

SO$_2$ and SiF$_2$ are isoelectronic, but near H–F calculations on both species show that *d*-orbital participation in the bonding is substantially less in SiF$_2$[189] than in SO$_2$.[190] Unless large basis sets are used for this type of species, too large *d*-orbital populations are often obtained. Wirsam has also studied SiF$_2$, including CI in the calculations.[191]

The series BeF$_2$, BF$_2$, CF$_2$, NF$_2$, and OF$_2$ are all known and a comparison of similar quality (DZ + P basis) SCF calculations and computed properties has appeared.[192] Thomson and Brotchie have also carefully studied BF$_2$,[193] particularly the equilibrium geometry of X^2A ($\angle$FBF = 120°), and the A^2B_1, $^2A_2$, and $^2B_1$ excited states, the latter with minimal GTO basis sets. Hyperfine coupling constants for B are reasonably well reproduced for the RHF wavefunctions, but those for F are a factor of two too small. The $^2B_2$ state also has an acute angle but is predicted to be bound.[194]

NF$_2$ was also studied by the same authors and by Brown and Williams.[195] Better agreement was obtained for the coupling constants when *d*-orbital optimization was carried out by Hinchcliffe and Cobb.[196]

There have been several calculations on difluorides of Group IIA elements, and Coulson has reviewed the bond angles in these molecules.[197] BeF$_2$ and MgF$_2$ are linear in the ground state but the energy required to deform the molecules is small.[198] The excited states of BeF$_2$ and MgF$_2$ have also been studied.[199] The Walsh energy diagram for ZnF$_2$ has been investigated theoretically by Yarkony and Schaefer.[200]

[184] S. V. O'Neil, H. F. Schaefer, tert., and C. F. Bender, *J. Chem. Phys.*, 1973, **59**, 3608.
[185] F. P. Billingsley and C. Trindle, *J. Phys. Chem.*, 1972, **76**, 2995.
[186] P. J. Hay, *J. Chem. Phys.*, 1973, **58**, 4706.
[187] P. Cremaschi and M. Simonetta, *Theor. Chim. Acta*, 1974, **34**, 175.
[188] D. C. McCain and W. E. Palke, *J. Chem. Phys.*, 1972, **56**, 4957.
[189] C. Thomson, *Theor. Chim. Acta*, 1973, **32**, 93.
[190] B. Roos and P. Siegbahn, *Theor. Chim. Acta*, 1971, **21**, 368; P. D. Dacre and M. Elder, *ibid.*, 1972, **25**, 254.
[191] B. Wirsam, *Chem. Phys. Letters*, 1973, **22**, 360.
[192] S. Rothenberg and H. F. Schaefer, tert., *J. Amer. Chem. Soc.*, 1973, **95**, 2095.
[193] C. Thomson and D. A. Brotchie, *Chem. Phys. Letters*, 1972, **16**, 573; *Theor. Chim. Acta*, 1973, **32**, 101.
[194] C. Thomson and D. A. Brotchie, *Mol. Phys.*, 1974, **28**, 301.
[195] R. D. Brown, F. R. Burden, B. T. Hart, and G. R. Williams, *Theor. Chim. Acta*, 1973, **28**, 399.
[196] A. Hinchcliffe and J. C. Cobb, *Chem. Phys.*, 1974, **3**, 271.
[197] C. A. Coulson, *Israel J. Chem.*, 1973, **11**, 683.
[198] J. L. Gole, A. K. Q. Siu, and E. F. Hayes, *J. Chem. Phys.*, 1973, **58**, 857.
[199] J. L. Gole, *J. Chem. Phys.*, 1973, **58**, 869.
[200] D. R. Yarkony and H. F. Schaefer, tert., *Chem. Phys. Letters*, 1973, **15**, 514.

Reference 4 gives more examples of this type of molecule, and we mention finally calculations on PF_2,[201] and a large calculation on SCl_2, where d-functions were shown to be more important on S than on Cl.[202]

ABC Molecules.—We have already mentioned the TiCO calculations.[141] LiON and LiNO, and FNO and FON, were studied with a small basis set by Peslak *et al.*[179] Minimal basis calculations on NSF have been reported,[203] and more recently a large DZ basis calculation on NSF and SSO,[204] with particular reference to the interpretation of photoelectron spectroscopy measurements. A series of more extensive calculations on FNO by Pulay and co-workers showed that force constants can be reliably obtained using even a $(5s2p)$ Gaussian basis set.[205]

4 *Ab initio* Investigations on Tetra-atomic Molecules

The past two years have seen a considerable number of increasingly accurate calculations on these molecules, especially including correlation.

Hydrides AH_3.—The question of the stability of $HeH_3{}^+$ is not yet resolved, despite fairly extensive CI calculations by Benson and McLaughlin,[206] since the more accurate calculations predict it to be unbound, whereas VB–CI[207] predicts a weakly bound species.

The dimerization energy of BH_3 has been extensively investigated by IEPA calculations and predicted to be 0.057 hartree.[208] The equilibrium geometry and force constants have also been computed.[209] Other work on BH_3 has appeared.[210]

There have been several recent studies on CH_3, culminating in a very large basis set calculation of the optimum geometry.[211] Kohnishi and Morukuma[212] have thoroughly investigated the hyperfine coupling constants in CH_3 using various CI wavefunctions, and IEPA calculations have also been reported.[213] The ions $CH_3{}^-$ and $CH_3{}^+$ have also been studied, the former, which is pyramidal, being investigated with respect to the inversion barrier. The origin of the barrier has been ascribed to polarization function influence and correlation effects by different authors.[213,214]

201 J. C. Cobb and A. Hinchcliffe, *Chem. Phys. Letters*, 1974, **24**, 75.
202 B. Solouki, P. Rosmus, and H. Bock, *Chem. Phys. Letters*, 1974, **26**, 20.
203 R. L. Dekock, D. Lloyd, A. Breeze, G. A. D. Collins, D. W. J. Cruickshank, and H. J. Lempka, *Chem. Phys. Letters*, 1972, **14**, 525.
204 P. Rosmus, P. D. Dacre, B. Solouki, and H. Bock, *Theor. Chim. Acta*, 1974, **35**, 129.
205 W. Sawodny and P. Pulay, *J. Mol. Spectroscopy*, 1974, **51**, 135.
206 M. J. Benson and D. R. McLaughlin, *J. Chem. Phys.*, 1973, **56**, 1322.
207 R. D. Poshusta and V. P. Agrawal, *J. Chem. Phys.*, 1973, **59**, 2477.
208 M. Gelus, A. Ahlrichs, and W Kutzelnigg, *Chem. Phys. Letters*, 1971, **7**, 503.
209 M. Gelus and W. Kutzelnigg, *Theor. Chim. Acta*, 1973, **28**, 103.
210 M. E. Schwartz and L. C. Allen *J. Amer. Chem. Soc.*, 1970, **92**, 1466; J. Paldus, J. Cizek, and I. Shavitt, *Phys. Rev.*, 1972, **5A**, 50.
211 R. E. Kari and I. G. Csizmadia, *Internat. J. Quantum Chem.*, 1972, **6**, 401.
212 H. Kohnishi and K. Morukuma, *J. Amer. Chem. Soc.*, 1972, **94**, 5603.
213 F. Driessler, R. Ahlrichs, V. Staemmler, and W. Kutzelnigg, *Theor. Chim. Acta*, 1973, **30**, 315.
214 A. J. Duke, *Chem. Phys. Letters*, 1973, **21**, 275.

Several investigations of NH_3 and PH_3 have appeared. A large STO basis near H–F calculation reproduced the barrier well,[215] and Laws *et al.* have computed a large number of one-electron properties.[216] Pulay and Meyer[217] have evaluated the force constants for a variety of basis sets. An alternative set of *d*-functions has been used by Gerloff *et al.*[218] Dejardin and co-workers have carried out MCSCF calculations, including 51 configurations, and have analysed the resulting wavefunctions.[219,220] Localized orbitals have been obtained by Wilhite and Whitten from both SCF and CI wavefunctions and these authors have also compared NH_3 and PH_3.[221,222] The influence of *d*-orbitals on the angle is much less in NH_3, although they are important in the σ-bonding. Lehn and Munsch[223] have investigated PH_3 using a variety of basis sets. Bond functions have been shown to be an efficient means of introducing polarization into the basis set by calculations on H_2O, NH_3, H_2O_2, and N_2H_4.[224]

There have been various investigations on H_3O^+, and once again it was shown that polarization functions have to be included in the basis set to obtain a pyramidal structure.[225] The inversion barrier was predicted to be only *ca.* 0.003 hartree. Lishka and Dyczmons[226] in an IEPA calculation, however, obtained a barrier substantially larger (*ca.* 0.005 hartree).

Among second-row hydrides, some calculations on PH_3 have already been mentioned, and SiH_3 and its ions have been investigated, in both ground and excited states, by Wirsam.[227] Aarons *et al.* have studied a variety of similar open-shell species, including SiH_3.[228]

AB₃ Molecules.—BF_3 and BCl_3 have been further studied,[229] as has CF_3,[212] which like CCl_3 is pyramidal.[228] For ACl_3 molecules relatively small basis sets have to be used, but SiF_3 and $SiCl_3$ both turn out to be pyramidal.[230] Several other calculations are discussed in ref. 4. Finally we note a rather large calculation on FeF_3 in which the deficiencies of small basis set calculations are pointed out.[231] It should thus be emphasized that *ab initio* calculations on large molecules must employ reasonably sized basis sets, and the usefulness of *isolated* minimal basis set calculations is questionable.

[215] R. M. Stevens, *J. Chem. Phys.*, 1971, **55**, 1725.
[216] E. A. Laws, R. M. Stevens, and W. N. Lipscomb, *J. Chem. Phys.*, 1972, **56**, 2029.
[217] P. Pulay and W. Meyer, *J. Chem. Phys.*, 1972, **57**, 3337.
[218] M. Gerloff, E. Ady, and J. Brickmann, *Mol. Phys.*, 1973, **26**, 561.
[219] P. Dejardin, E. Kochanski, and A. Veillard, *Chem. Phys. Letters*, 1972, **15**, 248.
[220] P. Dejardin, E. Kochanski, A. Veillard, B. Roos, and P. Siegbahn, *J. Chem. Phys.*, 1973, **59**, 5546.
[221] D. L. Wilhite and J. L. Whitten, *J. Chem. Phys.*, 1973, **58**, 948.
[222] J. D. Petke and J. L. Whitten, *J. Chem. Phys.*, 1973, **59**, 4855.
[223] J. M. Lehn and B. Munsch, *Mol. Phys.*, 1972, **23**, 91.
[224] J. O. Jarvie, A. Rauk, and C. Edmiston, *Canad. J. Chem.*, 1974, **52**, 2778.
[225] P. A. Kollman and C. F. Bender, *Chem. Phys. Letters*, 1973, **21**, 271.
[226] H. Lishka and V. Dyczmons, *Chem. Phys. Letters*, 1973, **23**, 167.
[227] B. Wirsam, *Chem. Phys. Letters*, 1973, **18**, 578.
[228] L. J. Aarons, I. H. Hillier, and M. F. Guest, *J.C.S. Faraday II*, 1974, **70**, 167.
[229] D. Goutier and L. A. Burnelle, *Chem. Phys. Letters*, 1973, **18**, 460.
[230] M. F. Guest, I. H. Hillier, and V. R. Saunders, *J.C.S. Faraday II*, 1972, **68**, 867.
[231] R. W. Land, J. W. Hunt, and H. F. Schaefer, tert., *J. Amer. Chem. Soc.*, 1973, **95**, 4517.

Miscellaneous Tetra-atomic Molecules.—H_4 is dealt with below. Both N_4 and P_4 have been investigated by Hillier's group,[232] and P_4 by Brundle *et al.*[233] As expected, N_4 is not stable, although P_4 is predicted to be.

An extensive investigation of basis set dependence has been reported in the case of H_2O_2[234] and the results have been compared with Veillard's earlier calculations.[235] H_2S_2 has also been studied.[236]

N_2H_2 is usually assumed to have a ^1A_g ground state. However, Wagniere found the A_2 state to be lower in energy,[237] but the geometry was not optimized in this work. The correlation energy in C_2H_2 has been investigated by Duben *et al.*,[238] and SCF calculations on Si_2H_2 have been reported by Wirsam.[239]

Among the relatively few A_2B_2 molecules studied, we may note a study of the possible forms of F_2C_2,[240] where the acetylene was shown to be more stable. Both C_2O_2 and $C_2O_2{}^+$ are bound, C_2O_2 having a $^3\Sigma^-$ ground state.[241] Various AB-AB dimers have been discussed elsewhere.[4]

Formaldehyde continues to receive much attention from theoreticians. A good account of the electronic spectrum is given by SCF-CI calculations,[242] but the $^1A\,\pi\pi^*$ state energy has been in dispute. A recent calculation using the MCSCF procedure puts the state 0.41 hartree above the ground state.[243] H_2CS has also been recently investigated.[244] Space does not permit mention of other work on tetra-atomic species, and we refer the reader to ref. 4 for a more extensive discussion.

5 *Ab initio* Calculations of Potential Energy Surfaces

Several of the papers referred to above have dealt with this problem and a very thorough and comprehensive review by Bader and Gangi in ref. 4 contains a detailed discussion of work up to mid-1974. In this section discussion is restricted to a few representative calculations which represent the state of the art in this field. Calculations of the total energy of a polyatomic molecule or molecules as

[232] M. F. Guest, I. H. Hillier, and V. R. Saunders, *J.C.S. Faraday II*, 1972, **68**, 2070.

[233] C. R. Brundle, N. A. Kuebler, M. B. Robin, and H. Basch, *Inorg. Chem.*, 1972, **11**, 20.

[234] J. P. Ranck and H. Johansen, *Theor. Chim. Acta*, 1972, **24**, 334.

[235] A. Veillard, *Theor. Chim. Acta*, 1970, **18**, 21; A. Veillard and H. Demuynck, *Chem. Phys. Letters*, 1970, **4**, 476.

[236] M. Schwartz, *J. Chem. Phys*, 1969, **51**, 4182.

[237] G. Wagniere, *Theor. Chim. Acta*, 1973, **31**, 269.

[238] A. J. Duben, L. Goodman, H. O. Pamuk, and O. Sinanŏglu, *Theor. Chim. Acta*, 1973, **30**, 177; S. Y. Chu, I. Ozkan, and L. Goodman, *J. Chem. Phys.*, 1974, **60**, 1268.

[239] B. Wirsam, *Theor. Chim. Acta*, 1972, **25**, 169.

[240] O. P. Strausz, R. Norstrom, A. C. Hopkinson, M. Schoenborn, and I. G. Csizmadia, *Theor. Chim. Acta*, 1973, **29**, 183.

[241] N. F. Beebe and J. R. Sabin, *Chem. Phys. Letters*, 1973, **24**, 389.

[242] R. J. Buenker and S. D. Peyerimhoff, *J. Chem. Phys.*, 1970, **53**, 1368; S. D. Peyerimhoff, R. J. Buenker, W. F. Kammer, and H. Hsu, *Chem. Phys. Letters*, 1971, **8**, 129; J. L. Whitten, *J. Chem. Phys.*, 1972, **56**, 5458; W. H. Fink, *J. Amer. Chem. Soc.*, 1972, **94**, 1073, 1078; D. M. Hayes and K. Morukuma, *Chem. Phys. Letters*, 1972, **12**, 539.

[243] S. R. Langhoff, S. T. Elbert, C. F. Jackels, and E. R. Davidson, *Chem. Phys. Letters*, 1974, **29**, 247; N. C. Baird and J. R. Swenson, *J. Phys. Chem.*, 1973, **77**, 277.

[244] P. J. Bruna, S. D. Peyerimhoff, R. J. Buenker, and P. Rosmus, *Chem. Phys.*, 1974, **3**, 35.

a function of the positions of the nuclei in order to obtain a P.E. surface are expensive computationally, but much useful information is obtainable from suitable systems.

Calculations at the SCF Level.—It is now well established that near Hartree–Fock calculations can give reliable geometry predictions and, even for basis sets of DZ quality, the same holds true provided polarization functions are included. However, the incorrect dissociation behaviour leads to a correlation energy error which is not constant over a wide range of intermediate distances in general. However, if a molecule dissociates to closed-shell products, *i.e.* there is no change in the number of electron pairs during the dissociation process, the asymptotic behaviour of the wavefunction is correct; E_c does not vary much with R and the H–F and true P.E. curves are approximately parallel. In these cases the H–F method is capable of yielding semi-quantitative P.E. curves with geometrical parameters characterizing minima in the surface accurate to *ca.* 1—2%, energy barriers accurate to *ca.* 0.001—0.003 hartree (1—2 kcal mol^{-1}), and energies of reaction within 0.008—0.016 hartree (5—10 kcal mol^{-1}).

Bader and Gangi give details of many earlier studies on such systems, the first of these being Clementi's calculation on $NH_3 + HCl$,[245] which surprisingly has not been re-investigated with polarization functions in the basis set.

The reaction of H_2 with HeH^+ gives He and $H_3{}^+$, and the work of Benson and McLaughlin[246] on this system showed the minimum energy pathway to exhibit C_{2v} symmetry, and predicted an exothermicity of 0.096 hartree with no barrier or local minima along the path. The corresponding CI calculations confirmed that the H–F and correlated surfaces were probably closely parallel.

Calculations by Kutzelnigg and co-workers provided further evidence for the close parallelism of the surfaces in the case of $LiH_2{}^+$.[247] Lester[248] has also studied this system.

A second class of systems A + B, where A possesses a closed shell and B a half-filled shell, also have the proper dissociation behaviour for diatomic species. When A and/or B is diatomic or polyatomic, the H–F method cannot yield a surface parallel to the true P.E. surface. A reaction of this type studied by Schaefer *et al.*[249] was $H_2 + Cl \rightarrow H + HCl$. The calculated SCF barrier for the linear pathway of 0.042 hartree (26.2 kcal mol^{-1}) was far too large. A similar calculation on $H_2 + F_2 \rightarrow H + HF$ showed the same general overestimation of the barrier height[250] (see below).

Several S_N2 reactions have been studied in order to find the minimum energy pathway between reactants and products. Full details are to be found in ref. 4, and as an example we consider

$$F^- + CH_3F \rightarrow [F\text{---}CH_3\text{---}F]^- \rightarrow CH_3F + F^- \qquad (4)$$

[245] E. Clementi, *J. Chem. Phys.*, 1967, **46**, 3851.
[246] M. J. Benson and D. R. McLaughlin, *J. Chem. Phys.*, 1972, **56**, 1322.
[247] W. Kutzelnigg, V. Staemmler, and C. Hoheisel, *Chem. Phys.*, 1973, **1**, 27.
[248] W. A. Lester, *J. Chem. Phys.*, 1970, **53**, 1511; 1971, **54**, 3171.
[249] S. Rothenberg and H. F. Schaefer, tert., *Chem. Phys. Letters*, 1971, **10**, 565.
[250] S. V. O'Neil, P. K. Pearson, and H. F. Schaefer, tert., *J. Chem. Phys.*, 1973, **58**, 1126.

studied by Veillard *et al.*[251] and Duke and Bader,[252] both groups using large basis sets and polarization functions. Both predicted an intermediate rather than a transition state without polarization functions, and found the linear $[FCH_3F]^-$ complex of D_{3h} symmetry and a reaction barrier of ~ 0.011—0.12 hartree (7—8 kcal mol^{-1}) with polarization functions. One feature found in this and other S_N2 reactions is the delayed departure of the leaving group.

Schaefer *et al.*[253] have studied the isomerization

$$CH_3NC \rightleftharpoons CH_3CN \tag{5}$$

which is believed to exhibit slow intramolecular vibrational relaxation. They did not find a minimum in the potential path when the P.E. surface was computed as a function of both the CH_3 rotational angle and the CNC angle. Various other interaction potentials have been reviewed by Yarkony *et al.*[254]

There are, however, two frequently used techniques which enable one to correct for the wrong dissociation behaviour of the H–F wavefunction without involving large CI. These are (1) the UHF method in which the single determinant is built up of spin orbitals with different spatial parts for different spins, and (2) correcting the RHF single determinant by adding enough determinants, usually only a few, to give the correct asymptotic behaviour. This type of calculation is reviewed in detail for $O(^3P$ or $^1D) + H_2$ by Bader[4] and we refer the reader to this review for details of the reaction

$$C + H_2 \rightarrow H_2O \tag{6}$$

The contributions of the additional configurations are usually only really important for large internuclear distances. The symmetrical insertion of $O(^3P)$ into H_2 gives a barrier of *ca.* 0.127 hartree (80 kcal mol^{-1}) and predicts a linear stable intermediate, and the energy minimum on this path is a saddle point which does not represent a stable nuclear configuration of triplet H_2O.

Murrell and co-workers[255] have examined the P.E. surface for the symmetrical insertion of CH_2 (singlet and triplet) into H_2 ($^1\Sigma_g^+$), with similar results to those quoted above. Baskin *et al.*[256] have studied the abstraction reaction

$$CH_2 + H_2 \rightarrow CH_3 + H \tag{7}$$

and among other studies of this type we mention $CH_3^+ + H_2 \rightarrow CH_5^+$,[257] $CH_4 + H \rightarrow CH_5$,[258] and the dissociation pathways for $H_3O \rightarrow H_2O + H$[259]

[251] G. Berthier, D. J. David, and A. Veillard, *Theor. Chim. Acta*, 1969, **14**, 369; A. Dedieu and A. Veillard, *Chem. Phys. Letters*, 1970, **5**, 328; *J. Amer. Chem. Soc.*, 1972, **94**, 6730.

[252] A. J. Duke and R. F. W. Bader, *Chem. Phys. Letters*, 1971, **10**, 631.

[253] D. H. Liskow, C. F. Bender, and H. F. Schaefer, tert., *J. Amer. Chem. Soc.*, 1972, **94**, 5178; *J. Chem. Phys.*, 1972, **57**, 4509.

[254] D. R. Yarkony, S. V. O'Neil, and H. F. Schaefer, tert., *J. Chem. Phys.*, 1974, **60**, 855.

[255] J. N. Murrell, J. B. Pedley, and S Durmaz, *J.C.S. Faraday II*, 1973, **69**, 1370.

[256] C. P. Baskin, C. F. Bender, C. W Bauschlicher, jun., and H. F. Schaefer, tert., *J. Amer. Chem. Soc.*, 1974, **96**, 2709.

[257] M. F. Guest, J. N. Murrell, and J. B. Pedley, *Mol. Phys.*, 1971, **20**, 81.

[258] S. Ehrenson and M. D. Newton, *Chem. Phys. Letters*, 1972, **13**, 24.

[259] R. A. Gangi and R. F. W. Bader, *Chem. Phys. Letters*, 1971, **11**, 216.

and $NH_4 \rightarrow NH_3 + H$.[260] One interesting conclusion is that, of the species CH_5, H_3O, and NH_4, only the last-named requires more than 0.032 hartree (20 kcal mol^{-1}) for its decomposition.

Additions of atoms to double bonds, such as $S(^3P, {}^1D)$ to ethylene, has been studied by Csizmadia's group,[261] and Buenker[262] and co-workers have carried out very detailed studies of the electrocyclic transformations between cyclic and open-chain hydrocarbons. Horsley *et al.*[263] have also studied the full 21-dimensional hypersurface of the isomerization of cyclopropane.

Before we consider explicit calculations designed to include as large a fraction of the correlation as possible, we should mention an important series of papers dealing with the solvation process with particular reference to the hydration of simple ions. P.E. surfaces for H_2O in the field of Li^+, Na^+, and K^+, and F^- and Cl^- were studied using a large GTO basis set and a single-determinant wavefunction.[264] Later papers dealt with an extensive investigation of the H_2O dimer.[265] Other workers have also investigated solvation, notably Pullman's group,[266] using, however, rather smaller basis sets.

Calculations including Extensive Electron Correlation.—We consider in this section those calculations designed to include enough CI to ensure that the reaction surface is similarly described along the whole path. In principle, full CI calculations can achieve this, but in practice various lesser CI calculations have to be carried out.

Minimal CI ensures the inclusion of those configurations necessary to ensure proper dissociation, as exemplified above, and we now turn to the intermediate, or truncated CI results. The problem of selection of configurations is a difficult one, and is dealt with elsewhere. Transformation to natural orbitals is now commonly used, or alternative procedures such as those proposed by Krauss[267] or Bender and Davidson.[268] The alternative to conventional CI is the MCSCF procedure as given in detail by Wahl and co-workers.[8] The IEPA method, which is referred to several times above, is also a useful method for small systems.

The simplest exchange reaction

$$H + H_2 \rightarrow H_2 + H \tag{8}$$

has been thoroughly reviewed elsewhere,[4] and the most recent work on this system culminates in Liu's surface, which is probably very close to the true surface. The

[260] J. Pelletier and R. F. W. Bader, unpublished results, cited in ref. 4.
[261] O. P. Strausz, H. E. Gunning, A. S. Denes, and I. G. Csizmadia, *J. Amer. Chem. Soc.*, 1972, **94**, 8317.
[262] K. Hsu, R. J. Buenker, and S. D. Peyerimhoff, *J. Amer. Chem. Soc.*, 1971, **93**, 2117; 1972, **94**, 5639; R. J. Buenker, S. D. Peyerimhoff, and K. Hsu, *ibid.*, 1971, **93**, 5005.
[263] J. A. Horsley, Y. Jean, C. Moser, L. Salem, R. M. Stevens, and J. S. Wright, *J. Amer. Chem. Soc.*, 1972, **94**, 279.
[264] H. Kistenmacher, H. Popkie, and E. Clementi, *J. Chem. Phys.*, 1973, **58**, 5627, and earlier papers in this series.
[265] H. Kistenmacher, H. Popkie, E. Clementi, and R. O. Watts, *J. Chem. Phys.*, 1974, **60**, 4455.
[266] M. Dreyfus, B. Maigret, and A. Pullman, *Theor. Chim. Acta*, 1970, **17**, 109.
[267] C. Edmiston and M. Krauss, *J. Chem. Phys.*, 1966, **45**, 1833.
[268] C. F. Bender and E. R. Davidson, *J. Phys. Chem.*, 1966, **70**, 2675.

surface for the

$$H_2 + D_2 \rightarrow 2HD \tag{9}$$

problem is still not completely understood despite a large amount of effort. The problem is that the experimental results from shock-tube experiments suggest a barrier height which is low compared to any yet computed from various possible transition states. The most extensive recent work of Silver and Stevens[270] still has not resolved this question.

The more complicated system $HeH^+ + H_2 \rightarrow He + H_3^+$ has been studied, where the best calculations indicate no barrier to reaction.[246]

The extensive series of calculations of Schaefer and co-workers on the reactions

$$F + H_2 \rightarrow FH + H \tag{10}$$

and

$$H + F_2 \rightarrow HF + H \tag{11}$$

probably indicate accurately the current state of the art in this field. A large (DZ + P) basis set was contracted and a large CI calculation carried out a geometry search carried out over a large number of points (*ca.* 300) with the result that the linear path was found to be of lowest energy and d and p polarization basis functions are necessary to obtain a reasonable barrier height for the first reaction.[271]

The 19-electron problem $H + F_2 \rightarrow HF + F$ was then studied,[250] using a DZ–GTO contracted basis, the results on the above reaction indicating this basis should be capable of yielding qualitatively accurate results. 555 configurations were included and once again the linear approach was favoured. It is expected that DZ + P calculations will improve the computed barrier height, which is less than the experimental value.

Further calculations on

$$Li + F_2 \rightarrow LiF + F \tag{12}$$

have been reported.[272,273] Pearson *et al.* carried out an SCF calculation and also a two-configuration calculation. The minimum energy path was found for the collinear reaction. A deep well (*ca.* 0.05 hartree) was found for a C_{2v} symmetry LiFLi species.

[269] B. Liu, *J. Chem. Phys.*, 1973, **58**, 1925.
[270] D. M. Silver and R. M. Stevens, *J. Chem. Phys.*, 1973, **59**, 3378.
[271] C. F. Bender, P. K. Pearson, S. V. O'Neil, and H. F. Schaefer, tert., *J. Chem. Phys.*, 1972, **56**, 4626; C. F. Bender, S. V. O'Neil, P. K. Pearson, and H. F. Schaefer, tert., *Science*, 1972, **176**, 1412.
[272] G. G. Balint-Kurti, *Mol. Phys.*, 1973, **25**, 393; G. G. Balint-Kurti and M. Karplus, *Chem. Phys. Letters*, 1971, **11**, 203.
[273] P. K. Pearson, W. J. Hunt, C. F. Bender, and H. F. Schaefer, tert., *J. Chem. Phys.*, 1973, **58**, 5358.

Although most workers in this field have used SCF–CI calculations, Basch[274] has examined the use of MCSCF in studying the reaction $2CH_2 \rightarrow C_2H_4$, and a few authors have used VB–CI wavefunctions (see ref. 4).

To summarize, we can see that the past two years have seen an impressive gain in accuracy in P.E. surface studies and it is clear that more chemically interesting results will be forthcoming in this area during the next few years.

[274] H. Basch, *J. Chem. Phys.*, 1971, **55**, 1700.

3 Atomic and Molecular Photoassociation

By B. STEVENS
*Department of Chemistry, University of South Florida,
Tampa, Florida 33620, U.S.A.*

1 Introduction

The tragic death of Professor Th. Förster in 1974 coincided with the twentieth anniversary of his publication of an article[1] entitled 'A Concentration Reversal of Fluorescence' in which the origin of a blue structureless fluorescence band observed in concentrated solutions of pyrene in benzene was attributed to 'an excited double molecule... formed from one excited and one unexcited pyrene molecule (which) decomposes after deactivation.' A key observation was the concentration independence of the absorption spectrum, indicating that molecular association follows the act of light absorption, and the term 'excimer' has been adopted to distinguish this product of photoassociation from an excited dimer produced by excitation of a relatively stable dimeric ground state.[2] As with the development of triplet-state spectroscopy, the phenomenon of photoassociation was recognized in atomic systems many years earlier[3] and emission from the mercury 'excimer' Hg_2^* was undoubtedly present in the light source used by Förster to excite pyrene.

Since 1954 it has been recognized that photoassociation is a general property of aromatic hydrocarbons (unless sterically hindered) and has been extended, largely due to the work of Weller and of Mataga, to include exciplexes formed by non-identical molecules as distinct from excited electron donor–acceptor (EDA) complexes stable in the ground state. The quantum-mechanical description of excimer and exciplex binding forces in terms of exciton and charge-resonance contributions is parametrized by donor ionization potentials and acceptor electron affinities much as the ionic character of covalent bonds is related to atomic electronegativities.

The development of this field is marked by the appearance of excimer and exciplex as index keywords and the organization of an International Exciplex Conference at the University of Western Ontario in May, 1974. A detailed account of the field is presented by Birks in his monumental treatise[4] and by

[1] Th. Forster and K. Kasper, *Z. phys. Chem.*, 1954, **1**, 275.
[2] *E.g.* T. Kajiwara, R. W. Chambers, and D. R. Kearns, *Chem. Phys. Letters*, 1973, **22**, 37.
[3] S. Mrozowski, *Z. Physik*, 1937, **106**, 458.
[4] J. B. Birks, 'Photophysics of Aromatic Molecules', Wiley, London and New York, 1970.

$$M + Q + h\nu_f$$

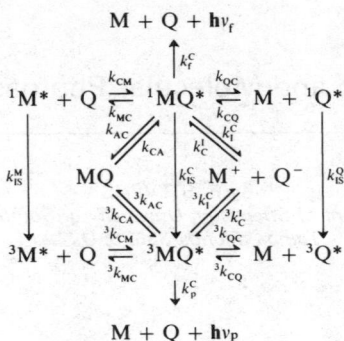

Figure 1 *Photoassociation and exciplex relaxation scheme*

Mataga and Kubota,[5] while reviews by Förster[6] and others[7,8] have appeared in recent years. The complexity and consequences of photoassociation are illustrated by the scheme in Figure 1 where evidence exists for each of the processes involved (although not usually in the same system) and this is presented as a basis for reporting some novel processes and different systems. These include atomic photoassociation, excimer and exciplex triplet states, photodimerization and photoaddition, exciplexes of molecular oxygen, excited dimers and complexes, and practical applications.

2 Atomic Systems

Despite the expectations that atomic photoassociation will be termolecular in the gas phase and that non-radiative relaxation of the diatomic product will be of less importance than it is in molecular complexes, the study of atomic systems may continue to provide precedent for the interpretation of molecular phenomena.

The emission spectrum of mercury vapour contains at least two broad structureless bands assigned to transitions (1) and (2) which decay exponentially, following

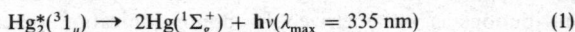

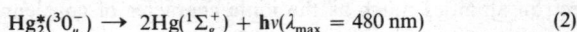

$$Hg_2^*(^3 1_u) \rightarrow 2Hg(^1\Sigma_g^+) + h\nu(\lambda_{max} = 335 \text{ nm}) \tag{1}$$

$$Hg_2^*(^3 0_u^-) \rightarrow 2Hg(^1\Sigma_g^+) + h\nu(\lambda_{max} = 480 \text{ nm}) \tag{2}$$

pulsed electron excitation, with a common lifetime of 14 μs; however, as the mercury vapour pressure is increased to several atmospheres the 480 nm peak shifts to 457 nm and the decay becomes non-exponential. Eckstrom *et al.*[9]

[5] N. Mataga and T. Kubota, 'Molecular Interactions and Electronic Spectra', Marcel Dekker, New York, 1970.
[6] Th. Forster, *Angew. Chem. Internat. Edn.*, 1969, **8**, 333.
[7] B. Stevens, *Adv. Photchem.*, 1971, **8**, 161.
[8] M. Ottolenghi, *Accounts Chem. Res.*, 1973, **6**, 153.
[9] D. J. Eckstrom, R. M. Hill, D. C. Lorents, and H. H. Nakarno, *Chem. Phys. Letters*, 1973, **23**, 112.

attribute the blue shift to effective vibrational relaxation of the emitter and interpret the hyperbolic decay component in terms of the (triplet–triplet?) excimer annihilation process (3) with a rate constant of the order of 10^{-11} cm^3 s^{-1} (the

$$Hg_2^* + Hg_2^* \rightarrow Hg_2^{**} + 2Hg \qquad (3)$$

collision frequency is of the order 10^{-10} cm^3 s^{-1}). However, since vibrational stabilization of Hg_2^* should lead to a red-shift of the associated emission band, the high-pressure band at 457 nm may originate from the annihilation product Hg_2^{**}, which by analogy with the same process in molecular systems could be the excimer singlet state [see process (10) below].

Jortner and co-workers[10] have reported that the excimer spectra of argon ($\lambda_{max} \simeq 128$ nm) and of xenon ($\lambda_{max} \simeq 172$ nm) are virtually identical in the gas, liquid, and solid phases (at similar temperatures) and show that the increase in band width with temperature (in the gas phase) is consistent with harmonic vibrational frequencies of 140 cm^{-1} (Xe$_2^*$) and 310 cm^{-1} (Ar$_2^*$). The emitting state is predominantly $^3\Sigma_u^+$, formed from the long-lived $^3P_2$ atomic state, which has been monitored in absorption in the region 810—995 nm by Oka et al.[11] for Ne$_2^*$, Ar$_2^*$, and Kr$_2^*$. In the absence of any pressure effects over the range 200—1400 Torr, these authors conclude that the respective decay constants of 0.15, 0.31, and 2.83 µs^{-1} describe the purely radiative relaxation. Spectroscopic evidence has been reported[12] for KrXe* ($\lambda_{max} \simeq 158$ nm) and KrAr* ($\lambda_{max} \simeq 135$ nm) in the liquid phase and for NeHe* in absorption,[13] whereas XeAr*, XeNe*, KrNe*, and ArNe* do not appear to be stable with respect to dissociative processes $^3k_{MC}$ or $^3k_{QC}$ of Figure 1. A broad structureless emission band with a maximum at 275 nm observed[14] from mercury vapour excited at 254 nm in the presence of xenon has been assigned to $^3HgXe^*$ whereas no comparable bands were observed in the presence of He, Ne, Ar, or Kr; since these have higher ionization potentials than Xe, it appears that $Hg6^3P_0$ acts as electron acceptor in $^3HgXe^*$.

From an extensive examination of the emission characteristics of charge-transfer complexes (exciplexes) of $Hg6^3P_0$ with such molecular electron donors as H_2O, NH_3, alcohols, and primary amines in the gas phase, Phillips and his group[15] found that the wavelength maximum of the structureless emission band (in the region 285–360 nm) undergoes a red shift with reduction in ionization potential of the donor. The emission quantum efficiencies are high for the simple

[10] O. Chesnovsky, B. Raz, and J. Jortner, *Chem. Phys. Letters*, 1972, **15**, 475.
[11] T. Oka, K. V. S. Rama Rao, J. L. Redpath, and R. F. Firestone, *J. Chem. Phys.*, 1974, **61**, 4740.
[12] O. Cheshnovsky, A. Gedanken, B. Raz, and J. Jortner, *Chem. Phys. Letters*, 1973, **22**, 23.
[13] Y. Tanaka and K. Yoshino, *J. Chem. Phys.*, 1972, **57**, 2964.
[14] C. G. Freeman, M. J. McEwan, R. F. C. Claridge, and L. F. Phillips, *Chem. Phys. Letters*, 1970, **6**, 482.
[15] C. G. Freeman, M. J. McEwan, R. F. C. Claridge, and L. F. Phillips, *Trans. Faraday Soc.*, 1971, **67**, 67, 2004; R. H. Newman, C. G. Freeman, M. J. McEwan, R. F. C. Claridge, and L. F. Phillips, *ibid.*, p. 1360.

donors (0.70 for $HgNH_3^*$ and 0.19 for HgH_2O^*) but generally decrease rapidly with donor complexity due largely to donor reactions with $Hg6^3P_1$, the precursor of $Hg6^3P_0$. An interesting feature of $Hg6^3P_0$ exciplex emission is the red shift of the maximum intensity wavelength with increasing donor pressure which these authors ascribe to vibrational relaxation through well-defined levels of the complex. In the case of NH_3 this shift amounts to *ca.* $1700\ cm^{-1}$ as the NH_3 pressure is increased to 760 Torr, which provides a lower limit of 5 kcal mol^{-1} for the exciplex dissociation energy; the binding energy limits for Hg–amine complexes are somewhat lower than this. At relatively low pressures of NH_3, the phase-shift difference between the $Hg6^3P_1$ and $HgNH_3^*$ emission is pressure dependent, reflecting the rate of exciplex formation in processes (4) and (5) and

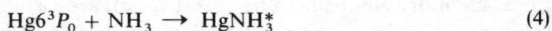

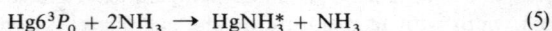

$$Hg6^3P_0 + NH_3 \rightarrow HgNH_3^* \tag{4}$$

$$Hg6^3P_0 + 2NH_3 \rightarrow HgNH_3^* + NH_3 \tag{5}$$

providing rate-constant values of

$$k_4 = 3.2 \times 10^{-13}\ cm^3\ molecule^{-1}\ s^{-1}$$

$$k_5 = 2.3 \times 10^{-30}\ cm^6\ molecule^{-2}\ s^{-1}$$

The constant residual phase shift at high pressures indicates that NH_3 does not quench the exciplex and is consistent with an exciplex radiative lifetime of 1.86 μs.

A different interpretation of the $HgNH_3^*$ emission red shift has been presented by Callear and Connor[16] who note that this reaches a maximum value in the presence of added N_2 or CF_4 (due to vibrational relaxation) which is considerably less than that produced by an increase in NH_3 pressure alone. The latter is attributed to the successive formation of a series of complexes $Hg(NH_3)_n^*$ with $n = 1$—4 (and possibly 5 and 6), and by an iterative procedure these authors deconvolute the emission profiles to obtain dissociation constants for the equilibrium (6) of the order of 10^{19} molecules cm^{-3}. A mean binding energy

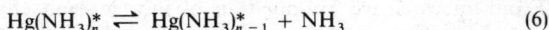

$$Hg(NH_3)_n^* \rightleftharpoons Hg(NH_3)_{n-1}^* + NH_3 \tag{6}$$

of *ca.* 9 kcal mol^{-1} for each NH_3 molecule in $Hg(NH_3)_2^*$ is estimated from the high-energy threshold emission frequencies of the separated emission profiles. An interesting feature of this work, which may be of relevance to molecular systems, is the appearance of a green structureless emission band ($\lambda_{max} \simeq 520$ nm) which increases in intensity with pressure of Hg vapour and which is attributed[17] to emission from Hg_2^* formed in process (7). Emission from the Cd^3P_0 exciplex

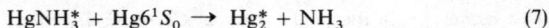

$$HgNH_3^* + Hg6^1S_0 \rightarrow Hg_2^* + NH_3 \tag{7}$$

with NH_3 has been reported[18] with $\lambda_{max} \simeq 430$ nm (red-shifted by 7400 cm^{-1} from the Cd^3P_1 line) and a radiative lifetime of 0.49 μs.

[16] A. B. Callear and J. H. Connor, *Chem. Phys. Letters*, 1972, **13**, 245.
[17] See, however, L. F. Phillips, *Chem. Phys. Letters*, 1973, **21**, 28.
[18] P. D. Morten, C. G. Freeman, M. J. McEwan, R. F. C. Claridge, and L. F. Phillips, *Chem. Phys. Letters*, 1972, **16**, 148.

3 Photoassociation of Singlet States

Excimer–Exciplex Fluorescence.—Unambiguous evidence of photoassociation (Process k_{CM} of Figure 1) is provided by the appearance of a broad structureless red-shifted fluorescence band as the concentration of Q (or M in the case of an excimer) is increased; this is accompanied by a reduction in fluorescence intensity of the directly excited molecule in the absence of any concentration dependence of the absorption spectrum. The radiative lifetime is considerably longer than that of the excited molecular precursor since the associated transition is symmetry forbidden, and additional criteria are provided by a negative temperature coefficient of intensity at higher temperatures due to an increase in rate constant k_{MC} of the dissociative feedback process, and, in the case of the exciplex, a reduction in intensity with increase in solvent dielectric constant, which promotes ionic dissociation (rate constant k_I^C).

The overall quantum yield of exciplex fluorescence reflects both the intrinsic yield given by equation (8), where k_i^C is the rate constant of the ith non-radiative

$$\gamma_F^C = k_F^C/(k_F^C + \sum_i k_i^C) \tag{8}$$

exciplex relaxation process, and the photoassociation efficiency, which if $k_{MC} \simeq 0$ is given by equation (9). Thus if the lifetime τ_M of the molecular precursor $^1M^*$

$$\phi_{CM} = k_{CM}\tau_M[Q]/(1 + k_{CM}\tau_M[Q]) \tag{9}$$

is short and [Q] (or [M] for an excimer formation) is limited by solubility, photoassociation will be restricted particularly in viscous solvents at low temperatures where the diffusional process (rate constant k_{CM}) is slow.

Various ingenious techniques have been exploited to increase the photoassociation efficiency. Johnson and Offen[19] have used the negative activation volume of photoassociation to record the elusive orange excimer fluorescence of perylene in cyclohexane at high pressures and reported an associated lifetime of 72 ns at 77 K ($\tau_M = 6.6$ ns) which is reduced to 13.4 ns at 296 K by a competing non-radiative process with an activation energy of 670 cm^{-1} and a frequency factor of 1.7×10^9 s^{-1} characteristic of k_{IS}^C.

The very short fluorescence lifetime, τ_M, and low emission yields of heterocyclic compounds are usually attributed to efficient intersystem crossing from the lowest $n\pi^*$ singlet state. In the case of quinoline and isoquinoline the energy of this state is raised above that of the singlet $\pi\pi^*$ state in hydrogen-bonding solvents with the result that τ_M is increased, and at higher concentrations in ethanol at 160 K excimer fluorescence bands are observed[20] red-shifted by 6500 and 6700 cm^{-1}, respectively, from the molecular 0–0 bands; no corresponding changes in the absorption spectra with concentration were noted.

An increase in local concentration (and photoassociation efficiency) may be effected in several ways, the simplest of which conceptually is use of the pure

[19] P. C. Johnson and H. W. Offen, *Chem. Phys. Letters*, 1973, **18**, 258.
[20] R. P. Blaunstein and K. S. Gant, *Photochem. and Photobiol.*, 1973, **18**, 347.

solid phase where 'sandwich' pairs of molecules may be present as the structural unit of the crystal or as defect sites induced by pressure or rapid sublimation. In this way Muller *et al.*[21] have excited a broad structureless emission band ($\bar{v}_{max} = 15\,500\,\text{cm}^{-1}$) in a sublimed tetracene film at 110 K, characteristic of excimer fluorescence with a lifetime of 65 ns ($\tau_M = 6.4$ ns). Although this lifetime is independent of temperature, the emission intensity is thermally quenched with an activation energy of 0.12 eV, corresponding to the energy deficiency for singlet (molecular) exciton fission into two triplet excitons. The estimated value of $2 \times 10^{10}\,\text{s}^{-1}$ for the rate constant of excited-pair formation is believed to describe the frequency of energy transfer from initially excited molecules to dimer site traps.

An effective increase in local concentration is also achieved by 1,3-distribution of aryl groups in a hydrocarbon (or other polymer) chain. From a study of the time-resolved intramolecular excimer fluorescence of 1,3-bis(α-naphthyl)propane in ethanol–glycerol mixtures, El-Bayoumi *et al.*[22] obtained a radiative lifetime of 720 ns, which is significantly longer than the value of 500 ns reported for the excimer of 1-methylnaphthalene in ethanol. The suggestion that the methylene chain locks the molecule in a perfect sandwich excimer configuration, restricting the torsional oscillations necessary to induce the symmetry-forbidden transition, is consistent with an apparent red shift of $\simeq 1000\,\text{cm}^{-1}$ in λ_{max} relative to the 1-methylanthracene excimer band peak. The relatively low values of the viscosity-dependent rate constants k_{CM} and k_{MC} indicate that rotation about the methylene chain limits the rates of photoassociation and dissociation in this intramolecular system. Chandross *et al.*[23] have also observed the intramolecular excimer fluorescence of 1,3-bis(α-naphthyl)propane produced by photochemical cleavage of the photoadduct in a rigid glass at 77 K, and noted that the excimer band is removed by thermal recycling, which presumably allows rotation about the methylene chain.

The generation of isolated sandwich pairs by photolysis of solutions of the photodimer or photoadduct in a rigid low-temperature glass has also been used by Chandross,[24] who developed this technique, to record the naphthalene–anthracene exciplex emission from the photochemically cleaved adduct of 1-(9-anthryl)-3(1-naphthyl)propane; the exciplex fluorescence peak is red-shifted by 3200 cm^{-1} from the 0–0 band of 9-methylanthracene (or by 10 000 cm^{-1} from the 0–0 band of 1-methylnaphthalene) and is characterized by a quantum yield of 0.37 and a lifetime of 33 ns at 77 K. A similar emission band is exhibited by single crystals[25] of this cyclo-adduct following photochemical cleavage, where, however, the lifetime is increased to 80 ns (at 60 K), indicating that some relaxation of the sandwich pair along the repulsive ground-state potential is accommodated

[21] H. Muller, H. Bassler, and G. Vaubel, *Chem. Phys. Letters*, 1974, **29**, 102.
[22] P. Avouris, J. Kordas, and M. A. El-Bayoumi, *Chem. Phys. Letters*, 1974, **26**, 373.
[23] E. A. Chandross and C. J. Dempster, *J. Amer. Chem. Soc.*, 1970, **92**, 704.
[24] E. A. Chandross and A. H. Schiebel, *J. Amer. Chem. Soc.*, 1973, **95**, 1671.
[25] J. Ferguson, A. W.-H. Mau, and M. Puza, *Mol. Phys.*, 1974, **27**, 377; **28**, 1457, 1467.

by a hydrocarbon glass. Other aspects of this work are reported in appropriate sections below.

Triplet–triplet annihilation [process (10)] generates excited singlet and ground-state molecules in close proximity, which may subsequently photoasso-

$$^3M^* + {}^3M^* \rightarrow {}^1M^* + M \tag{10}$$

ciate in a re-encounter; this accounts for the increased ratio DEF/DF of excimer/molecular fluorescence intensities in the delayed emission spectrum relative to this ratio in the prompt fluorescence spectrum, and for the reduction in this ratio as the solvent viscosity is increased at lower temperature. Alternatively, triplet–triplet annihilation may produce the excimer directly in the competitive process (11). Tachikawa and Bard[26] have shown that the DEF/DF ratio of pyrene and

$$^3M^* + {}^3M^* \rightarrow {}^1MM^* \tag{11}$$

of 1,2-benzanthracene in cyclohexane at room temperature is essentially independent of applied magnetic field up to 8 kG, consistent with a common origin of both emitting species. Since excimer dissociation is thermally restricted at low temperatures where DF ≫ DEF, these findings, which contradict the results of a previous study of magnetic field effects by Wyrsch and Labhart,[27] support the exclusive operation of process (10) as the primary annihilation event.

Indirect evidence for process (10) as the primary annihilation event is provided by the interesting report by Nickel[28] that this leads to emission from higher singlet states S_n of benzanthracene ($n = 1, 2,$ or 3) and fluoranthene ($n = 1, 2, 3,$ or 4) in liquid paraffin; the absence of corresponding excimer bands may be attributed to the very rapid radiative decay of $S_n (n > 1)$ which effectively competes with re-encounter association.

Excited and unexcited molecules may also be generated in close proximity by the doublet–doublet annihilation process (12) (rate constant $k_c^{\dagger}$ of Figure 1)

$$^2M^- + {}^2Q^+ \rightarrow {}^{1,3}M^* + Q \tag{12}$$

$$^2M^- + {}^2Q^+ \rightarrow {}^{1,3}MQ^* \tag{13}$$

where M* may be a triplet state if electron transfer enthalpy is insufficient to produce $^1M^*$, in which case process (12) may be followed by (10) or (11). Again, direct association [process (13)] must also be considered, and the resulting emission is described as chemiluminescence (CL) if the radical ions are prepared chemically or electrogenerated chemiluminescence (ECL) if they are produced electrochemically. Following a study of the CL of 9-methylanthracene (M) and tri-*p*-tolylamine (Q) Weller and Zachariasse[29] concluded that the primary process (13) is followed by singlet exciplex emission which competes with thermal

[26] H. Tachikawa and A. J. Bard, *Chem. Phys. Letters*, 1974, **26**, 568.
[27] D. Wyrsch and H. Labhart, *Chem. Phys. Letters*, 1971, **8**, 217.
[28] B. Nickel, *Chem. Phys. Letters*, 1974, **27**, 84.
[29] A. Weller and K. Zachariasse, *Chem. Phys. Letters*, 1971, **10**, 590.

dissociation (process k_{MC} of Figure 1) responsible for the observed fluorescence of M. However, Tachikawa and Bard[26] find that the reaction enthalpy computed from electrochemical data for this system is insufficient to produce $^1MQ^*$ in process (13) whereas magnetic field effects implicate intermediary triplet–triplet annihilation in the formation of both $^1M^*$ and $^1MQ^*$. These authors[26] suggested that $^3M^*$ is formed in process (12) and subsequently undergoes process (10), as in the pyrene–tetramethylenephenylenediamine system, but the origin of $^1MQ^*$, if this is indeed responsible for the broad emission band at 520 nm, is less certain.

Balzani's group[30] have reported the appearance of a well-characterized structureless emission band ($\lambda_{max} = 650$ nm) as naphthalene is added to solutions of the cis-dichlorobis-(1,10-phenanthroline)iridium(III) cation; this is attributed to the first 'inorganic' exciplex in which naphthalene adopts a sandwich configuration with a phenanthroline ligand with an estimated binding energy of 4 kcal mol^{-1}; however, the exciplex spectrum is not exhibited by 1,10-phenanthroline in the presence of naphthalene.

The first report of molecular excimer fluorescence in the vapour phase concerns 1-azabicyclo[2,2,2]octane, which exhibits a structureless emission band ($\lambda_{max} = 375$ nm) red-shifted by 12 400 cm^{-1} from the molecular 0–0 fluorescence band. Halpern[31] found that this low-energy band is enhanced by the addition of n-hexane, and analysed the data in terms of a photoassociation mechanism to obtain an associated lifetime of 440 ns and an equilibrium constant (k_{CM}/k_{MC}) of 2.4×10^5 dm^{-3} mol^{-1}.

Excimer–Exciplex Absorption.—The observation of excited singlet states in absorption, facilitated by the development of excitation pulses and detection techniques with resolution in the nanosecond region, has been extended to excimers and exciplexes. Assignments are usually based on intensity variations with temperature and concentration, and may be confirmed by the coincidence of relaxation times in absorption and emission, although this is not always possible. Thus Thomas and co-workers,[32] using pulsed radiolysis, find transient absorption bands in the xylenes, mesitylene, and pseudocumene at 600 nm which are enhanced on cooling and are attributed to singlet excimer absorption; the exponential decay is characterized by lifetimes of ca. 20 ns, and molar extinction coefficients of the order 10^4 dm^3 mol^{-1} cm^{-1} are calculated from the total singlet G-values and the photoassociation equilibrium constants. An absorption band at 1100 nm in the transient spectrum of anthracene in benzene following pulsed radiolysis has been assigned by Rodgers[33] to the anthracene excimer with a lifetime of <3 ns.

[30] R. Ballardini, G. Varani, L. Moggi, and V. Balzani, J. Amer. Chem. Soc., 1974, **96**, 7123.

[31] A. M. Halpern, J. Amer. Chem. Soc., 1974, **96**, 4392.

[32] R. V. Bensasson, J. T. Richards, T. Gangwer, and J. K. Thomas, Chem. Phys. Letters, 1972, **14**, 430.

[33] M. A. J. Rodgers, Chem. Phys. Letters, 1972, **12**, 612.

A pyrene excimer absorption band ($\lambda_{max} = 490$ nm) has been reported by Ottolenghi's group[34] using nitrogen laser excitation; the associated lifetime is reduced by an increase in excitation intensity due to excimer–excimer annihilation analogous to process (3). The assignment has been confirmed by Slifkin *et al.*[35] using modulation excitation spectroscopy.

The assignments of more complex exciplex absorption spectra are facilitated by the recognition that certain bands are due to localized excitation of donor and acceptor radical ions [processes (14) and (15)] exhibited by ground-state

$$^1(M^-Q^-) + h\nu \rightarrow {}^1(M^{+*}Q^-) \quad (14)$$

$$^1(M^-Q^-) + h\nu \rightarrow {}^1(M^+Q^-*) \quad (15)$$

EDA complexes. Thus peaks at 540, 580, and 620 nm in the biphenyl(BP)– tetramethylphenylenediamine (TMPD) exciplex spectrum are assigned[36] to TMPD$^+$, whereas those at 410 and 650 nm are attributed to BP$^-$. Additional bands originate in reverse CT transitions (16) and (17) whereas others may be

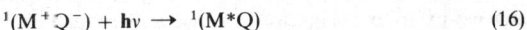

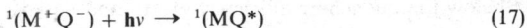

$$^1(M^+Q^-) + h\nu \rightarrow {}^1(M^*Q) \quad (16)$$

$$^1(M^+Q^-) + h\nu \rightarrow {}^1(MQ^*) \quad (17)$$

associated with states of donor or acceptor which are inaccessible by one-photon absorption. The reader is referred to the excellent review by Ottolenghi[8] for further details of this important area of molecular spectroscopy.

An interesting consequence of exciplex absorption is ionic dissociation leading to transient photoconductivity in a biphotonic process, as exemplified by the naphthalene(N)–TMPD system excited by a nitrogen laser ($\lambda = 337.1$ nm) in non-polar solvents.[36] The second-power dependence of photoconductivity on

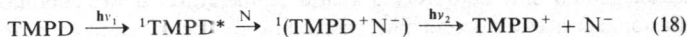

$$\text{TMPD} \xrightarrow{h\nu_1} {}^1\text{TMPD}^* \xrightarrow{N} {}^1(\text{TMPD}^+\text{N}^-) \xrightarrow{h\nu_2} \text{TMPD}^+ + \text{N}^- \quad (18)$$

light intensity is consistent with the sequence (18), where the energy of the dissociated ion pair exceeds that of the laser quantum. In solvents of higher dielectric constant the energy of the ion pair is reduced and the photoconductivity varies linearly with laser excitation intensity.[37]

Kinetic Consequences of Photoassociation.—In the absence of direct spectroscopic evidence, photoassociation may be inferred from the negative temperature coefficient of the yield of a bimolecular photochemical or photophysical process due to the energy dependence of the competing feedback process (k_{MC} of Figure 1). An example is provided by the photochemical addition of *trans*-stilbene to olefins described by Saltiel *et al.*[38] Lewis and Ware[39] have shown at some length how the feedback process may introduce discrepancies between steady-state and

[34] C. R. Goldschmidt and M. Ottolenghi, *J. Phys. Chem.*, 1970, **74**, 2041.
[35] M. A. Slifkin and A. O. Al-Chalabi, *Chem. Phys. Letters*, 1973, **20**, 211.
[36] A. Alchalal, M. Tamir, and M. Ottolenghi, *J. Phys. Chem.*, 1972, **76**, 2229.
[37] Y. Taniguchi, Y. Nishima, and N. Mataga, *Bull. Chem. Soc. Japan*, 1972, **45**, 764.
[38] J. Saltiel, J. T. D'Agostino, O. L. Chapman, and R. D. Lura, *J. Amer. Chem. Soc.*, 1971, **93**, 2804.
[39] C. Lewis and W. R. Ware, *Mol. Photochem.*, 1973, **5**, 261.

time-dependent quenching constants which can be analysed to obtain lifetimes and binding energies of excimers–exciplexes even when these are non-fluorescent, and two papers[40] have recently appeared from Ware's laboratory describing the application of these methods to real systems.

Seliger et al.[41] have pointed out that exciplex binding energies can only be obtained from the temperature dependence of fluorescence intensity quotients if the photoassociation and feedback frequencies greatly exceed the relaxation constants of the emitting species, and suggested that the independence of this quotient on concentration of added quencher (e.g. O_2) be adopted as a criterion for this condition. These authors question the findings of Mataga et al.[42] that the photoassociation enthalpy of the pyrene–dimethylaniline exciplex is reduced from 6.9 kcal mol^{-1} in hexane ($\varepsilon = 1.9$) to 2.3 kcal mol^{-1} in pyridine ($\varepsilon = 12.3$) on the grounds that equilibrium is not established. An increase in association enthalpy with dielectric constant reported by Koizumi et al.[43] is to be expected.

If it is accepted that the exchange transfer of electronic excitation energy is accommodated by processes k_{CM} and k_{QC} (Figure 1), then any radiative or non-radiative exciplex relaxation process will reduce the sensitization efficiency of Q* below the quenching efficiency of M* and may provide indirect evidence of photoassociation. However, discrepancies of this type do not seem to have been reported for energy transfer in the singlet manifold.

4 Excimer–Exciplex Triplet States

Spectroscopic Evidence.—Since exciton interactions are not expected to contribute significantly to binding energies of the excimer triplet state, these will be considerably smaller than the singlet-state binding energy, and excimer phosphorescence should only be observed at low temperatures under conditions where diffusional processes permit photoassociation. The question then arises as to whether the triplet excimer is produced by intersystem crossing from the excimer singlet state (process k_{IS}^C) or directly by photoassociation in the triplet manifold (process $^3k_{CQ}$); in either case the effects of trace amounts of quenching or luminescent impurities cannot be overlooked, and several authors have preferred to eliminate these by examining rigid systems containing molecular pairs of the appropriate geometry.

Following a useful summary of the evidence for excimer triplet states, Averis and El-Bayoumi[44] reported the phosphorescence of 1,3-diphenylpropane in isopentane, which shifts from a toluene-like (molecular) spectrum at 77 K to a broader band ($\lambda_{max} = 420$ nm) at 115 K and is replaced by the excimer fluorescence at higher temperatures. These authors ascribe the 420 nm emission band to excimer phosphorescence, red-shifted by ca. 5000 cm^{-1} from the molecular

[40] W. R. Ware, D. Watt, and J. D. Holmes, J. Amer. Chem. Soc., 1974, **96**, 7853; W. R. Ware, J. D. Holmes, and D. R. Arnold, ibid., p. 7861.
[41] R. J. McDonald and B. K. Selinger, Mol. Photochem., 1971, **3**, 99.
[42] T. Okada, H. Matsui, H. Oohari, H. Matsumoto, and N. Mataga, J. Chem. Phys., 1968, **49**, 4717.
[43] S. Murata, H. Kokubun, and M. Koizumi, Z. phys. Chem. (Frankfurt), 1970, **70**, 47.
[44] P. Averis and M. A. El-Bayoumi, Chem. Phys. Letters, 1973, **20**, 59.

0–0 phosphorescence band, and ascribe its formation to photoassociation of the molecular triplet state. Yokoyama *et al.*[45] report the presence of a broad band ($\lambda_{max} = 459$ nm) in the phosphorescence spectrum ($\tau = 32$ ms) of poly-(3,6-dibromo-N-vinylcarbazole) in rigid 2-methyltetrahydrofuran at 77 K. This exhibits a linear dependence on light intensity, and in the absence of photochemical decomposition is assigned to the (intramolecular) excimer phosphorescence which is red-shifted by 2200 cm^{-1} from the 0–0 phosphorescence band of the 3,6-dibromo-N-ethylcarbazole molecule chosen as a model monomer; this red shift may be compared with the value of 3000 cm^{-1} observed for the corresponding excimer fluorescence.

Rigid low-temperature solutions of the photochemically cleaved photodimer of 1,3-bis(α-naphthyl)propane exhibit excimer fluorescence and a structured phosphorescence, shifted to the red of naphthalene phosphorescence by 250 nm. After thermal recycling only molecular fluorescence and phosphorescence is observed, leading Chandross and Dempster[23] to the conclusion that the naphthalene triplet-state photoassociation is prevented by steric requirements or thermal instability. Ferguson and co-workers[46] concluded that the phosphorescence quantum yield of anthracene sandwich pair produced by dimer cleavage must be $<1\%$ of the molecular phosphorescence yield.

The structureless phosphorescence emission band at 13 800 cm^{-1} exhibited by single crystals of pyrene at room temperature has also been observed from powdered pyrene after the sample was subjected to high pressures (15 ton cm^{-2}) to create lattice defects and is attributed by Langelaar *et al.*[47] to defect excimers lying 2000 cm^{-1} below the triplet exciton energy in the crystal. Perhaps the most conclusive evidence for triplet excimers has been reported by El-Sayed's group,[48] who used the phosphorescence microwave double-resonance technique to determine the principal magnetic axes of the triplet state responsible for excimer emission in a hexachlorobenzene crystal at 1.6 K; the zero-field transition frequencies are $|D| + |E| = 4.665$ GHz, $|D| - |E| = 3.692$ GHz and $2|E| = 0.972$ GHz.

Using the modulation excitation technique, Slifkin *et al.*[49] observed broad new structureless absorption bands in very concentrated solutions of pyrene, benzanthracene, and dibenzanthracene, attributed to the products of triplet-state photoassociation. In this respect it is of interest to note the thesis of Hoytink *et al.*[50] which ascribes the short concentration-dependent lifetimes of aromatic hydrocarbon triplet states in solution to self-quenching *via* photoassociation rather than to impurity quenching. Following a detailed treatment of the concentration and temperature dependence of these lifetimes, these authors

[45] M. Yokoyama, M. Funaki, and H. Mikawa, *J.C.S. Chem. Comm.*, 1974, 372.
[46] J. Ferguson, A. W.-H. Mau, and J. M. Morris, *Austral. J. Chem.*, 1973, **26**, 91.
[47] O. L. Gijzeman, J. Langelaar, and J. D. W. VanVoorst, *Chem. Phys. Letters*, 1970, **5**, 269; 1971, **11**, 526.
[48] M. A. El-Sayed, C. T. Lin, and R. Leyerle, *Chem. Phys. Letters*, 1974, **25**, 457.
[49] M. A. Slifkin and A. O. Al-Chalabi, *Chem. Phys. Letters*, 1973, **20**, 211; 1974, **29**, 110.
[50] J. Langelaar, G. Jansen, R. P. H. Rettschnick, and G. J. Hoytink, *Chem. Phys. Letters*, 1971, **12**, 86.

found that the triplet excimer formation efficiency is of the order 10^{-4} for four different fluor molecules, and is attended by an energy barrier of *ca.* 400 cm^{-1} which excludes triplet-state photoassociation at 77 K. Lindquist *et al.*[51] have attributed the concentration quenching of the pentacene triplet state to the formation of a dimeric species which undergoes ground-state dissociation with an activation energy of 4600 cm^{-1} and frequency factor of 10^{11} s^{-1}, consistent with a lifetime of 36 ms at 298 K.

Ottolenghi *et al.*[52] found that the decay of the ^{1}(TMPD$^+$N$^-$)* exciplex absorption in non-polar solvents generates the naphthalene triplet state at the same rate. The overall process may be regarded as an intersystem crossing to the exciplex triplet state (k_{IS}^C) which rapidly dissociates ($^3k_{MC}$). This 'slow' intersystem crossing from the thermalized exciplex is distinguished from a 'fast' process which leads to the absorption of aromatic triplet states in the presence of NN-diethylaniline immediately after the laser flash and prior to any significant decay of the singlet exciplex; this is attributed to intersystem crossing from the vibrationally unrelaxed exciplex. Since both 'slow' and 'fast' intersystem crossing populate the acceptor triplet state, the yield of the latter is temperature dependent, as in the case of isolated molecules where intersystem crossing to different triplet states above and below the singlet state is believed to take place.

Whitten and co-workers[53] have assigned new transient absorption spectra from anthracene and metalloporphyrin triplet states in the presence of nitro- and chloro-aromatic quenchers at high concentration to triplet exciplexes of 1:1 and 1:2 stoicheiometry. These may result from exciplexes of different geometry, described by Chandross *et al.*[54] as sandwich pair and localized pair, the latter being responsible for non-linear Stern–Volmer quenching at high quencher concentrations due to the formation of 1:2 exciplexes which shift λ_{max} (of exciplex fluorescence) to longer wavelengths.

Kinetic Evidence.—In support of an intersystem crossing (process k_{IS}^C) in the excimer, various workers have reported non-radiative relaxation of singlet excimers with rate constants of the order expected for spin-prohibitive transitions, *i.e. ca.* 10^9 s^{-1}. Thus Offen *et al.*[19] found that the temperature dependence of the perylene excimer fluorescence decay constant is characterized by a frequency factor of 1.7×10^9 s^{-1}, whereas Cundall's group[55] reported a temperature dependence of the non-radiative relaxation of the 1-methylnaphthalene excimer in the form

$$k_{NR}/s^{-1} = 5 \times 10^6 + 3.2 \times 10^{11} \exp(-3400\,K/T)$$

with the reasonable suggestion that these numerical terms describe intersystem crossing and internal conversion, respectively. This same group reported an

[51] C. Hellner, L. Lindquist, and P. C. Roberge, *J.C.S. Faraday II*, 1972, **68**, 1928.
[52] N. Orbach, J. Novros, and M. Ottolenghi, *J. Phys. Chem.*, 1973, **77**, 2831; C. R. Goldschmidt, R. Potashnik, and M. Ottolenghi, *ibid.*, 1971, **75**, 1025.
[53] J. K. Roy, F. A. Carroll, and D. G. Whitten, *J. Amer. Chem. Soc.*, 1974, **96**, 6349.
[54] G. N. Taylor, E. A. Chandross, and A. H. Schiebel, *J. Amer. Chem. Soc.*, 1974, **96**, 2693.
[55] R. B. Cundall and L. C. Pereira, *Chem. Phys. Letters*, 1972, **15**, 383; R. B. Cundall, L. C. Pereira, and D. A. Robinson, *ibid.*, 1972, **13**, 253.

increase in benzene triplet yield with benzene concentration to 0.56 in pure benzene, from which an analysis based on the increased intersystem-crossing yield in the excimer provides an intrinsic value of $\geqslant 0.71$ for this parameter.

If the transfer of triplet energy involves photoassociation (processes $^3k_{CM}$ and $^3k_{QC}$ of Figure 1) any competing relaxation of the exciplex triplet state will reduce the rate (and yield) of excited acceptor formation below the rate of donor quenching. To the Reporter's knowledge this has only been observed[56] for the energy donor–acceptor system of biacetyl–benzil, where the donor quenching constant and acceptor sensitization constants are 900 and 390 $dm^3\ mol^{-1}\ s^{-1}$ respectively. This energy transfer deficiency provides indirect evidence for an intermediate exciplex triplet state which may undergo chemical or physical relaxation, and it may be operating in other systems where an energy-transfer efficiency of unity is assumed to compute either triplet acceptor absorption coefficients[57] or donor intersystem crossing yields.[58]

5 Oxygen Exciplexes

The interpretation of bimolecular quenching of electronically excited molecules, in terms of primary photoassociation with the quenching species followed by exciplex relaxation should logically apply to oxygen as an effective quenching molecule. However, the photoassociation scheme in this case is more complex than that shown in Figure 1 owing to the triplet nature of the $O_2(^3\Sigma_g^-)$ ground state, the existence of several low-lying singlet states ($O_2\ ^1\Delta_g$ at 8000 cm^{-1} and $O_2\ ^1\Sigma_g^+$ at 13 000 cm^{-1} above $O_2\ ^3\Sigma_g^-$), and the relatively high electron affinity of $O_2\ ^3\Sigma_g^-$; thus the lowest state of the oxygen exciplex (or oxciplex) is a singlet and of the (dissociated) ground state is a triplet.

The evidence for oxygen exciplexes with aromatic molecules is largely mechanistic, although the excited complexes produced by absorption at high oxygen pressures in the classic work of Evans[59] are almost certainly those which result from photoassociation; moreover, Ottolenghi and co-workers[60] have observed the radical cation of pyrene following oxygen quenching of the triplet state in solvents of high dielectric constant, and the new low-energy emission bands observed by Kasha *et al.*[61] from air-saturated dye solutions may originate from an oxygen exciplex state.

Mechanistically, the evidence is largely based on kinetic studies[62] of the sensitized photoaddition of molecular oxygen to an aromatic hydrocarbon M such as anthracene, which can act as its own sensitizer. A wealth of evidence supports the intermediary role of $O_2\ ^1\Delta_g$, which appears to be formed solely by oxygen quenching of the sensitizer triplet state, and the kinetic observations are

[56] H. H. Richtol and A. Belorit, *J. Chem. Phys.*, 1966, **45**, 35.
[57] R. Bensasson and E. J. Land, *Trans. Faraday Soc.*, 1971, **67**, 1904.
[58] R. B. Cundall and W. Tippett, *Trans. Faraday Soc.*, 1970, **66**, 350.
[59] *E.g.* D. F. Evans, *J. Chem. Soc.*, 1961, 2566.
[60] R. Potashnik, C. R. Goldschmidt, and M. Ottolenghi, *Chem. Phys. Letters*, 1971, **9**, 424.
[61] D. E. Brabham and M. Kasha, *Chem. Phys. Letters*, 1974, **29**, 159.
[62] B. Stevens, *Accounts Chem. Res.*, 1973, **6**, 90.

$$M(S_1) + O_2(^3\Sigma) \longrightarrow {}^3\Gamma_{ct}(S_1{}^3\Sigma)$$

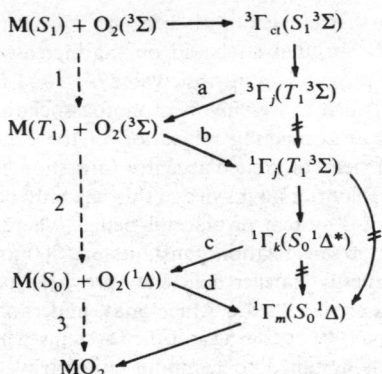

Figure 2   *The suggested role of oxciplex states in the self-sensitized photoperoxidation of an aromatic hydrocarbon* M

(Reproduced by permission from *Chem. Phys. Letters*, 1974, **27**, 157)

consistent with the sequential operation of processes 1—3 of Figure 2 rather than direct addition of $O_2(^3\Sigma)$ to either the excited singlet or triplet states of the aromatic hydrocarbon M. This has been rationalized[63] in terms of electronic relaxation of the intervening states Γ_i of the oxciplex, indicated in Figure 2 by solid arrows, which may involve dissociation and recombination to overcome the spin barrier ${}^3\Gamma_j \rightarrow {}^1\Gamma_j$ (processes a and b) or symmetry restriction attending the internal conversion ${}^1\Gamma_k \rightarrow {}^1\Gamma_m$ (processes c and d).

A complete photoassociation scheme should include the state ${}^3\Gamma_h(T_1{}^1\Sigma)$, which may lie above or below Γ_j, together with ${}^5\Gamma_j(T_1{}^3\Sigma)$ and the ground state ${}^3\Gamma_0(S_0{}^3\Sigma)$ but as yet even less is known of the behaviour of the first two of these.

6 Photoaddition

From the purely photochemical standpoint process k_A^C, leading to the formation of photoadduct (or photodimer), is the important exciplex (excimer) relaxation route. The intermediary role of the exciplex in photoaddition is confirmed experimentally in at least one case but is more often a conceptual convenience in systems where exciplex fluorescence is not observed.

A novel system has been reported by Chandross and Schiebel,[64] who found that 1-(9-anthryl)-3-(1-naphthyl)propane undergoes an intramolecular photo-addition at low concentrations $(2 \times 10^{-5}\,\mathrm{mol\,dm^{-3}})$ in methylcyclohexane, but at higher concentrations $(> 10^{-3}\,\mathrm{mol\,dm^{-3}})$ forms an intermolecular photo-dimer of two anthracene moieties. Chandross and Dempster[23] also report that the intramolecular photodimerization of α-naphthyl substituents at the 1 and 3 positions of an alkane chain competes with excimer emission and is reversible,

[63] B. Stevens, *J. Photochem.*, 1974, **3**, 393.
[64] E. A. Chandross and A. H. Schiebel, *J. Amer. Chem. Soc.*, 1973, **95**, 611, 1671.

whereas the same substituents at 1,2- and 1,4-positions are photochemically unreactive and exhibit no spectroscopic evidence of intramolecular photoassociation.

Ferguson and co-workers[25] found that the thermal quenching of the low-temperature intramolecular exciplex emission of 1-(9-anthryl)-3-(1-naphthyl)-propane is due to photoaddition since both processes are characterized by the same frequency factor and an activation energy of $600\,cm^{-1}$; this must be regarded as conclusive evidence for the role of the exciplex intermediate in this case, and it may be instructive to examine the orbital correlations for the excimer–photodimer system using anthracene A, in which the excimer is believed to be of ^1L_a origin, as a model. The correlation diagram shown in Figure 3 is restricted to the HOMO (Ψ_7) and LUMO (Ψ_8) orbitals of two anthracene molecules which transform as b_{1g} and b_{3u} under symmetry operations of the D_{2h} point group. In the symmetrical sandwich excimer configuration of the same symmetry these split into dimer orbitals $\Psi_7 + \Psi_7$ (b_{2u}), $\Psi_7 - \Psi_7$ (b_{1g}), $\Psi_8 + \Psi_8$ (a_{1g}), and $\Psi_8 - \Psi_8$ (b_{3u}) as shown. Correlation of these with the photodimer orbitals is facilitated by the correlation of $\Psi_1 - \Psi_6$ and $\Psi_8 - \Psi_{14}$ Π-orbitals of each anthracene molecule with symmetry-adapted linear combinations $\Psi_i \pm \Psi_i$ ($i = 1, 2,$ or 3) and $\Psi_j \pm \Psi_j$ ($j = 4, 5,$ or 6) of the benzene ring Π-orbitals in the photodimer; this restricts further examination to symmetry-adapted linear combinations of σ-orbitals located at the 9,9'- and 10,10'-positions in the photodimer which transform as a_{1g} ($\sigma_c + \sigma_c$), b_{2u} ($\sigma_c - \sigma_c$), b_{3u} ($\sigma_c^* + \sigma_c^*$), and b_{1g} ($\sigma_c^* - \sigma_c^*$), under the same symmetry operations. As is evident from the orbital occupancies shown in Figure 3, the excimer correlates with a singly-excited state

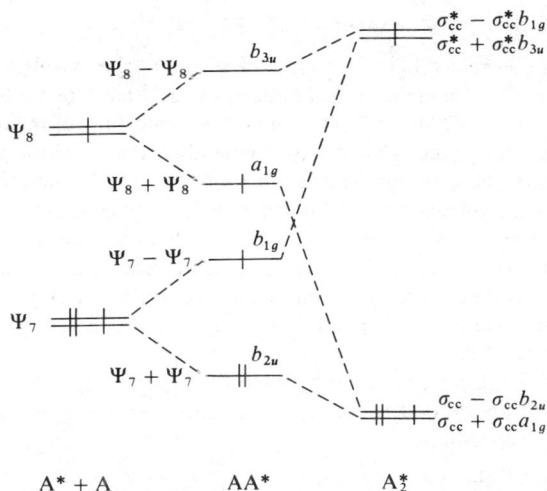

Figure 3 *Orbital correlation diagram for the photodimerization of anthracene* A *via the excimer intermediate* AA*

of the photodimer, which unless subject to electronic relaxation along the reaction profile must undergo internal conversion or bimolecular quenching in the product configuration.

In this respect it is interesting to note that Saltiel and Townsend[65] have reported an increase in the quantum yield of anthracene photodimerization in the presence of conjugated dienes which also quench the anthracene fluorescence. Of the alternative processes

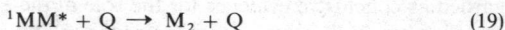

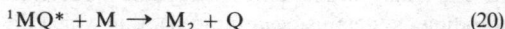

$$^1MM^* + Q \rightarrow M_2 + Q \tag{19}$$

$$^1MQ^* + M \rightarrow M_2 + Q \tag{20}$$

these authors found that the latter is quantitatively consistent with their data; however, the dependence of yield on anthracene concentration was not examined and exciplex emission was not observed, although the exciplex lifetime must exceed that of the anthracene singlet state (*ca.* 5 ns) if process (20) is to offset the quenching of this state by the diene.

Campbell and Liu[66] have reported the general diene catalysis of 9-phenyl-anthracene photodimerization which was previously believed to be sterically hindered, and following a detailed analysis of mechanisms involving either process (19) or process (20), they have shown that only the former is consistent with the observed linear dependence of reciprocal quantum yield on reciprocal concentration of 9-phenylanthracene. A possible implication of these results is re-formation of the excimer from the excited dimer in the absence of diene quenching, facilitated by the small energy barrier to photochemical addition reported by Ferguson *et al.*;[25] in this case symbolic differentiation of excimer and excited dimer in the sequence (21) may be desirable.

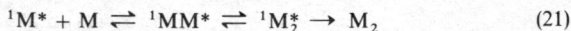

$$^1M^* + M \rightleftharpoons {}^1MM^* \rightleftharpoons {}^1M_2^* \rightarrow M_2 \tag{21}$$

Compelling evidence for the intermediary role of the exciplex in the photo-addition of phenanthrene to dimethyl fumarate and maleate is provided by the work of Creed and Caldwell,[67] who found that quenching of the exciplex fluorescence by electron donors quantitatively parallels the reduction in yield of the photoproducts; these authors suggest the term 'exterplex' to describe the excited termolecular intermediate for which precedent is reported in atomic systems [process (6), $n = 2$]. Sasaki *et al.*[68] have found that the photochemical addition of cycloheptatriene to anthracene competes with anthracene photodimerization and that the adduct/dimer yield quotient is reduced in heavy-atom solvents; these authors assume the formation of a singlet exciplex intermediate but the results may be consistent with photodimerization (of anthracene) in the triplet manifold. This is accommodated by the orbital correlation scheme in Figure 3, where it also appears that triplet–triplet annihilation could lead directly to the photodimer ground state.

[65] J. Saltiel and D. E. Townsend, *J. Amer. Chem. Soc.*, 1973, **95**, 6140.
[66] R. O. Campbell and R. S. H. Liu, *Mol. Photochem.*, 1974, **6**, 207.
[67] D. Creed and R. A. Caldwell, *J. Amer. Chem. Soc.*, 1974, **96**, 7369.
[68] T. Sasaki, K. Kanematsu, and K. Hayakawa, *J. Amer. Chem. Soc.*, 1973, **95**, 5632.

Suggested chemical reactions of the exciplex are not restricted to addition. Thus Gutierrez and Whitten[69] find that the quantum yield of *trans–cis* isomerization of 1,2-bis-(4-pyridyl)ethylene (BPE) in acetonitrile is reduced by the addition of bromoethane, which also quenches the PBE fluorescence to a different extent; this is attributed to the formation of a relatively long-lived BPE–solvent exciplex which interacts with bromoethane to release the *cis*-BPE isomer. Proton transfer in, the exciplex is proposed by Bowman *et al.*[70] in connection with the photoaddition of acrylonitrile to indene, and by Yang and Libman[71] in their mechanism for the 9,10-addition of secondary amines to anthracene in benzene, the quantum yield of which is reduced by deuteriation of the amine.

The reverse of photoaddition (process k_c^A of Figure 1) may be expected to generate molecular excited states and is the basis of mechanisms proposed for chemiluminescence in certain systems. Thus Turro and Lechtken[72] have demonstrated that tetramethyl-1,2-dioxetan thermally decomposes to two molecules of acetone, one of which is in the lowest triplet state, whereas in the oxalate–hydrogen peroxide system an energetic dioxetandione (a CO_2 dimer) has been suggested as a key intermediate[73] which can sensitize emission from fluorescent acceptors with simultaneous self-dissociation.

7 Excited Dimers and Complexes

The absence both of vibrational structure in the excimer fluorescence spectrum and of a corresponding band in the absorption spectrum is consistent with a repulsive interaction potential between the unexcited molecular components in the excimer configuration. However, the introduction of energy barriers to molecular separation, as in the crystalline state or in rigid glass solutions of photochemically cleaved photodimers, may produce an interaction potential minimum leading to characteristic dimer absorption bands which by definition are associated with an excited dimer as the upper state, although the resulting emission is usually indistinguishable from excimer fluorescence if the dimer pair has a sandwich configuration. In this case the topochemical requirements for photodimerization are satisfied and the red excimer fluorescence of crystalline sandwich pairs of anthracene is thermally quenched.[25]

Since the 'exact' sandwich pair configuration is probably the least stable in the ground state, a partial relaxation of the energy barriers to rotation and dissociation by softening the glass or photochemical disruption of the crystal lattice will lead to more stable pair configurations with different absorption and emission characteristics and photochemical properties. These have been

[69] A. R. Gutiérrez and D. G. Whitten, *J. Amer. Chem. Soc.*, 1974, **96**, 7129.
[70] R. M. Bowman, T. R. Chamberlain, C. W. Huang, and J. J. McCullough, *J. Amer. Chem. Soc.*, 1974, **96**, 692.
[71] N. C. Yang and J. Libman, *J. Amer. Chem. Soc.*, 1973, **95**, 5783.
[72] N. J. Turro, P. Lechtken, N. E. Schore, G. Schuster, H. C. Steinmetzer and A. Yekta, *Accounts Chem. Res.*, 1974, **7**, 97; N. J. Turro and P. Lechtken, *J. Amer. Chem. Soc.*, 1972, **94**, 2886; *Mol. Photochem.*, 1974, **6**, 95.
[73] H. Gusten and E. F. Ullman, *Chem. Comm.*, 1970, 28.

studied by Ferguson and co-workers,[25] who prepared molecular pairs by photo-chemical cleavage of photodimers in the crystalline state and in rigid glass solutions or by the controlled softening of glassy solutions of anthracene and derivatives at low temperatures. In the latter case[74] the appearance of new absorption spectra with isosbestic points is attributed to the successive formation of thermally unstable dimers and tetramers with well-defined spectral properties and short (ca. 5 ns) fluorescence lifetimes; the existence of in-phase and out-of-phase dimer components is established from the polarized fluorescence excitation spectrum in the $26\,000$ cm^{-1} region, indicating an angle of ca. 60° between the short in-plane molecular axes.

The behaviour of sandwich dimer pairs formed by dianthracene photocleavage in a rigid matrix at 77 K is marked by discontinuities in their fluorescence characteristics with progressive softening of the glass.[46] Thus the characteristic excimer emission maximum first shifts slightly to the blue and the lifetime changes from 200 to 100 ns, following which the lifetime is reduced to 5 ns and broad structure appears in the emission band; if the latter is associated with an excited dimer, as these authors suggest, then this must be distinguished from the excimer not only in the association–excitation sequence. Conceivably the 100 ns component could originate from an unrelaxed excited state of the photodimer (Figure 3) which may be in photochemical equilibrium with the excimer if the activation energy is low (cf. 600 cm^{-1} for anthracene–naphthalene photoaddition), since the radiative transition from this correlated state should also be forbidden.

The spectroscopic identity of the exciplex and excited (EDA) complex has been proposed by Itoh[75] following an examination of the emission characteristics of both species for a series of 9,10-dicyanoanthracene–alkylnaphthalene systems. The exciplex was formed following acceptor excitation in 3-methylpentane at room temperature whereas absorption in the CT band at 77 K was used to excite the EDA complex. A similar conclusion was reached by Shirota et al.[76] for arene–fumaronitrile complexes in non-polar solvents. However, depending on the donor–acceptor properties of other systems, the exciplex may be identifiable as the unexcited EDA complex.

8 Applications

Studies of photoassociation in aromatic polymer chains have been used to estimate the distribution of 'excimer-forming sites' and the rate of interconversion of chain conformations resulting from C—C bond rotation. An excellent summary of the extensive work in this area is given in a recent paper by Frank and Harrah.[77]

[74] J. Ferguson, A. W.-H. Man, and J. M. Morris, *Austral. J. Chem.*, 1973, **26**, 103.
[75] M. Itoh, *Chem. Phys. Letters*, 1974, **24**, 551; **26**, 505.
[76] Y. Shirota, I. Tsuhui, and H. Mikawa, *Bull. Chem. Soc. Japan*, 1974, **47**, 991.
[77] C. W. Frank and L. A. Harrah, *J. Chem. Phys.*, 1974, **61**, 1526.

Recent work concerning the role of excimers and exciplexes in biochemical systems is included in recent reviews by Song.[78] Of interest here is the use of pyrene excimer fluorescence as an optical probe to determine coefficients of lateral diffusion in the hydrophobic regions of membranes.[79]

Since, by definition, photoassociation leads to population inversion and excimer–exciplex emission is significantly red-shifted from the molecular absorption spectrum, the feasibility of stimulated emission from these systems has been recognized for some time. Unfortunately radiative transition probabilities from the sandwich configuration are relatively small and success appears to have been achieved only when the molecular components adopt a coplanar configuration due to hydrogen or other intramolecular bonding and the transition becomes allowed. Thus Srinavasan *et al.*[80] reported laser emission from concentrated solutions of alkylaminocoumarins in the presence of an acid and suggested that the emitting state is a protonated coplanar excited dimer of these dipolar molecules. Mataga *et al.*[81] have summarized the possibilities of exciplex laser systems and reported laser action from the intramolecular exciplex emission of p-(9'-anthryl)-NN-dimethylaniline in non-polar solvents; however, the structure and absorption spectrum of this molecule do not seem to warrant use of the term 'intramolecular exciplex'. An analysis[82] of the laser capabilities of molecular mercury has been referred to recently.[9]

[78] P.-S. Song, *Photochem. and Photobiol.*, 1973, **18**, 531; P.-S. Song and H. Baba, *ibid.*, 1974, **20**, 527.
[79] H.-J. Galla and E. Sackman, *Biochem. Biophys. Acta*, 1974, **339**, 103.
[80] R. Srinivasan, R. J. von Gutfeld, C. S. Angadiyavar, and R. W. Dreyfus, *Chem. Phys. Letters*, 1974, **25**, 537.
[81] N. Bakashima, N. Mataga, C. Yamanaka, R. Ide, and S. Misumi, *Chem. Phys. Letters*, 1973, **18**, 386.
[82] D. C. Lorents, R. M. Hill, and D. J. Eckstrom, 'Molecular Metal Lasers', Semiannual Technical Report No. 1, Contract Nooo-14-72-C-0478 Stanford Research.

4 Interactions between Molecules and Electrons of Low Energy

By I. C. WALKER

Department of Chemistry, University of Stirling

1 Introduction

High-energy electrons have long been used by chemists to explore molecular structure; witness mass spectroscopy and electron diffraction. Recent years have seen the development of techniques for handling electrons of low energy, and it is now apparent that they also can reveal a wealth of information on molecular structure, not readily accessible by other means. It will be the aim of this Report to introduce, to chemists, this relatively novel field of study, through those recent results and developments which impinge on areas of chemical interest. In keeping with the philosophy of the Annual Reports, the hope is to provide an article palatable to the general reader. Only selected references are given. This has meant the omission of some very fine detailed work on simple systems, where these are judged primarily of interest to physicists. Also, a brief outline of fundamental concepts is included, although these are, in general, available in textbooks.[1,2] Different aspects of electron–molecule interactions have been reviewed in recent years; reference will be made to these, as appropriate, in the text.

2 Basic Concepts

A. Definitions.—Electron-impact studies are concerned with sorting out what happens when an electron of low energy meets a molecule. To begin with, one can attempt to classify the interactions into resonant and non-resonant processes. In the former, the electron is directly scattered by the target molecule, with or without excitation of molecular energy states. In this latter, scattering proceeds *via* formation and decay of a short-lived negative ion; this is a resonant process, because, apart from any other requirements, in order to be captured the electron must have just the right energy to be accommodated in an orbital of the target molecule. Not surprisingly, such a classification is not clear-cut; in practice,

[1] H. S. W. Massey, 'Electronic and Ionic Impact Phenomena', Vols. I and II, 2nd edn. Oxford University Press, 1969.
[2] J. B. Hasted, 'Physics of Atomic Collisions', 2nd edn., Butterworths, London, 1972.

the negative ion resulting from electron capture may be too short-lived to be distinguished from direct potential scattering.

The language of electron-impact experiments is that of scattering theory, and events are described in terms of collision cross-sections rather than rate constants or extinction coefficients. Figure 1 is a schematic diagram of a typical experimental set-up. A beam of electrons, of energy E_i and intensity I_0, is directed through the target gas. The attenuation of the electron beam is given by equation (1), where n is the molecule number density, l the path length, and Q,

$$I_t = I_0 \exp(-nQl) \tag{1}$$

which may be seen to have the dimensions of area, is a collision cross-section. It is proportional to the probability that an electron be scattered in travelling unit distance through the gas at unit number density, and is the sum of the cross-sections for all contributing scattering processes [equation (2)], where Q_{el} is

$$Q = Q_{el} + \sum_i Q_i \tag{2}$$

the elastic scattering cross-section; the summation includes all inelastic processes – those in which the electron transfers energy to (or from) internal degrees of freedom of the molecule. These may be identified by collection and energy analysis of the scattered electrons. The scattering angle θ is an important experimental parameter; a knowledge of the angular distribution of the scattered electrons helps one to deduce details of the electron–molecule interaction. So a differential cross-section $I(\theta)$ is defined [equation (3)], giving the probability of scattering into the solid angle $d\Omega = \sin\theta\, d\theta\, d\phi$, where ϕ is the azimuthal scattering angle.

$$Q = \int_0^{2\pi} \int_0^{\pi} I(\theta) \sin\theta\, d\theta\, d\phi \tag{3}$$

B. Theoretical.—It has been suggested that electron-scattering processes can be treated theoretically using only classical mechanics, and indeed some remarkable results have been claimed using classical concepts.[3] However, conventional approaches use quantum theory, which recognizes that the electron wavelength ($\lambda = \sqrt{154/E}$ Å, where E is in eV) cannot be neglected for energies in the eV region. Firstly, any electron will have an angular momentum, J, about the scattering centre, and this must be quantized, as shown in equation (4). A wave

$$J = \sqrt{l(l+1)}\hbar \tag{4}$$

for which $l = 0$ is an s-wave, one for which $l = 1$ is a p-wave, and so on. Equation (3) implies also that the electron-impact parameter, b, is quantized, as shown in equation (5), where m_e is the electron mass and v_e its velocity. An exact description

$$J = m_e v_e b \tag{5}$$

[3] M. Gryzinski, *Phys. Rev. Letters*, 1970, **24**, 45.

of the scattering process requires, of course, solution of the appropriate wave equation. The resulting eigenfunction Ψ must have the asymptotic form of equation (6), where k, the electron wavenumber, is $2\pi/\lambda$. The first term here

$$\Psi = \exp(ikz) + r^{-1}\exp(ikr)f(\theta) \qquad (6)$$

represents the incoming electron wave, the second term the outgoing scattered wave. The problem is to evaluate $f(\theta)$, the amplitude of the scattered wave; equation (7) shows how it is related to the differential cross-section, $I(\theta)$. It turns

$$I(\theta) = |f(\theta)|^2 \qquad (7)$$

out that $I(\theta) = A^2 + B^2$, where A and B are defined by equation (8), P_l are the

$$A = \tfrac{1}{2k}\Sigma(2l + 1)[\cos 2\eta_l - 1]P_l(\cos\theta) \qquad (8a)$$
$$B = \tfrac{1}{2k}\Sigma(2l + 1)\sin 2\eta_l P_l(\cos\theta) \qquad (8b)$$

Legendre polynomials, and η_l are phase shifts introduced by the scattering molecule into the incident electron wave-trains. An observer detects scattering through these phase shifts. It follows [cf. equation (3)] that Q can be defined by equation (9). The summations are over all the partial waves that 'see' the molecule.

$$Q = \frac{4\pi}{k^2}\Sigma(2l + 1)\sin^2\eta_l \qquad (9)$$

If the electron velocity is very low, so that $b = J/m_e v_e$ is large, even for a p-wave ($l = 1$), then only the s-wave will be scattered (in a head-on collision) and the cross-section will be given by equation (10). Equations (8) illustrate that scattering

$$Q_0 = \frac{4\pi}{k^2}\sin^2\eta_0 \qquad (10)$$

need not be isotropic; the angular distribution of the scattered electrons depends on which waves are scattered.

There is a class of experiments, electron swarm experiments, in which cross-sections are not measured directly, but deduced from the behaviour of a swarm or pulse of electrons pushed through the gas by a d.c. electric field. The measurable parameters depend on the momentum transfer cross-section, Q_D, a cross-section weighted according to the anisotropy of collisions and defined by equation (11).

$$Q_D = \int_0^\pi \int_0^{2\pi} I(\theta)(1 - \cos\theta)\sin\theta\, d\theta\, d\phi \qquad (11)$$

$$Q_D = \frac{4\pi}{k^2}\Sigma(l + 1)\sin^2(\eta_l - \eta_{l+1}) \qquad (12)$$

In terms of phase shifts, Q_D may be expressed as equation (12). These equations may be generalized to include inelastic effects.

The task confronting the theoretician is to calculate the phase shifts. For electrons of low energy, no one theory is universally applicable. Several different approaches are currently applied, and then only to the simplest systems. Discussion of these is outside the scope of this article but reference to them will be made in the text, as appropriate. At high electron energies, the cross-sections are calculable using the Born approximation. Briefly, then, the electron–molecule interaction is taken simply as the Coulomb interaction between incident and bound electrons, and it is assumed that the electron wave is negligibly disturbed by the molecule. For inelastic scattering, this approximation demonstrates that, at high incident energies, the molecule does not distinguish between the electron and a passing photon, so that scattering is predominately in the forward direction, and any molecular transitions excited are optically allowed. As the electron energy is reduced and one looks at large scattering angles, optically forbidden transitions gain over those that are electric-dipole-allowed. Spin-forbidden electronic transitions, excited through electron exchange, are particularly enhanced at low energy. Happily, these conclusions of the Born approximation, confirmed experimentally, may be extrapolated to low energies, where the theory is no longer valid. An electron energy-loss spectrum recorded at low incident energies and large scattering angles may be dominated by optically forbidden transitions. One other result of the Born approximation appears to be generally applicable; the relative differential cross-sections for excitation of different vibrational levels of a molecular transition are invariant with incident electron energy E_i and angle θ, and equal the Franck–Condon factors. This last point is an important one; if, in a transition, departures from the Franck–Condon factors appear as either E_i or θ is varied, then one must suspect contributions from an underlying transition.[4]

It is the practice to describe inelastic processes in terms of f, the generalized oscillator strength. For a transition, of energy loss W, f is defined by equation (13), where K is defined by equation (14) and k_i and k_f are initial and final wave-numbers, respectively. Equation (13) allows one to define f by equation (15),

$$f = (W/2)(k_i/k_f)K^2 I(\theta) \qquad (13)$$

$$K^2 = (k_i^2 + k_f^2 - 2k_i k_f \cos \theta) \qquad (14)$$

$$f = 2W\varepsilon\varepsilon^*/K^2 \qquad (15)$$

where ε is a transition matrix element. In the limit of high impact energy and zero scattering angle, the generalized oscillator strength reduces to the optical oscillator strength. This allows a direct comparison of electron-scattering results with those from optical spectroscopy. Lassettre has reviewed this aspect of electron-scattering spectroscopy.[5] The generalized oscillator strength appears in a useful equation [equation (16)] recently derived, relating the energy of a triplet state, E_T, to that of the corresponding singlet state E_S.[6] W is the excitation

$$E_S - E_T = (\pi g W)^{-1} \int_0^\infty K^2 f \, dK \qquad (16)$$

energy of the singlet state, g its degeneracy, and f its generalized oscillator strength. This equation, applicable where the Born approximation is valid, has been tested for several states in helium and carbon monoxide for which both $(E_S - E_T)$

[4] S. Trajmar, J. K. Rice, and A. Kuppermann, *Adv. Chem. Phys.*, 1970, **18**, 15.

[5] E. N. Lassettre, *Canad. J. Chem.*, 1969, **47**, 1733.

[6] E. N. Lassettre and M. A. Dillon, *J. Chem. Phys.*, 1973, **59**, 4778.

and f are known. It offers the interesting prospect of positioning a triplet state from experimental measurements on the corresponding singlet at high electron energies.

C. Experimental Considerations.—Common to all electron spectroscopy experiments is the need to know the energy of the incident electrons. This is most frequently achieved by passing the electrons, usually from a thermionic source, through electrostatic and/or magnetic fields, tuned to transmit electrons of a single energy. Favourite monochromators are the $127°$ electrostatic selector and the spherical electrostatic selector. These have been fully described in the literature.[7] With such instruments, workable electron beams of 30—50 meV energy spread are relatively easily obtained (1 eV = $1.60199_6 \times 10^{-19}$ J molecule^{-1}). 'Workable' is an important qualification in describing electron beams. Beams of low energy are very sensitive to space-charge effects, surface imperfections, and various 'relaxation' processes, not yet understood, all of which serve to broaden the energy distribution. So, although a nice beam may emerge from the monochromator, one cannot assume that it will survive into the collision region, particularly in the presence of molecular gases which have pronounced effects on surface potentials and hence on the electron beam. One somewhat novel and essentially simple monochromator has been developed which avoids some of these problems. In this, the trochoidal electron monochromator, electron energy selection results from passage of the electrons through mutually perpendicular electrostatic and magnetic fields.[8] The latter is in the direction of the incident electron beam and 'contains' the low-energy electrons, so that the

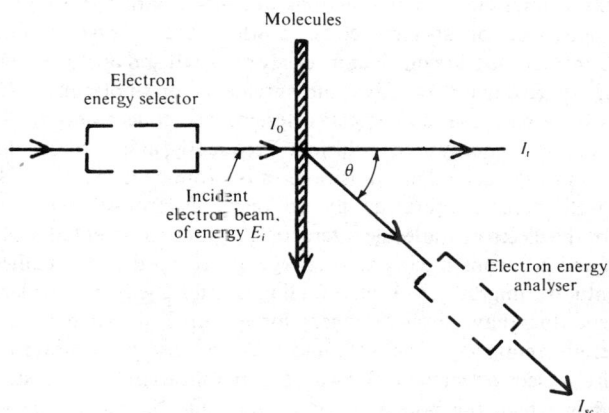

Figure 1 *Schematic diagram of an electric scattering experiment. An electron beam, of energy E_i, intensity I_0, crosses a molecular beam, or passes through a chamber containing the sample gas at low pressure. The scattered electrons, of intensity I_{sc}, have energy E_f; θ is the scattering angle. I_t is the intensity of the transmitted electron current*

[7] J. B. Hasted, *Contemporary Phys.*, 1973, **14**, 357.
[8] A. Stamatovic and G. J. Schulz, *Rev. Sci. Instr.*, 1970, **41**, 423.

monochromator can be used from zero energy upwards, with a resolution of about 20 meV, at best. The magnetic field, which must penetrate through to the collision region, does, however, mean that the scattering angle, θ (Figure 1), is not measurable.

High-resolution spectra at low energies have been obtained without mono-chromation of the electron beam, by time-of-flight analysis.[9] Single electrons which have drifted through a long path-length, L, are detected and registered according to their respective times of flight, t. For any electron, velocity = L/t. Electron energy is thus determined from two accurately measurable parameters L and t. Measurements are made without and with gas (under single-collisions conditions) in the flight path, and thus cross-sections are determined simultaneously for a range of electron energies.

Another technique which avoids the use of monochromators produces electrons through photoionization.[10] A recently described source generates electrons by photoionization of a metastable $(^1D_2)$ barium beam with a He–Cd laser. The resulting photoelectrons have 17 meV kinetic energy and a calculated energy spread of less than 1 meV. It is not easy to verify such an energy spread experimentally, but it seems that in the final beam it is about 6 meV, and most of this is ascribable to Doppler broadening in the target beam and to potential gradients across the collision volume. This is something of a breakthrough in electron-beam technology.

The electrostatic and magnetic monochromators can, of course, be used to analyse the energy of the scattered electrons, if this is desired. In practice, the analyser end of an electron-scattering system is simpler to design than the input end, because inter-electron interactions are less troublesome after scattering. (Hence, photoelectron spectrometers, in which the energies of photo-ejected electrons are analysed, are much more easily designed and operated than electron-scattering spectrometers.) Many measurements on molecular excitation by electrons have been made in systems designed to collect only those electrons which lose all their energy in the collision. Frequently, in such threshold excitation spectra, optically forbidden transitions are prominent (Section 2B). In the trapped-well method, initiated, like so many electron-scattering techniques, by Schulz, the electron–molecule interaction region is a potential 'well' such that the scattered electrons having zero energy are trapped in the collision region, and eventually migrate to a surrounding scattered-electron collector.[7] This experiment does have disadvantages; for example, negative ions, as well as electrons, are retained in the well, and it is not easy to separate negative-ion current from electron current. However, a modification to the system has been described in which the potential barrier defining the exit to the well can be modulated.[11] This overcomes the negative-ion problem as well as improving the effective resolution of the trap.

[9] G. C. Baldwin and S. I. Friedman, *Rev. Sci. Instr.*, 1967, **38**, 519.
[10] A. C. Gallagher and G. York, *Rev. Sci. Instr.*, 1974, **45**, 662.
[11] F. W. E. Knoop, H. H. Brongersma, and A. J. H. Boerboom, *Chem. Phys. Letters*, 1970, **5**, 450.

A completely new technique for obtaining high-resolution threshold excitation spectra has been developed by Cvejanović and Read.[12] A weak electric field penetrates into the collision chamber, producing a potential distribution which has the effect of extracting from the collision region only those electrons whose energy is within about 16 meV of zero. It would be nice to see some molecules studied in this system.

3 'Direct' Scattering

A. Elastic and Total Collision Cross-sections.—The equations of Section 2B indicate that total molecular collision cross-sections might be expected to vary markedly with electron energy, even for simple potential scattering. While it is too much to hope that molecular electron distributions be established from electron-scattering data, one can begin to try, by correlation of experiment and theory, to identify the dominant electron–molecule interactions.

The most direct measurement of total collision cross-sections is in a single-collision experiment where I_t is measured at different gas pressures (Figure 1). Log I_t [*cf.* equation (1)] as a function of molecule number density, n, gives Q. This kind of experiment has disadvantages, particularly if absolute cross-sections are demanded. Firstly, n must be measured accurately, and this is not a trivial exercise in the relevant pressure region, 10^{-3}—10^{-4} Torr (760 Torr = 101 325 N m^{-2}). Most usually, capacitance manometers are used, but such instruments must be calibrated individually.[13] Further, the path length, l, is not easily defined. Finally, electrons scattered through small angles may enter the trans-mitted-current collector. However, if carefully done, this kind of experiment is still useful, at least for measuring relative cross-sections.

An alternative and completely different technique is the molecular-beam recoil experiment. In this, scattering is detected by recoil of the target molecule, not of the electron. One obvious advantage of this is that instead of the molecule number density, the electron number density is required, and this is easily measured as electron current. A possible disadvatage is that, because of the great difference between molecule and electron momenta, the 180° centre-of-mass scattering angle is compressed into a few degrees of apparatus angle. (Contrast the electron-scattering situation, where centre-of-mass and laboratory scattering angles are effectively the same). This means that in some cases the transformation from apparatus to centre-of-mass angle is not simple. Recent workers describe a fairly sophisticated kinematic analysis of beam recoil data.[14] This is one of a series of papers on scattering of electrons by alkali-metal halides.[15] These molecules are well chosen. They have very large dipole moments (~ 10 D $\approx$ 3.33×10^{-29} C m). The long-range electron–dipole interaction gives cross-sections which are very high and should be amenable to theoretical treatment.

[12] S. Cvejanović and F. H. Read, *J. Phys. (B)*, 1974, **7**, 1180.
[13] G. C. Baldwin and M. R. Gaertner, *J. Vacuum Sci. Technol.*, 1973, **10**, 215.
[14] M. G. Fickes and R. C. Stern, *J. Chem. Phys.*, 1974, **60**, 4710.
[15] R. C. Slater, M. G. Fickes, W. G. Becker, and R. C. Stern, *J. Chem. Phys.*, 1974, **61**, 2290.

In fact, the first Born approximation has been applied to the scattering of very low-energy electrons from polar molecules, the justification being that most of the cross-section comes in small contributions from distant and thus only slightly perturbed electrons. The Born approximation gives a cross-section that is purely inelastic, involving rotational motion [equation (17)], where m is the electron–molecule reduced mass, μ is the molecular dipole moment (assumed to be a point dipole), J is the molecular rotational quantum number, and $\underline{J}$

$$I(\theta) = \frac{4}{3}\left(\frac{me\mu}{\hbar^2}\right)^2 \frac{k_f}{k_i} \frac{\underline{J}}{2J+1}|k_f - k_i|^{-2} \tag{17}$$

is either J (for the transition $J \rightarrow J - 1$) or $J + 1$ (for $J \rightarrow J + 1$). Other transitions contribute nothing to $I(\theta)$. In the beam recoil experiments, cross-sections are extracted from measurements using model 'Born-type' cross-sections with adjustable parameters, which may be chosen to fit the measured data; that is they are model-dependent. Total cross-sections so obtained are about a factor of two smaller than those calculated from equation (17), while associated momentum transfer cross-sections differ from calculated values by as much as a factor of ten, below 1 eV. A modification to the simple Born approach produces better agreement with experiment. This gives the molecule a core of finite size, arising out of the repulsive forces between incident and molecular electrons which are assumed to be so great that the electron can never penetrate the molecule.[16] Cross-sections for Na_2 and K_2 have also been obtained in molecular beam recoil experiments.[17]

Very low-energy total collision cross-sections have been estimated from time-of-flight measurements (Section 2C). An early paper, describing the time-of-flight spectrometer in detail, reported cross-sections for helium and argon.[9] A recent one examines electron–nitrogen scattering for 0.3—1.6 eV.[18] At 0.9—1.5 eV, the total collision cross-section equals, numerically, the momentum transfer cross-section obtained from analysis of swarm experiments. This means that, within this energy range, the scattering is isotropic; at lower energies Q is less than Q_D, indicating predominance of back-scattering [equation (11)].

These results vindicate the electron-swarm work. In the swarm experiments, a pulse of electrons is pushed through the gas, at high number density, N, by a d.c. electric field E. The swarm very quickly establishes its velocity distribution, which depends only on E/N. Transport coefficients (diffusion coefficients, drift velocity) are measurable as functions of E/N. These, of course, depend on the nature of the collisions between the electrons and the gas molecules. However, the evaluation of cross-sections consistent with the measured parameters is not easy.

The most satisfactory analytical technique was pioneered by Phelps et al. In this, the Boltzmann transport equation, appropriate to the particular system,

[16] M. R. H. Rudge, J. Phys. (B), 1974, 7, 1323.
[17] T. M. Miller and A. Kasdan, J. Chem. Phys., 1973, 59, 3913.
[18] G. C. Baldwin, Phys. Rev. (A), 1974, 9, 1225.

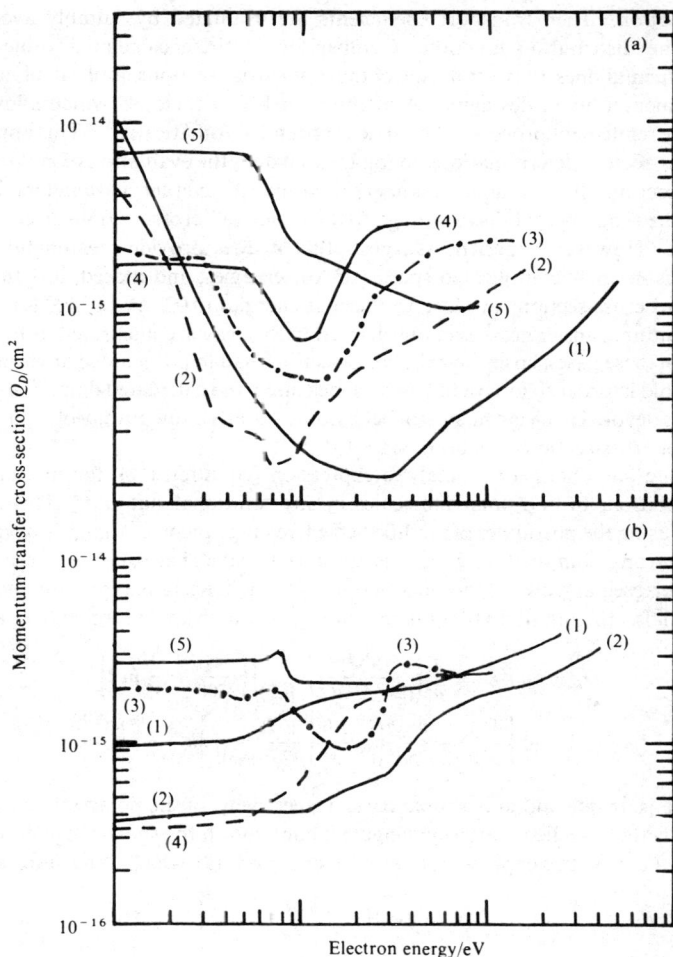

Figure 2 *Momentum transfer cross-sections, Q_D, as a function of electron energy, for some hydrocarbons.*

(a) *Saturated molecules* (C_nH_{2n+2});
    (*1*) *methane*, CH_4[a]   (*2*) *ethane*, C_2H_6[b]
    (*3*)*propane*, C_3H_8[b]   (*4*) *butane*, C_4H_{10}[b]
    (*5*) *neopentane*, $C(CH_3)_4$[c]
(b) (*1*) *acetylene*, C_2H_2[d]   (*2*) *ethylene*, C_2H_4[d]
    (*3*) *propylene*, C_3H_6[b]   (*4*) *cyclopropane* C_3H_6[b]
    (*5*) *but-1-ene*, C_4H_8[c]

[a] Ref. 19; [b] ref. 20; [c] I. C. Walker, unpublished results; [d] C. W. Duncan and I. C. Walker, *J.C.S. Faraday II*, 1972, **68**, 1800

is solved numerically, using assumed cross-sections, to give the electron energy distribution. Then transport coefficients are evaluated by suitably averaging over the distribution function. Comparison of these calculated values with experimental ones allows tailoring of the input cross-sections until calculated and experimental quantities agree. A method has been developed which allows the lengthy refinement process to be carried through automatically.[19] This approach to cross-section determinations, involving as it does the evaluation of microscopic cross-sections from a small number of measured average parameters, suffers from the fundamental disadvantage that the derived cross-sections need not be unique. However, experience suggests that it does provide meaningful cross-sections in fairly complicated species at low energies, and indeed, it is the only route to cross-sections at close to thermal energies (0.028 eV, at 300 K). Some momentum transfer cross-sections derived in this way are illustrated in Figure 2. In all of these gases, large inelastic cross-sections had to be included at vibrational threshold in order to force a fit between measured and calculated data.[20] A recent book is devoted to electron-swarm behaviour,[21] and a bibliography of momentum transfer cross-sections has been compiled.[22]

Chang has obtained an analytic expression [equation (18)] for the diffusion cross-sections of non-polar molecules in any rotational state J.[23] This is produced using the postulates of modified effective range theory, originally developed for electron–atom scattering. In this, the partial wave phase-shifts [equations (8)] are expressed as power series in k and $\ln k$ for the long-range electron–molecule potentials. In equation (18), q is the molecular quadrupole moment, α_0 and α_2

$$Q_D = 4\pi a^2 \left\{ 1 + \frac{q^2}{a^2} \left(\frac{J(J+1)}{(2J-1)(2J+3)} \right) \left(\frac{481}{2700} + \frac{2}{45} \ln 2 \right) \right.$$
$$\left. + \left[\frac{4\pi\alpha_0}{5a} + \frac{\pi\alpha_2 q}{25a^2} \left(\frac{J(J+1)}{(2J-1)(2J+3)} \right) \right] k + \cdots \right\} \qquad (18)$$

are the isotropic and anisotropic parts, respectively, of the polarizability tensor, and a is the so-called scattering length; it contains all the short-range potentials. When J is zero, this expression reduces to equation (19), which is the more familiar

$$Q_D = 4\pi a^2 \left[1 + \frac{4\pi\alpha_0 k}{5a} + \cdots \right] \qquad (19)$$

equation for electron–atom scattering. For para-hydrogen, 99.5 % of which is in the $J = 0$ level, equation (19) applies. A good fit to measured momentum transfer cross-sections is obtained for $\alpha_0 = 5.5\, a_0$. This compares with $5.44\, a_0$ measured independently and $5.41\, a_0$ calculated. The scattering length is $1.26\, a_0$; the cross-section extrapolates to $4\pi a^2$ at zero energy. (Atomic units are used; $m = e = \hbar = a_0 = 1$.) Analysis for other gases is less straightforward because of the

[19] C. W. Duncan and I. C. Walker, *J.C.S. Faraday II*, 1972, **68**, 1514.
[20] C. W. Duncan and I. C. Walker, *J.C.S. Faraday II*, 1974, **70**, 577.
[21] L. G. H. Huxley and R. W. Crompton, 'The Diffusion and Drift of Electrons in Gases', John Wiley, New York, 1974.
[22] Y. Itikawa, Argonne National Laboratory Report, ANL-7939, 1972.
[23] E. S. Chang, *Phys. Rev. (A)*, 1974, **9**, 1644.

rotational dependence of the cross-sections, but it seems clear that the term in q^2/a^2 is correct, and that Q_D does depend on the rotational population.

Differential cross-sections contain more information than total collision cross-sections, but few such measurements have been made at low energies. Truhlar *et al.* have studied simple diatomic species at intermediate energies. They attempt to reproduce experimental differential cross-sections using a Born-type approximation, with different model interaction potentials. The energy region they scan is too low for the Born approximation to be strictly valid, but, in carbon monoxide, for example, one potential (the sum of dipole, quadrupole, and polarization interactions), chosen by Crawford and Dalgarno to give correct momentum transfer cross-sections at 0.03 eV, produces good differential cross-sections at $10—80$ eV; *i.e.* a model potential, 'calibrated' at thermal energies, can account for intermediate energy results.[24]

That theory and experiment are closely interdependent in this field of study is illustrated by He. Momentum transfer cross-sections, believed good to 2%, have been evaluated for helium, from swarm data. The calculated value is still 5% lower than that measured. The discrepancy still has to be sorted out, and this for the second most simple scattering system.[25]

B. Electronic and Vibrational Excitation.—One aspect of electron scattering which has obvious implications for our understanding of chemical processes is inelastic scattering with electronic excitation. As indicated above, when the energy of the incident electron is low, optically forbidden transitions may be excited. In particular, spin-forbidden transitions are accessible through excitation with electron exchange, and this affords a reliable means of positioning triplet states of molecules. Excitation with exchange requires the electron to associate fairly intimately with the molecule, so that scattering tends to be isotropic; in a spin-allowed transition, scattering is predominately about $\theta = 0°$. This means that a spin-forbidden transition is recognizable because its differential cross-section, relative to that of a spin-allowed transition, increases with scattering angle. Also, the cross-sections for excitation of spin-forbidden transitions tend to peak close to threshold energy while those for spin-allowed transitions are maximal at higher energies.[4]

In Table 1 some low-lying triplet levels of unsaturated molecules, recorded in electron scattering, are listed. Most of the assignments are supported by calculations. In some cases (notably acetylene), information on the triplet manifold comes solely from electron scattering. Where triplets have been located optically, in general, the O_2 enhancement technique, pioneered by Evans, has been used. In this, a spectrum is recorded in the presence of several atmospheres of O_2 – this is hazardous, to say the least. An alternative technique simply uses very long path lengths to ensure detectable absorption, but this is very sensitive to impurities. In contrast, an electron energy-loss spectrum is recorded at low gas pressures, so is not sensitive to impurities, and, close to threshold, a spectrum may be dominated by triplets.

[24] D. G. Truhlar, W. Williams, and S. Trajmar, *J. Chem. Phys.*, 1972, **10**, 4307.
[25] D. K. Gibson, R. W. Crompton, and G. Cavalleri, *J. Phys. (B)*, 1973, **6**, 1118.

Table 1 *Triplet states in some unsaturated molecules. The energy/eV refers to the position of maximum intensity*

Molecule	T_1	T_2	T_3	Method and reference
$CH_2{=}CH_2$	4.3 ($^3B_{1u}$)	—	—	electron trap, *a, b*
$MeCH{=}CH_2$	4.35	—	—	
$EtCH{=}CH_2$	4.25	—	—	
cis-$MeCH{=}CHMe$	4.3	—	—	
trans-$MeCH{=}CHMe$	4.3	—	—	
$Me_2C{=}CHMe$	4.2	—	—	
$Me_2C{=}CMe_2$	4.1	—	—	
$CHF{=}CH_2$	4.4	—	—	electron energy-loss, *c*
$CF_2{=}CH_2$	4.63	—	—	
cis-$CHF{=}CHF$	4.28	—	—	
trans-$CHF{=}CHF$	4.18	—	—	
$CHF{=}CF_2$	4.43	—	—	
$CF_2{=}CF_2$	4.68	—	—	
s-trans-butadiene	3.22 ($^3B_{2u}$)	4.91 (^3A_g)	—	electron energy-loss, *d*
trans-hexa-1,3,5-triene	2.66 (^3B_u)	4.2 (^3A_g)	—	modulated trap, *e*
$HC{\equiv}CH$	5.3 ($^3\Sigma_u^+$)	6.0 ($^3\Delta_u$)	8.01 ($^3\Pi_u$)	trap, *f*
$MeC{\equiv}CH$	5.2	5.8	—	
$EtC{\equiv}CH$	5.3	5.8	—	
$MeC{\equiv}CMe$	5.0	5.8	—	
acetone	4.16	5.3–6.1	—	electron energy-loss, *g*

a D. F. Dance and I. C. Walker, *Proc. Roy. Soc.*, 1973, **A334**, 259; *b* D. F. Dance and I. C. Walker, unpublished results; *c* M. J. Coggiola, O. A. Mosher, W. M. Flicker, and A. Kuppermann, *Chem. Phys. Letters*, 1974, **27**, 14; *d* O. A. Mosher, W. M. Flicker, and A. Kuppermann, *J. Chem. Phys.*, 1973, **59**, 6502; *e* F. W. E. Knoop and L. J. Oosterhoff, *Chem. Phys. Letters*, 1973, **22**, 247; *f* D. F. Dance and I. C. Walker, *J.C.S. Faraday II*, 1974, **70**, 1426; *g* W. M. St. John, R. C. Estler, and J. P. Doering, *J. Chem. Phys.*, 1974, **61**, 763.

Simpler species which have been studied are carbon dioxide[26] and nitrous oxide.[27] In the former an energy-loss spectrum recorded for incident energies very close to threshold and over a range of scattering angles has been unravelled, with the aid of theoretical calculations on excited states, to give transitions assigned to excitation of $^3\Sigma_u^+$, $^{1,3}\Sigma_u^-$, $^{1,3}\Pi_g$ and $^{1,3}\Delta_u$ states between 7 and 10 eV; likewise for N_2O, where a number of new low-lying states are claimed, starting at about 5 eV. No electronic levels below 5 eV are apparent. This contradicts earlier threshold work which positions triplet states at 3.8 and 4.4 eV.[28] The spectrum of water has been explored extensively. An explanation for one controversial

[26] R. I. Hall, A. Chutjian, and S. Trajmar, *J. Phys.* (*B*), 1973, **6**, L264.
[27] R. I. Hall, A. Chutjian, and S. Trajmar, *J. Phys.* (*B*), 1973, **6**, L365.
[28] M.-J. Hubin-Franskin and J. E. Collin, *Bull. Soc. roy. Sci. Liège*, 1971, **5–8**, 361.

aspect of its spectrum has been offered by Lassettre and Huo.[29] Most of the recorded spectra for H_2O show a weak feature corresponding to an energy loss of about 4.5 eV, which experimentalists liked to assign to the lowest triplet state, $\tilde{a}\,^3B_1$. This would imply a bound state; the dissociation limits are 5.11 eV for $OH(^2\pi) + H(^2S)$ and 5.03 eV for $O(^3P) + H_2(^1\Sigma_g{}^+)$. However, repeated computations have not been able to produce a suitably attractive potential for this triplet level. Lassettre and Huo have now pointed out that, in each of the experiments, H^- could have been transmitted to the scattered electron collector to give a signal at about 4.5 eV.[29] The $\tilde{a}\,^3B_1$ state probably lies at 7.2 eV. This is near the calculated position. Also, measurements on the optical oscillator strength of the corresponding singlet, at 7.4 eV, for 300, 400, and 500 eV incident energies, lead, through equation (16) to[30] equation (20). A position of 7.2 eV

$$E_s - E_T = 0.58 \pm 0.42 \tag{20}$$

for the $^3B_1$ state is just within this range. Another H_2O triplet has been located at 9.81 eV.[31]

Results for triplet states in a number of other molecules are cited in a recent review of electron-scattering spectroscopy.[32]

At the other end of the spectrum, electron energy-loss is useful for sorting out high-energy transitions. These include excitation of Rydberg states. A Rydberg-excited state is one where the excited electron finds itself in an orbital sufficiently far from the positive ion core to see it as atom-like. The energies of Rydberg states are then given by the formulae (21), where A is an ionization energy, R is

$$W = A - [R/(n - \delta)^2] \tag{21}$$

the Rydberg constant, and δ the Rydberg correction or quantum defect. For molecules built up from atoms of the first Period, $\delta < 0.1$ for states derived from nd electrons, 0.3—0.5 for np electrons, and 0.9—1.2 for ns electrons.

The study of these excited states in molecules has been somewhat neglected because they frequently correspond to the vacuum-u.v. region of the spectrum and also because, as the ionization limit is approached, transitions to Rydberg-excited states become weaker and are difficult to sort out from strong valence transitions in the same spectral region. In electron scattering, Rydberg states are frequently strongly excited. The interpretation of electron-scattering results is helped by photoelectron spectroscopy. Firstly, photoelectron spectroscopy gives accurate values for ionization energies. Secondly, because Rydberg orbitals do not contribute to bonding, the vibrational structure associated with a Rydberg transition is similar to the vibrational spacing of the positive ion core, and this is also observed in photoionization. Thus it is likely that photoelectron and electron-scattering spectroscopy will combine to give useful results in this area.

[29] E. N. Lassettre and W. M. Huo, *J. Chem. Phys.*, 1974, **61**, 1703.
[30] E. N. Lassettre and A. Skerbele, *J. Chem. Phys.*, 1974, **60**, 2466.
[31] S. Trajmar, W. Williams, and A Kuppermann, *J. Chem. Phys.*, 1973, **58**, 2521.
[32] I. C. Walker, *Chem. Soc. Rev.*, 1974, **3**, 467.

Oxygen is a case in point; only a small number of Rydberg states are easily studied in photoabsorption. Electron-impact work, aided by *ab initio* calculations and photon absorption measurements, has located the first members ($n = 3$) of the Rydberg series, which converge to the O_2^+ ($\tilde{X}\ ^2\Pi_g$) state (removal of a π^* electron).[33] The lowest of these is the $^3\Pi_g$ Rydberg state, which is the result of excitation of the π_g (π^*) electron to the $3s\sigma_g$ level; it appears, in electron impact, as structure on top of the Schumann–Runge continuum ($\tilde{B}\ ^3\Sigma_u^- \leftarrow \tilde{X}\ ^3\Sigma_g^-$). The spacing and relative intensities of this vibrational structure indicate that the internuclear distance of this $^3\Pi_g$ (R) state is greater than that of the positive ion to which it is converging. Further, the spacing between the vibrational levels 1 and 2 in this R state is anomalously large compared to that between levels 0 and 1. This is taken as evidence that the R state is crossed by a repulsive potential-curve between $v' = 1$ and 2. This same anomalous spacing has been observed for the negative ion O_2^{*-}, in which two electrons are accommodated in the $3s\sigma_g$ orbital[34] (see Section 4).

The assignment of the electronic spectrum of acetone is still a matter of some debate. The lowest singlet transition, seen as the $\tilde{A}$ band, of vertical transition energy 4.4 eV, is $n_0 \rightarrow \pi^*$, but there has been considerable speculation as to the origin of all higher transitions. An electron-energy-loss spectrum has identified higher members of three Rydberg series with quantum defects of 1.03, 0.81, and 0.315.[35] The lower members of these extend into the spectral regions of the $\tilde{B}$, $\tilde{C}$, and $\tilde{D}$ bands, which explains why they could not be classified in terms of valence transitions. Work on naphthalene likewise emphasizes the need to include Rydberg configurations for any theoretical analysis of its spectrum.[36] Rydberg series have been assigned in alkyl derivatives of water and aliphatic carbonyl compounds, and substituent effects on Rydberg orbital energies have been discussed using Taft σ^* values.[37] Electron-impact and photoelectron spectra of cyclopropenone have been interpreted with the aid of SCF MO calculations.[38] Removal of either an oxygen lone-pair electron or an olefinic electron causes significant change in the structure of the ion relative to the ground-state molecule, indicating that these electrons are delocalized in the molecule. A number of Rydberg states are observed.

Peart and Dolder claim to have prepared, in its ground electronic state, H_3^+,

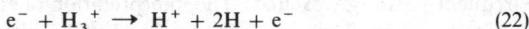

$$e^- + H_3^+ \rightarrow H^+ + 2H + e^- \tag{22}$$

the simplest polyatomic species.[39] In the reaction (22) the threshold for H^+ production is ~ 15 eV, which can be equated with the energy of the first excited

[33] D. C. Cartwright, W. J. Hunt, W. Williams, S. Trajmar, and W. A. Goddard, *Phys. Rev.* (*A*), 1973, **8**, 2436.

[34] L. Sanché and G. J. Schulz, *Phys. Rev.* (*A*), 1972, **6**, 69.

[35] R. H. Huebner, R. J. Celotta, S. R. Mielczarek, and C. E. Kuyatt, *J. Chem. Phys.*, 1973, **59**, 5434.

[36] R. H. Huebner, S. R. Mielczarek, and C. E. Kuyatt, *Chem. Phys. Letters*, 1972, **16**, 464.

[37] W.-C. Tam and C. E. Brion, *J. Electron Spectroscopy*, 1974, **3**, 467.

[38] W. R. Harshbarger, N. A. Kuebler, and M. B. Robin, *J. Chem. Phys.*, 1974, **60**, 345.

[39] B. Peart and K. T. Dolder, *J. Phys.* (*B*), 1974, **7**, 1567.

electronic state of H_3^+, ^3E, while further enhancement of H^+ production places the ^1E level at about 19.25 eV.

Calculations of cross-sections for electronic excitation of molecules by electrons of low energy are few. Chung and Lin report calculated excitation functions for eleven electronic states of carbon monoxide (singlet and triplet) from threshold to 1000 eV.[40] For the molecular wavefunctions, Gaussian-type orbitals are used and a Born-type approximation is employed. This is, of course, a high-energy approximation, and so good results are not to be expected at low energies. It also turns out that the experimental data are inadequate for an assessment of the accuracy of the calculations

Electrons can excite almost all vibrational modes of molecules. Swarm results are consistent with large inelastic cross-sections for i.r.-active vibrations, close to threshold;[20] this could happen through interaction of the electron with the instantaneous molecular dipole. A variety of other interactions (electron–quadrupole, electron–induced dipole, *etc.*) can cause excitation of i.r.-inactive vibrational modes.[41]

4 Resonance Scattering

The negative ions familiar to chemists are those formed by attachment of an electron to a ground-state molecule having a positive electron affinity. Being bound, they survive for sufficiently long to be detected in a negative-ion mass spectrometer. A study of electron scattering reveals negative ions where the electron is associated with excited molecular states. It is not unusual for a molecule which has a negative electron affinity in the ground state to have excited states with positive electron affinities. Also, negative ions lying *above* the parent molecular state are detectable in scattering work. Not surprisingly, these negative ions have relatively short lifetimes, and are frequently referred to as electron–molecule resonances. The discovery of these resonances (the first, firmly identified by Schulz, in 1963, in helium) caused a flurry of activity among physicists, and, in fact, is probably directly responsible for much of the current popularity of atomic and molecular physics. Resonance phenomena were familiar to physicists; they are widespread in sub-nuclear particles. The mathematical and physical properties of resonances have therefore been developed within the framework of nuclear physics. Schulz has summarized our present understanding of resonances in atoms and diatomic molecules.[42] Briefly, a resonance which lies above its parent state is a *shape* resonance, so-called because the trapping is the result of the shape of the effective interaction potential; the electron is held within a potential barrier generated by a combination of attractive forces and repulsive (centrifugal) forces. The repulsive potential, which requires the electron to have angular momentum, is lacking for s-waves, and this places

[40] S. Chung and C. C. Lin, *Phys. Rev. (A)*, 1974, **9**, 1954.
[41] H. T. Davis and L. D. Schmidt, *Chem. Phys. Letters*, 1972, **16**, 260.
[42] G. J. Schulz, *Rev. Mod. Phys.*, 1973, **45**, 423.

a symmetry restriction on the molecular orbitals which can trap the electron. The other kind of resonance is described as Feshbach resonance. A Feshbach resonance results when the electron enters an orbital of an excited molecule, to give a *bound* negative ion. Usually, a Feshbach resonance lies within 1 eV of its parent molecular state. These resonances are, of course, not stationary states; they decay exponentially with time, as shown in equation (23), where

$$|\psi_n|^2 \propto \exp\left(-\Gamma_n t/\hbar\right) \tag{23}$$

ψ_n is the resonance wavefunction and Γ_n its linewidth. Its lifetime τ equals $\hbar/\Gamma_n$. In general, Feshbach resonances are the longer-lived and may survive for sufficiently long to have well-developed vibrational structure. The lifetimes of shape resonances vary considerably. At the one extreme, they may be so short-lived as to be barely detectable; at the other, for a system with a sufficiently attractive potential within a high barrier, a shape resonance may merge into a bound, conventional negative ion.

In the presence of resonance scattering, the phase shift has two components, a potential one, η_{lp}, and a resonance one, η_{lr}, which are related by equation (24).

$$\eta_l = \eta_{lp} + \eta_{lr} \tag{24}$$

So a resonance may produce structure in the total cross-section, at the resonance energy. Also, a resonance may decay into lower-lying excited states [equation (25)]

$$M + e_1 \rightarrow (M^{*-}) \rightarrow M^* + e_2 \tag{25}$$

and so be detectable as enhancement of particular inelastic cross-sections at the resonance energy. For short-lived resonances (10^{-14}—10^{-15} s) any associated structure in the total collision cross-section is broad—and may not be distinguishable from any broad structure that is due to potential scattering. It may then be profitable to study the resonances in inelastic scattering. The different fates of a resonance are summarized in equations (26).

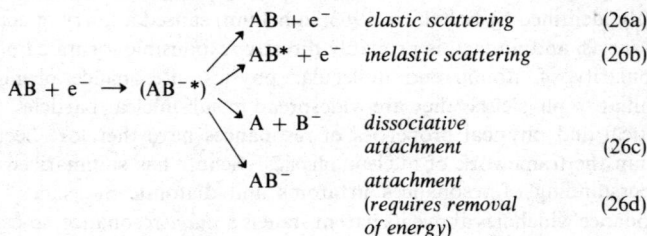

$$AB + e^- \rightarrow (AB^{-*})$$

$AB + e^-$	elastic scattering	(26a)
$AB^* + e^-$	inelastic scattering	(26b)
$A + B^-$	dissociative attachment	(26c)
AB^-	attachment (requires removal of energy)	(26d)

Table 2 gives properties of some molecular resonances seen as fine structure in the electron current transmitted through the sample gas; electron energy was selected in a trochoidal electron monochromator.[43] Each of these resonances (except perhaps that for SO_2) is a shape resonance in which the electron enters the

[43] L. Sanché and G. J. Schulz, *J. Chem. Phys.*, 1973, **58**, 479.

Table 2 *Some properties of shape resonances in polyatomic molecules*[a]

Molecule	Energy/eV	Vibrational mode[b]	Vibrational spacing (0—1)/eV	Symmetry
CO_2	3.14 (onset)	100	0.138	$^2\Pi_u$
NO_2	0.14 (onset)	100	0.130[c]	$^1A_1$
		010	0.065	—
SO_2	2.87 (onset)	100	0.093[d] or 0.128	—
N_2O	2.34 (max.)	—	—	$^2\Sigma$
H_2S	2.30 (max.)	—	—	$^2A_1$
C_2H_4	1.76 (max.)	—	—	$^2B_{3u}$
C_6H_6	1.14 (onset)	C—C sym. str.	0.123	$^2E_{2u}$

[a] L. Sanché and G. J. Schulz, *J. Chem. Phys.*, 1973, **58**, 479; [b] The three numbers represent the quantum numbers for symmetrical stretching, bending, and unsymmetrical stretching, respectively; [c] These values were obtained by extrapolating the spacing of the vibrational progressions to below zero energy, stopping at the known electron affinity, 2.38 eV; [d] The value depends on the identity of the resonance. If the observed structure corresponds to highly excited levels of a ground-state shape resonance, then the spacing is 0.128 eV. If the resonance is associated with an excited molecule, 0.093 eV applies; see text.

lowest available molecular orbital in the ground-state molecule. SO_2 is anomalous. It has a positive electron affinity (E.A. ~ 1.1 eV), *i.e.* the negative ion formed when an electron goes into the lowest available orbital is stable with respect to the neutral ground-state molecule. One then expects to see resonance structure, corresponding to formation of vibrationally excited SO_2^-, starting for electrons of energy close to zero volts. The structure, in fact, does not start until ~ 2.8 eV. Perhaps, SO_2^- in its lower vibrational levels is so long-lived (sharp) that it is not detectable in this experiment, while at higher energies the lifetime decreases, making the resonance visible. Alternatively, the resonance may be associated with an excited electronic level of SO_2. Benzene appears to have a second shape resonance at 4—6 eV, in addition to the one tabulated; this is interpreted in terms of a negative ion in which the electron enters the highest vacant π^* molecular orbital. Substitution in the benzene ring removes the degeneracy of the lowest π^* orbitals so that substituted benzenes have two shape resonances below 2 eV;[44] the same is true of N-heterocyclic molecules.[45] At higher energies, each of the molecules in Table 2 (except benzene) shows sharp structure arising from Feshbach resonances associated with Rydberg-excited molecular states. Only one such resonance is apparent in ethylene, lying about 0.5 eV below the first Rydberg excited state (the R state). This same resonance was independently identified in a threshold excitation experiment.[46] The lowest Rydberg excited state of acetylene, likewise has an electron affinity of *ca.* 0.5 eV.[47]

[44] L. G. Christophorou, D. L. McCorkle, and J. G. Carter, *J. Chem. Phys.*, 1974, **60**, 3779.
[45] M. N. Pisanias, L. G. Christophorou, J. G. Carter, and D. L. McCorkle, *J. Chem. Phys.*, 1973, **58**, 2110.
[46] D. F. Dance and I. C. Walker, *Proc. Roy. Soc.*, 1973, **A334**, 259.
[47] D. F. Dance and I. C. Walker, *J.C.S. Faraday II*, 1974, **70**, 1426.

Bound negative ions are not apparent in the alkyl-substituted ethylenes and acetylenes.

Of the polyatomic shape resonances in Table 2, that in CO_2 has been most intensively studied.[48] At 3.8 eV, vibrational excitation of CO_2 is possible, both directly and *via* the resonance; in the former case, scattering is predominantly in the forward direction. The resonance can thus be nicely examined through large-angle scattering with vibrational excitation. Energy-loss spectra recorded at 90° scattering angle and around 3.8 eV incident energy show resonance excitation of $n00$ (symmetric stretch) and $n10$ (symmetric stretch plus bend) vibrational modes. For any particular energy-loss, the cross-section as a function of energy shows structure whose position and spacing depend on incident energy and also the particular vibrational modes excited. This irregularity in the structure has also been seen in the N_2 shape resonance (~ 2.3 eV), and there successfully interpreted in terms of a resonance of 'intermediate' lifetime, the 'boomerang' model. (Were the negative ion long-lived, the structure would correspond to its vibrational levels.) The lowest lying CO_2^- ion is bent in its equilibrium configuration and also has a longer equilibrium C—O bond length than the neutral, ground-state molecule. Therefore, at the instant of attachment, the CO_2^-, in order to attain its equilibrium configuration, begins to stretch, rapidly, and, more slowly, to bend. If the ion survives for about 10^{-14} s, then it has time to complete about one complete stretching vibration but only part of a bending cycle. The stretching motion will therefore encompass many highly vibrationally excited stretching levels of the ground-state molecule but only a few low-lying bending vibrations, as is observed. Angular-distribution measurements support the notion of two distinct energy-loss processes (in addition to direct excitation).[49] Symmetry considerations show that the angular distributions of electrons resonantly scattered by molecules depend on the vibronic symmetry of the molecular states involved, and for CO_2^- the theoretically predicted distributions conform to those measured.[50] This same negative-ion state of CO_2 has been detected in a mass spectrometer with a lifetime of some microseconds.[51] In this case it was produced by bombardment of succinic anhydride or maleic anhydride with electrons. In these organic species, the CO_2 entity is 'bent', and the long lifetime against detachment is explicable if there is an unfavourable Franck–Condon overlap between the bent ion and the linear neutral parent and if the potential-energy curve of the negative ion lies below that of the ground-state neutral species at 134°. It is, of course, about 3.8 eV above at 180°. Shape resonances in more complicated molecules are discussed in a recent review article.[32]

Detailed measurements on resonances in diatomics and atoms continue. A high-resolution study on oxygen using a time-of-flight spectrometer has resolved the doublet structure of the $^2\Pi_g O_2^-$ ion.[52] This experiment confirms

[48] M. J. W. Boness and G. J. Schulz, *Phys. Rev.* (*A*), 1974, **9**, 1969.
[49] D. Danner, thesis, Physikalisches Institut der Universitat Freiberg, 1970.
[50] D. Andrick and F. H. Read, *J. Phys.* (*B*), 1971, **4**, 389.
[51] C. D. Cooper and R. N. Compton, *Chem. Phys. Letters*, 1972, **14**, 29.
[52] J. E. Land and W. Raith, *Phys. Rev.* (*A*), 1974, **9**, 1592.

an accidental coincidence between the $v = 3$ level of O_2 ($\tilde{X}\ ^3\Sigma_g^-$) and the $v = 8$ level of O_2^-. Vibrational excitation of O_2 at energies between 4 and 15 eV is resonance-enhanced.[53] Two low-lying resonances, $^4\Sigma_u^-$ (shape) and $^2\Sigma_u$ (valence Feshbach) appear to be accessible in this region.[54]

It has been established that any molecule or radical whose dipole moment exceeds 1.625 D (1 D = 3.33 $\times$ 10^{-30} C m) has a positive vertical electron affinity; the additional electron is bound by the molecular dipole.[55] Sharp structure has been observed in the elastic and vibrational cross-sections of the ground electronic state of carbon monoxide at energies coincident with the $v = 0$, 1, and 2 levels of the $\tilde{a}\ ^3\Pi$ state, a valence excited state ($\sigma \rightarrow \pi^*$).[56] No comparable structure is detectable in the energy range of the equivalent $\tilde{B}\ ^3\Pi_g$ state of isoelectronic nitrogen. The explanation offered is that the electric dipole moment of the excited CO state (1.38 D) temporarily binds the incoming electron. The $\tilde{B}\ ^3\Pi_g$ state of nitrogen does not have a permanent dipole moment.

A common energy-calibration point in electron-impact work is the helium resonance (He$^-$: $1s,2s^2$). Two recent measurements place this resonance at 19.367 $\pm$ 0.009 eV[57] and 19.361 $\pm$ 0.009 eV.[12] The first of these also evaluates the s-wave phase shift at the resonance energy, from measurements on the elastic differential cross-section at 90° scattering angle, where there is no contribution from p-wave scattering [equations (8)]; this gives $\eta_{or} = 106° \pm 3°$. The resonance width is given as 9 $\pm$ 1 meV, which is in good agreement with previous estimates. Incidentally, as suggested by Gibson and Dolder,[58] this helium resonance offers a means of making measurements absolute, independent of a knowledge of gas pressures (see Section 3A). Differential measurements give access to the phase shifts at the resonance energy [equations (8)]. These can then be used to calculate absolute cross-sections. This calibration technique may be applied to other simple systems where differential measurements are feasible.

5 Stable Negative Ions

A. Experimental Methods for Measuring Electron Affinities.—Electron scattering does give some information on the structure of stable negative ions, where these appear as nuclear excited resonances, but a complete description of these negative ions, including perhaps the most important fundamental property – the electron affinity – requires measurements on the ground-state negative ion. That this is not easy is manifest in the fact that accurate electron affinities are at present available for only a handful of species. Most of these have been obtained within the past three or so years, the result of the development of beam techniques, in particular those based on photodetachment processes (27). The measured

[53] S. F. Wong, M. J. W. Boness, and G. J. Schulz, *Phys. Rev. Letters*, 1973, **31**, 969.
[54] M. Krauss, D. Neumann, A. C. Wahl, G. Das, and W. Zemke, *Phys. Rev. (A)*, 1973, **7**, 69.
[55] O. H. Crawford, *Mol. Phys.*, 1973, **26**, 139.
[56] S. F. Wong and G. J. Schulz, *Phys. Rev. Letters*, 1974, **33**, 134.
[57] S. Cvejanović, J. Comer, and F. H. Read, *J. Phys. (B)*, 1974, **7**, 468.
[58] J. R. Gibson and K. T. Dolder, *J. Phys. (B)*, 1969, **2**, 1180.

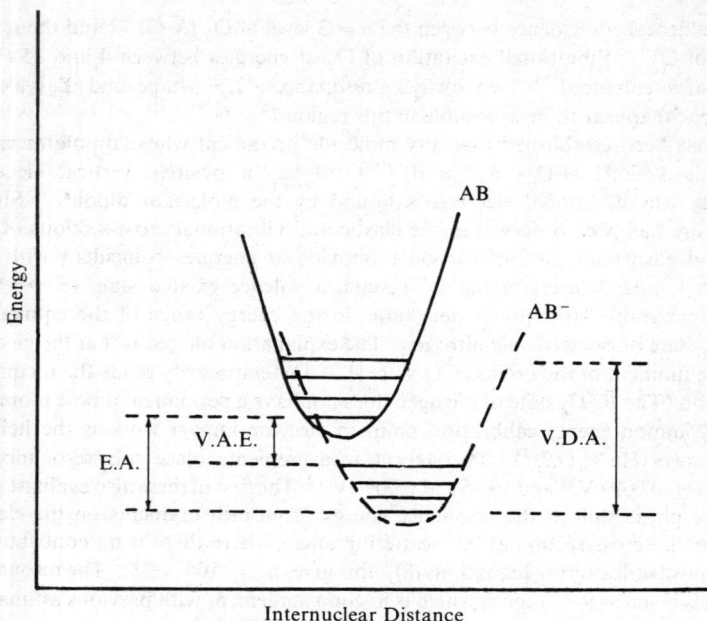

Figure 3 *Energy terms associated with negative ions. E.A. is the electron affinity, V.A.E. is the vertical attachment energy, and V.D.A. is the vertical detachment energy. Recent electron-affinity determinations concentrate on measurement of detachment-energy spectra*

$$AB^- + h\nu \rightarrow AB + e^- \qquad (27)$$

quantity is the photon energy needed to detach the electron from the negative ion. For a molecule, the electron affinity is the detachment energy between the rotational ground state of the negative ion in its zeroth vibrational state and the corresponding rotational ground state of the neutral molecule in its zeroth vibrational state. This, and other measurable parameters, are illustrated in Figure 3. A good description of photodetachment experiments is given in a review article.[59] Now, particularly with laser sources, they are producing unambiguous electron affinities for atoms and simple molecules. Three somewhat different experiments are currently employed. In the first, a linearly polarized laser beam of fixed frequency crosses the negative-ion beam, prepared in a discharge.[60] The ejected electrons are directed into a hemispherical electrostatic analyser, where they are analysed for energy E_e. Then E.A. = $h\nu - E_e$. In this experiment, a current maximum is recorded at energy E_e; the maximum can be precisely located with reference to the electron signal from a species of known

[59] B. Steiner, 'Case Studies in Atomic Collision Physics II', ed. E. W. McDaniel and M. R. C. McDowell, North-Holland Publishing Company, Amsterdam, 1972, p. 483.
[60] M. W. Siegel, R. J. Celotta, J. L. Hall, J. Levine, and R. A. Bennett, *Phys. Rev. (A)*, 1972, **6**, 607.

electron affinity, say O^- (E.A. $= 1.462^{+0.003}_{-0.007}$ eV). A molecular ion will usually give a detachment spectrum, corresponding to transitions between different vibrational levels of ionic and neutral species (Figure 3). The different vibrational levels must be identified, of course, to give the electron affinity, as defined. This same experiment has the useful facility of giving the angular distributions of ejected electrons by rotation of the laser polarization rather than by changing the geometry of the collision region.

In a second experiment, a fast negative-ion beam is crossed by a beam from a continuously tuneable laser.[61] Detachment is detected by directing the resulting fast neutral species at the cathode of a multiplier and monitoring the secondary electron current. The signal, proportional to the cross-section for neutral atom production, is a step function of the energy. This particular experiment allows exploration of the cross-section near threshold. This is expected to be a function of the forces between the product particles.[62] For atoms, $Q \propto E^{(2l+1)/2}$, where l is the orbital angular momentum of the detached electron and E is its energy. Experiment has shown that this threshold law applies only very close to threshold, within 5 meV in Se[61] and 50 meV in heavier species. This emphasizes the errors inherent in extrapolating cross-section data to threshold over wide energy ranges. In molecules, the threshold behaviour will be modified by additional long-range electron–molecule interactions, and this is not yet fully understood. Detachment using a tuneable laser has provided evidence for excited electronic states of C_2^-.[63] The electron affinity of C_2 is ca. 3.5 eV, one of the highest known electron affinities. Irradiation of C_2^- with a laser over the wavelength range 5300—5450 Å (2.5—2.2 eV) detaches an electron. The results are consistent with a two-photon process in which electronic excitation to a bound electronically excited state is followed by electron detachment from this intermediate state. These measurements establish that the ground state of C_2^- is $^2\Sigma_g^+$. Bound electronic excited states are not known, in the gas phase, for any other molecular negative ion.

A third photodetachment experiment, using a laser source, can be carried out in an ion–cyclotron mass spectrometer.[64] Negative ions are generated in the resonance cell, in which a tuneable laser is directed along the cell axis. Photodetachment is detected by monitoring the negative-ion concentration in the mass spectrometer; the signal is a step function of the laser frequency.

All of these experiments can, of course, be performed with conventional sources. However, it seems likely that the best electron-affinity measurements will continue to be obtained using such photodetachment procedures with laser beams. The experimental procedures are fully described in the cited papers.

Two other beam techniques are also being applied to the determination of electron affinities. Both are based on the measurement of energy-threshold

[61] H. Hotop, T. A. Patterson, and W. C. Lineberger, *Phys. Rev. (A)*, 1973, **8**, 762.

[62] T. F. O'Malley, *Phys. Rev. (A)*, 1965, **137**, 1668.

[63] W. C. Lineberger and T. A. Patterson, *Chem. Phys. Letters*, 1972, **13**, 40.

[64] K. Smythe and J. I. Brauman, *J. Chem. Phys.*, 1972, **56**, 1132.

processes producing the negative ion in question. In endothermic negative-ion charge-transfer the reaction is:[65]

$$X^- + AB \rightarrow X + AB^- \qquad (28)$$

A negative-ion beam whose energy can be varied is directed through the sample gas, AB, and the threshold energy, E_{th}, for production of AB^- is measured. Then equation (29) is valid. This experiment requires a knowledge of the electron

$$E.A.(AB) = E.A.(X) - E_{th} \qquad (29)$$

affinity of X as well as a means of establishing the threshold energy. This last requirement is not easy. Experimentally, the cross-section as a function of energy displays tailing and curvature, which can obscure the true onset. This tailing depends on the thermal velocity distribution of the neutral gas molecules (Doppler broadening) and the shape of the true threshold function. This can be arrived at by substituting a calculated energy distribution into an assumed threshold function and adjusting the threshold function to get a match with the experimental curve. In addition, any one value of electron affinity should be deduced from several different charge-transfer reactions and using negative-ion reactants produced in different sources. Considering the difficulties associated in evaluating the data, the results achieved for small molecules in this experiment seem good, and stand up to comparison with those from photodetachment work.

The other charge-transfer reaction is collisional ionization:[66]

$$X + AB \rightarrow X^+ + AB^- \qquad (30)$$

where X is commonly caesium. A Cs beam of variable energy passes through the sample gas and, as before, the threshold energy for reaction (30) is identified. If AB and AB^- are in their ground energy levels, equation (31) holds. Cs is a

$$E.A.(AB) = I.P.(Cs) - E_{th} \qquad (31)$$

popular component in molecular beams, and so the physics of the interaction leading to the electron transfer is fairly well understood. In this experiment, again, the energy threshold is obscured by the essential lack of resolution, and unfolding techniques must be used to produce a true cross-section curve from that observed. Details of the data-handling procedures have been described.[64] Calculations suggest that the Cs^+ acts as a very efficient third body, removing excess vibrational energy from the negative ion, so that equation (31) applies. However, absolute values appear less good than those from endothermic charge-transfer reactions; where comparison with reliable values is possible, collisional ionization values appear somewhat high (Table 3).

B. Measured Electron Affinities.—The electron-affinity values in Tables 3 and 4 are limited mainly to those recently measured using tested techniques. For some

[65] B. M. Hughes, C. Lifshitz, and T. O. Tiernan, *J. Chem. Phys.*, 1973, **59**, 3162.
[66] S. J. Malley, R. N. Compton, H. C. Schwanter, and V. E. Anderson, *J. Chem. Phys.*, 1973, **59**, 4125.

Table 3 *Selected molecular electron affinities*

Ion	E.A./eV	Method	Ref.	Comments
O_2^-	0.440 ± 0.008	Pa	a	see text
O_3^-	~ 2.0	CT	b	
S_2^-	1.663 ± 0.040	Pa	c	see text
F_2^-	3.08 ± 0.1	CT	d	
Cl_2^-	2.32 ± 0.1	CT	e	
	2.38 ± 0.1	CT	d	
Br_2^-	2.62 ± 0.2	CT	e	
	2.51 ± 0.1	CT	d	
I_2^-	2.42 ± 0.2	CT	e	
	2.58 ± 0.1	CT	d	
IBr^-	2.7 ± 0.2	CT	d	
NH^-	0.38 ± 0.03	Pa	c	
OH^-	$1.829 \begin{smallmatrix} +0.010 \\ -0.014 \end{smallmatrix}$	Pa	c	
	1.825 ± 0.002	Pb	f	see text
OD^-	1.823 ± 0.002	Pb	f	
SH^-	2.301 ± 0.001	Pc	g	
SeH^-	2.21 ± 0.03	Pc	h	
NO^-	$0.024 \begin{smallmatrix} +0.01 \\ -0.005 \end{smallmatrix}$	Pa	i	
	0.015 ± 0.1	CT	e	
	0.1 ± 0.1	CI	j	
	0.025	ES	k	
N_2O^-	-0.15 ± 0.1	CI	j	
NO_2^-	2.36 ± 0.10	Pb	l	see text
	2.5 ± 0.1	CI	j	
	2.28 ± 0.1	CT	e	
	2.38 ± 0.06	CT	m	
NH_2^-	0.779 ± 0.037	Pa	e	
	0.744 ± 0.022	Pc	n	
PH_2^-	1.25 ± 0.03	Pc	o	
AsH_2^-	1.27 ± 0.03	Pc	n	
C_2H^-	2.21 ± 0.4	CT	e	
SiH_3^-	$\leqslant 1.44 \pm 0.03$	Pc	p	
GeH_3^-	$\leqslant 1.74 \pm 0.04$	Pc	p	
SO_2^-	1.097 ± 0.036	Pa	c	see text
	0.99 ± 0.1	CT	e	
CS_2^-	0.5 ± 0.2	CT	e	
SF_5^-	$\geqslant 2.8 \pm 0.1$	CT	q	
	$\geqslant 2.8 \pm 0.2$	CI	r	
SF_6^-	$\geqslant 0.6 \pm 0.1$	CT	q	
	$0.54 \begin{smallmatrix} +0.1 \\ -0.17 \end{smallmatrix}$	CI	r	
TeF_6^-	$3.34 \begin{smallmatrix} +0.1 \\ -0.17 \end{smallmatrix}$	CI	r	
(hexafluorobenzene)$^-$	$\geqslant 1.8 \pm 0.3$	CT	q	
(octafluorotoluene)$^-$	$\geqslant 1.7 \pm 0.3$	CT	q	
(perfluorocyclohexene)$^-$	$\geqslant 1.4 \pm 0.3$	CT	q	
(nitrobenzene)$^-$	$\geqslant 0.7 \pm 0.2$	CT	q	
(tetracyanoethylene)$^-$	2.03 ± 0.07	Pd	s	
(ethyl nitrene)	$\leqslant 1.87 \pm 0.16$	Pc	t	

Table 3—continued

Ion	E.A./eV	Method	Ref.	Comments
(maleic anhydride)$^-$	1.4 ± 0.2	CI	u	
(cyclopentadienide)$^-$	$\leqslant 1.84 \pm 0.03$	Pd	v	
(methylcyclopentadienide)$^-$	$\leqslant 1.67 \pm 0.04$	Pc	v	
(tetrafluorosuccinic anhydride)$^-$	0.5 ± 0.2	CI	w	
(hexafluoroglutaric anhydride)$^-$	1.5 ± 0.2	CI	w	

Abbreviations: Pa photodetachment with a fixed frequency laser, Pb photodetachment with a tuneable laser, Pc photodetachment in an ion–cyclotron resonance mass spectrometer, Pd photodetachment with conventional optical sources, CT charge transfer [equation (28)], CI collisional ionization [equation (30)], ES electron scattering; [a] R. J. Celotta, R. A. Bennett, J. L. Hall, M. W. Siegel, and J. Levine, *Phys. Rev. (A)*, 1972, **6**, 631; [b] J. A. Rutherford, B. R. Turner, and D. A. Vroom, *J. Chem. Phys.*, 1973, **58**, 5267; [c] R. J. Celotta, R. A. Bennett, and J. L. Hall, *J. Chem. Phys.*, 1974, **60**, 1740; [d] W. A. Chupka, J. Berkowitz, and D. Gutman, *J. Chem. Phys.*, 1971, **55**, 2724; [e] ref. 65; [f] H. Hotop and T. A. Patterson, *J. Chem. Phys.*, 1974, **60**, 1806; [g] J. R. Eyler and G. H. Atkinson, *Chem. Phys. Letters*, 1974, **28**, 217; [h] K. C. Smyth and J. I. Brauman, *J. Chem. Phys.*, 1972, **56**, 5993; [i] ref. 60; [i] ref. 66; [k] P. Burrow, *Chem. Phys. Letters*, 1974, **26**, 265; [l] E. Herbst, T. A. Patterson, and W. C. Lineberger, *J. Chem. Phys.*, 1974, **61**, 1300; [m] D. B. Dunkin, F. C. Fehsenfeld, and E. E. Ferguson, *Chem. Phys. Letters*, 1972, **15**, 257; [n] K. Smyth and J. I. Brauman, *J. Chem. Phys.*, 1972, **56**, 4620; [o] ref. 64; [p] K. J. Reed and J. I. Brauman, *J. Chem. Phys.*, 1974, **61**, 4830; [q] C. Lifshitz, T. O. Tiernan, and B. M. Hughes, *J. Chem. Phys.*, 1973, **59**, 3182; [r] R. N. Compton and C. D. Cooper, *J. Chem. Phys.*, 1973, **59**, 4140; [s] L. E. Lyons and L. D. Palmer, *Chem. Phys. Letters*, 1973, **21**, 442; [t] J. H. Richardson, L. M. Stephenson, and J. I. Brauman, *Chem. Phys. Letters*, 1974, **25**, 321; [u] R. N. Compton and P. W. Reinhardt, *J. Chem. Phys.*, 1974, **60**, 2953; [v] J. H. Richardson, L. M. Stephenson, and J. I. Brauman, *J. Chem. Phys.*, 1973, **59**, 5068; [w] C. D. Cooper and R. N. Compton, *J. Chem. Phys.*, 1974, 60, 2424.

diatomic species the measured data allow construction of a potential-energy curve for the negative ion, giving access to structural parameters for the ion. For $O_2{}^-$, $r_e = 134.1 \pm 1$ pm and $\omega_e = 1089$ cm^{-1} (corresponding values for the O_2 ground state are 120.74 pm and 1580.36 cm^{-1}). For NO$^-$, $r_e = 125.8 \pm 1$ pm and $\omega_e = 1470$ cm^{-1} (corresponding values for NO are 115.08 pm and 1904 cm^{-1}). A number of experiments on the hydroxyl radical confirm that the internuclear distance and vibration frequency of the negative ion are very close to those of the neutral parent radical: $r_e(\mathrm{OH}^-) = r_e(\mathrm{OH}) \pm 0.1$ pm and $\omega_e(\mathrm{OH}^-) = \omega_e(\mathrm{OH}) - (51 \pm 74)$ cm^{-1}. This is consistent with placement of the extra electron in a non-bonding orbital. The vibration frequency of $S_2{}^-$ is 725.68 cm^{-1} ($h\nu_{\mathrm{vib}} = 0.065$ eV) and the symmetric stretch frequency of $SO_2{}^-$ is 988.37 cm^{-1} ($h\nu_{\mathrm{vib}} = 0.122$ eV). The difference in electron affinities between OH and OD is believed to be real and almost entirely due to differences in position of the rotational states in OH ($^2\Pi_{\frac{3}{2}}$) and OD ($^2\Pi_{\frac{3}{2}}$). The electron affinities of the halogens are not known precisely, but do increase in the order $Cl_2 < Br_2 \approx I_2 < F_2$. The electron affinity of NO_2, long disputed, is at last established as ~ 2.36 eV. Further, the photodetachment experiments have indicated an isomer of $NO_2{}^-$, perhaps a peroxy isomer, with an electron affinity of *ca.* 1.8 eV. Calculations seem to support the existence of such a species. The contrasting behaviour of

Table 4 *Selected atomic electron affinities*

Ion	E.A./eV	Method	Ref.	Comments
Li⁻	0.620	Pa	a	
	0.61 ± 0.05	Pd	b	
Na⁻	0.548	Pa	a	
	0.543	Pb	a	
	0.53 ± 0.05	Pd	b	
K⁻	0.5012	Pa	a	
	0.50 ± 0.05	Pd	b	
Rb⁻	0.486	Pa	a	see text
	0.4859	Pb	a	
	0.48 ± 0.05	Pd	b	
Cs⁻	0.470	Pa	a	
	0.472	Pb	a	
	0.47 ± 0.05	Pd	b	
Ge⁻	1.20 ± 0.1	Pd	b	
Sn⁻	1.25 ± 0.1	Pd	b	
P⁻	0.77 ± 0.05	Pd	b	
As⁻	0.80 ± 0.05	Pd	b	
Sb⁻	1.05 ± 0.05	Pd	b	
Bi⁻	0 9—1.2	Pd	b	
O⁻	1.462 $^{+\,0.003}_{-\,0.007}$	Pa		see ref. c
S⁻	2.0772 ± 0.0005	Pb	d	
Se⁻	2.0206 ± 0.003	Pb	e	
Te⁻	1.9 ± 0.15	Pd	b	
Cr⁻	0.66 ± 0.05	Pd	b	
Cu⁻	1.226 ± 0.01	Pa	c	
Ag⁻	1.303 $^{+\,0.007}_{-\,0.011}$	Pa	c	
Au⁻	2.3086 ± 0.0007	Pb	f	
Pt⁻	2.128 ± 0.002	Pb	f	

Abbreviations are as for Table 3; [a] J. A. Patterson, H. Hotop, A. Kasdan, D. W. Norcross, and W. C. Lineberger, *Phys. Rev. Letters*, 1974, **32**, 189; [b] D. Feldmann, R. Rackwitz, E. Heinicke, and H. J. Kaiser, *Phys. Letters*, 1973, **45A**, 404; [c] H. Hotop, R. A. Bennett, and W. C. Lineberger, *J. Chem. Phys.*, 1973, **58**, 2373; [d] W. C. Lineberger and B. Woodward, *Phys. Rev. Letters*, 1970, **25**, 424; [e] H. Hotop and T. A. Patterson, *Phys. Rev. (A)*, 1973, **8**, 762; [f] H. Hotop and W. C. Lineberger, *J. Chem. Phys.*, 1973, **58**, 2379.

SF_6 and TeF_6 towards low-energy electrons is interesting. As indicated in Table 3, TeF_6 has a very large electron affinity and yet it has a very small attachment rate constant. The converse holds for SF_6.

The photodetachment experiments of Patterson *et al.* on the alkali metals have revealed, in addition to reliable electron-affinity values, strong resonances just below the first excited electronic states, $^2P_{\frac{1}{2}}$ and $^2P_{\frac{3}{2}}$. In Rb, the lower resonance has an estimated width of about 150 μeV, by far the narrowest resonance yet detected.

Comparison of the electron-affinity data of Table 4 with those in a recent compilation[59] shows that, until now, the best estimates of many of these electron affinities were calculated values. In particular, some of the empirical methods, where electron affinities are extrapolated from known ionization energies, have given remarkably good results.[59] The exact quantum calculation of electron affinities is a challenging problem. Any electron affinity is the small difference between two large quantities, the energy of neutral species and ion, respectively. So, for example, correlation effects, which are sometimes negligible in Hartree–Fock computations, cannot be ignored in electron-affinity calculations. A review of the theoretical methods that have been adopted is contained in a paper which also details a new approach to the problem.[67] In this, ionization energies (which include electron affinities) are determined directly without recourse to separate evaluation of energies and wavefunctions of the neutral and ionic species. For OH, it gives an electron affinity within 0.1 eV of the measured value.[68] An alternative, direct calculation using the one-particle Green-function method has also been applied to the evaluation of molecular electron affinities. It gives -0.24 eV[69] for the electron affinity of the methyl radical; this has not yet been measured.

Uncertainties associated with experimental electron affinities are frequently due to uncertainty about the energy state of the ion under study. Negative-ion formation through electron attachment [equations (26c) and (26d)] is an active area of study, not treated here. A chapter of a recent book covers this topic.[70]

6 Conclusion

This article has concentrated exclusively on electron–molecule systems in the gas-phase. In this area there are now available experimental techniques which could be almost routinely applied to the exploration of molecules of chemical interest. It is hoped that this Report has indicated the usefulness of such work. At the other extreme, the study of electrons in condensed phases is attracting increasing attention. It would seem appropriate to look also at the changing behaviour of electrons from dilute gases, through high-pressure gases, to liquids in which a good deal is already known about the physical and chemical behaviour of solvated electrons.[71] Some work in this area has been done in the measurement of electron drift velocities in gases at up to 50 atm. pressure, when the drift velocity becomes a function of gas pressure. Two, not necessarily mutually exclusive, explanations have been offered. Firstly, at moderate pressures, electron trapping, through resonance formation, could impede the electron's progress through the gas; at such pressures, the cumulative effect of the very many delaying encounters would be measurable.[72] At very high pressures, a more plausible explanation is that the electron sees a homogeneous scattering medium, approaching the condensed-phase situation. This could account for observed

[67] J. Simons and W. D. Smith, *J. Chem. Phys.*, 1973, **58**, 4899.
[68] W. D. Smith, T.-T. Chen, and J. Simons, *Chem. Phys. Letters*, 1974, **27**, 499.
[69] L. S. Cederbaum and W. von Messen, *Phys. Letters*, 1974, **47A**, 199.
[70] L. G. Christophorou, 'Atomic and Molecular Radiation Physics', Wiley–Interscience, New York, 1971, Ch. 6.
[71] See *e.g.* D. C. Walker, *Quart. Rev.*, 1967, **21**, 79; M. Anbar, *ibid.*, 1968, **22**, 578.
[72] R. W. Crompton and A. G. Robertson, *Austral. J. Phys.*, 1971, **24**, 543.

effects.[73] Of course, under these conditions, resonances arising from electron trappings by aggregates of molecules might be feasible. More recently, Christophorou *et al.* have looked at electron attachment to sample gases, in the presence of very high pressures of inert molecules. They conclude from measurements on benzene that it must have a small, *positive* electron affinity;[74] other electron-scattering experiments have not indicated this (see, for example, Table 2).

In any event, it seems likely that the free electron will find increasing applications as a 'reactant' in chemistry laboratories.

[73] W. Legler, *Phys. Letters*, 1970, **31A**, 129.
[74] L. G. Christophorou and R. E. Goans, *J. Chem. Phys.*, 1974, **60**, 4244.

5 Heavy-atom Kinetic Isotope Effects

By A. MACCOLL

Department of Chemistry, University College, London, WC1H 0AJ

1 Introduction

To the best knowledge of the reporter this subject has not as such been previously reported on, although specific results of the application of the techniques have been referred to from time to time. For this reason it is proposed to give a general account of the theory and practice of the technique, so that the reader interested in applying it, or assessing the results obtained, will have key references. A useful starting point is the indexed bibliography produced by Stern and Wolfsberg and published by the National Bureau of Standards.[1] This lists papers published since the earliest entry until 1965 in the case of experimental work, and 1968 in the case of reviews and theoretical papers.

The plan of the present article is to summarize first the terms used, and the theoretical treatment. This is then followed by a section on experimental methods and on the interpretation of the experimental results. Finally, an account is given of applications of the method in the cases of carbon, nitrogen, oxygen, sulphur, and chlorine.

One of the challenging problems of chemical kinetics is the specification of the properties of the transition state of a chemical reaction. The problem may be approached at either a topological or a metrical level; in the latter case one requires, as well as the geometrical structure, the mechanical properties. By topological level is meant the discussion of whether, for example, a transition state is cyclic or acyclic or which end of a bidentate group attacks a molecule. Such is the province of physical organic chemistry, although the metrical aspects can be hinted at when the degree of bond breaking or bond forming is discussed. The requirement of knowing the geometry and the mechanical properties must be met if an attempt is to be made at predicting quantitatively the rate of reaction. Thus in the case of unimolecular gas-phase reactions (of molecules, radicals, or ions) the rate constant, as a function of internal energy, is given by

$$k(e) = \frac{\sigma}{h} \left\{ \frac{(I_A I_B I_C)^+}{I_A I_B I_C} \right\}^{\ddagger} \frac{W(e - e_0)}{\rho(e)} \tag{1}$$

[1] M. J. Stern and M. Wolfsberg, 'Heavy Atom Kinetic Isotope Effects. An Indexed Bibliography', N.B.S. Special Publication 349, U.S. Dept. of Commerce, Washington D.C., 1972.

where e_0 is the critical energy, e the internal energy, σ the reaction path degeneracy, $W(e - e_0)$ the number of states of the activated complex with energy less than or equal to $e - e_0$, and $\rho(e)$ the density of states for the normal molecule at energy e. $I_A I_B I_C$ and $(I_A I_B I_C)^+$ are the products of the principal moments of inertia for the ground and transition states, respectively. To evaluate both the density and the number of states a knowledge of the vibrational frequencies in both the ground and transition states is required. For a reaction of a molecule, the ground-state frequencies are usually known; those for the transition state have to be estimated. On the other hand, for the reaction of a radical (e.g. decomposition) or of an ion (e.g. electron impact fragmentation) the frequencies for the ground state are usually unknown, as are those of the transition state, so the problem really becomes difficult. Any evidence pertaining to the properties of the transition state must per se be valuable. In the case of ground-state molecules, spectroscopic techniques, including isotopic substitution, can lead to a complete structural definition, together with an identification of the vibrational frequencies. This suggests that the effects of isotopic substitution upon reaction rates can in principle do the same for transition states, which is the underlying reason for the studies described herein.

The term 'heavy-atom kinetic isotope effect' refers to the effect for atoms other than hydrogen, mainly carbon, nitrogen, oxygen, sulphur, and chlorine. The atomic weights and percentage abundances are shown in Table 1. The important

Table 1   *Isotopic masses and percentage abundances*

Element	Atomic mass (C = 12.000 00)	Percentage abundance
^{12}C	12.000 00	98.892
^{13}C	13.003 36	1.108
^{14}N	14.003 07	99.645
^{15}N	15.000 11	0.355
^{16}O	15.994 91	99.759
^{17}O	16.999 13	0.037
^{18}O	17.999 16	0.204
^{32}S	31.972 07	95.018
^{33}S	32.971 46	0.750
^{34}S	33.967 86	4.215
^{36}S	35.967 09	0.017
^{35}Cl	34.968 85	75.529
^{37}Cl	36.965 90	24.471

approximation that can be made in this case is $m^* - m \ll m, m^*$ and $v - v^* \ll v$, v^*, where the asterisk refers to the heavy isotope.

The first person to realize the importance of, and to formalize the treatment of kinetic isotope effects was Bigeleisen[2] following upon the formalization of isotopic equilibria by Bigeleisen and Goeppert-Mayer.[3] The theory will be dealt with

[2] J. Bigeleisen, *J. Chem. Phys.*, 1949, **17**, 675.
[3] J. Bigeleisen and M. Goeppert-Mayer, *J. Chem. Phys.*, 1947, **15**, 261.

in a subsequent section. Experimentally, most workers have used isotopic species of natural abundance, and made precise measurements of the isotopic ratio as a function of percentage reaction. To do this, a double-collector mass spectrometer is necessary, which enables a comparison of the ion beam from the minor isotopic species to be backed off against a fraction of the ion beam from the major isotopic species *e.g.* $[^{13}CO_2{}^+]/[^{12}CO_2{}^+]$. For technical reasons, such as discrimination in the mass spectrometer, absolute ratio measurements are not attempted; rather the ratio for the unknown is compared with that for a standard. This is quite suitable for kinetic studies, since the ratio for the reactant (or product) at time t can be compared with that of the reactant at zero time or a product (provided it contains all the isotopic species) at infinite time. Because of the experimental skills required and because twin-collector mass spectrometers have only recently become available commercially, work in this field has, in the past, been restricted to a relatively small number of schools. Since it is a method of considerable power, it can be anticipated that increasing efforts will be made in this field. This is especially true in the area of gas-phase kinetics, where the behaviour of an isolated molecule can be examined without the complicating co-operative effects of the solvent inherent in studies of reactions in solution.

2 Some Terms and Definitions

The isotope effect is measured as k/k^*, where the asterisk refers to the heavier species. If $k/k^* > 1$, the effect is said to be *direct* or *normal*; if $k/k^* < 1$ it is said to be *inverse*. If the heavy atom is directly concerned in bond breaking or bond making in the transition state, the effect is said to be *primary*; if not the effect is *secondary*. Consider a species such as malonic acid, for which there are four possible decarboxylation reactions (Scheme 1). k_3/k_2 measures the *intramolecular*

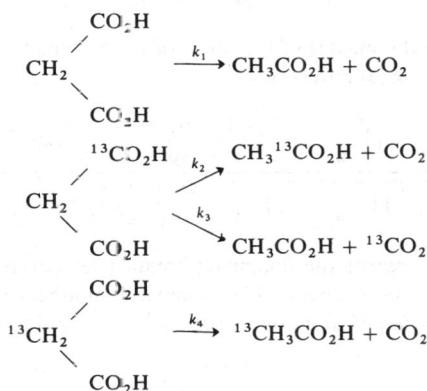

Scheme 1

kinetic isotope effect and k_1/k_2 the *primary intermolecular* kinetic isotope effect. k_1/k_4 measures the *secondary intermolecular* kinetic isotope effect.

3 Theoretical Treatment

Bigeleisen[2] started with the absolute rate theory equation

$$k = \kappa \frac{C^{\ddagger}}{C_A C_B \ldots} \left(\frac{kT}{2\pi M^{\ddagger}}\right)^{\frac{1}{2}} \frac{1}{\delta} \tag{2}$$

where the symbols have their usual significance. The ratio of the rate constants is then given by

$$\frac{k}{k^*} = \frac{\kappa}{\kappa^*} \frac{Q^{\ddagger}}{Q^{**}} \left(\frac{M^{**}}{M^{\ddagger}}\right)^{\frac{1}{2}} \tag{3}$$

This equation finally reduces to

$$\frac{k}{k^*} = \frac{\kappa}{\kappa^*} \frac{\left(\dfrac{M^*}{M}\right)^{\frac{1}{2}}}{\left(\dfrac{M^{**}}{M^{\ddagger}}\right)^{\frac{1}{2}}} \frac{\left(\dfrac{I_A^* I_B^* I_C^*}{I_A I_B I_C}\right)^{\frac{1}{2}}}{\left(\dfrac{I_A^{\ddagger *} I_B^{\ddagger *} I_C^{\ddagger *}}{I_A^{\ddagger} I_B^{\ddagger} I_C^{\ddagger}}\right)^{\frac{1}{2}}} \frac{\dfrac{s}{s^*}}{\dfrac{s^{\ddagger}}{s^{**}}} \frac{\prod_i^{3n-6} \dfrac{1-\exp(-u_i)}{1-\exp(-u_i^*)}}{\prod_i^{3n^{\ddagger}-7} \dfrac{1-\exp(-u_i^{\ddagger})}{1-\exp(-u_i^{**})}} \frac{\exp \sum_i^{3n-6}(u_i-u_i^*)/2}{\exp \sum_i^{3n^{\ddagger}-7}(u_i^{\ddagger}-u_i^{**})/2} \tag{4}$$

by substituting the usual expressions for the partition functions. In equation (4), the M's are the molecular weights, the I's the moments of inertia and the u's are given by $u_i = \mathbf{h}v_i/kT$, where the v_i's are the vibration frequencies. The s's are the symmetry numbers. For isotopically substituted molecules the Teller–Redlich product rule[4] holds, namely

$$\left(\frac{M}{M^*}\right)^{\frac{1}{2}} \left(\frac{I_A I_B I_C}{I_A^* I_B^* I_C^*}\right)^{\frac{1}{2}} = \prod_i^n \frac{m_i}{m_i^*} \prod_i^{3n-6} \frac{v_i}{v_i^*} \tag{5}$$

where the m's are the masses of the atom or atoms which are isotopically substituted. Introducing (5) into (4) gives

$$\frac{k}{k^*} = \frac{v^{\ddagger}}{v^{**}} \frac{\dfrac{s}{s^*}}{\dfrac{s^{\ddagger}}{s^{**}}} \frac{\prod_i^{3n-6} \dfrac{u_i^*}{u_i}}{\prod_i^{3n^{\ddagger}-6} \dfrac{u_i^{**}}{u_i^{\ddagger}}} \cdot \frac{\prod_i^{3n-6} \dfrac{1-\exp(-u_i)}{1-\exp(-u_i^*)}}{\prod_i^{3n^{\ddagger}-7} \dfrac{1-\exp(-u_i^{\ddagger})}{1-\exp(-u_i^{**})}} \cdot \frac{\exp \sum_i^{3n-6}(u_i-u_i^*)/2}{\exp \sum_i^{3n^{\ddagger}-7}(u_i^{\ddagger}-u_i^{**})/2} \tag{6}$$

where the $v^{\ddagger}$'s represent the imaginary frequencies corresponding to motion along the reaction co-ordinate. The symmetry numbers can be removed by multiplying both sides of equations (4) and (6) by $(s^{\ddagger}/s^{**})/(s/s^*)$ to yield

$$\frac{ks^{\ddagger}s^*\kappa^*}{k^*s^{**}s\kappa} = HRR$$

[4] O. Redlich, *Z. Phys. Chem.*, 1935, **B28**, 371.

where HRR is the rate constant ratio on the harmonic oscillator, rigid rotator approximation.† Equations (4) and (6) can also be written

$$HRR = MMI \times EXC \times ZPE \qquad (4a)$$

$$HRR = \frac{v^{\ddagger}}{v^{*\ddagger}} \times VP \times EXC \times ZPE \qquad (6a)$$

In these expressions, MMI is the mass-moment of inertia factor, EXC the vibrational excitation factor, ZPE the zero point energy factor, and VP the vibrational frequency product factor.‡ In the case of heavy atom kinetic isotope effects, Bigeleisen introduced the quantity $\Delta u_i = u_i - u_i^*$, and using the fact that $\Delta u_i \ll u_i$, expanded equation (6) to yield

$$HRR = \frac{v^{\ddagger}}{v^{*\ddagger}} \left\{ 1 + \sum_i^{3n-6} G(u_i)\Delta u_i - \sum_i^{3n\ddagger-7} G(u_i^{\ddagger})\Delta u_i^{\ddagger} \right\} \qquad (7)$$

where $G(u) = (1/2) - (1/u) + [1/(e^u - 1)]$, and had previously been tabulated by Bigeleisen and Goeppert-Mayer.[3] An alternative form of equation (4), appropriate to the case of unimolecular gas-phase reactions, has been suggested by Christie.[7] It is

$$HRR = MI \left[1 + \tfrac{1}{2} \sum_i^{3n-6} (\coth u_i/2)\,\Delta u_i - \tfrac{1}{2} \sum_i^{3n-7} (\coth u_i^{\ddagger}/2)\,\Delta u_i^{\ddagger} \right] \qquad (8)$$

where

$$MI = \frac{\left(\dfrac{I_A^{\ddagger} I_B^{\ddagger} I_C^{\ddagger}}{I_A^{*\ddagger} I_B^{*\ddagger} I_C^{*\ddagger}} \right)^{\ddagger}}{\left(\dfrac{I_A I_B I_C}{I_A^{*} I_B^{*} I_C^{*}} \right)^{\ddagger}}$$

In order to apply these formulae, it is necessary to know the geometry of the ground and transition states and also the vibrational frequencies. In the cases of equations (4), (6), and (8), the expression for HRR can be factored into a temperature-independent factor (TIF) and a temperature-dependent factor (TDF).

The above formula can be simplified at low and high temperatures. For as $T \to 0$, $u \to \infty$ and $EXC \to 1$, equations (4) and (6) become

$$HRR = MMI \times ZPE \qquad (9)$$

$$HRR = \frac{v^{\ddagger}}{v^{*\ddagger}} \times VP \times ZPE \qquad (10)$$

---

† This seems a better use of the letters than "harmonic rate ratio" used by Van Hook.[5]
‡ This nomenclature was introduced by Wolfsberg and Stern.[6]

[5] W. A. Van Hook, in 'Isotope Effects in Chemical Reactions', ed. C. J. Collins and N. S. Bowman, ACS monograph No 167, Von Nostrand Rheinhold, New York, 1970, p. 11.
[6] M. Wolfsberg and M. J. Stern. *Pure Appl. Chem.*, 1964, **8**, 225.
[7] Private communication. See also J. R. Christie, W. D. Johnson, A. G. Loudon, A. Maccoll, and L. M. N. Machacek, *J.C.S. Faraday I*, submitted for publication

Because of the approximations made in deducing equation (8) it is not amenable to proceeding to the limit $T \to 0$. Again at high temperatures, $u \to 0$, ZPE $\to 1$ and EXC $\to (VP)^{-1}$ and equations (4) and (6) become

$$HRR = \frac{MMI}{VP} \tag{11}$$

$$HRR = \frac{v^{+}}{v^{*+}} \tag{12}$$

The high-temperature limit of equation (8) is

$$HHR = MI \tag{13}$$

A further form of the high-temperature approximation is given by expanding equation (7) in powers of T^{-1}. This yields

$$\ln HHR = \ln \frac{v^{+}}{v^{*+}} + \frac{1}{24}\left(\frac{h}{2\pi kT}\right)^{2} \sum_{i} (a_{ii} - a_{ii}^{+})(m_{i}^{-1} - m_{i}^{*-1}) \tag{14}$$

where $a_{ii} - a_{ii}^{+}$ is the difference in the diagonal element of the cartesian force constant between the ground and the transition state. Alternatively

$$\ln HRR = \ln \frac{v^{+}}{v^{*+}} + \frac{1}{24}\left(\frac{h}{2\pi kT}\right)^{2}\left\{\sum_{i,j} f_{ij}(g_{ij} - g_{ij}^{*}) - \sum_{i,j} f_{ij}^{*}(g_{ij}^{+} - g_{ij}^{*+})\right\} \tag{15}$$

where the f_{ij} and g_{ij} are the elements of the Wilson F and G matrices.[8] To extend the range of application of equations (14) and (15) to lower temperatures, Bigeleisen and Wolfsberg[9] defined a parameter $\bar{\gamma} = 12G(u_i)/u_i$, and replaced the factor $1/24$ by $\bar{\gamma}/24$ and $\bar{\gamma}^{+}/24$, where $\bar{\gamma}$ is an average value for the given molecule, $\bar{\gamma}^{+}$ being that for the activated complex. This approximation depends upon the fact that γ_i is a slowly varying function of u_i. Wolfsberg and Stern[6] examined the validity of this approximation, concluding that it is only useful for making qualitative predictions of isotope effects.

The question arises as to the calculation of v^{+}/v^{*+}, and two approaches have been used. The first derives from the Slater theory,[10] in which bond rupture occurs when a bond reaches a critical extension. In this case

$$\frac{v^{+}}{v^{*+}} = \left(\frac{\mu}{\mu^{*}}\right)^{\frac{1}{2}} \tag{16}$$

where $\mu = M_A M_B/(M_A + M_B)$, A and B being the two atoms concerned in the bond. Alternatively, Bigeleisen and Wolfsberg[9] consider the two fragments produced by bond rupture, in which case the atomic masses in equation (16) become replaced by the fragment masses. More realistically the vibrational problem can be solved on the basis of an assumed potential-energy surface for the

[8] E. B. Wilson, jun., J. C. Decius, and P. C. Cross, 'Molecular Vibrations', McGraw-Hill, New York, 1955.
[9] J. Bigeleisen and M. Wolfsberg, *Adv. Chem. Phys.*, 1958, **1**, 15.
[10] J. C. Slater, 'Theory of Unimolecular Reactions', Methuen, London, 1959.

transition states, and the ratio of the imaginary frequencies for crossing the col calculated.

It has already been noted that, in order to treat any specific case, the geometry of both the ground state and the transition state must be known, as must the vibration frequencies for each of the states. The moments of inertia of the ground state can readily be obtained from structural data; those for the transition state will contain parameters which can be varied according to the model. It should be noted that, using equation (4) or (8), the terms MMI and MI are temperature-independent, and so, in principle, should enable transition-state geometry to be decided. Unfortunately the correspondence will not be one to one, in that a range of parameters will give rise to the same value of MMI or MI. The frequencies for the two species in the ground state can be determined experimentally or calculated by the Wilson method from a knowledge of the force constants. It will be found that only a limited number of these will change significantly upon isotopic substitution. For the transition state, the frequencies or force constants have to be assumed. This part of the model will be reflected in the temperature-dependent factor of the rate constant ratio.

Williams and Taylor,[11] in their studies of t-butyl chloride, have synthesized t-$C_4H_9{}^{35}Cl$ and t-$C_4H_9{}^{37}Cl$ and measured their spectra. They have also used Wilson's *FG* method to calculate the frequencies. Frequencies for t-$C_4H_9{}^{35}Cl$

Table 2   *Vibrational frequencies and shifts in* t-C_4H_9Cl

Degeneracy	Observed/cm^{-1}		Calculated/cm^{-1}	
	v(t-$C_4H_9{}^{35}Cl$)	Δv	v(t-$C_4H_9{}^{35}Cl$)	Δv
1				
2	301	$\leqslant 1.5$	301.0	1.48
1	372	3.8 ± 0.5	372.2	4.16
2	408	—	408.0	0.10
1	585	2.6 ± 0.1	585.0	2.71
1	818	0.5 ± 0.3	818.1	0.72
2	1210	$\leqslant 0.2$	1210.1	0.20

and the corresponding shifts are shown in Table 2. By these methods, a complete treatment of the calculation of the rate constant ratio can be made.

Stern and Wolfsberg[6,12] have described a 'cut-off' technique that simplifies the calculation. They point out that the computer time required for the solution of the vibrational problem varies roughly as the third power of the number of atoms. Thus for a molecule of any degree of complexity, the computer time required may be prohibitive. According to equation (15), it is permissible 'to omit (cut off) portions of the molecule that do not include the atoms necessary to specify all the internal co-ordinates which both involve the isotopic position and have associated with them force constant or geometry changes between reactant and transition state'. The results of calculations of these authors on the decomposition of malonic acid (*vide infra*) are shown in Table 3.[12]

[11] R. C. Williams and J. W. Taylor, *J. Amer. Chem. Soc.*, 1973, **95**, 1710.
[12] M. J. Stern and M. Wolfsberg, *J. Chem. Phys.*, 1966, **45**, 4105.

Table 3 *Cut-off calculations for* $HOOCCH_2{}^{13}COOH \rightarrow [HOOCCH_2\cdots{}^{13}COOH]^*$

	$\dfrac{k_1/2k_3}{v_{1L}^{\ddagger}/v_{2L}^{\ddagger}}$	Individual factors at			
Model reaction	300 K	VP	EXC	ZPE	$v_1^{\ddagger}/v$
$HOOCCH_2{}^{13}COOH \rightarrow [HOOCCH_2\cdots{}^{13}COOH]^b$	1.0245	0.9946	1.0004	1.0296	1.005
$C-CH_2{}^{13}COOH \rightarrow [C-CH_2\cdots{}^{13}COOH]$	1.0244	0.9962	0.9994	1.0290	1.003
$^{45}Y-CH_2{}^{13}COOH \rightarrow [^{45}Y-CH_2\cdots{}^{13}COOH]^c$	1.0244	0.9944	1.0004	1.0298	1.005
$C-{}^{13}COOH \rightarrow [COO^{13}COOH]$	1.0241	0.9977	0.9991	1.0273	1.002
$^{59}Y-{}^{13}COOH \rightarrow [^{59}Y\cdots{}^{13}COOH]^d$	1.0265	0.9938	0.9992	1.0338	1.006
$C-CH_2-{}^{13}C \rightarrow [C-CH_2\cdots{}^{13}C]$	1.0340	0.9740	1.0027	1.0587	1.026
$C-{}^{13}C \rightarrow [C\cdots{}^{13}C]$	1.0334	0.9806	1.0005	1.0533	1.019

(a) MMI in all cases is unity, since reactant and transition-state geometries are identical. (b) Complete calculation. Geometry and force constants given in footnote a of Table 8, of ref. 12. (c) ^{45}Y is an atom with mass of 45 a.m.u., representing COOH. (d) ^{59}Y is an atom with mass of 59 a.m.u., representing HOOCCH$_2$.

A further example of cut-off calculations comes from Fry *et al.*[13] These authors investigated ^{14}C and ^{37}Cl isotope effects for the reaction

$$Y + R^*CH_2X \rightarrow [Y\cdots R^*CH_2\cdots X]^+ \rightarrow R^*CH_2Y + X$$

The models used were (1) the complete molecule, (2) inclusion of three ring-carbon atoms, and (3) inclusion of only one ring-carbon atom.

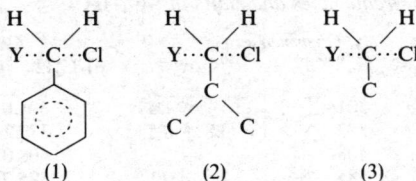

The calculations were performed for various bond orders of the $Y-C$ (n_1) and $C-Cl$ (n_2) bonds. The corresponding force constants were taken to be

Table 4 *Calculated ^{14}C (upper) and ^{37}Cl (lower) isotope effects at 30 °C*

n_1	Complete model (1)	Cut-off models (2)	(3)
0.9	1.03764	1.03894	1.04531
	1.00348	1.00208	1.00048
0.7	1.05576	1.05750	1.06907
	1.01077	1.00924	1.00607
0.5	1.05473	1.05619	1.06788
	1.01702	1.01582	1.01265
0.3	1.03922	1.04022	1.04939
	1.02123	1.02064	1.01784
0.1	1.01090	1.01158	1.01689
	1.02451	1.02386	1.02493

13 L. B. Sims, A. Fry, L. T. Netherton, J. C. Wilson, K. D. Reppond, and S. W. Crook, *J. Amer. Chem. Soc.*, 1972, **94**, 1364.

$F_{ab} = n_{ab}F_{ab}^0$ and $F_{abc} = n_{ab}n_{bc}F_{cbc}^0$ for a typical stretch and bend frequency. The n_{ij} are the bond orders. Constant total bonding was assumed in the reaction *i.e.* $n_1 + n_2 = 1$. The results are shown in Table 4, where Y is taken to be O as in hydrolysis.

An interesting difference is found between the effects for ^{14}C and ^{37}Cl. The latter effect increases monotonically as the bond order of the C—Cl bond increases, whereas the former effect goes through a maximum. This behaviour has not as yet been reported experimentally.

The reaction

$$A + BC \rightarrow AB + C$$

has been discussed in some detail by Bigeleisen.[14]

The anomalous temperature dependence of carbon kinetic isotope effects has been examined theoretically by Yankwich and his collaborators.[15] Crossover refers to a change-over from a situation in which $k/k^* > 1$ to one in which $k/k^* < 1$, or *vice versa* as the temperature is changed. They used a modification of the Wilson FG matrix method to solve the vibrational problem and explore the effects of cross-terms in the potential function. They concluded that crossover does not occur 'when the reaction co-ordinate contains but a single element', but can occur if more than one element is involved. Such crossover has been reported for small-molecule isotopic exchange reactions by Stern *et al.*[16]

Following a series of papers on the calculation of the Bigeleisen–Mayer function by use of a finite orthogonal polynomial expansion,[17—20] Bigeleisen and Ishida[21] have developed simple expressions for the calculation of isotope effects consequent upon end-atom substitution. The effects can be calculated from a knowledge of the atomic masses and the stretching and bending force constants. The solution of the vibrational secular equations is not necessary. Further calculations by this method are promised.

Kidd and Yankwich[22] have considered the effect of curvature of the potential-energy surface along the reaction co-ordinate at the transition state upon heavy-atom kinetic isotope effects. The results are analysed in terms of the TIF and TDF.

The work so far described regards the reacting molecule as being in the gas phase. Medium effects have been investigated by Keller and Yankwich[23] using a cell model developed by Stern, Van Hook, and Wolfsberg[24] for the study of vapour pressure isotope effects. The model predicts that medium-induced

[14] J. Bigeleisen, *Pure Appl. Chem.*, 1964, **8**, 217.
[15] T. T. S. Huang, W. J. Kass, W. E. Buddenbaum, and P. E. Yankwich. *J. Phys. Chem.*, 1968, **72**, 4431.
[16] M. J. Stern, W. Spindel, and E. U. Monse, *J. Chem. Phys.*, 1968, **48**, 2908.
[17] J. Bigeleisen and T. Ishida, *J. Chem. Phys.*, 1968, **48**, 1311.
[18] J. Bigeleisen and T. Ishida, *Adv. Chem. Ser.*, 1969, **89**, 192.
[19] J. Bigeleisen, T. Ishida, and W. Spindel, *Proc. Nat. Acad. Sci. U.S.A.*, 1970, **67**, 113.
[20] J. Bigeleisen, T. Ishida, and W. Spindel, *J. Chem. Phys.*, 1971, **55**, 5021.
[21] J. Bigeleisen and T. Ishida, *J. Amer. Chem. Soc.*, 1973, **94**, 6155.
[22] R. W. Kidd and P. E. Yankwich, *J. Chem. Phys.*, 1973, **59**, 2723.
[23] J. H. Keller and P. E. Yankwich, *J. Amer. Chem. Soc.*, 1973, **95**, 4811, 7968.
[24] M. J. Stern, W. A. Van Hook, and M. Wolfsberg, *J. Chem. Phys.*, 1963, **39**, 3179.

isotope effects will be negligible unless medium–reactant interactions are so strong that gross rate effects would be observable. Keller and Yankwich[25] subsequently introduced a structured medium model, as distinct from the cell or continuous medium model. A non-linear triatomic molecule is taken as the reactant, but it is influenced by the attachment and coupling of one or two mass points, representing the effects of the medium. The effects on the TDF calculated in this fashion are greater than those calculated on the cell model. Also substantial effects are produced on the TIF which have no counterparts on the cell model. The authors suggest that a comparison of related inter- and intra-molecular kinetic isotope effects would be useful in discovering the conditions for which a given model should be used.

4 Experimental Methods

The magnitude of the heavy-atom kinetic isotope effect can be gauged from model calculations[26] based on a diatomic molecule (Table 5). It is seen that the effects

Table 5 *Estimates of heavy-atom kinetic isotope effects* (k/k^*)

	Temperature/°C		
System	25	100	200
$^{12}C^{12}C/^{12}C^{13}C$	1.0548	1.0444	1.0363
$^{12}C^{14}N/^{12}C^{15}N$	1.0436	1.0354	1.0290
$^{12}C^{16}O/^{12}C^{18}O$	1.0675	1.0547	1.0447
$^{12}C^{32}S/^{12}C^{34}S$	1.0184	1.0152	1.0128
$^{12}C^{35}Cl/^{12}C^{37}Cl$	1.0141	1.0118	1.0101

are very small. It is possible in principle to prepare isotopically pure specimens of A and A*, and by using conventional kinetic techniques determine k and k^*. However, the observed effect is not greatly different from the experimental error and so it is much better to use a comparative technique. To this end a twin-collector mass spectrometer is essential.[27] Those commercially available are usually small-radius permanent magnetic analysers, with electrostatic scanning. Two types of twin collector are possible, the first consisting of a slit and a collector (Figure 1a), the main beam being collected on the slit, and the minor beam on the

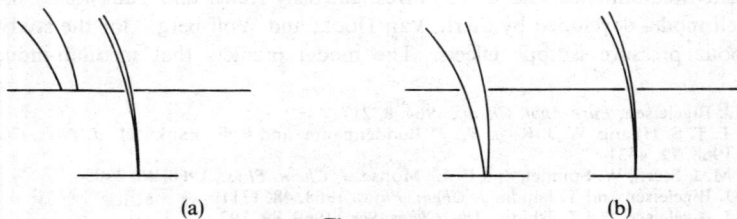

(a)   (b)
Figure 1

[25] J. H. Keller and P. E. Yankwich, *J. Amer. Chem. Soc.*, 1974, **96**, 2303.
[26] V. J. Shiner and W. E. Buddenbaum, in M.T.P. International Review of Science, Physical Chemistry, Series Two, Volume 5, ed. A. Maccoll, Butterworths, London, 1975.
[27] J. H. Beynon, 'Mass Spectrometry and its Applications to Organic Chemistry', Elsevier, Amsterdam, 1960.

collector; the second (Figure 1b) has a pair of slits and collectors, which have to be positioned for a given isotopic species. The disadvantage of the first system is that beams other than the main beam will also be collected on the slit, and of the second that the collector system must be repositioned when the isotopic species is charged. In either case a double inlet system is used, preferably with time-controlled magnetic valves, so that the ratio can be determined alternately for the unknown sample and the standard. If the ratio for the reactant is determined at various degrees of reaction the obvious standard is the reactant itself. If, on the other hand, the ratio for the product is determined, provided all the isotopic species goes into the product investigated, then the product at completion of reaction is the appropriate standard. Thus either $r = (A^*/A)/(A_0^*/A_0)$ or $r' = (P^*/P)/(P_\infty^*/P_\infty)$ are the observed quantities.

In order to test the mass spectrometer, measurements should be made on zero enrichment. To this end, 10 sets of 10 differences of ratios, with the same sample admitted to each inlet system, are measured. The mean and the standard deviation can then be calculated. Reproducibility should be within 0.01 % or better, depending on the instrument. The results of such a test on a GEC–AEI MS20 are shown in Table 6. In all cases the observed values are within the permitted limits.

Table 6   *The zero enrichment test*

Atom	Ratio (zero enrichment)
$^{13}C/^{12}C(CO_2)$	$+0.0014 \pm 0.0079$
$^{15}N/^{14}N(N_2)$	$+0.0016 \pm 0.0089$
$^{37}Cl/^{35}Cl(CH_3Cl)$	$+0.0062 \pm 0.013$
(C_2H_5Cl)	-0.0079 ± 0.010

It is usually desirable to convert the compound to be analysed into a small gaseous molecule before analysis. Great care must be taken to ensure that isotopic fractionation does not occur during the preparation of the sample. Also if the compound under investigation contains more than one atom of the type that is isotopically substituted care must be taken to ensure that the compound analysed contains only the atom of interest, and none of the other atoms of the same type. Alternatively, the assumption of isotopic homogeneity can be made, namely that the isotopic label has an equal probability of occurrence at each possible position. Thus in the case of malonic acid (Scheme 1, p. 79) the labelled carbon can occur either at one of the carbonyl positions, or in the methylene group. In phenyldiazonium compounds, the labelled nitrogen can occur either adjacent to or remote from the ring.

Beynon[27] has discussed the determination of isotopic ratios in some detail. For ^{13}C, conversion into carbon dioxide is the accepted method:[28] for ^{15}N, conversion into nitrogen.[29] In the case of ^{18}O, oxygen gas can be used,[30] but

[28] D. D. Van Slyke and J. Folch, *J. Biol. Chem.*, 1940, **136**, 509; F. Pregl and J. Grant, 'Quantitative Organic Analysis', Churchill, London, 1951.

[29] D. Rittenberg and L. Ponticorvo, *J. Appl. Radiation and Isotopes*, 1956, **1**, 208.

[30] C. C. Sweeley, W. H. Elliot, I. Fries, and R. Ryhage, *Analyt. Chem.*, 1966, **38**, 1549.

more conveniently the oxygen can be converted into water and equilibrated with carbon dioxide, which is then measured.[31] In the case of sulphur, the compound can conveniently be converted into sulphur dioxide, and measured as such.[32] For the halogens, the elemental substances or the hydrides have the disadvantage of producing corrosion and long-lived memory effects, and so the compound is usually converted into methyl chloride.[33] A recent paper recommends the use of negative ion mass spectrometry for determining $^{37}Cl/^{35}Cl$ ratios.[34]

5 Interpretation of Experimental Results

This subject has been fully treated by Bigeleisen and Wolfsberg[9] and by Melander[35] and will be summarized here since the results obtained are fundamental to the method. Considering the kinetics of a system undergoing a thermal

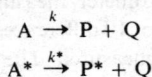

$$A \xrightarrow{k} P + Q$$
$$A^* \xrightarrow{k^*} P^* + Q$$

Scheme 2

unimolecular isotopic competitive reaction (Scheme 2), the rates of decomposition of isotopically substituted reactants are

$$\frac{-dA}{dt} = kA, \qquad \frac{-dA^*}{dt} = k^*A \qquad (17a,b)$$

Thus

$$\frac{dA}{dA^*} = \frac{k}{k^*}\frac{A}{A^*}$$

or

$$\frac{k}{k^*} = \frac{\ln A/A_0}{\ln A^*/A_0^*} \qquad (18)$$

In this form equation (18) is of little use, since the observed quantities are $(A^*/A)/(A_0^*/A_0)$ or $(P^*/P)/(P_\infty^*/P_\infty)$ depending upon whether the isotopic ratio is determined on the remaining reactant or on the products.

Solution of the differential equations (17a,b) gives $\ln A/A_0 = -kt$, $\ln A^*/A_0 = -k^*t$. For the fraction undecomposed $(1 - f)$:

$$(1 - f) = \frac{A + A^*}{A_0 + A_0^*} = \frac{A^* R_0(1 + R_f)}{A_0^* R_f(1 + R_0)} \qquad (19)$$

[31] C. R. McKinney, J. M. McCrea, S. Epstein, H. A. Allen, and H. C. Urey, *Rev. Sci. Instr.*, 1950, **21**, 724; L. Freedman and J. Bigeleisen, *J. Chem. Phys.*, 1950, **18**, 1325.

[32] A. G. Harrison and H. G. Thode, *Trans. Faraday Soc.*, 1957, **53**, 1648; V. Agarwala, E. E. Rees, and H. G. Thode, *Canad. J. Chem.*, 1965, **43**, 2802.

[33] J. W. Hill and A. Fry, *J. Amer. Chem. Soc.*, 1962, **84**, 2763.

[34] J. W. Taylor and E. P. Grimsrud, *Analyt. Chem.*, 1969, **41**, 805.

[35] L. Melander, 'Isotope Effects on Reaction Rates', The Ronald Press, New York, 1960.

Then equation (18) can be written

$$\frac{k}{k^*} = \frac{\log\left[(1 - f)S\right]}{\log\left[r(1 - f)S\right]} \tag{20}$$

where $R_f = A^*/A$, $R_0 = A_0^*/A_0$, $r = R_f/R_0$ and $S = (1 + R_0)/(1 + R_f)$. The factor S is essentially a correction term and can be estimated with sufficient accuracy by using the literature value for A_0^*/A_0, or it can be measured, from which $R_f = rR_0$ can be substituted into equation (20). The latter is the relevant equation for calculating k/k^* where isotopic analysis is made on the substrate.

In the other case of interest, namely where isotopic analysis is made on the product of reaction, the set of differential equations for product formation can be written

$$\frac{dP}{dt} = k(P_\infty - P), \qquad \frac{dP^*}{dt} = k^*(P_\infty^* - P^*) \tag{17c,d}$$

or by division

$$\frac{dP}{dP^*} = \frac{k}{k^*} \frac{(P_\infty - P)}{(P_\infty^* - P)^*}$$

which on solution gives

$$\frac{k}{k^*} = \frac{\log\left(1 - P/P_\infty\right)}{\log\left(1 - P^*/P_\infty^*\right)} \tag{21}$$

The fraction which has undergone reaction can be defined as

$$f = \frac{P}{P_\infty} \frac{(1 + P^*/P)}{(1 + P_\infty^*/P_\infty)} = \frac{P(1 + R_f')}{P_\infty r'(1 + R_\infty')} \tag{22}$$

where $R_f' = P^*/P$, $R_\infty' = P_\infty^*/P_\infty$ and $r' = R_f'/R_\infty'$. Then since

$$P/P_\infty = f\frac{(1 + R_\infty')}{(1 + R_f')} \quad \text{and} \quad P^*/P_\infty^* = r'f\frac{(1 + R_\infty')}{(1 + R_f')}$$

$$\frac{k}{k^*} = \frac{\log\left[1 - f(1 + R_\infty')/(1 + R_f')\right]}{\log\left[1 - r'(1 + R_\infty')/(1 + R_f')\right]} \tag{23}$$

or

$$\frac{k}{k^*} = \frac{\log\left[1 - fS'\right]}{\log\left[1 - r'fS'\right]} \tag{24}$$

where $S' = (1 + R_\infty')/(1 + R_f')$ and the value of $R_\infty' = R_0$ is taken from the literature. S' can be calculated by the substitution $R_f' = r'R_\infty'$ or measured.

For reactions involving elements such as carbon, oxygen, nitrogen, and sulphur, for which one isotopic species is present to the extent of only a few percent, equations (20) and (24) reduce to

$$k/k^* = \log(1 - f)/\log[r(1 - f)] \tag{25a}$$

$$k/k^* = \log(1 - r'f)/\log(1 - f) \tag{25b}$$

Even for molecules containing heavy isotopes for which R_0 or $R'_\infty \simeq 1$, equations (17a) and (17b) can be employed without appreciable loss of accuracy. For if $R_f = (1 - \delta)R_0$ then

$$(1 + R_0)/(1 + R_f) = (1 + R_0)/[1 + R_0(1 - \delta)]$$

$$\simeq (1 - \delta R_0)/(1 + R_0)$$

$$\simeq 1 + \delta R_0/(1 + R_0)$$

$$\simeq 1$$

If f is small, then equation (25b) reduces further to

$$\frac{k}{k^*} = r' \tag{26}$$

as can be seen by expanding the logarithms. This implies making measurements on the initial product. Bigeleisen and Allen[36] and Jones[37] have made estimates of the accuracy with which k/k^* can be derived from equations (25a) and (25b).

Carbon Isotope Effects.—These are of very great importance in the investigation of organic reaction mechanisms, since organic reactions of necessity involve change of bonding at a carbon atom. By far the most widely investigated reaction is the decarboxylation of diacids such as malonic acid. Bigeleisen and Friedman[38] first reported results on this system using the molten phase; $k_4/k_3 = 1.0204$ and $k_1/2k_3 = 1.037$ at 138 °C, the rate constants being defined on p. 79. Results obtained by other authors are shown in Table 7. In addition some measurements

Table 7 ^{13}C *Isotope effects in the decarboxylation of malonic acid*

k_4/k_3	1.020	1.021	1.028	1.030, 1.027	1.027
$k_1/2k_3$	1.037	1.036	—	—	1.040
k_1/k_2	—	—	—	—	1.017
Temp./°C	138	138–199	140	140	150
Ref.	38	39, 40	41	42, 43	44

[36] J. Bigeleisen and T. L. Allen, *J. Chem. Phys.*, 1951, **19**, 760.
[37] W. M. Jones, *J. Chem. Phys.*, 1951, **19**, 78.
[38] J. Bigeleisen and L. Friedman, *J. Chem. Phys.*, 1949, **17**, 998.
[39] J. G. Lindsay, A. N. Bourns, and H. G. Thode, *Canad. J. Chem.*, 1951, **29**, 192.
[40] J. G. Lindsay, A. N. Bourns, and H. G. Thode, *Canad. J. Chem.*, 1952, **30**, 163.
[41] P. E. Yankwich and E. C. Stivers, *J. Chem. Phys.*, 1953, **21**, 61.
[42] P. E. Yankwich and A. L. Promislow, *J. Amer. Chem. Soc.*, 1954, **76**, 4648.

were done with ^{14}C. The early measurements by different workers were inconsistent, but Yankwich *et al.*[43] finally reported for k_4/k_3, 1.0285 (^{13}C) and 1.0545 (^{14}C), the difference being in agreement with that expected from theory. One value for k_1/k_2 (^{14}C), 1.076 (154 °C)[45] is not in agreement with the ^{13}C value for this ratio reported in Table 7. Bigeleisen and Wolfsberg[9] have made calculations assuming that a carbon-stretching frequency is lost in the transition state using equations (7) and (16) together with the fragment mass approximation for the calculation of v^+/v^{*+}. The results were $k_1/2k_3 = 1.029$ and $k_4/k_3 = 1.025$, in reasonable agreement with the experimental results. However, a concerted six-centre mechanism (Scheme 3) yielding the enol form of acetic acid is also a

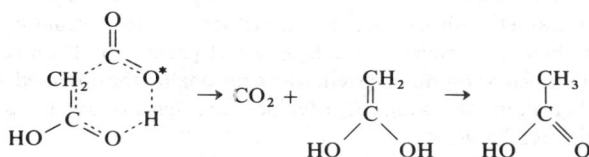

Scheme 3

possibility. The interpretation is rendered less certain by a lack of knowledge of the molecularity of the reaction and the extent of cooperative interaction between the molecules in the liquid phase.

An elegant series of papers arose out of a study of ^{13}C isotope effects in the solvolysis of 1-bromo-1-phenylethane. Stothers and Bourns[46] investigated nucleophilic substitution at the saturated carbon atom on the assumption that kinetic isotope effects might provide a useful criterion of mechanism since, on a simplistic view, the main change in the S_N1 mechanism is bond cleavage, whereas in the S_N2 mechanism both bond cleavage and bond formation are of importance, suggesting that the kinetic isotope effect might be greater for the S_N1 reaction than for the S_N2. However, while their work was in progress, Bender *et al.* had shown[47] that in the S_N2 reaction of methyl iodide with several tertiary amines gave $^{12}k/^{13}k = 1.09$—1.14, whereas hydrolysis of t-butyl chloride in dioxan, a classic S_N1 reaction, gave $^{12}k/^{13}k = 1.03$. Bender *et al.* in consequence concluded that ^{13}C kinetic isotope effects were of little use in differentiating between the S_N1 and S_N2 mechanisms.

Stothers and Bourns developed an ingenious method for determining $[^{13}C]/[^{12}C]$ at the carbon atom being substituted and found $^{12}k/^{13}k = 1.0065$ for methanolysis at 25 °C and 1.0064 for methanolysis at 45 °C. These rather

[43] P. E. Yankwich, A. L. Promislow, and R. F. Nystrom, *J. Amer. Chem. Soc.*, 1954, **76**, 5893.
[44] A. G. Loudon, A. Maccoll, and D. Smith, *J.C.S. Faraday I*, 1973, **69**, 894.
[45] G. A. Ropp and V. F. Raaen, *J. Amer. Chem. Soc.*, 1952, **74**, 4992.
[46] J. B. Stothers and A. N. Bourns, *Canad. J. Chem.*, 1960, **38**, 923.
[47] M. L. Bender and D. F. Hoeg, *Chem. and Ind.*, 1957, 463; *J. Amer. Chem. Soc.*, 1957, **79**, 5649; G. J. Buist and M. L. Bender, *J. Amer. Chem. Soc.*, 1958, **80**, 4308.

surprising results were interpreted as indicating a strengthening of the bonding of the isotopic carbon with the ring in the transition state. This view received support from subsequent work[48] by Kresge et al.[49] relating to strengthening of the $Ar-C^+$ bond in the triphenylmethyl carbonium ion. Stothers and Bourns also claimed that their results supported the fragment model of Bigeleisen for calculating heavy-atom kinetic isotope effects.

In a subsequent paper, these two authors investigated[49] the ^{13}C effect for the bimolecular reaction of 1-bromo-1-phenylethane and benzyl bromide with alkoxide ions in alcoholic solution. The former compound gave $^{12}k/^{13}k = 1.0032$ for ethoxide ion in ethanol and the latter 1.0531 for reaction with methoxide ion in methanol. The former value was later corrected to 1.0321. The authors concluded that the kinetic isotope effect provides a very sensitive means of obtaining information about the transition state. A recent paper[50] by Bron reports calculations on the benzyl bromide system which rationalize the observed differences in ^{13}C and D isotope effects and their temperature dependences in S_N1 and S_N2 and borderline mechanisms.

Bron and Stothers[51] then took up the matter of the temperature dependence of $^{12}k/^{13}k$ for the alcoholysis of 1-bromo-1-phenylethane. They found mean values of 1.0018, 1.0044, and 1.0064 at 0, 25, and 45 °C, respectively. Of considerable interest is the increasing isotope effect with increasing temperature. They calculated the $^{12}k/^{13}k$ ratio on the basis of equation (7), calculating the Δv_i on the diatomic approximation $(v_i/v_i^*) = (\mu_i^*/\mu_i)^{\frac{1}{2}}$, a method which may overestimate the effect. The results suggested $^{12}k/^{13}k = 0.9924$ (0 °C) and $= 0.9931$ (45 °C). Thus an inverse effect was predicted as against the direct effect observed. This may also be due to the solvolysis proceeding through an intimate ion-pair transition state.

For the bimolecular displacements of benzyl bromide and 1-bromo-1-phenylethane, relatively large temperature effects were found, namely for the former compound 1.0578 (-23 °C) and 1.0531 (0 °C) and for the latter 1.0359 (0 °C) and 1.0321 (25 °C).[52] These compounds show the expected temperature dependence, i.e. $^{12}k/^{13}k$ decreasing as the temperature is increased. This contrasts with the behaviour of 1-bromo-1-phenylethane in alcoholysis, where the opposite effect is reported. A further paper[53] describes the effect of p-substitution on the alcoholysis of 1-bromo-1-(p-substituted phenyl) ethanes. The results are shown in Table 8.

The general conclusion to be drawn from this work is that low ^{13}C isotope effects are to be predicted for S_N1 reactions as compared with S_N2. Also, electron-donating groups at the reaction centre tend to stabilize the activated complex (more bond formation) leading to lower isotope effects. In fact, as Table 8 shows, the effect for Me is inverse.

[48] A. J. Kresge, N. N. Lichtin, K. N. Rao, and R. E. Weston, J. Amer. Chem. Soc., 1965, 87, 437.
[49] J. B. Stothers and A. N. Bourns, Canad. J. Chem., 1962, 40, 2007.
[50] J. Bron, Canad. J. Chem., 1974, 52, 903.
[51] J. Bron and J. B. Stothers, Canad. J. Chem., 1968, 46, 1435.
[52] J. Bron and J. B. Stothers, Canad. J. Chem., 1968, 46, 1825.
[53] J. Bron and J. B. Stothers, Canad. J. Chem., 1969, 47, 2506.

Table 8 $^{12}k/^{13}k$ *for the alcoholysis of 1-bromo-1-(p-substituted phenyl)-ethanes*

	MeOH		EtOH	
p-*Substituent*	0 °C	25 °C	0 °C	25 °C
Me	0.9995	—	1.0005	—
H	1.0005	1.0065	1.0018	1.0044
Br	1.0127	1.0113	—	—

Another means of studying nucleophilic displacement reactions is by labelling the nucleophile, as was first done by Nair,[54] who reported $^{12}k/^{13}k = 1.005$. Lynn and Yankwich[55] studied the effect in the attack of cyanide ions upon methyl iodide, reporting $^{12}k/^{13}k = 1.0149$ (11.4 °C), thus confirming the S_N2 character of this reaction.

Little work has been done on ^{13}C isotope effects in gas-phase reactions. The structural isomerization of cyclopropane to propylene, probably the most widely investigated unimolecular reaction, was studied by Sims and Yankwich,[56] over a pressure range 1—760 mmHg and a temperature range 450—519 °C. At 760 mmHg $^{12}k/^{13}k = 0.995$ exp (19.1/RT) values being 1.012 (513.8 °C) and 1.014 (450.3 °C). The pressure dependence of the isotope effect was explained, at least qualitatively, on the basis of the transition state proposed by Simons and Rabinovitch,[57] involving ring relaxation with hindered rotation of the enol methylene groups. Wettaw and Sims[58] have reported an investigation of the isomerization of methyl isocyanide to acetonitrile, from 10—760 mmHg, at 226 °C. Values of $^{12}k/^{13}k$ of 1.018 at 760 mmHg, falling to 1.011 at 10 mmHg, were reported. The average value of the effects for $^{13}CH_3NC$ and $CH_3N^{13}C$ was related to the individual isotope effects, which were calculated on the basis of Rice–Ramsperger–Kassel–Marcus theory. Good agreement was obtained with a cyclic transition state suggested by Rabinovitch.[59]

An interesting gas-phase application of ^{14}C was carried out by Bigley and Thurman[60] in their investigation of the pyrolysis of 2,2-dimethyl-4-phenylbut-3-enoic acid (Scheme 4). At 278 °C, $k/k^* = 1.035$ (^{14}C), and at 286 °C, k^H/k^D was

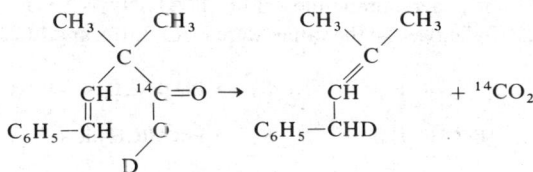

Scheme 4

[54] P. M. Nair, *Diss. Abs.*, 1957, **17**, 1469.
[55] K. R. Lynn and P. E. Yankwich, *Chem. and Ind.*, 1960, 117; *J. Amer. Chem. Soc.*, 1961, **83**, 53.
[56] L. B. Sims and P. E. Yankwich, *J. Phys. Chem.*, 1967, **71**, 3459.
[57] J. W. Simons and B. S. Rabinovitch, *J. Phys. Chem.*, 1964, **68**, 1322.
[58] J. F. Wettaw and L. B. Sims, *J. Phys. Chem.*, 1968, **72**, 3440.
[59] F. W. Schneider and B. S. Rabinovitch, *J. Amer. Chem. Soc.*, 1962, **84**, 4215; F. J. Fletcher, B. S. Rabinovitch, K. W. Watkins, and D. J. Locker, *J. Phys. Chem.*, 1966, **70**, 2823.
[60] D. B. Bigley and J. C. Thurman, *J. Chem. Soc. (B)*, 1967, 941; *Tetrahedron Letters*, 1967, 2377.

2.87, the deuterium being substituted in the OH group. The results were taken to support the six-centre mechanism.

Nitrogen Isotope Effects.—Decomposition of substituted phenyldiazonium salts has been the subject of a number of studies. Early work by Lewis and Insole[61] compared the rates of decomposition of $CH_3C_6H_4{}^{14}N_2{}^+BF_4{}^-$ and $CH_3C_6H_4{}^{15}N^{14}N^+BF_4{}^-$ (>99%) and found $^{14}k/^{15}k = 1.019$. Brown and Drury[62] measured the rate ratios for a number of substituted compounds using two different nucleophiles, at two low temperatures. Values ranged from 1.043 to 1.047. Later work by Lewis et al.[63] suggested that the decomposition in aqueous solution was essentially a bimolecular attack of water on the diazonium cation. Loudon et al.[64] have presented evidence to the effect that the reaction is essentially unimolecular. This is consistent with the observed values of $^{14}k/^{15}k$.

Ayrey, Bourns, and Vyas[65] investigated the nitrogen isotope effect in the reaction of the 2-phenyltrimethylammonium ion with ethoxide ion in ethanol. A value of 1.017 was found for $^{14}k/^{15}k$ at 60 °C. For the reaction of the same substrate with hydroxide ions in water, Bourns and Smith[66] reported $^{14}k/^{15}k = 1.009$ at 97 °C. These values are relatively large and indicate a considerable amount of C—N weakening in the transition state. Values of 1.014, 1.015, and 1.011 for p-CH_3O, p-H, and p-Cl substituted substrates were reported.[66]

Bourns et al. have continued work in this field, the latest paper[67] reporting values of 1.0137, 1.0133, 1.0114, and 1.0088 for p-OCH_3, p-H, p-Cl, and p-CF_3 in the reactions of $XC_6H_4(CH_2)_2\overset{+}{N}(CH_3)_3$ with sodium ethoxide in ethanol.

A striking result of the effect of the stereochemistry of the substrate upon the kinetic isotope effect has been reported.[68] For the decomposition of trimethyl-*cis*-2-phenylcyclohexylammonium iodide the isotope effect was 1.012, whereas that for the *trans* compound was only 1.002. These values are correlated with the fact that the former compound can undergo *trans*-elimination, whereas the latter can react only by *cis*-elimination.

Seltzer and his co-workers[69] have investigated $^{14}k/^{15}k$ ratios for the decomposition of some azo-compounds $PhCH(CH_3)N{=}NR$. For R = methyl, isopropyl, and α-phenylethyl the values were 1.013, 1.015, and 1.023. Two possible routes are

$$[PhCH(CH_3)\cdots N{\overset{-}{=}}N{-}R] \rightarrow Ph\dot{C}HCH_3 + \dot{N} = NR$$
$$[PhCH(CH_3)\cdots N{\overset{-}{=}}N\cdots R] \rightarrow PhCH(CH_3)R + N_2$$

[61] E. S. Lewis and J. M. Insole, *J. Amer. Chem. Soc.*, 1964, **86**, 34.
[62] L. L. Brown and J. S. Drury, *J. Chem. Phys.*, 1965, **43**, 1688.
[63] E. S. Lewis, L. D. Hartung, and B. M. McKay, *J. Amer. Chem. Soc.*, 1969, 91, 419.
[64] A. G. Loudon, A. Maccoll, and D. Smith, *J.C.S. Faraday I*, 1973, **69**, 899.
[65] G. Ayrey, A. N. Bourns, and V. A. Vyas, *Canad. J. Chem.*, 1963, **41**, 1759.
[66] A. N. Bourns and P. J. Smith, *Proc. Chem. Soc.*, 1964, 366.
[67] P. J. Smith and A. N. Bourns, *Canad. J. Chem.*, 1974, **52**, 749.
[68] G. Ayrey, E. Buncel, and A. N. Bourns, *Proc. Chem. Soc.*, 1961, 458.
[69] S. Seltzer and F. T. Dunne, *J. Amer. Chem. Soc.*, 1965, **87**, 2628; S. Seltzer and S. G. Mylonakis, *J. Amer. Chem. Soc.*, 1967, **89**, 6584.

On the basis of the above evidence and from a study of the [^{13}C]methyl and [^{2}H$_3$]methyl compounds, it was concluded that whereas the methyl compound decomposed by the two-step process and the α-phenylethyl compound by the one-step process, the isopropyl compound decomposed by a mixture of the two.

A nitrogen isotope effect study of an aromatic nucleophilic substitution reaction has been reported.[70] Hydrolysis of *p*-nitroaniline in aqueous sodium hydroxide was shown to be second order and Arrhenius parameters were reported. No nitrogen isotope effect was found: (^{14}N^{14}N/^{14}N^{15}N) = 136.09 ± 0.015 at 10% reaction and 136.14 ± 0.11 at 100% reaction. It was concluded that C—N stretching was unimportant in the transition state.

Oxygen Isotope Effects.—Hart and Bourns [71] investigated the $^{16}k/^{18}k$ ratio in the displacement of phenoxide ion by piperidine in phenyl 2,4-dinitrophenyl ether

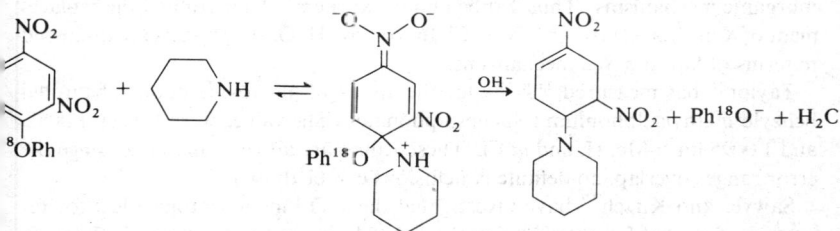

Scheme 5

(Scheme 5). When the concentration of base is high the first step is rate-controlling and as the bonding at O is not appreciably altered, the isotope effect would be expected to be small. At low base concentrations, however, the second step should be rate-controlling and a normal isotope effect for bond rupture should be observed. With increasing concentrations of base from 0.005 to 0.149, the observed values of $^{16}k/^{18}k$ were 1.0109, 1.0070, and 1.0024, confirming the proposed mechanism.

Oxygen kinetic isotope effects have also been used in the study of the methanolysis of aryl benzoates. Mitton and Schowen[72] found $^{16}k/^{18}k = 1.018$ and 1.024 for X = Br and H in Scheme 6. The effect is very large, despite the fact that

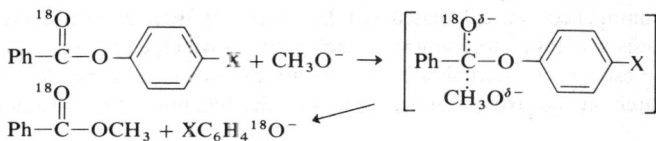

Scheme 6

[70] G. Ayrey and W. A. Wylie, *J. Chem. Soc.* (*B*), 1970, 738.
[71] C. R. Hart and A. N. Bourns, *Tetrahedron Letters*, 1966, 2995.
[72] C. G. Mitton and R. L. Schowen, *Tetrahedron Letters*, 1968, 5803.

it is essentially a secondary effect. However, the transition state suggested does imply a considerable change in bonding to O, and so it could be considered as a primary effect.

Mention has been made already of the use of $^{16}k/^{18}k$ measurements in a study of gas-phase decarboxylation. Ropp and Guillory[73] have investigated the ^{18}O isotope effect on the rate of photochemical oxidation of formic acid by chlorine in the gas phase. The ratio obtained, 1.002, caused the authors to conclude that the abstraction of the hydroxyl hydrogen by chlorine atoms makes no appreciable contribution to the reaction mechanism.

A large ^{18}O isotope effect has been reported by Goldstein[74] in the thermal dissociation of acetyl peroxide. The value was 1.023 at 45 °C. He suggested that C—C cleavage must accompany O—O cleavage to a large extent, on the basis of model calculations.

Not surprisingly, ^{18}O isotope effects have been investigated in the field of inorganic mechanisms. Thus Taube and co-workers[75] have studied the replacement of X in $[Co(NH_3)_5X]^{2+}$ (X = Cl, Br, or I) by H_2O. The results are discussed in terms of S_N1 and S_N2 mechanisms.

Taylor[76] has measured $^{16}k/^{18}k$ for the attack of phenoxide upon substituted benzyldimethylsulphonium toluenesulphonates. The values were 1.0074, 1.0082, and 1.0095 for p-Me, H, and m-Cl. These values are all very small and, since the error ranges overlap, no definite conclusions can be drawn.

Sawyer and Kirsch[77] have investigated the ^{18}O kinetic isotope effect for reactions of methyl formate($[^{18}O]$methoxyl). Their results are shown in Table 9.

Table 9 ^{18}O *Kinetic isotope effects for methoxy-labelled methyl formate*

Reaction	$^{16}k/^{18}k$
Acid-catalysed hydrolysis	1.0009 ± 0.0004
Alkaline hydrolysis	1.0091 ± 0.0004
General base-catalysed hydrolysis	1.0115 ± 0.0002
Hydrazinolysis	1.0048 ± 0.0006
Hydrazinolysis	1.0621 ± 0.0008

In the case of hydrazinolysis, the two values refer to conditions under which the formation of the tetrahedral intermediate and its breakdown are rate-controlling. I.r. studies yielded methyl rocking and formyl-C-methoxyl-O stretching frequencies of 1162.3, 1144.8 and 1208.3, 1187.4 cm^{-1}, respectively. From these the maximum effect was calculated as 1.052. The very small effect observed for acid catalysis was taken to confirm a mechanism in which equilibrium protonation of the ester is counterbalanced by the normal kinetic isotope effect for attack of water on the oxocarbonium ion. For the remaining four reactions, it was

[73] G. A. Ropp and W. A. Guillory, *J. Phys. Chem.*, 1961, **65**, 1496.
[74] M. J. Goldstein, *Tetrahedron Letters*, 1964, 1601.
[75] F. A. Posey and H. Taube, *J. Amer. Chem. Soc.*, 1957, **79**, 252; M. Green and H. Taube, *Inorg. Chem.*, 1963, **2**, 948.
[76] L. H. Taylor, PhD Thesis, Massachusetts Institute of Technology, 1963.
[77] C. B. Sawyer and J. F. Kirsch, *J. Amer. Chem. Soc.*, 1973, **95**, 7375.

concluded that the order of the acyl carbon–methoxyl oxygen bond is only slightly reduced in the transition state.

A very careful study has been made by Margolin and Samuel[78] of the ^{13}C and ^{18}O kinetic isotope effects in the decarbonylation of benzylformic acid in concentrated sulphuric acid. The original Hammett mechanism[79] is shown in Scheme 7. Ropp[80] measured the ^{13}C isotope effect and found $^{12}k/^{13}k = 1.039$,

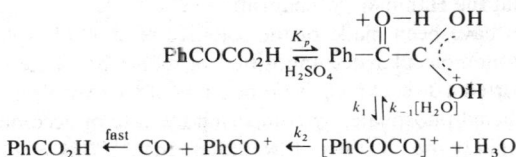

$$PhCOCO_2H \underset{H_2SO_4}{\overset{K_p}{\rightleftharpoons}} Ph-C-C\begin{matrix} +O-H & OH \\ \vdots \\ OH \end{matrix}$$

$$k_1 \big\uparrow\big\downarrow k_{-1}[H_2O]$$

$$PhCO_2H \overset{fast}{\leftarrow} CO + PhCO^+ \overset{k_2}{\leftarrow} [PhCOCO]^+ + H_3O$$

Scheme 7

a value inconsistent with the Hammett scheme. A mechanism which makes C—C rupture rate-controlling is shown in Scheme 8.

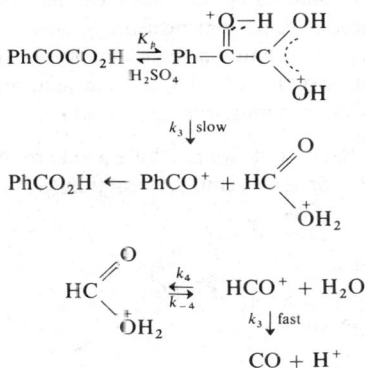

$$PhCOCO_2H \underset{H_2SO_4}{\overset{K_p}{\rightleftharpoons}} Ph-C-C\begin{matrix} +O-H & OH \\ \vdots \\ OH \end{matrix}$$

$$k_3 \big\downarrow slow$$

$$PhCO_2H \leftarrow PhCO^+ + HC\begin{matrix} O \\ \diagup\diagup \\ \diagdown \\ OH_2 \end{matrix}$$

$$HC\begin{matrix} O \\ \diagup\diagup \\ \diagdown \\ OH_2 \end{matrix} \underset{k_{-4}}{\overset{k_4}{\rightleftarrows}} HCO^+ + H_2O$$

$$k_3 \big\downarrow fast$$

$$CO + H^+$$

Scheme 8

Margolin and Samuel calculated the oxygen isotope effects on the basis of the $^{18}O/^{16}O$ ratio in the product carbon monoxide. There are three ^{18}O effects, intermolecular ($2k_3/k_1 = 0.981$), primary intramolecular ($k_3/k_2 = 0.986$), and secondary intramolecular ($2k_2/k_1 = 0.995$). For carbon, the effect may be on the carboxyl carbon ($^{13}k/^{12}k = 0.947$) or on the carbonyl carbon ($^{13}k/^{12}k = 1.000$). By a kinetic analysis of the two schemes including the reversible exchange of oxygen between the carbonyl group and solvent water $(1, -1)$ (Scheme 7) or $(4, -4)$ (Scheme 8), the authors deduced the following rate equations:

Scheme 7: $\qquad k_{exp} = k_1 k_2/(k_{-1}[H_2O] + k_2) \times K_p h_0^2$

Scheme 8: $\qquad k_{exp} = k_3 K_p h_0^2$

[78] Z. Margolin and D. Samuel, *Chem. Comm.*, 1970, 802.
[79] L. P. Hammett, 'Physical Organic Chemistry', McGraw-Hill, New York, 1940, p. 253.
[80] G. A. Ropp, *J. Amer. Chem. Soc.*, 1960, **82**, 842.

where h_0 is the Hammett acidity function. The equations indicate that, when C—O bond rupture is rate-limiting (Scheme 7), an isotope effect of the carbonyl carbon can only be expected when there is oxygen exchange with the solvent. However, when C—C rupture is rate-limiting (Scheme 8) there should be a carbon isotope effect of the carbonyl carbon. In 99.5—100% sulphuric acid at 0 °C, no exchange was observed during decarbonylation. The observed results show conclusively that the Hammett mechanism is correct.

Many studies have been made of the role of the monophosphate ion intermediate in the unimolecular decomposition of monophosphate esters. Gorenstein[81] has reported a direct kinetic isotope study of the hydrolysis of the dianion of 2,4-dinitrophenyl phosphate, by comparing the rate of decomposition of the ^{16}O ester with that of the ^{18}O ester. It was found that $^{16}k/^{18}k = 1.0204 \pm 0.0044$ in the temperature range 39–55 °C and the pH range 4.4—8.0. Bunton *et al.*[82] and Kirby and Varvoglis[83] have shown that the hydrolysis probably proceeds through P—O cleavage. The present observation of a large ^{18}O effect is taken by the author to indicate substantial P—O bond breaking in the transition state. Preliminary results reported suggest that no isotope effect has been found for dibenzyl-2,4-dinitrophenyl phosphate, a compound which must hydrolyse through addition–elimination or by a direct $S_N2(P)$ mechanism, a result in marked contrast to that observed for the unimolecular process.

Sulphur Isotope Effects.—Bader and Bourns[84] were able to distinguish between two suggested mechanisms for the Tschugaeff reaction (Scheme 9). For ^aS and

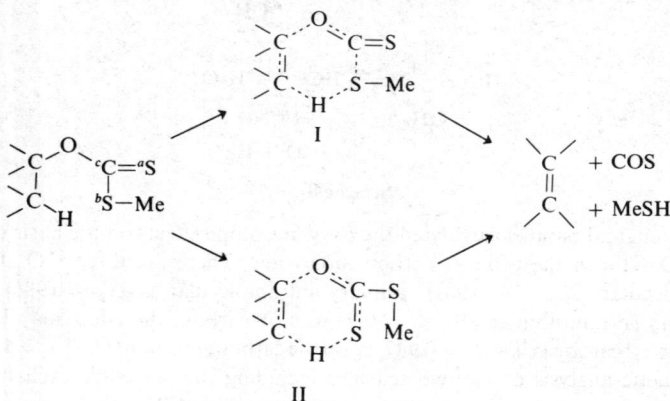

Scheme 9

^bS the effects were 1.0086 and 1.0021, respectively, while for ^cC, $^{12}k/^{13}k = 1.0004$. Thus it would appear that there is a large bonding change at ^aS, a small one at

[81] D. G. Gorenstein, *J. Amer. Chem. Soc.*, 1972, **94**, 2523.
[82] C. A. Bunton, D. R. Llewellyn, K. G. Oldham, and C. A. Vernon, *J. Chem. Soc.*, 1958, 3574.
[83] A. J. Kirby and A. G. Varvoglis, *J. Amer. Chem. Soc.*, 1967, **89**, 415.
[84] R. F. W. Bader and A. N. Bourns, *Canada J. Chem.*, 1961, **39**, 348.

^{b}S, and almost none at ^{c}C. On this basis I was ruled out, and II confirmed as the structure of the transition state.

An interesting effect of solvent upon a sulphur isotope effect has been reported by Cockerill and Saunders[85] in the reaction of hydroxide ion with 2-phenylethyldimethylsulphonium bromide. In pure water $^{32}k/^{34}k = 1.0074$ decreasing to 1.0011 in an aqueous solution containing 7 mol l^{-1} of Me$_2$SO. These authors also studied the β-deuterium effect and found this to pass through a maximum as the Me$_2$SO content increased. They interpreted these results as indicating a reagent-like transition state, with less weakening of both the C—S and α-C—H bonds as the Me$_2$SO increased.

The bromodesulphonation of sodium p-methoxybenzenesulphonate (Scheme 10) (and of potassium 1-methylnaphthalene-4-sulphonate) has been investigated by Baliga and Bourns.[86] For low concentrations of bromide ion, the first step

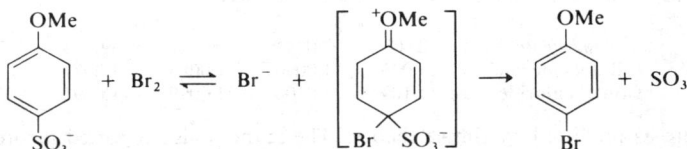

Scheme 10

is rate-controlling, the species in the bracket is an intermediate, and $^{32}k/^{34}k = 1.0032$. In the presence of added bromide ion, the first step becomes reversible, the rate falls and the second step becomes rate-controlling. The isotope effect increases to 1.0127 at 0.03 M bromide and 1.0173 at 0.5 M bromide.

Chlorine Isotope Effects.—As chlorine is a well-established leaving group in nucleophilic substitution, $^{35}k/^{37}k$ values have proved useful in determining mechanism. Following on the original investigation of Bartholemew, Brown, and Lounsbury,[87] Hill and Fry[38] investigated the effect in reactions of benzyl and substituted benzyl chlorides with various nucleophiles. These authors observed that, for those reactions following first-order kinetics, $^{35}k/^{37}k \simeq 1.008$, whilst for those following second-order kinetics the value was 1.006. They suggested that the value of $^{35}k/^{37}k$ might be used to differentiate between S_N1 and S_N2 reactions. Grimsrud and Taylor[89] investigated nucleophilic displacements at a saturated carbon atom by varying both the nucleophilic power of the attacking group and the electron-donating power of the p-substituent in a series of benzyl chlorides with a view to testing the predictions of Thornton.[90] In the

[85] A. F. Cockerill and W. H. Saunders, *J. Amer. Chem. Soc.*, 1967, **89**, 4985.
[86] B. T. Baliga and A. N. Bourns, *Canad. J. Chem.*, 1966, **44**, 363.
[87] R. M. Bartholemew, F. Brown, and M. Lounsbury, *Canad. J. Chem.*, 1954, **32**, 979; *Nature*, 1954, **174**, 133.
[88] J. W. Hill and A. Fry, *J. Amer. Chem. Soc.*, 1962, **84**, 2763.
[89] E. P. Grimsrud and J. W. Taylor, *J. Amer. Chem. Soc.*, 1970, **92**, 739.
[90] E. R. Thornton, 'Solvolysis Mechanisms', Ronald Press, New York, 1964.

case of attack by $C_6H_5S^-$ and CH_3O^-, $^{35}k/^{37}k$ decreased in the order $CH_3O > H > NO_2$, these being the p-substituents. Again, for the variation of the nucleophile, in the case of p-nitrobenzyl chloride and t-butyl chloride the isotope effect was smaller for n-$C_4H_9S^-$ than for $C_6H_5S^-$. Finally, for the reactions of the same substrates the isotope effect was greater for n-$C_4H_9S^-$ than for CH_3O^-, and greater for $C_6H_5S^-$ than for $C_6H_5O^-$. The fact that the oxide nucleophiles show consistently smaller effects than the sulphur analogues suggests that the oxides are the stronger nucleophiles, a view not consistent with the rate data $k_{RS^-} \gg k_{RO^-}$. This contradiction was discussed in terms of the basicity, polarizability and solvation effects on the nucleophile.

A very careful study of temperature effects on $^{35}k/^{37}k$ in the solvolysis of n- and t-butyl chlorides was made by Taylor and co-workers.[91] The results are shown in Table 10. In each case, log $^{35}k/^{37}k$ is inversely proportional to tem-

Table 10 $^{35}k/^{37}k$ *for solvolysis of the butyl chlorides*

Substrate	0 °C	20 °C	40 °C	60 °C
n-Butyl chloride	1.0096	1.0090	1.0084	1.0079
t-Butyl chloride	1.0109	1.0106	1.0099	1.0095

perature as predicted by simple theory. The authors also reported theoretical calculations based upon equation (7). In fact it was assumed that only the ground-state C—Cl stretch was lost in the transition state, so that

$$\text{TDF} = 1 + G(u_i)\Delta u_i$$

where $u_i = hv_i/kT$, v_i being the C—Cl stretching frequency. They concluded that 'the temperature dependence of the kinetic isotope effect is best evaluated from consideration of the ground-state configuration plus information on the number of isotopically important ground-state vibrational frequencies for a given reactant'.

In a later series of papers, Williams and Taylor[11,92] have initiated a thorough study of models for use in calculating chlorine kinetic isotope effects. They investigated[11] the isotopic shifts of the C—Cl frequencies for a series of alkyl chlorides and compared them (in the case of t-butyl chloride) with values calculated by the Wilson *FG* matrix method (see p. 82). These authors also applied[92] equation (4) to the calculation of the kinetic isotope effect for t-butyl chloride solvolysis. The ground-state geometry and the ground-state force field (leading to vibration frequencies) are known; the corresponding data for the transition state were assumed for a series of models and $^{35}k/^{37}k$ was calculated. The most probable transition state for methanolysis was C—Cl = 1.89 Å, C—CH_3 = 1.50 Å, and $z(CH_3) = -0.16$ Å relative to the central carbon atom. These values, together with an assumed force field, gave values of $^{35}k/^{37}k$ of 1.010 87 and 1.009 51 at 10 and 60 °C, respectively. The experimental values were 1.0109_7 and 1.0095_3.

[91] C. R. Turnquist, J. W. Taylor, E. P. Grimsrud, and R. C. Williams, *J. Amer. Chem. Soc.*, 1973, **95**, 4133.
[92] R. C. Williams and J. W. Taylor, *J. Amer. Chem. Soc.*, 1974, **96**, 3721.

Graczyk and Taylor[93] have used kinetic isotope effects to investigate the participation of ion-pairs in the reactions of p-methoxybenzyl chloride in aqueous acetone. This was done by competition between water and sodium azide. It was found that the observed effect increases by $>30\%$ as the azide concentration is varied from 0—0.25 mol l^{-1}. This observation is shown to be consistent with the ion-pair theory but not with simultaneous S_N1 and S_N2 processes. Raaen *et al.* have discussed isotope criteria in relation to the problem of ion-pair participation.[94]

Sims *et al.*[13] have made model calculations for S_N2 reactions, assuming variable degrees of bond breaking and bond forming in the transition state. They found that, while the $^{12}C/^{14}C$ effect goes through a maximum at *ca.* 50% breaking, the $^{35}Cl/^{37}Cl$ effect increases monotonically. So far no experiments have been reported which confirm this prediction.

Up to the present no chlorine kinetic isotope effects have been reported in gas-phase reactions. However, current experiments by Maccoll and Machacek[95] are designed to determine the variation of $^{35}k/^{37}k$ over the series ethyl, isopropyl, and t-butyl chlorides. Results to date show that the effect is very much larger for t-butyl chloride than for ethyl chloride.

[93] D. G. Graczyk and J. W. Taylor, *J. Amer. Chem. Soc.*, 1974, **96**, 3255.
[94] V. F. Raaen, T. Juhlke, F. J. Brown, and C. J. Collins, *J. Amer. Chem. Soc.*, 1974, **96**, 5928.
[95] A. Maccoll and L. M. N. Machacek, unpublished work.

6 The Thermochemistry of Organometallic Compounds

By W. V. STEELE
Department of Chemistry, University of Stirling
Stirling FK9 4LA, Scotland

1 Introduction

The thermochemistry of organometallic compounds is still a field which is sparse in accurate data; therefore reliable information on the strength of carbon–metal bonds is still meagre. This sorry state of affairs has been caused by a number of difficulties not usually encountered in the thermochemical investigation of organic compounds. The standard enthalpy of formation of an organic compound containing C, H, O, and N is usually obtained by conventional oxygen bomb calorimetry in a static-bomb calorimeter.

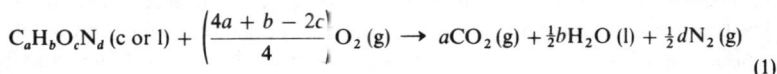

$$C_aH_bO_cN_d \text{ (c or l)} + \left(\frac{4a + b - 2c}{4}\right) O_2 \text{ (g)} \rightarrow aCO_2 \text{ (g)} + \tfrac{1}{2}bH_2O \text{ (l)} + \tfrac{1}{2}dN_2 \text{ (g)}$$

(1)

The standard enthalpies of formation of the products of equation (1), ΔH_f° $(CO_2, g) = -393.51 \pm 0.13$ kJ mol^{-1} and ΔH_f° $(H_2O, l) = -285.830 \pm 0.042$ kJ mol^{-1}, are both accurately known, so the standard enthalpy of formation of the organic compound can be calculated by a simple Hess's cycle. However, when a metallic atom is present, e.g. $M(CH_3)_4$ in the tetra-alkyls of Group IVB organometallic compounds, the products are usually found to contain not only CO_2 (g), H_2O (l), and MO_2, the standard oxide of the metal, but also a number of other inorganic compounds in varying amounts. These are usually other oxides, free metal, carbonates, nitrates, and unburnt carbon admixed with the metal. Such mixtures would need to be analysed accurately before the final state for the combustion calorimetry could be defined. This, if possible, would be time-consuming and would need to be repeated for each combustion, since every combustion gives a different ratio of products. An example, quoted by Good and Scott,[1] is the study of the combustion of tetraethyl-lead carried out by Knowlton in a static-bomb calorimeter. Here the products were found to contain PbO, PbO_2, $PbCO_3$, and $Pb(NO_3)_2$ admixed with unburnt Pb. The interpretation of the measured energy of combustion with respect to these products led to a value for the enthalpy of formation of tetraethyl-lead which was in error by no less than 170 kJ mol^{-1}.

[1] W. D. Good, D. W. Scott and G. Waddington, *J. Phys. Chem.*, 1956, **60**, 1090.

103

2 Rotating-bomb Calorimetry

During the past two decades the introduction by Sunner[2] of rotating-bomb combustion calorimetry has helped to overcome some of the difficulties encountered in static-bomb calorimetry. The rotating-bomb combustion calorimeter combines the functions of a combustion calorimeter and a solution calorimeter in the same instrument. By introducing a suitable solvent the products of combustion can be dissolved immediately after formation, to give a well-defined final solution, of uniform composition, in the bomb. The choice of solvent is restricted somewhat since it must not react with the interior of the bomb, and it is advantageous not to produce any gaseous products on dissolving the solids produced. Similarly, if an alkali is used as solvent, it should be of sufficient strength to dissolve completely the carbon dioxide produced. The experimentalist must exercise considerable artifice, exploring such variables as sample size, container material, auxiliary substance used to promote combustion, oxygen pressure, amount and nature of liquid reagent, and crucible size, shape, and mass in order to find the conditions which ensure a thermodynamically definable state after combustion.

In conventional bomb calorimetry of compounds containing C, H, O, and N, Washburn[3] devised a series of standard corrections to convert the measured energy released in the combustion reaction to that which would be released under idealized isothermal conditions. To set up such a correction procedure for each solvent used in the rotating-bomb calorimetry of organometallic compounds is a massive task, so some sort of comparison experiment is required. The normal type of comparison experiment is set out below. In it, an oxide of the metal, whose enthalpy of formation is well defined, is dissolved in the solvent, after combustion of a standard organic compound, to produce as accurately as possible the same amount of carbon dioxide and the same evolution of energy as in the combustion of the organometallic compound. Within limits, combinations of benzoic acid with a hydrocarbon oil usually give the desired amounts of energy and carbon dioxide. If both the evolution of energy and the production of carbon dioxide cannot be made to match, the latter at least should be made to do so. After correction for the energy of combustion of the organic sample, the results of the comparison experiments yield a value of the enthalpy of solution of the oxide (plus water from combustion of the organic sample) in the initial bomb solution to form the final solution. By combining the result with that found in the main experiments it is possible to obtain an enthalpy of combustion of the organometallic compound which depends only on the enthalpies of formation of water and carbon dioxide and on the enthalpy of formation of the oxide actually used in the comparison experiments. A comparison experiment of this type makes a separate standard-state correction unnecessary, and so eliminates errors which would arise through inadequate information on the quantities (e.g. the solubility and enthalpy of solution of CO_2 in the final solution)

[2] S. Sunner, *Svensk. kem. Tidskr.*, 1950, **58**, 71.
[3] E. M. Washburn, *J. Res. Nat. Bur. Stand.*, 1933, **10**, 525.

that would have to be used in reducing the bomb contents to standard states. It also removes uncertainties about the enthalpies of formation of the combustion products under the conditions in which they are produced in the bomb.

The most troublesome problem in the combustion calorimetry of organometallic compounds by rotating-bomb calorimetry is that of incomplete combustion. Incomplete combustion can either mean that not all of the sample has burned or that the carbon is not oxidized completely. The attainment of complete combustion is an art, and may be obtained by either higher or, in some cases, *lower* oxygen pressures or a change of crucible size and shape. Some organometallic compounds detonate on combustion, *e.g.* tetramethyl-lead, while others, *e.g.* lead oxalate, do not sustain combustion. In both cases the addition of an auxiliary substance of known energy of combustion, in a suitable amount, sustains smooth, complete combustion. Where solid unburnt residue is produced, a correction for incomplete combustion has to be applied. The mass of the residue must be determined with an accuracy of ± 0.05 mg, or even better in careful work. The energy of combustion of the deposit will depend on the composition and physical state of the residue and may be greater or less than that of carbon in its standard state. If sufficient sample is present for analysis, its composition can be found and a suitable correction made, but if not, the experimentalist must select the value he considers appropriate. Usually an equimolar amount of compound and carbon is selected. Sometimes complete combustion has to be sacrificed in order to obtain the desired form of solid products that can dissolve in the solvent in the bomb.

A fuller account of the methods and problems attached to the combustion calorimetry of organometallic compounds by rotating-bomb calorimetry can be found in ref. 4. Details of the methods used in specific cases are given below in the discussion of the standard enthalpies of formation of metal alkyls.

3 Reaction Calorimetry

Other methods used for the determination of the standard enthalpies of formation of organometallic compounds lack the specificity of the combustion reaction. Enthalpies of reaction can be determined directly by calorimetry or indirectly from the temperature dependence of the equilibrium constant, or from a knowledge of the equilibrium constant and the entropy change at one temperature.

The application of non-combustion thermochemistry to organometallic compounds has been widespread. The determination of the enthalpies of formation of transition-metal complexes has become a massive field of its own and will not be covered at all in this Report. During the past decade, considerable effort has gone into the development of calorimetric techniques suited to the study of complex formation, and the output of new thermochemical data on these compounds has been considerable. The whole field, until 1970, has been extensively reviewed by Ashcroft and Mortimer in their book.[5]

---

[4] 'Experimental Thermochemistry', ed. H. A. Skinner, Wiley–Interscience, New York, 1962, Vol. II.

[5] S. J. Ashcroft and C. T. Mortimer, 'Thermochemistry of Transition Metal Complexes', Academic Press, London, 1970.

The most common types of reactions studied calorimetrically are usually classified as 'hydrogenation', 'halogenation', and 'hydrolysis'. The design and operation of a calorimeter suitable for any one of the above types of reaction depends on many variables, such as the nature of the compounds to be studied, *e.g.* whether they are gaseous or solid, and on the energy output in the reaction. The general principles involved in such constructions can be found in ref. 4. A number of these calorimeters have been developed by Skinner's group at

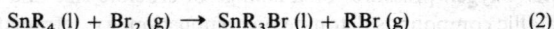

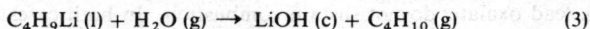

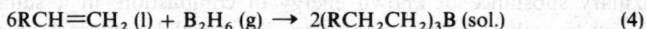

$$SnR_4 (l) + Br_2 (g) \rightarrow SnR_3Br (l) + RBr (g) \qquad (2)$$

$$C_4H_9Li (l) + H_2O (g) \rightarrow LiOH (c) + C_4H_{10} (g) \qquad (3)$$

$$6RCH{=}CH_2 (l) + B_2H_6 (g) \rightarrow 2(RCH_2CH_2)_3B (sol.) \qquad (4)$$

Manchester, and they include ones used to study reactions (2)—(4).[6—8] The main difficulty in the reaction calorimetry of organometallic compounds, particularly with the more reactive ones of Groups I, II, and III, is the violence of the reactions, which are difficult to control. *Carson et al.*[9] measured the energy evolved in the hydrolysis of dimethylzinc in aqueous ether solution [equation (5)] and in the decomposition of the same compound by dilute sulphuric

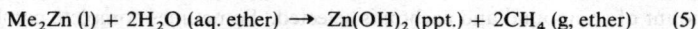

$$Me_2Zn (l) + 2H_2O (aq. ether) \rightarrow Zn(OH)_2 (ppt.) + 2CH_4 (g, ether) \qquad (5)$$

acid [equation (6)]. The large uncertainties in the enthalpies of both reactions

$$Me_2Zn (l) + H_2SO_4 (aq.) \rightarrow ZnSO_4 (aq.) + 2CH_4 (g) \qquad (6)$$

arose because of the ill-defined nature of the precipitated $Zn(OH)_2$ in the former case and in the side-reactions which arose because of the violence of the main reaction in the latter case.

Data for enthalpies of formation derived by the use of reaction calorimetry are discussed in Section 5.

4 Enthalpies of Vaporization or Sublimation

The standard enthalpy of formation of any compound in the solid or liquid state depends on both the chemical binding forces within the molecule and the forces between the molecules. For the discussion of the chemical binding forces only, as for example in the determination of bond-energy terms, it is necessary to remove the intermolecular forces from consideration. This can be achieved by conversion of ΔH_f° (l) to ΔH_f° (g), the standard enthalpy of formation in the gas phase, using equations (7) and (8). Although progress in the determination of

$$\Delta H_f^\circ (g) = \Delta H_f^\circ (l) + \Delta H^\circ \text{ (vaporization)} \qquad (7)$$

$$\Delta H_f^\circ (g) = \Delta H_f^\circ (c) + \Delta H^\circ \text{ (sublimation)} \qquad (8)$$

[6] J. B. Pedley, H. A. Skinner, and C. L. Chernick, *Trans. Faraday Soc.*, 1957, **53**, 1612.
[7] P. A. Fowell and C. T. Mortimer, *J. Chem. Soc.*, 1961, 3793.
[8] A. E. Pope and H. A. Skinner, *J. Chem. Soc.*, 1963, 3704.
[9] A. S. Carson, K. Hartley, and H. A. Skinner, *Trans. Faraday Soc.*, 1949, **45**, 1159.

the standard enthalpies of formation of organometallic compounds is moving ahead, albeit slowly, this is not the case with enthalpies of vaporization and sublimation. In this area the data are virtually non-existent. General methods of determining enthalpies of vaporization and sublimation have been otlined by Cox and Pilcher.[10] In the case of organometallic compounds the low vapour pressures and the possibility of aggregate molecules in the gas phase are problems which need to be taken into account in a reliable determination of ΔH° (vap. or sub.). The second of these possibilities is one which has not been taken into account too often in the past. The wide discrepancies in the temperature dependence of the sublimation pressures of $SiPh_4$ reported by McCauley and Smith[11] and by Calle and Kana'an[12] have been attributed to incomplete degassing in the former case.

A novel method of determining the enthalpy of sublimation of diphenylmercury is due to Carson *et al.*[13] who measured the vapour pressure over a short range of temperature near 298 K by using the Knudsen effusion method on a sample of Ph_2Hg labelled with the radioactive isotope ^{203}Hg. However, using a similar technique, Carson, Copper, and Stranks[14] studied tritium-labelled tetraphenyltin and tetraphenyl-lead, obtaining a value of 66.32 kJ mol^{-1} for the enthalpy of sublimation of the former. Keiser and Kana'an[15] have redetermined ΔH° (sub.) for tetraphenyltin, using the techniques of simultaneous measurement of the torsional recoil and of the rate of mass effusion, and they obtained a vaue of 161.1 kJ mol^{-1}. The latter study gives a more reliable value for this quantity, since the entropy of sublimation in the former case is impossibly low. The probable reason for the widely different values of the enthalpies of sublimation is that at the lower temperatures used by Carson *et al.* the species effusing from the Knudsen cell was polymeric in nature, although this has yet to be proved. Using the mass-effusion and torsional-recoil methods simultaneously, it is possible to calculate the vapour pressure by both methods, the results agreeing when the correct molecular weight is used for the species present in the gas phase. The agreement is excellent when the molecular weight of the monomer is used in Kana'an's work. Further examples of the use of this method to determine the enthalpies of sublimation of organometallic compounds are given in the next section. Reliable information on the enthalpies of vaporization and sublimation of other compounds awaits the construction and operation of other experimental systems of this type.

5 Reliable Thermochemical Data on Organometallic Compounds

The compilation of reliable values for the enthalpies of formation of organometallic compounds is not an easy job for the non-thermochemist. Often he is

[10] J. D. Cox and G. A. Pilcher, 'Thermochemistry of Organic and Organometallic Compounds', Academic Press, London, 1970.
[11] J. A. McCauley and N. O. Smith, *J. Chem. Thermodynamics*, 1973, **5**, 31.
[12] L. M. Calle and A. S. Kana'an *J. Chem. Thermodynamics*, 1974, 935.
[13] A. S. Carson, D. R. Stranks, and B. R. Wilmshurst, *Proc. Roy. Soc.*, 1958, **A244**, 72.
[14] A. S. Carson, R. Cooper, and D. R. Stranks, *Trans. Faraday Soc.*, 1962, **58**, 2125.
[15] D. Keiser and A. S. Kana'an, *J. Phys. Chem.*, 1969, **73**, 4264.

confronted with two or more values, all of which are different, and is unable to distinguish between them. Such an example is the standard enthalpy of formation of dibenzenechromium; here two values exist in the literature; Fischer, Cotton, and Wilkinson[16] gave ΔH_f° $C_{12}H_{12}Cr$ (c) = 89.1 $\pm$ 33 kJ mol^{-1}, and Fischer and Schreiner[17] found a value of ΔH_f° $C_{12}H_{12}Cr$ (c) = 213.0 $\pm$ 12 kJ mol^{-1}. Both these values were obtained by static-bomb calorimetry, but insufficient details are given, particularly in the latter case, for an objective decision to be made on the relative merit of the results. It is probable that neither are reliable in view of the unreliable results normally obtained by static measurements.

Cox and Pilcher,[10] in their comprehensive review of the enthalpies of formation of organic and organometallic compounds, tabulate critically the data available to 1966. The tabulation is particularly good as it gives not only the results of the thermochemical measurements, but also some indications of how the measurements were made and a realistic error limit on the values. Many of the values listed there bring up to date the earlier review by Skinner,[18] who gives even more detail. Here we review the data available up to December 1974 Group by Group. The reader is referred to ref. 10 to obtain values where these are relevant.

Group IA.—Very few reliable thermochemical data exist on the organometallic compounds of this Group. Cox and Pilcher summarize all the data available at present. It is worth noting that the data have been obtained by static-bomb calorimetry, and therefore their reliability is questionable. The enthalpies of vaporization quoted are for the formation of polymeric vapours in the gas phase, and accurate bond-dissociation energies for these compounds are still in doubt.

Group IIA.—Cox and Pilcher select only one piece of work on the hydrolysis dicyclopentadienylmagnesium for inclusion in their compilation. Since then, the only measurements on the enthalpies of formation of the organometallic compounds of this Group are due to Holm.[18a] He has measured the enthalpies of reaction of 17 alkyl halides with magnesium metal in diethyl ether, using a steady-state heat-flow calorimeter. He measured the enthalpies of the reaction

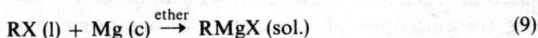

$$RX\ (l) + Mg\ (c) \xrightarrow{ether} RMgX\ (sol.) \tag{9}$$

(9), correcting for the side-reactions (10) and (11).

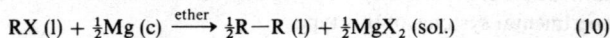

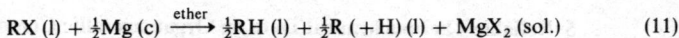

$$RX\ (l) + \tfrac{1}{2}Mg\ (c) \xrightarrow{ether} \tfrac{1}{2}R-R\ (l) + \tfrac{1}{2}MgX_2\ (sol.) \tag{10}$$

$$RX\ (l) + \tfrac{1}{2}Mg\ (c) \xrightarrow{ether} \tfrac{1}{2}RH\ (l) + \tfrac{1}{2}R\ (+H)\ (l) + MgX_2\ (sol.) \tag{11}$$

The enthalpies of reaction of the hydrogen halides with Grignard reagents in ether were also measured. The results were used to calculate the enthalpies of formation of the Grignard reagents in ether along with their relative

[16] A. K. Fischer, F. A. Cotton, and G. Wilkinson, *J. Phys. Chem.*, 1959, **63**, 154.
[17] E. O. Fischer and S. Schreiner, *Chem. Ber.*, 1958, **91**, 2213.
[18] H. A. Skinner, in 'Advances in Organometallic Chemistry', Academic Press, London, 1964, Vol. 2, p. 49.
[18] (a) T. Holm, *J. Organometallic Chem.*, 1973, **56**, 87.

bond-dissociation energies. Absolute values of the dissociation energies cannot be calculated in the absence of both the standard enthalpies of formation of the RMgX compounds and their enthalpies of sublimation. Holm concluded that, in relative terms, the R—MgX bond strength is constant for various primary alkyl groups but is 50 kJ mol^{-1} higher for Me and Ph groups. It is lower by 21, 22, and 62 kJ mol^{-1} for isopropyl, s-butyl, and t-butyl, respectively. Benzyl and allyl groups gave values which were approximately 33 kJ mol^{-1} weaker than the primary groups. These results follow that which is normally observed in the variation of bond dissociation energies with alkyl groups in organometallics (see section 6).

Group IIB.—The thermochemistry of the organometallic compounds of Zn, Cd, and Hg has been well documented by Cox and Pilcher.[10] The results come from two main sources; static-bomb calorimetry and reaction calorimetry.

In the case of the zinc alkyls the uncertainties in the final values obtained both by static-bomb calorimetry and hydrolysis reaction calorimetry were high, owing to the lack of a well-defined thermodynamic end state and the violence of the reaction. A similar situation exists for the dialkyls of cadmium.

The study of the thermochemistry of the organo-compounds of mercury has been undertaken in the main by Skinner's group at Manchester. In the early fifties a series of papers tackled the determination of ΔH_f° for R_2Hg and RHgX compounds by various methods. In this case static combustion bomb calorimetry can be used with great effect to determine the standard enthalpies of combustion, and hence formation, since the mercury for the main part does not burn, and only small traces of red HgO were formed on the bomb fittings. All these early results are tabulated by Cox and Pilcher. More recently, Carson and Wilmshurst[19] have remeasured ΔH_f° Ph_2Hg (c) by the static-bomb method to try and remove the uncertainty which existed over its value. Their result ΔH_f° Ph_2Hg (c) $= 282.8 \pm 7.9 \text{ kJ mol}^{-1}$, agrees well with that obtained by Skinner and Fairbrother.[20] The value now seems to be well established at $281.6 \pm 4.2 \text{ kJ} \text{ mol}^{-1}$. The novel method used in the determination of the enthalpy of sublimation of this compound has already been mentioned in Section 4.

Group IIIA.—Cox and Pilcher[10] tabulate data on 60 organometallic compounds of boron. The majority of the trialkylboron values listed were obtained either by static-bomb calorimetry or by measuring the enthalpy of addition of diborane to an olefin in 'monoglyme' solvent, e.g. reaction (12). The static-bomb com-

$$B_2H_6 \text{ (g)} + 6C_2H_4 \text{ (g)} \rightarrow 2B(C_2H_5)_3 \text{ (l)} \tag{12}$$

bustion measurements are complicated by the fact that the boron burns to form boric oxide, which becomes largely hydrated by the water formed and which is thermodynamically ill-defined. A residue, which has been shown to be a mixture of B_4C together with boron and carbon, is also formed. Good and Månsson[21]

[19] A. S. Carson and B. R. Wilmshurst, *J. Chem. Thermodynamics*, 1971, **3**, 251.
[20] D. M. Fairbrother and H A. Skinner, *Trans. Faraday Soc.*, 1956, **52**, 956.
[21] W. D. Good and M. Månsson, *J. Phys. Chem.*, 1966, **70**, 97.

have developed a rotating-bomb calorimetric technique to solve these problems in order to determine accurate values for the enthalpies of formation of boron compounds. The method involved the combustion of the boron-containing compound mixed with a fluorine-containing combustion promoter, vinylidene fluoride polymer. The combustion bomb initially contained an aqueous solution of hydrofluoric acid in such concentration that the boron appeared in the reaction products in a homogeneous solution. Comparison experiments were used to minimize errors from the inexact reductions to standard states (see Section 2). In these experiments benzoic acid and/or hydrocarbon oil were used to produce the same amount of CO_2 and temperature rise as in the main combustions. The bomb contained initially an aqueous solution of HF and HBF_4, which, on dilution with the water formed by the combustion of the mixture, gave a solution of nearly the same amount and concentration as in the main experiments. The enthalpies of formation of orthoboric acid, trimethylamineborane, and diammoniumdecaborane were determined. The energy of combustion of trimethylamineborane can be represented by the reaction scheme of reactions (13) and (14).

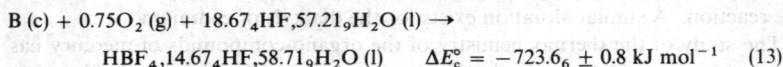

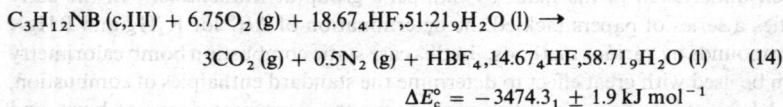

$$B (c) + 0.75O_2 (g) + 18.67_4HF,57.21_9H_2O (l) \rightarrow$$

$$HBF_4,14.67_4HF,58.71_9H_2O (l) \quad \Delta E_c^\circ = -723.6_6 \pm 0.8 \text{ kJ mol}^{-1} \quad (13)$$

$$C_3H_{12}NB (c,III) + 6.75O_2 (g) + 18.67_4HF,51.21_9H_2O (l) \rightarrow$$

$$3CO_2 (g) + 0.5N_2 (g) + HBF_4,14.67_4HF,58.71_9H_2O (l) \quad (14)$$

$$\Delta E_c^\circ = -3474.3_1 \pm 1.9 \text{ kJ mol}^{-1}$$

Combining reactions (13) and (14) with data for the enthalpies of formation of CO_2 (g) and H_2O (l) and the enthalpy of dilution of aqueous HF, the enthalpy of formation of trimethylamineborane was found to be ΔH_f° $C_3H_{12}NB$ (c, III) = $-142.4_2 \pm 4.6_0$ kJ mol^{-1}. This method has not been applied to other boron organometallic compounds, and the values for many remain suspect. Pedley[22] has produced an internally consistent analysis by computer of the thermochemical data on boron organometallic compounds and other boron compounds.

The enthalpies of formation of aluminium alkyls in the literature are internally inconsistent. They have nearly all been determined by static-bomb calorimetry, with all the defects that that entails. Values reported for Et_3Al differ by as much as 84 kJ mol^{-1}. The most reliable piece of work in the field appears to be that of Mortimer and Sellers,[23] who studied the reaction between Me_3Al and acetic acid in toluene [reaction (15)] and obtained ΔH_f° $Me_3Al(l) = -150.6 \pm 6.7$ kJ mol^{-1}.

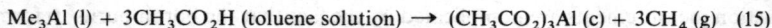

$$Me_3Al (l) + 3CH_3CO_2H \text{ (toluene solution)} \rightarrow (CH_3CO_2)_3Al (c) + 3CH_4 (g) \quad (15)$$

Recently, Smith[24] has attempted to resolve the conflict in published enthalpies of formation of aluminium alkyls by use of a one-constant equation (the

[22] J. B. Pedley, M. F. Guest, and M. Horn, *J. Chem. Thermodynamics*, 1969, **1**, 345.
[23] C. T. Mortimer and P. W. Sellers, *J. Chem. Soc.*, 1963, 1978.
[24] M. B. Smith, *J. Organometallic Chem.*, 1974, **76**, 171.

'Displacement Rule') to formulate a self-consistent set of values. However, such a set of values can only be tentative, and the design of a rotating-bomb calorimetric set-up suitable for the combustion of these compounds is required to produce valid data. Smith states that the measurement of the enthalpies of acid hydrolysis appears to be particularly appropriate for aluminium compounds.

Group IIIB.—The results from determinations of the standard enthalpies of formation of the alkyl organometallic compounds of gallium, indium, and thallium are also meagre. The only addition to the values quoted in ref. 10 is the determination of the enthalpies of formation of Et_3Ga, Bu_3Ga, and Bu_3^iGa by Rabinovich, Kol'yakova, and Zorina[24a] by static-bomb calorimetry. They give little experimental detail, and the reliability of the values is questionable. The results given are as follows:

$$Et_3Ga \quad \Delta H_f^o \text{ (l)} = -117 \pm 4 \text{ kJ mol}^{-1} \quad \Delta H_f^o \text{ (g)} = -75 \text{ kJ mol}^{-1} \quad (16)$$

$$Bu_3Ga \quad \Delta H_f^o \text{ (l)} = -280 \pm 4 \text{ kJ mol}^{-1} \quad \Delta H_f^o \text{ (g)} = -222 \text{ kJ mol}^{-1} \quad (17)$$

$$Bu_3^iGa \quad \Delta H_f^0 \text{ (l)} = -289 \pm 4 \text{ kJ mol}^{-1} \quad \Delta H_f^o \text{ (g)} = -234 \text{ kJ mol}^{-1} \quad (18)$$

Group IVB.—The difficulty of obtaining reliable thermochemical data on organosilicon compounds has often been noted. A recent paper[25] on the bond-energy terms for Group IVB organometallic compounds does not include any values for alkylsilanes, stating the reason as lack of reliable data. Quane[26] has combined data from two recent experimental studies; the electron-impact work of Potzinger and Lampe[27] and the determination of enthalpies of combustion of methyl chlorosilanes by Hajiev and Agarunov,[28] to obtain values for silicon bond-energy terms which reproduce the input data within ± 12 kJ mol^{-1}.

The breakthrough in the field of obtaining reliable data on silicon organometallic compounds is due to Good *et al.*,[29] who determined the enthalpy of formation of hexamethyldisiloxane by a similar method to that described above for boron compounds. They showed that the combustion of the compound in the presence of $\alpha\alpha\alpha$-trifluorotoluene resulted in the formation of gaseous silicon tetrafluoride in place of silica, and a complete, clean combustion resulted. Water placed initially in the bomb dissolved the silicon tetrafluoride to produce a homogeneous solution of hexafluorosilicic acid in excess hydrofluoric acid. The energy of formation of the hexafluorosilicic acid was determined in a separate experiment in which elemental silicon mixed with a sample of polyvinylidene fluoride of known energy of combustion was burned, with a solution of hydrofluoric acid placed initially in the bomb. It was thus possible to determine the

[24a] I. B. Rabinovich, G. M. Kol'yakova, and E. N. Zorina, *Doklady Akad. Nauk S.S.S.R.*, 1973, **209**, 616.
[25] A. S. Carson, P. G. Laye, J. A. Spencer, and W. V. Steele, *J. Chem. Thermodynamics*, 1970, **2**, 659.
[26] D. Quane, *J. Phys. Chem.*, 1971, **75**, 2480.
[27] P. Potzinger and F. W. Lampe, *J. Phys. Chem.*, 1970, **74**, 719.
[28] S. N. Hajiev and M. J. Agarunov, *J. Organometallic Chem.*, 1970, **22**, 305.
[29] W. D. Good, J. L. Lacina, B. L. DePrater, and J. P. McCullough, *J. Phys. Chem.*, 1964, **68**, 579.

enthalpy of formation of the hexamethyldisiloxane with respect to the particular sample of elemental silicon used in the measurements.

Iseard, Pedley, and Treverton,[30] using a similar method, have recently determined the standard enthalpies of formation of tetraethylsilicon, hexamethyldisiloxane, and hexamethyldisilane. The agreement between the two values for hexamethyldisilane is excellent, and augurs well for the use of this method in the future. The enthalpies of formation determined by the Sussex group were:

$$\Delta H_f^{\circ} \ Me_6Si_2O \ (l) = -812.5 \pm 19.7 \ kJ \ mol^{-1} \tag{19}$$

$$\Delta H_f^{\circ} \ Me_6Si_2 \ (l) = -403.3 \pm 7.1 \ kJ \ mol^{-1} \tag{20}$$

$$\Delta H_f^{\circ} \ Et_4Si \ (l) = -277.8 \pm 18.8 \ kJ \ mol^{-1} \tag{21}$$

The accurate determination of the enthalpies of formation of a number of organogermanium compounds has appeared in the literature since the publication of ref. 10. The work by Bills and Cotton[31] on the determination of the enthalpy of formation of tetraethylgermanium by rotating-bomb calorimetry has led the way in this field. They used 10% hydrofluoric acid as solvent to dissolve the products of combustion. Since then, Carson *et al.*[32] at Leeds have determined the standard enthalpies of formation of tetraphenylgermanium, tetrabenzylgermanium, and hexaphenyldigermanium by rotating-bomb calorimetry, using 1M-potassium hydroxide as solvent. Small unburnt residues, which produced CO_2 on heating in oxygen, were observed, but these were so small that their exact composition was difficult to determine. The results were calculated on the assumption that this residue was a 50% mixture of carbon and unburnt compound. The resultant derived enthalpies of formation are:

$$\Delta H_f^{\circ} \ Ph_4Ge \ (c) = 281.2 \pm 13.8 \ kJ \ mol^{-1} \tag{22}$$

$$\Delta H_f^{\circ} \ Ph_6Ge_2 \ (c) = 446.4 \pm 10.5 \ kJ \ mol^{-1} \tag{23}$$

$$\Delta H_f^{\circ} \ Bz_4Ge \ (c) = 219.7 \pm 10.5 \ kJ \ mol^{-1} \tag{24}$$

Kana'an[33] has determined the sublimation pressures and the associated thermochemical quantities for tetraphenylgermanium and hexaphenyldigermanium by simultaneous measurement of the torsional recoil and rate of mass effusion. He obtained values of:

$$\Delta H_{sub}^{\circ} \ Ph_4Ge = 156.9 \pm 4.6 \ kJ \ mol^{-1} \tag{25}$$

$$\Delta H_{sub}^{\circ} \ Ph_6Ge_2 = 209.2 \pm 4.2 \ kJ \ mol^{-1} \tag{26}$$

and combining these values with those in equations (22) and (23), respectively,

[30] B. S. Iseard, J. B. Pedley, and J. A. Treverton, *J. Chem. Soc. (A)*, 1971, 3071.
[31] J. L. Bills and F. A. Cotton, *J. Phys. Chem.*, 1964, **68**, 806.
[32] (a) G. P. Adams, A. S. Carson, and P. G. Laye, *Trans. Faraday Soc.*, 1969, **65**, 113; (b) A. S. Carson, E. M. Carson, P. G. Laye, J. A. Spencer, and W. V. Steele, *ibid.*, 1970, **66**, 2459.
[33] A. S. Kana'an, *J. Chem. Thermodynamics*, 1974, **6**, 191.

we obtain:

$$\Delta H_f^\circ \, Ph_4Ge \, (g) = 438.1 \pm 14.2 \, kJ \, mol^{-1} \qquad (27)$$

$$\Delta H_f^\circ \, Ph_6Ge_2 \, (g) = 655.6 \pm 11.3 \, kJ \, mol^{-1} \qquad (28)$$

These values differ from those quoted by Carson *et al.*, who used estimated values for the enthalpies of sublimation. Using the above values for the enthalpies of sublimation, it is possible to estimate that the enthalpy of sublimation of tetra-benzylgermanium should be:

$$\Delta H_{sub}^\circ \, Bz_4Ge = 184 \pm 8 \, kJ \, mol^{-1} \qquad (29)$$

and hence

$$\Delta H_f^\circ \, Bz_4Ge \, (g) = 403.8 \, kJ \, mol^{-1} \qquad (30)$$

The static-bomb and reaction calorimetry of organotin compounds has been reviewed by Skinner,[18] and the values obtained are listed by Cox and Pilcher.[10] Static-bomb calorimetry appears to be able to give reliable values for tin organometallic compounds since the only products of combustion are SnO_2, CO_2, and H_2O, with only a small amount of unburnt tin and soot. However, in the case of tetraphenyltin the combustion proved difficult, owing to explosions.[34] Adams, Carson, and Laye[35] overcame this problem by using an aneroid calorimeter in their combustions. Owing to the smaller heat capacity of the system, they were able to use much smaller samples and still get a reasonable temperature rise. A silica chimney helped to obtain complete combustion. The agreement obtained between the two investigations was excellent. Carson *et al.*[36] have also determined the standard enthalpy of formation of hexaphenyl-ditin in the same calorimeter. They obtained:

$$\Delta H_f^\circ \, Ph_6Sn_2 \, (c) = 660.2 \pm 8.4 \, kJ \, mol^{-1} \qquad (31)$$

Keiser and Kana'an[15] have measured the enthalpies of sublimation of both tetraphenyltin and hexaphenylditin, obtaining values of:

$$\Delta H_{sub}^\circ \, Ph_4Sn = 161.1 \pm 5.0 \, kJ \, mol^{-1} \qquad (32)$$

$$\Delta H_{sub}^\circ \, Ph_6Sn_2 = 188.3 \pm 3.9 \, kJ \, mol^{-1} \qquad (33)$$

Combination of these values with those obtained for the standard enthalpies of formation of the compounds gives:

$$\Delta H_f^\circ \, Ph_4Sn \, (g) = 571.1 \pm 5.9 \, kJ \, mol^{-1} \qquad (34)$$

$$\Delta H_f^\circ \, Ph_6Sn_2 \, (g) = 848.5 \pm 9.2 \, kJ \, mol^{-1} \qquad (35)$$

As stated above, the static-bomb calorimetry of the organometallic compounds of lead gave values which were in error by up to 150 kJ mol^{-1}. Accurate thermochemical values for lead alkyls have been obtained by rotating-bomb calorimetry

[34] A. E. Pope and H. A. Skinner, *Trans. Faraday Soc.*, 1964, **60**, 1402.
[35] G. P. Adams, A. S. Carson, and P. G. Laye, *J. Chem. Thermodynamics*, 1969, **1**, 393.
[36] W. V. Steele, A. S. Carson, P. G. Laye, and J. A. Spencer, *J. Chem. Thermodynamics*, 1973, **5**, 477.

by Good et al.[1] The bomb contained nitric acid as solvent for the solid products of the combustions, and the final solution was a homogeneous one of lead nitrate in excess nitric acid. The bomb processes can be summarized by equations (36) and (37).

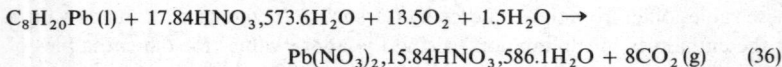

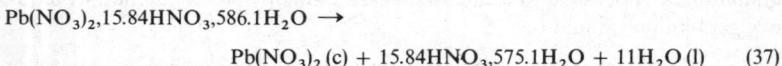

$$C_8H_{20}Pb\ (l) + 17.84HNO_3,573.6H_2O + 13.5O_2 + 1.5H_2O \rightarrow$$

$$Pb(NO_3)_2,15.84HNO_3,586.1H_2O + 8CO_2\ (g) \qquad (36)$$

$$Pb(NO_3)_2,15.84HNO_3,586.1H_2O \rightarrow$$

$$Pb(NO_3)_2\ (c) + 15.84HNO_3,575.1H_2O + 11H_2O\ (l) \qquad (37)$$

The derived enthalpies of formation are given in equations (38) and (39), using

$$\Delta H_f^\circ\ Me_4Pb\ (l) = 64.4 \pm 3.8\ kJ\ mol^{-1} \qquad (38)$$

$$\Delta H_f^\circ\ Et_4Pb\ (l) = 108.8 \pm 2.5\ kJ\ mol^{-1} \qquad (39)$$

the most recent value for the enthalpy of formation of solid lead nitrate.[36] Carson et al.[37] have determined the standard enthalpy of formation of tetra-phenyl-lead, using a similar method to that of Good but using yellow lead dioxide as the solid lead compound in the comparison experiments; lead nitrate is hygroscopic and requires careful handling in weighing. The following enthalpies of formation were derived from the results:

$$\Delta H_f^\circ\ Ph_4Pb\ (c) = 515.0 \pm 15.0\ kJ\ mol^{-1} \qquad (40)$$

$$\Delta H_f^\circ\ Ph_4Pb\ (g) = 709.6 \pm 20.0\ kJ\ mol^{-1} \qquad (41)$$

Group VB.—The values tabulated by Cox and Pilcher for the enthalpies of formation of the organophosphorus compounds all have large uncertainties attached to them. Recent work by Birley and Skinner[38] and Head and Lewis[39] on the enthalpy of formation of aqueous orthophosphoric acid should lead the way in obtaining more reliable values for all the compounds. Head and Lewis burnt white phosphorus in a rotating-bomb calorimeter; hydrolysis of the reaction products produced a mixture of acids. When the bomb was maintained at 323 K the product was orthophosphoric acid when perchloric acid was used as solvent. Their paper outlines methods for the detection of small amounts of the phosphorus acids in the presence of a large excess of the ortho-acid. It should be used as a model by anyone undertaking the determination of the enthalpies of these compounds by rotating-bomb calorimetry.

The static-bomb calorimetry of arsenic organometallic compounds is fraught with difficulty, as the products are a complex mixture of solid As_2O_3, As_2O_4, unburnt As, and carbon, together with an aqueous solution of As^{III} and As^V in varying concentrations within the bomb itself. The values obtained by this

[37] A. S. Carson, P. G. Laye, J. A. Spencer, and W. V. Steele, J. Chem. Thermodynamics, 1972, **4**, 783.

[38] G. I. Birley and H. A. Skinner, Trans. Faraday Soc., 1968, **64**, 2232.

[39] A. J. Head and G. B. Lewis, J. Chem. Thermodynamics, 1970, **2**, 701.

method must therefore be suspect Mortimer and Sellers[40] have used the rotating-bomb method to determine the standard enthalpy of formation of triphenylarsine. The solvent used in the combustions was sodium hydroxide, giving a final solution of sodium arsenate, sodium arsenite, sodium carbonate, and excess sodium hydroxide. After analysis of the solution, corrections were made for the enthalpies of formation of the sodium carbonate and the sodium arsenate.

The only values available for organo-antimony and -bismuth compounds were obtained by static-bomb calorimetry, and are suspect. Cox and Pilcher list uncertainty intervals of ± 16 kJ mol^{-1} for these compounds, and this is probably an under-estimate.

Group VIA and Group VIIA.—With the exception of the triacetylacetonates of chromium and manganese(II), whose enthalpies of formation have been determined by reaction calorimetry, only static-bomb calorimetry values exist for the organometallic compounds of both of these Groups.[10] A rotating-bomb method of determining the standard enthalpies of formation of manganese organometallic compounds has been developed by Good et al.,[41] but to date the only manganese compound studied has been dimanganese decacarbonyl. The solvent used in the rotating-bomb experiments to dissolve the products of combustion was a mixture of nitric acid and hydrogen peroxide, which converted all the manganese into Mn^{II} ions. A solution of manganese nitrate Mn(NO$_3$)$_2$, 10.3H$_2$O, encapsulated in a Pyrex ampoule, was used to reproduce the same final solution in the comparison experiments. The idealized combustion reaction can be represented by equation (42), and

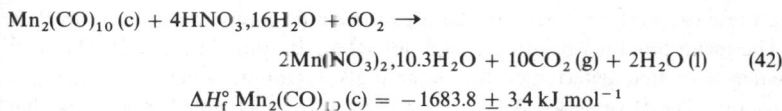

$$Mn_2(CO)_{10} (c) + 4HNO_3,16H_2O + 6O_2 \rightarrow$$

$$2Mn(NO_3)_2,10.3H_2O + 10CO_2 (g) + 2H_2O (l) \quad (42)$$

$$\Delta H_f^\circ \, Mn_2(CO)_{10} \, (c) = -1683.8 \pm 3.4 \text{ kJ mol}^{-1}$$

Group VIB.—The thermochemistry of organoselenium compounds is in its infancy. Barnes and Mortimer[42] have determined the standard enthalpies of formation of selenium dioxide and diphenyl selenide by rotating-bomb calorimetry, using deionized water as the solvent. The enthalpy of selenium dioxide in water was determined in separate experiments by reaction calorimetry. The following enthalpies of formation were determined:

$$\Delta H_f^\circ \, SeO_2 \, (c) = -226.4 \pm 2.1 \text{ kJ mol}^{-1} \quad (43)$$

$$\Delta H_f^\circ \, Ph_2Se \, (c) = 227.2 \pm 6.0 \text{ kJ mol}^{-1} \quad (44)$$

$$\Delta H_f^\circ \, Ph_2Se \, (g) = 291.0 \pm 8.4 \text{ kJ mol}^{-1} \quad (45)$$

Group VIII.—The data on the organometallic compounds of this Group have been summarized by Cox and Pilcher[10] in half a page of their tables, and the field is virtually bare of any accurate values.

[40] C. T. Mortimer and P. Sellers, *J. Chem. Soc.*, 1964, 1965.
[41] W. D. Good, D. M. Fairbrother, and G. Waddington, *J. Phys. Chem.*, 1958, **62**, 853.
[42] D. S. Barnes and C. T. Mortimer, *J. Chem. Thermodynamics*, 1974, **6**, 371.

6 Mean Metal Bond Dissociation Energies

The difference between the concepts of bond-energy terms and bond-dissociation energies is well known, and will not be explained here (see for example ref. 18). The enthalpies of formation of organometallic compounds can be used in conjunction with the enthalpies of formation of the free radicals and the enthalpies of formation of the gaseous metal atoms to calculate *mean* bond-dissociation energies of metal–carbon bonds in metallic alkyls and related compounds. If the enthalpy of process (46) can be calculated, then we have equation (47),

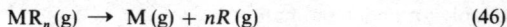

$$MR_n(g) \rightarrow M(g) + nR(g) \tag{46}$$

$$\Delta H^\circ = \Delta H_f^\circ M(g) + n \Delta H_f^\circ R(g) - \Delta H_f^\circ MR_n(g) \tag{47}$$

and the *mean* bond-dissociation energy, $\bar{D}(M—R)$, is $\Delta H^\circ/n$. In general, these *mean* bond-dissociation energies will differ from the actual metal–carbon dissociation energies $D(R_{n-1}M—R)$, which still have not received much attention and few measured values have been reported. An exception to this is the study of the individual bond-dissociation energies in mercury dialkyls, in which the inequality of the two dissociation energies was demonstrated. The whole field has been reviewed by Skinner.[18]

The *mean* bond-dissociation energies in the metal alkyls as defined above are listed in Table 3 and have been derived from the data in Tables 1 and 2 along with the enthalpies of formation of the metal alkyls from Cox and Pilcher[10] or given in Section 5. Values derived from static-bomb calorimetry are given in parentheses as these can only be tentative. However, certain trends are evident, and these can best be shown for the organometallic compounds of Group IVB.

The mean bond-dissociation energy for RMe_4 is approximately $30 \, kJ \, mol^{-1}$ greater than that determined for other n-alkyl groups, which stays virtually constant. The $\bar{D}$ value for RPh_4 is approximately $65 \, kJ \, mol^{-1}$ more than that determined for the normal alkyl groups. Also $\bar{D}(M—R)$ falls progressively as we descend the Group, i.e. $\bar{D}(C—R) > \bar{D}(Si—R) > \bar{D}(Ge—R) > \bar{D}(Sn—R) > \bar{D}(Pb—R)$.

The comments made in the last paragraph apply to a number of other values in Table 3 and parallel those made by Holm[18a] in his discussion of the relative bond-dissociation energies of $R—MgX$ compounds (see Section 5).

Cox and Pilcher[10] list values for bond-energy terms for organometallic compounds which reproduce the data reasonably well (within $\pm 12 \, kJ \, mol^{-1}$). It is therefore not proposed to set out these values here again, but the interested reader is referred to their book. It is also worth noting the scheme derived by Carson et al.[19,36,37] for the organometallic compound of Group IVB.

7 Conclusions

The thermochemistry of organometallic compounds has progressed in the past two decades since the advent of the rotating-bomb calorimeter. A number of workers, notably Good and Skinner, have pioneered methods of determining

Table 1 *Enthalpies of formation of gaseous atoms at* 298.15 K[a]

Atom	ΔH_f° (g)/kJ mol^{-1}	Atom	ΔH_f° (g)/kJ mol^{-1}
H	217.997 ± 0.006	Cu	338.5 ± 2.1
Li	161 ± 2	Zn	130.54 ± 0.21
Be	322 ± 2	Ga	272 ± 2
B	559 ± 15	Ge	283 ± 2
C	716.7 ± 0.4	As	288 ± 12
N	472.7 ± 0.4	Se	207.5 ± 4.2
O	249.17 ± 0.10	Br	111.88 ± 0.12
F	78.5 ± 10.0	Rb	79.5 ± 2.1
Na	108.43 ± 0.63	Sr	165 ± 2
Mg	143.68 ± 0.20	Y	420 ± 2
Al	332.6 ± 0.2	Zr	593 ± 10
Si	453.5 ± 12.6	Nb	753 ± 8
P	316.2 ± 0.6	Mo	659 ± 2
S	277.8 ± 0.3	Tc	661 ± 17
Cl	121.290 ± 0.008	Ru	650 ± 10
K	89.60 ± 0.21	Rh	558 ± 8
Ca	179 ± 2	Pd	373.2 ± 3.3
Sc	380.7 ± 2.1	Ag	284.9 ± 0.8
Ti	467.97 ± 1.72	Cd	111.56 ± 0.10
V	514.6 ± 1.0	In	236.73 ± 0.42
Cr	397 ± 2	Sn	300.7 ± 0.3
Mn	284.9 ± 1.7	Sb	265 ± 8
Fe	415 ± 7	Te	193 ± 8
Co	422.6 ± 0.8	I	106.762 ± 0.040
Ni	421.86 ± 0.29	Cs	78.6 ± 1.3
Ba	185 ± 2	Au	368 ± 4
Hf	619 ± 4	Hg	61.329 ± 0.054
Ta	782 ± 10	Tl	181.2 ± 2.1
W	851.4 ± 0.8	Pb	194.97 ± 0.84
Re	778.0 ± 2.7	Bi	208.8 ± 0.8
Os	790 ± 6	Ra	162 ± 10
Ir	669 ± 8	Th	597.5 ± 4.6
Pt	564.5 ± 2.1	U	536 ± 8

[a] All values are taken from 'Bond Energies. Data and Methodology' by W. V. Steele, Butterworths, London, in the press. In this monograph the values listed above are those accepted as the 'best' available from a review of literature data up to July 1974.

Table 2 *Enthalpies of formation of free radicals at* 298.15 K

Radical	ΔH_f°/kJ mol^{-1}	Reference
CH_3	145.6 ± 0.4	a
Et	106.7 ± 4.2	b
Pr^n	85.8 ± 7.5	b
Bu^n	67 ± 10	b
Pr^i	76 ± 7.5	b
Bu^i	54 ± 9	c
Bu^t	33 ± 6	c
Ph	324.3 ± 10.0	d
Bz	188 ± 13	e
cyclo-C_5H_5	188 ± 40	e

[a] W. A. Chupka, *J. Chem. Phys.*, 1968, **48**, 2337; [b] J. A. Kerr, *Chem. Rev.*, 1966, **66**, 465; [c] H. E. O'Neil and S. W. Benson, Chapter 17, in 'Free Radicals', ed. J. K. Kochi, Wiley–Interscience, 1973, vol. II; [d] G. A. Chamberlain and E. Whittle, *Trans. Faraday Soc.*, 1971, **67**, 2077; [e] J. S. Roberts and H. A. Skinner, *Trans. Faraday Soc.*, 1949, **45**, 339.

Table 3 *Mean bond-dissociation energies in* MR_n, $\bar{D}(M—R)/kJ\ mol^{-1}$ [a]

Bond	$\bar{D}/kJ\ mol^{-1}$	Bond	$\bar{D}/kJ\ mol^{-1}$
Li—Me	(248)	Ga—Bu	232 ± 8?
Li—Bu	(253)	Ga—Bu^i	223 ± 8?
Mg—cyclo-C_5H_5	191 ± 45	In—Me	168 ± 6
Zn—Me	183 ± 7	Si—Et	290 ± 25
Zn—Et	133 ± 10	Ge—Et	244 ± 8
Zn—Pr^n	145 ± 25	Ge—Pr	239 ± 8
Zn—Bu^n	157 ± 25	Ge—Ph	306 ± 16
Cd—Me	147 ± 4	Ge—Bz	183 ± 16
Cd—Et	111 ± 8	Sn—Me	226 ± 4
Hg—Me	124 ± 4	Sn—Et	193 ± 8
Hg—Et	100 ± 8	Sn—Pr	197 ± 8
Hg—Pr^n	99 ± 9	Sn—Pr^i	182 ± 10
Hg—Pr^i	86 ± 9	Sn—Bu^n	198 ± 10
Hg—Ph	157 ± 8	Sn—Ph	257 ± 10
B—Me	(373)	Pb—Me	168 ± 4
B—Et	(344)	Pb—Et	139 ± 6
B—Pr^n	(351)	Pb—Ph	196 ± 10
B—Pr^i	(346)	P—Me	(283)
B—Bu^n	(349)	P—Et	(229)
B—Bu^i	(334)	P—Ph	(319)
B—Ph	(466)	As—Me	(238)
Al—Me	(286)	As—Et	(184)
Al—Et	(272) or (245)[b]	As—Ph	280 ± 15
Ga—Me	252 ± 8	Cr—C_6H_6	(166)
Ga—Et	223 ± 8?	Mo—C_6H_6	(211)
Sb—Me	(224)	Se—Et	(238)
Sb—Et	(179)	Se—Ph	277 ± 10
Sb—Ph	(267)	Fe—cyclo-C_5H_5	288 ± 40
Bi—Me	(140)	Ni—cyclo-C_5H_5	237 ± 40
Bi—Et	(109)		
Bi—Ph	(200)		

[a] Values calculated from equation (47), Tables 1 and 2, and values of ΔH_f° (g) from ref. 10 or text; [b] Value calculated from ΔH_f° Et$_3$Al(g) $= -84\ kJ\ mol^{-1}$ from P. A. Fowler, Ph.D. Thesis, University of Manchester, 1961.

the standard enthalpies of formation of these compounds, but several important groups remain untackled, notably organo-aluminium and -phosphorus compounds. Kana'an's work on the determination of the enthalpies of sublimation of the tetraphenyls of Group IVB includes virtually the first such determinations in the field, and it is to be hoped that more values will be forthcoming soon. The basis of the field has been well founded, and it now needs a set of dedicated, scrupulous workers to continue the good work.

7 The High-temperature Thermodynamics of Inorganic Substances

By G. V. JAGANNATHAN, G. R. WOOLLEY, and P. A. H. WYATT
Department of Chemistry, The University of St. Andrews, Scotland

1 Introduction

The extension of chemical thermodynamics to reactions at high temperatures is of course by no means new, and is in principle straightforward. Indeed, the preference in the past for working within 100 K or so of room temperature reflected in many publications in the purely chemical literature (with notable exceptions) has obviously been dictated by experimental convenience, and many metallurgists have probably felt, with some justification, that they had a broader vision of the scope of chemical thermodynamics than the average chemist. Despite the experimental difficulties, substances which react at useful rates only at high temperatures have nevertheless frequently demanded practical attention because of their economic importance, and consequently much of the information on metals and other inorganic substances and their mixtures has been widely scattered through the specialist literature in subjects outside the arbitrary confines of mainstream chemistry. It is therefore particularly useful when books like F. D. Richardson's 'Physical Chemistry of Melts in Metallurgy' (in two volumes)[1] appear and provide an up-to-date key to reviews and other references in a well-defined and important area of this large field.

Although supplementary information about the behaviour of mixtures is often essential, it is still broadly true that the principal object of thermodynamic studies at high temperatures is to permit the calculation of yields of chemical reactions (McGlashan[2]) and for this purpose free-energy information on pure substances over wide temperature ranges will continue to be of paramount importance. While third-law determinations and statistical calculations continue, the free-energy values for many compounds still depend upon the measurement of a chemical or phase equilibrium at a high temperature, if only for checking internal consistency, and here a good feeling for the possibility of chemical complications, and even kinetic constraints, is obviously vital. Since dozens of elements are potentially important economically, and every element, in spite of

¹ F. D. Richardson, 'Physical Chemistry of Melts in Metallurgy', Volumes I and II, Academic Press, London, 1974.
² 'Chemical Thermodynamics', ed. M. L. McGlashan, (Specialist Periodical Reports), The Chemical Society, London, 1973, Vol. 1.

family resemblances, has largely its own distinctive chemistry, there will always be important questions of detail to settle experimentally even after the broad framework of consistent free-energy data throughout the Periodic Table has largely been set up. In particular, nonstoicheiometry will always bring special problems of its own wherever it occurs.

During the past two decades a good introduction to the general physico-chemical background and the high-temperature techniques involved has been provided by successive editions of 'Metallurgical Thermochemistry', by Kubaschewski, Evans, and Alcock,[3] and by the Discussions of the Faraday Society on 'The Physical Chemistry of Process Metallurgy' (No. 4, 1948), 'The Structure and Properties of Ionic Melts' (No. 32, 1961), and 'The Vitreous State' (No. 50, 1970). Calculations of reaction yields, vapour pressures, *etc.* have also been greatly facilitated by the replacement of laborious expressions involving temperature series by compilations of $-(G^{\ominus} - H_0^{\ominus})/T$ values or, often more usefully still, of $\Delta G_f^{\ominus}$ values over a wide range of temperatures, as in the JANAF tables.[4] In recent years, however, the whole field has expanded considerably, largely for two reasons. On the one hand, alongside the steady progress in the traditional fields of metallurgy, glass technology, and ceramics, considerable research effort has been directed towards investigations of the properties of materials in any way connected with nuclear power, rocket fuels, and semiconductors. On the other hand, advances in mass spectrometry and the development of the matrix-isolation technique (ironically taking advantage of very *low*-temperature properties) have produced a revolution in our knowledge of the constitution of the vapour phase over high-temperature melts and solids.

In preparing this Report, a few hundred references have been traced for the past two years alone, and it will clearly be impracticable and probably also pointless to cover all of them adequately, since they will in any case be familiar to specialists in the various fields. We have therefore designed our Report rather for those approaching this subject from the traditionally 'chemical' end and have aimed at giving an adequate introductory bibliography and a survey both of the standard methods of investigation and of the novel experimental techniques that are gaining favour. The selection of examples chosen to illustrate lines of research cannot be exhaustive but it will give some indication of the problems one can expect to meet.

2 Sources of Information

Between 1956 and 1966 seven reviews appeared of high-temperature chemical studies. The first four[5—8] covered mainly experimental techniques, while the

[3] O. Kubaschewski, E. L. Evans, and C. B. Alcock, 'Metallurgical Thermochemistry', 4th edn., Pergamon Press, Oxford, 1967.
[4] D. R. Stull and H. Prophet, 'Joint Army, Navy, and Air Force Thermochemical Tables', U.S. Govt. Printing Office, Washington, 2nd edn., 1971.
[5] L. Brewer and A. W. Searcy, *Ann. Rev. Phys. Chem.*, 1956, **7**, 259.
[6] J. L. Margrave, *Ann. Rev. Phys. Chem.*, 1959, **10**, 457.
[7] P. W. Gilles, *Ann. Rev. Phys. Chem.*, 1961, **12**, 355.
[8] R. F. Porter, *Ann. Rev. Phys. Chem.*, 1959, **10**, 219.

later ones[9-11] have, in addition, discussed the status of high-temperature topics[10] and have turned towards the correlation of results[9,11] and the discussion of structures and stabilities of species and the adequacy of theoretical treatments.[11] The dissociation energies of diatomic gases,[5,10] the attainment and measurement of high temperatures,[6] and the kinetics of high-temperature reactions[6] and heterogeneous equilibria[8] are among the topics dealt with.

The period since 1966 has seen the rapid growth of matrix-isolation spectroscopy, photoelectron spectroscopy, and, to a lesser extent, Mössbauer spectroscopy, all of which have potentially important contributions to make to the investigation of high-temperature species; however, the combination of mass spectrometry with the Knudsen effusion cell still seems to be the most widely used technique for this purpose. Reference will be made to some of these methods and their recent variants in this Report. A review of the field up to 1970 was published by Margrave *et al.*,[11] who referred to the experimental techniques available for the characterization of high-temperature species and discussed the conclusions which had been drawn about the stabilities and structures of a number of the latter.

More recently, the first Chemical Society Specialist Periodical Report on Chemical Thermodynamics[2] contains some excellent chapters of direct relevance to high temperatures. One, by Kubaschewski, Spencer, and Dench, is specifically on a restricted branch of this subject ('Metallurgical Thermochemistry at High Temperatures') and surveys the thermochemical quantities required and the experimental methods for their determination, while another, by Frankiss and Green, is on the 'Statistical Methods of Calculating Thermodynamic Functions' and sets out the requisite formulae in some detail, though it must be admitted that the examples chosen for the applications are almost entirely organic molecules. The most useful chapter for present purposes, however, is that of Herington on 'Thermodynamic Quantities, Thermodynamic Data, and their Uses', which must be one of the most comprehensive, yet manageable, guides to the literature in any subject. Theory, units, symbols, and measuring scales are all surveyed, along with an extensive bibliography of the available tabulations of thermodynamic data and a discussion of their uses.

To the Herington list of sources may now be added (i) a monograph by Mills[12] on inorganic sulphides, selenides, and tellurides, surveying the literature up to 1970 and listing the heats of formation, standard entropies, heat capacities, enthalpies, vapour pressures, and dissociation energies of over 700 compounds up to 2000 K; (ii) a special compilation by Horvath[13] of the physical properties of 31 compounds of major importance in the chemical industry, including in graphical form such properties as vapour pressure, density (orthobaric and

[9] J. Drowart and P. Goldfinger, *Ann. Rev. Phys. Chem.*, 1962, **13**, 459.
[10] R. J. Thorn, *Ann. Rev. Phys. Chem.*, 1966, **17**, 83.
[11] J. W. Hastie, R. H. Hauge, and J. L. Margrave, *Ann. Rev. Phys. Chem.*, 1970, **21**, 475.
[12] K. C. Mills, 'Thermodynamic Data for Inorganic Sulphides, Selenides and Tellurides', Butterworths, London, 1973.
[13] A. L. Horvath, 'Physical Properties of Inorganic Compounds', Arnold, London, 1974.

supercritical), latent heat of vaporization, thermal capacity, viscosity, thermal conductivity, surface tension, and solubility; and (iii) a book by Barin and Knacke[14] on inorganic substances, the review of which by Skinner[15] implies that it does not contain much data independent of the JANAF tables, Landolt–Börnstein, and 'Metallurgical Thermochemistry'.[3]

The second edition of the JANAF Thermochemical Tables is welcomed in a review by F. D. Rossini,[16] who makes a plea that the full notation $(G^{\ominus} - H_0^{\ominus})/T$ should be adhered to and not replaced by contractions like gef_T. (On this theme, the article following Rossini's review is a guide to the IUPAC recommendations on the presentation of thermodynamic results, prepared by Commission I.2 on thermodynamics and thermochemistry under the chairmanship of E. F. Westrum jun.) Further JANAF supplements, beyond the first 33 now incorporated in the second edition, are now published from time to time in J. Phys. Chem. Ref. Data (see, e.g., Vol. 3, No. 2, 1974). Apart from revisions for lead halides and titanium oxides, the programme has continued with new tables on cobalt chlorides, oxides of V, Nb, Ta, and Cr, selected carbides, nitrides, and alkaline-earth compounds; and it is proposed to follow on with refractory metal halides, polyatomic carbon gases, and sulphur–fluorine compounds.[17]

The International Council of Scientific Unions (ICSU) and the Committee on Data for Science and Technology (CODATA) set up in 1968 a Task Group on Key Values for Thermodynamics, with the object of producing an internationally agreed set of values for the thermodynamic properties of chemical species. Following upon the initial publication of the first report, a final list has now been produced[18] of recommended values in both J and thermochemical calories of $\Delta H_f^{\ominus}$ (298.15 K), $S^{\ominus}$ (298.15 K), and $[H^{\ominus}$ (298.15 K) $- H^{\ominus}(0)]$ for the atomic and molecular forms of the elements O, H, Cl, Br, I, N, C (with, for example, $\Delta H_f^{\ominus}$ (298.15 K) for C(g) at 716.67 kJ mol^{-1} or 171.29 kcal mol^{-1}), and the rare gases, plus H_2O (l and g), and gaseous HCl, HBr, HI, CO, and CO_2. It is stressed that these recommended values will not be entirely consistent with any other published set but the Task Group has as its ultimate goal the revision of the 298.15 K values for the whole field. Mention should also be made of the CODATA 'International Compendium of Numerical Data Projects',[19] the CATCH (Computer Analysis of Thermochemical Data) tables available from the School of Molecular Sciences of the University of Sussex, and of the quarterly current awareness bibliographies on the 'Higher Temperature Chemistry and Physics of Materials',[20] edited by J. J. Diamond and covering research at temperatures above 1000 °C. It is

[14] I. Barin and O. Knacke, 'Thermochemical Properties of Inorganic Substances', Springer-Verlag, Berlin/Heidelberg, 1973.
[15] H. A. Skinner, J. Chem. Thermodynamics, 1974, 6, 711.
[16] F. D. Rossini, J. Chem. Thermodynamics, 1972, 4, 509.
[17] M. W. Chase, personal communication.
[18] Codata Key Values, Part I, J. Chem. Thermodynamics, 1972, 4, 331.
[19] Codata, 'International Compendium of Numerical Data Projects', Springer-Verlag, Berlin, 1969.
[20] 'Bibliography on the High Temperature Chemistry and Physics of Materials', ed. J. J. Diamond, NBS Special Publ. 315-1, 1968 and onwards.

divided into two parts, (I) solids and liquids and (II) gases, and groups the references under 15 headings such as 'Phase equilibria above 1000 °C', 'Devices for achieving temperatures above 1500 °C, or 'Spectroscopy of interest to high temperature chemistry'.

When investigating certain high-temperature equilibria, *e.g.* by the Knudsen effusion technique, where the residence time of the molecules in the cell may be rather short, it is necessary to know which of the possible reactions have time to reach equilibrium. Kinetic information is then vital. Specifically high-temperature reaction rate data (up to 5000 K) for homogeneous gas reactions are critically surveyed by Baulch *et al.*,[21] with O.S.T.I. support, and a paper on current compilations and evaluations of kinetic data is also available.[22]

3 Enthalpy Measurements

The established methods of attaining and measuring high temperatures are well reviewed.[6,7,9] A new International Practical Temperature Scale was adopted in 1968 (IPTS 68), being higher by 1.24 K at 1000 °C and 5.9 K at 3000 °C than the older (IPTS 48) scale: further information is given by Herington.[2]

Improvements have been made in the traditional resistive and inductive heating of furnaces, but novel methods are constantly being introduced or revived,[9,23] such as arc and solar imaging, shock tubes[24] and lasers,[25] flames augmented by electric power, adiabatic expansion,[26] and the radio-frequency heating associated with electromagnetic levitation. The latter technique, with which temperatures above 2700 K are obtainable, is particularly suited to the study of reactive alloys at high temperatures and has been shown to give results for enthalpies of mixing and other thermodynamic quantities in good agreement with former methods for the liquid Fe–Ni system,[27] but it is clearly of restricted application to inorganic substances in general.

Calorimetry.—The various forms of calorimeter are well reviewed and classified,[2,3,28,29] and details of specifically high-temperature devices can be found in specialist books[3] and reviews.[29]

To judge from recent publications, the most popular method of determining thermal capacities and enthalpies is still drop calorimetry. Here a capsule containing the specimen is heated in a furnace to a determined temperature and then dropped into a calorimeter of known heat capacity kept near room temperature,

[21] D. L. Baulch, 'High Temperature Reaction Rate Data Reports', Department of Physical Chemistry, Leeds University, Leeds, England, 1970.
[22] L. H. Gevantman and D. Garvin, *Internat. J. Chem. Kinetics*, 1973, **5**, 213.
[23] Chemical Society Faraday Symposium, No. 8, 'High Temperature Studies in Chemistry', London, 1973.
[24] W. T. Rawlins and W. C. Gardiner, jun., *J. Phys. Chem.*, 1974, **78**, 497.
[25] R. T. Meyer, A. W. Lynch, and J. M. Freese, *J. Phys. Chem.*, 1973, **77**, 1083.
[26] J. Berkowitz, *J. Chem. Phys.*, 1972, **56**, 2766.
[27] K. C. Mills, K. Kinoshita, and P. Grieveson, *J. Chem. Thermodynamics*, 1972, **4**, 581.
[28] 'Physicochemical Measurements at High Temperatures', ed. J. O'M Bockris, J. L. White, and J. D. MacKenzie, Butterworths, London, 1959.
[29] 'Experimental Thermochemistry', Vol. II, ed. H. A. Skinner, Interscience–Wiley, London, 1962.

so that the enthalpy change of the sample between the two temperatures can be determined directly and the heat capacity from the rate of change with temperature of the enthalpy. In this way $[H^{\ominus}(T) - H^{\ominus}(298.15 \text{ K})]$ and $C_p^{\ominus}(T)$ {and hence sometimes $[S^{\ominus}(T) - S^{\ominus}(298.15 \text{ K})]$ and other functions} have been determined for $FeSe_2$ between 300 and 853 K by Svendsen;[30] for Na_2O up to 1300 K and Na(l) up to 1505 K by Fredrickson and Chasanov;[31] for UN up to 1700 K by Oetting and Leitnaker;[32] and for Pt and a Pt–Rh alloy (10 mass per cent Rh) between 400 and 1700 K,[33a] $UO_{2.25}$ up to 1600 K,[33b] and LiF, NaF, KF, RbF, and CsF from 500 to 1600 K,[33c] all by MacLeod. For the alkali-metal fluorides MacLeod[33c] also reports the enthalpies of fusion as 26.7 kJ mol^{-1} (LiF), 33.5 kJ mol^{-1} (NaF), 29.5 kJ mol^{-1} (KF), 25.9 kJ mol^{-1} (RbF), and 14.0 kJ mol^{-1} (CsF). With the exceptions of the KF value, which is higher than other published data, and that of CsF, which is considerably lower than the only other published value, these agree well with other determinations. Mar and Stout[34] have continued their studies of high-temperature enthalpies of compounds of high boron content, and Spedding et al.[35] report much useful data, tabulated at 100 K intervals from 100 to 1600 K, of ten rare-earth fluorides.

By adiabatic shield calorimetry, Grønvold[36] determines the following quantities for liquid Se: $C_p^{\ominus}$ (1000 K) = 35.62 J K^{-1} mol^{-1}, $[H^{\ominus}$ (1000 K) − $H^{\ominus}$ (298.15 K)] = 28.593 kJ mol^{-1}, and $[S^{\ominus}$ (1000 K) − $S^{\ominus}$ (298.15 K)] = 49.81 J K^{-1} mol^{-1}; and for hexagonal Se he finds the melting point to be (494.33 ± 0.02) K and the enthalpy of fusion (6159 ± 4) J mol^{-1}. With his co-workers[37] he is also investigating the thermodynamic properties (up to 900 K and above) of selenides and tellurides of Ni, Fe, and Cr, some of which are non-stoicheiometric compounds, and of alkali-metal–$MgCl_2$ double chlorides.[38] The Co and Ni tellurides undergo interesting structural changes, and the heat capacities of phases of various composition are reported by Mills.[39] A neat, high-temperature calorimeter (for 900—1800 K), which can serve either as an adiabatic or as a heat-flow device, is described by Malinsky and Claisse[40] for the study of metal systems. The authors claim that it is reliable, sensitive, and easy to operate, and have measured the enthalpy of formation at 1550 K and the heat capacity from 1100 to 1700 K of another non-stoicheiometric compound, $Cr_{0.47}Tm_{0.53}$. Recently, high-temperature microcalorimetry has been used by Campserveux and

[30] S. R. Svendsen, Acta Chem. Scand., 1972, 26, 3834.
[31] D. R. Fredrickson and M. G. Chasanov, J. Chem. Thermodynamics, 1973, 5, 485; ibid., 1974, 6, 629.
[32] F. L. Oetting and J. M. Leitnaker, J. Chem. Thermodynamics, 1972, 4, 199.
[33] (a) A. C. MacLeod, J. Chem. Thermodynamics, 1972, 4, 391; (b) ibid., p. 699; (c) J.C.S. Faraday I, 1973, 69, 2026.
[34] R. W. Mar and N. D. Stout, J. Chem. Thermodynamics, 1974, 6, 943.
[35] E. H. Spedding, B. J. Beaudry, D. C. Henderson, and J. Moorman, J. Chem. Phys., 1974, 60, 1578.
[36] F. Grønvold, J. Chem. Thermodynamics, 1973, 5, 525.
[37] F. Grønvold, Acta Chem. Scand., 1972, 26, 2085; F. Grønvold, N. J. Kveseth, and A. Sween, J. Chem. Thermodynamics, 1972, 4, 337; F. Grønvold; ibid., 1973, 5, 545.
[38] I. L. Holm, B. J. Holm, B. Rinnan, and F. Grønvold, J. Chem. Thermodynamics, 1973, 5, 97.
[39] K. C. Mills, J.C.S. Faraday I, 1974, 70, 2224.
[40] I. Malinsky and F. Claisse, J. Chem. Thermodynamics, 1973, 5, 615.

Gerdanian[41] for determining partial molar enthalpies of solution of O_2 in $CeO_{1.5}$ and CeO_2 at 1353 K, and the less common materials used in semiconductors are attracting interest as investigations over wider ranges of temperature become possible. For example, McMasters *et al.*[42] have reported enthalpies and Gibbs free energies of formation of the europium chalcogenides EuO, EuS, EuSe, and EuTe, which become ferromagnetic or antiferromagnetic at temperatures down to 20 K.

Flames.—Certain gaseous systems can be studied above 2000 K by injection of the reactants into the premixed constituent gases forming the flame and examining their subsequent changes by emission spectroscopy.[43] Kinetic aspects feature prominently in such studies, but equilibrium constants are also obtainable and hence standard enthalpies of formation.

Jensen and Jones[44] have used H_2–N_2–O_2 flames to examine gas-phase Al- and Fe- containing species. Aluminium is an important rocket fuel and consequently knowledge of its high-temperature chemistry is essential. It was found that when Al was present in the H_2–N_2–O_2 flame the main product was $Al(OH)_2$, although Fe produced FeOH. The following thermodynamic information was obtained (all species are gaseous):

$$Al + 2H_2O = Al(OH)_2 + 2H \tag{1}$$

$$\Delta H_0^\ominus = (-95 \pm 30)\,\text{kJ mol}^{-1}: \qquad K = 3.0\exp(4300\,\text{K}/T)$$

$$Al - OH = AlO + H \tag{2}$$

$$\Delta H_0^\ominus = (-172 \pm 20)\,\text{kJ mol}^{-1}; \qquad K = 0.67\exp(19\,200\,\text{K}/T)$$

$$Fe + H_2O = FeOH + H \tag{3}$$

$$\Delta H_0^\ominus = (110 \pm 20)\,\text{kJ mol}^{-1}; \qquad K = 66\exp(-16\,100\,\text{K}/T)$$

$$Fe + OH = FeO + H \tag{4}$$

$$\Delta H_0^\ominus = (22 \pm 20)\,\text{kJ mol}^{-1}; \qquad K = 0.67\exp(-3160\,\text{K}/T)$$

Enthalpies of formation from the *gaseous* elements were also derived: $Al(OH)_2(g)$, $(-1005 \pm 30)\,\text{kJ mol}^{-1}$; $AlO(g)$, $(-349 \pm 20)\,\text{kJ mol}^{-1}$; $FeOH(g)$, (69 ± 20) kJ mol^{-1}; $FeO(g)$, $(259 \pm 20)\,\text{kJ mol}^{-1}$. The derived AlO bond energy, $(596 \pm 20)\,\text{kJ mol}^{-1}$, is at variance with some former estimates. Similar studies on Ba by Jones and Broida[45] have produced information on the population of states of BaO, and shown that Ba gives rise to a greater proportion of charged species in the flame than Al.

A discussion of the formation of inorganic oxide aerosols of controlled dimensions and generated from anhydrous chlorides in H_2–O_2 flames is contained in

[41] J. Campserveux and P. Gerdanian, *J. Chem. Thermodynamics*, 1974, **6**, 795.
[42] O. D. McMasters, K. A. Gschneidner jun., E. Kaldis, and G. Sampietro, *J. Chem. Thermodynamics*, 1974, **6**, 845.
[43] (a) A. G. Gaydon, "Dissociation Energies and Spectra of Diatomic Molecules', Chapman and Hall, London, 1968. (b) A. G. Gaydon, 'The Spectroscopy of Flames', Chapman and Hall, London, 1974.
[44] (a) D. E. Jensen and G. A. Jones, *J.C.S. Faraday I*, 1972, **68**, 259; (b) *ibid.*, 1973, **69**, 1448.
[45] C. R. Jones and H. P. Broida, *J. Chem. Phys.*, 1974, **60**, 4369, 4377.

the Faraday Symposium Report on Fogs and Smokes.[46a] If of submicronic size, these particles exhibit unusual photocatalytic properties. For example, TiO_2[46] allows the catalytic photo-oxidation (in the u.v. range) of organic and inorganic compounds at ambient temperature.

Enthalpy Information from Other Sources.—A feature of the Jensen and Jones work mentioned above[44a] is their careful comparison of the second- and third-law estimates of the thermodynamic quantities involved. A similar comparison has been made in a closely related study by Farber, Srivastava, and Uy[47] of the vapour species over Al_2O_3 at temperatures near 2000 K. Estimates are given of the partial pressures of all the gaseous species Al, O, AlO, Al_2O, Al_2O_2, AlO_2, and (provisionally) Al_2O_3 between 1900 and 2600 K, and internally consistent heats of formation and equilibrium constants are derived. The vapour species were determined mass spectrometrically after emission in a collimated beam from an elongated orifice in an effusion cell constructed from alumina itself, and not from metals such as tungsten and molybdenum, reactions with which are demonstrated to account reasonably for the former discrepancies between mass spectrometric and weight-loss evidence. However, despite the internal consistency achieved separately in the completely independent flame and mass-spectrometric investigations, the derived thermodynamic values are not yet completely consistent between themselves. For example, the bond energy of AlO quoted by Farber *et al.*,[47] though higher than some earlier estimates, is still, at about 510 kJ mol^{-1}, significantly lower than the 600 kJ mol^{-1} figure mentioned by Jensen and Jones.[44a] Farber *et al.*[47b] discuss this discrepancy further and refer to the work of Newman and Page,[48] who also reported a 'high' value for the dissociation energy of AlO of 601 kJ mol^{-1} (second law) and 589 kJ mol^{-1} (third law) from a spectroscopic study similar to that of Jensen and Jones. The establishment of equilibrium concentrations in the flame is felt to be acceptable, although Jensen and Jones state that their H, OH, and O concentrations were slightly above the equilibrium values. The discrepancy is apparently more likely to lie in the intensity–concentration calibration of AlO, which was assumed to be the only Al-containing species present at 4200 K other than Al itself. If this is not the case, and the suboxides of Al are more stable at increased temperatures, this leads to an uncertain electronic *f* number for the 0—0 band and a consequent error in D(Al—O). Although one cannot draw further definite conclusions, it is interesting to note that Das *et al.*[49] have examined AlO theoretically and report (along with data on CN) a value of 4.24 eV (*i.e.* only 409 kJ mol^{-1}!) for D_e for the ground state of AlO. The next excited states are assigned D_e values of 3.70 eV (354 kJ mol^{-1}) and 3.55 eV (342 kJ mol^{-1}). Clearly, further high-temperature work on Al and its oxides should solve the problem.

[46] (a) F. Juillet, F. Lecomte, H. Mozzanega, S. J. Teichner, A. Thevenet, and P. Vergnon, *Chemical Society Faraday Symposium* No. 7, 'Fogs and Smokes', Swansea, 1973, p. 57; (b) A. P. George, R. D. Murley, and E. R. Place, *ibid.*, p. 63.

[47] (a) M. Farber, R. D. Srivastava, and O. M. Uy, *J.C.S. Faraday I*, 1972, **68**, 249; (b) *J.C.S. Faraday II*, 1972, **68**, 1388.

[48] R. N. Newman and F. M. Page, *Combustion and Flame*, 1971, **17**, 149.

[49] G. Das, T. Janis, and A. C. Wahl, *J. Chem. Phys.*, 1974, **61**, 1274.

4 Free-energy Measurements: Vapour Pressures

Vapour-pressure measurements of one form or another are still the most abundant source of direct equilibrium information at high temperatures. Despite their closer relationship to free energy, vapour pressures are conveniently determined at several temperatures and enthalpy values are thereby also derived, as mentioned above for the aluminium oxide system.[47] A somewhat arbitrary selection of 'second law' enthalpy information derived in this way therefore falls naturally into this section for illustrative purposes. Such second-law enthalpy values are in any case now commonly compared with third-law values to check the overall consistency of the thermodynamic interpretation. This has often been done in the investigations referred to below though it is not specifically mentioned here.

The well-established Knudsen effusion techniques acquired new versatility when combined with mass spectrometry in the 1960's and, more recently, with low-temperature matrix-isolation methods for characterizing the vapour species. These important developments have been well reviewed.[11,50,51] It has always been necessary to characterize the vapour species before the measured vapour pressure could be quantitatively interpreted, and even with purely mechanical methods, such as weight loss and torsion effusion, ingenious combinations of results could in principle give some indication of the molecular weight of the effusing species. However, as a result of the much greater power of the newer techniques both to determine the species present and to estimate their concentrations, not only has a wealth of new vapour-pressure information accumulated, but there has also been a complete revolution in our ideas about the degree of chemical complication of the species which can exist in the vapour phase.

Mass Spectrometry.—Our first example of unexpectedly complicated vapour species also illustrates the problems that can be introduced by having a temporary embarrassment of information from the modern highly sensitive instruments. In a mass-spectrometric investigation of the species evaporating from $V_2O_5-WO_3$ mixtures at 1255—1465 K, evidence was obtained for the existence of more than 60 ions by Gilles and co-workers,[52] who identified many previously unknown ternary species of high molecular weight such as $V_2W_2O_{11}^+$ and $V_3W_2O_{13}^+$, some of which disappeared, however, as effusion progressed and the composition of the sample approached the $VO_2/WO_2 + WO_3$ solid-solution region. The system was unfortunately too complex to derive thermodynamic results. By contrast, the vapour species over solid and liquid $WOCl_4$ is regarded as $WOCl_4$ by Enghag and Staffansson,[53] who determined the vapour pressure by a transpiration method and also reported heats of fusion and sublimation of 10.13 and 72.22 kJ mol^{-1}, respectively, at the melting point (478 K) and of vaporization, 62.09 kJ mol^{-1}, at 501 K.

[50] 'The Characterization of High Temperature Vapours', ed. J. L. Margrave, Wiley, New York, 1967.
[51] 'Vibrational Spectroscopy of Trapped Species', ed. H. E. Hallam, Wiley, London, 1973.
[52] S. L. Bennett, S. S. Lin, and P. W. Gilles, *J. Phys. Chem.*, 1974, **78**, 266.
[53] P. Enghag and L. Staffansson, *Acta Chem. Scand.*, 1972, **26**, 1067.

Other examples of recently identified species are SiON, detected by Muenow[54] (who has also investigated Ge_2N and GeSiN), and HfN, which was identified above 2800 K by Kohl and Stearns,[55a] who report $\Delta H_0^{\ominus} = (60.5 \pm 30)$ kJ mol^{-1} for the reaction:

$$HfN = Hf(g) + \tfrac{1}{2}N_2(g) \tag{5}$$

together with free-energy values at 2885 and 2969 K and deduce $D(Hf-N) = (531 \pm 30)$ kJ mol^{-1}. Continuing their studies of the thermodynamics of aluminium-containing molecules and of the vaporization of metal carbides, the same authors[55b] report dissociation energies of AlC_2, Al_2C_2, and $AlAuC_2$ obtained with the use of a tantalum effusion cell.

Two papers report the presence of carbon species in the vapour when certain substances are heated in graphite Knudsen cells. Thus Hildenbrand[56] found gaseous SCF_2 in addition to SF and SF_2 on investigating the effect at 1500 K on a graphite cell of SF_6, which he was led to examine because of its possible use as an electron scavenger in plasmas and also as a source of fluorine atoms for use in lasers. Standard enthalpies of formation of all three gaseous species are given along with S—F bond dissociation energies. Guido and Gigli[57a,b] have similarly identified the species CeSiC effusing from a graphite-lined cell containing $CeSi_2$. The same authors describe, in the more recent paper cited, the use of a double-oven technique for the study of gaseous GaCN in the range 1398—1783 K and give tables of $-(G_T^{\ominus} - H_0^{\ominus})/T$ and enthalpies.

The sublimation of graphite itself has been further investigated by Zavitsanos and Carlson[58] using a technique involving r.f. heating of the Knudsen cell, which they show to be convenient and compatible with high-temperature time-of-flight mass spectrometry. With 450 kHz heating at 25 kW, temperatures in excess of 2700 K are possible, and all the vapour species C_1, C_2, C_3, and C_4 were detected in the range 2320—3000 K. The partial pressures of the C_2 and C_3 species over graphite varied in this temperature range according to the equations:

$$\log_{10}[P(C_2)/N\ m^{-2}] = 14.977 - 44\,230\ K/T \tag{6}$$

$$\log_{10}[P(C_3)/N\ m^{-2}] = 14.950 - 40\,670\ K/T \tag{7}$$

and the entropy change for vaporization to C_3 is quoted as 190.6 J K^{-1} mol^{-1} at 2740 K. The temperature range has been extended still further to 4000 K and above by Meyer et al.,[25] who have investigated reactions with H_2, O_2, and CH_4 of carbon species (C_n) from the laser-induced evaporation of graphite and tantalum carbide.

Among several recent papers by Gingerich and co-workers[59] there is a study of the reactions between carbon and phosphorus in a graphite-lined Ta cell and

54 D. W. Muenow, J. Phys. Chem., 1973, 77, 970; J. Chem. Phys., 1974, 60, 3382.
55 (a) F. J. Kohl and C. A. Stearns, J. Phys. Chem., 1974, 78, 273; (b) ibid., 1973, 77, 136.
56 H. D. Hildenbrand, J. Phys. Chem., 1973, 77, 897.
57 (a) M. Guido and G. Gigli, J. Chem. Phys., 1973, 59, 3437; (b) ibid., 1974, 60, 721.
58 P. D. Zavitsanos and G. A. Carlson, J. Chem. Phys., 1973, 59, 2966.
59 J. Kordis and K. A. Gingerich, J. Chem. Phys., 1973, 58, 5058; ibid., 1973, 58, 5141; K. A. Gingerich and G. D. Blue, ibid., 1973, 59, 185; D. L. Cocke and K. A. Gingerich, ibid., 1974, 60, 1958: J. Kordis and K. A. Gingerich, J. Phys. Chem., 1973, 77, 700; K. A. Gingerich, D. L. Cocke, and J. Kordis, ibid., 1974, 78, 603.

there are others of equilibria involving gaseous Sb_2, Sb_3, Sb_4, SbP, SbP_3, and P_2; of AlAu; of TiRh, Rh_2, and Ti_2Rh; of Eu_2 and EuAg; and of AsP and BiP.

Free-energy, enthalpy, and bond-energy values are given for many of the compounds and, in the case of the europium compounds, the authors test the Pauling polar compound model (in which respect see also the recent work of Neubert and Zmbov[60] on CuLi, AgLi, and AuLi). They also give methods of estimating the heats of formation of as yet unknown diatomic lanthanides. A different europium compound has been studied by Hariharan and Eick,[61] who find the sublimation vapour pressure of $Eu^{II}Se$ over the range 1808—2131 K to be

$$\log_{10}(P/N\ m^{-2}) = (8.53 \pm 0.096) - (2.26 \pm 0.02) \times 10^4\ K/T \qquad (8)$$

using W and Mo Knudsen cells and time-of-flight mass spectrometry. The dissociation energies of the gaseous oxides of another rare-earth metal, CeO_2 and Ce_2O_2, are reported by Piacente and co-workers,[62] who also measured the vapour pressure of rubidium in the range 402—551 K and the dissociation energy of the Rb_2 molecule. Ackermann and Rauh,[63] who in their later papers report several free-energy and enthalpy results for the oxides Y_2O_3, YO, ZrO, and HfO, have determined the vapour pressures over the pure solid or liquid metals Th, Hf, and Zr, which fit the following equations in the ranges stated:

$$\log_{10}[P(Th,l)/N\ m^{-2}] = (11.032 \pm 0.098) - (29\ 769 \pm 219)\ K/T \qquad (9)$$
$$at\ 2020—2500\ K;$$

$$\log_{10}[P(Hf,l)/N\ m^{-2}] = (11.312 \pm 0.085) - (30\ 446 \pm 240)\ K/T \qquad (10)$$
$$at\ T > 2464\ K;$$

$$\log_{10}[P(Hf,s)/N\ m^{-2}] = (11.862 \pm 0.072) - (31\ 801 \pm 153)\ K/T \qquad (11)$$
$$at\ 1940—2464\ K;$$

$$\log_{10}[P(Zr,l)/N\ m^{-2}] = (11.548 \pm 0.081) - (29\ 944 \pm 240)\ K/T \qquad (12)$$
$$at\ 2134—2550\ K.$$

The standard heats of sublimation at 298 K are rather similar: Th (597.5 ± 4.6), Hf (620.9 ± 5.0), and Zr (560.0 ± 5.0) kJ mol^{-1}. The gaseous oxides ThO and ThO_2 have been investigated recently in the range 1700—1900 K by Hildenbrand and Murad.[64]

Among the Group V elements, elementary arsenic has been studied in the vapour phase over solid $MoAs_2$, Mo_2As_3, GaAs, and InAs by Murray, Pupp, and Pottie.[65] Excellent agreement between second- and third-law estimates is reported and the equilibrium constant for the reaction $As_4(g) \rightleftharpoons 2As_2(g)$ is derived with values varying from 4×10^{-4} atm at 1048 K to 1.1×10^{-7} atm at 807 K. Two simple oxide species of phosphorus, PO and PO_2, have had their atomization energies determined as (593.0 ± 8) kJ mol^{-1} and (1086.2 ± 11)

[60] A. Neubert and K. F. Zmbov, *J.C.S. Faraday I*, 1974, **70**, 2219.
[61] A. V. Hariharan and H. A. Eick, *J. Chem. Thermodynamics*, 1974, **6**, 373.
[62] V. Piacente, G. Bardi, L. Malaspina, and A. Desideri, *J. Chem. Phys.*, 1973, **59**, 31; V. Piacente, G. Bardi, and L. Malaspina, *J. Chem. Thermodynamics*, 1973, **5**, 219.
[63] R. J. Ackermann and E. G. Rauh, *J. Chem. Thermodynamics*, 1972, **4**, 521; *ibid.*, 1973, **5**, 331; *J. Chem. Phys.*, 1974, **60**, 2266.
[64] D. L. Hildenbrand and E. Murad, *J. Chem. Phys.*, 1974, **61**, 1232.
[65] J. J. Murray, C. Pupp, and R. F. Pottie, *J. Chem. Phys.*, 1973, **58**, 2569; *J. Chem. Thermodynamics*, 1974, **6**, 123.

kJ mol^{-1}, respectively, and the equilibrium constants have been determined in the range 1200—2500 K for some of their reactions with Y, Gd, and Sn, all in the gas phase, by Drowart *et al.*[66a]

Uy and Drowart[66b] find dissociation energies, free-energy functions, and equilibrium constants at 1200—1700 K for the reactions of compounds of Al and Cu with the Group VI elements S, Se, and Te. Rosenqvist and Tungesvik[67]

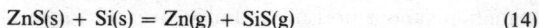

$$2ZnS(s) = 2Zn(g) + S_2(g) \qquad (13)$$

$$ZnS(s) + Si(s) = Zn(g) + SiS(g) \qquad (14)$$

have examined the equilibria (13) and (14), Farber and Srivastava[68] the vaporization of VN and V, and, very recently, Sigai and Wiedemeier[69] the activities of CdSe and MnSe in their solid solutions.

Gas-phase Negative Ions in Mass Spectrometry.—Negative ions attract attention both when they can be found as equilibrium vapour species at the very high temperatures (above 2000 K) now accessible to investigation and when they can be used to determine electron affinities. In such investigations Srivastava, Uy, and Farber[70] employ a dual vacuum-chamber mass spectrometer–furnace assembly and enhance the formation of negative oxide ions by adding preheated KCl from a second boiler to take advantage of the excellent electron-donor properties of the alkali metal. Equilibrium constants are then determined for such reactions as (15) and (16) at 1623—2100 K, whence electron affinities are

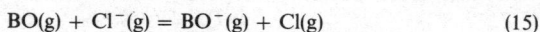

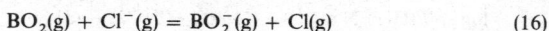

$$BO(g) + Cl^-(g) = BO^-(g) + Cl(g) \qquad (15)$$

$$BO_2(g) + Cl^-(g) = BO_2^-(g) + Cl(g) \qquad (16)$$

derived as (300.8 ± 8.4) kJ mol^{-1} for BO and (344.3 ± 12.55) kJ mol^{-1} for BO$_2$. A similar investigation with the corresponding Al compounds, in the somewhat higher temperature range 2080—2222 K, leads to electron affinities of (354.8 ± 12.55) kJ mol^{-1} for AlO and (396.2 ± 12.55) kJ mol^{-1} for AlO$_2$ (greater than that of Cl) and has already been mentioned in connection with the discrepancy between the mass spectrometric and flame spectroscopic results for the AlO(g) heat of formation and bond energy.[47]

Margrave and his co-workers[71] have recently measured the heat of formation of GeF$_3$(g) and its electron affinity from the appearance potential of the negative ion in the usual way. They emphasize the importance of determining the kinetic energies of the fragments from the electron impact since allowance for this excess energy makes an appreciable difference to the estimated heats of formation of

[66] (*a*) J. Drowart, C. E. Myers, R. Szwarc, A. V. Auwera-Mahieu, and O. M. Uy, *J.C.S. Faraday II*, 1972, **68**, 1749; (*b*) O. M. Uy and J. Drowart, *Trans. Faraday Soc.*, 1971, **67**, 1293.

[67] T. Rosenqvist and K. Tungesvik, *Trans. Faraday Soc.*, 1971, **67**, 2945.

[68] M. Farber and R. D. Srivastava, *J.C.S. Faraday I*, 1973, **69**, 390.

[69] A. G. Sigai and H. Wiedemeier, *J. Chem. Thermodynamics*, 1974, **6**, 983.

[70] R. D. Srivastava, O. M. Uy, and M. Farber, *Trans. Faraday Soc.*, 1971, **67**, 2941.

[71] J. L. Wang, J. L. Margrave, and J. L. Franklin, *J. Chem. Phys.*, 1974, **60**, 2158.

GeF_4, GeF_3, and $GeF_3{}^-$. The present values of the bond-dissociation energies of the Group IV fluorides are summarized in a table.[71]

Negative ions are also being considered as intermediates in the pyrolytic decomposition of inorganic substances. From e.s.r. evidence, Harrison and Ng[72] deduce that $NO_3{}^{2-}$ occurs as an intermediate in the decomposition of strontium and barium nitrates, while Cordes and Smith[73] find evidence for $ClO_3{}^-$ when $LiClO_4$ decomposes to $LiCl$ and O_2 (with some ClO_2 also formed) and compare the behaviour of $LiClO_4$ with that of other alkali-metal perchlorates. Tang and Fenn[74] sublimed NH_4ClO_4 and examined the vapour by time-of-flight mass spectrometry. This substance is of some importance since it is used as the oxidizer in most solid-fuel rockets and contributes about half the total weight of most ballistic missiles. Tang and Fenn do not consider negative ions at this stage, but they do report that quite different results were obtained depending upon the form of the solid. Whereas a single crystal loses NH_3 and $HClO_4$ on heating, a vapour of *lower* molecular weight is obtained from a compressed powder. The authors suggest that sublimation processes may be classified as associative, dissociative, or destructive and regard the behaviour of the compressed solid as indicative of decomposition in the vapour after initial dissociation sublimation. If this dependence of vapour constitution upon the state of aggregation of the solid proves at all common, it points to a variable that will have to be taken into account in projected work and may possibly be a factor in explaining some of the discrepancies which arise from time to time between the results of different groups of investigators.

Matrix Isolation.—While the use of matrix isolation as a technique for the trapping, identification, and estimation of concentration of vapour species is now well established,[50,51,75] there have been few recent studies specifically directed towards this application, most of the activity having been channelled into i.r. and other purely spectroscopic investigations of the species conveniently isolated in that way[76] and of the reactions between such species in the matrix itself. (For recent examples see the work of Spiker and Andrews on reactions of alkali metals with N_2O;[77a] with NO;[77b] of Rb and Cs with O_2;[77c] of NiF_2 and $NiCl_2$ with CO, N_2, and O_2;[77d] and of CO with CaF_2, CrF_2, MnF_2, and ZnF_2;[77d] and the work of Bos, Ogden, and Orgee[77e] on $Ge + O_2$ vapour reactions). From that point of view, vapour-pressure cells are simply sources of supply of the required molecular beams, but the results of such work will have a direct

[72] L. G. Harrison and H. N. Ng, *J.C.S. Faraday I*, 1973, **69**, 1432.

[73] H. F. Cordes and S. R. Smith, *J. Phys. Chem.*, 1974, **78**, 773, 776.

[74] S. P. Tang and J. B. Fenn, *J. Phys. Chem.*, 1973, **77**, 940.

[75] A. J. Barnes and H. E. Hallam, *Quart. Rev.*, 1969, **23**, 392; H. E. Hallam, *Ann. Reports*, 1970, **67**, (*A*), 117.

[76] S. D. Gabelnick, G. T. Reedy, and M. G. Chasanov, *J. Chem. Phys.*, 1974, **60**, 1167.

[77] (*a*) R. C. Spiker, jun., and L. Andrews, *J. Chem. Phys.*, 1973, **58**, 702, 713; (*b*) L. Andrews and D. E. Tevault, *J. Phys. Chem.*, 1973, 77, 1640, 1646: (*c*) R. R. Smardzewski and L. Andrews, *ibid.*, p. 801; L. Andrews, J. T. Hwang, and C. Trindle, *ibid.*, p. 1065; (*d*) D. A. Van Leirsburg and C. W DeKock, *ibid.*, 1974, **78**, 134; (*e*) A. Bos, J. S. Ogden, and L. Orgee, *ibid.*, p. 1763.

bearing on the high-temperature field both by indicating which species to look for in experimental work (and supplying a technique for their identification and measurement) and by providing the detailed molecular information for a reliable statistical mechanical calculation of their thermodynamic functions. For example, in the absence of anharmonicity constant data, a simple harmonic model must be used, and this could introduce significant errors at high temperatures. Margrave and co-workers[11] review the situation and list geometries and frequencies for a wide range of oxides and halides. Herzberg[78] and Gaydon[43] give spectroscopic data on many species present in high-temperature environments. A new evaluation of the molecular constants of 16 bent symmetric SY_2-type molecules is given by Thirugnanasambandam and Mohan.[79] Since many of the species trapped in matrices often contain only 2 or 3 atoms, they are attracting the interest of the theoretical chemists. For example, O'Neil *et al.*[80] report an *ab initio* calculation on the LiO_2 radical giving a bond angle of 44.5° in the ground state. Theoretical treatments of small molecules are discussed by Thomson elsewhere in this volume.

Many references to the spectra of matrix-trapped species could be cited, but only one or two of the recent technical developments will be mentioned here. Schoch and Kay[81] recommend a triode sputtering source for substances which are difficult to vaporize and have used it to prepare samples of Ag, Ta, W, and Mo trapped in Ar and Xe for spectroscopy; and Perutz and Turner[82] compare the properties of matrices prepared by 'slow spray on' (SSO) and pulsed matrix isolation (PMI) methods. Most of the isolated species up to the present have been examined by i.r. spectroscopy. As a recent example of the combination of i.r. with Mössbauer spectroscopy, we cite the study of SnO, Sn_2P_2, Sn, Sn_2, and higher polymers of both in N_2 matrices by Bos *et al.*,[83] who give references to earlier work in the past three or four years. They point out that the Mössbauer technique is better than other forms of spectroscopy for the study of low-temperature diffusion and reactivity because the area of the absorption peaks is comparatively independent of the chemical environment of the resonance-active nuclei.

Photoelectron Spectroscopy.—Conventional photoelectron spectrometers are too insensitive to deal with the low concentrations of vapour species commonly encountered over inorganic substances at high temperatures, but Berkowitz has increased the detection efficiency by increasing the acceptance angle with a cylindrical mirror arrangement and can thereby operate satisfactorily with an effective sample gas pressure in the ionization region of the order of 10^{-2} N m^{-2}. In this way, photoelectron spectra have been mapped out and interesting inferences about bonding have been drawn for the high-temperature vapour species

[78] G. Herzberg, 'The Spectra and Structures of Simple Free Radicals', Cornell Univ. Press, Ithica, New York, 1971.
[79] P. Thirugnanasambandam and S. Mohan, *J. Chem. Phys.*, 1974, **61**, 470.
[80] S. V. O'Neil, H. F. Schaefer tert., and C. F. Bender, *J. Chem. Phys.*, 1973, **59**, 3608.
[81] F. Schoch and E. Kay, *J. Chem. Phys.*, 1973, **59**, 718.
[82] R. N. Perutz and J. J. Turner, *J.C.S. Faraday II*, 1973, **69**, 452.
[83] A. Bos, A. T. Howe, B. W. Dale, and L. W. Becker, *J.C.S. Faraday II*, 1974, **70**, 440; A. Bos and A. T. Howe, *ibid.*, p. 451.

TlCl, TlBr, and TlI[84a] (all of which were chosen because they were previously known to give rise to simple monatomic vapour species); Group III monohalides InCl, InBr, and InI;[84b] monomer and dimer photoelectron spectra of TlF;[84c] the halides of Cs;[84d] the trihalides of In and Ga;[84e] and the halides (except F) of Zn, Cd, and Hg.[84f]

Static, Dynamic, and Simple Effusion Methods.—Most of the classical vapour-pressure methods have been adapted at some time or other for high-temperature work. Elliott *et al.*[85] have described an isopiestic balance for measurements between 939 and 1037 K and used it to determine cadmium vapour pressures over Cd–Au alloys. More recently, transpiration methods have been used by De Maria and Piacente[86] in determining the vapour pressure of Ca from 1126 to 1300 K, for which equation (17) is valid, and Sr from 1086 to 1310 K, for which

$$\log_{10}(P/N\ m^{-2}) = (9.94 \pm 0.13) - (8550 \pm 158)\,K/T, \tag{17}$$

$$\log_{10}(P/N\ m^{-2}) = (9.76 \pm 0.14) - (7720 \pm 161)\,K/T, \tag{18}$$

equation (18) applies, and by Battat *et al.*[87] on systems involving Ga, As, HCl, and H_2O after first testing their novel apparatus with water at the lower temperatures and lead at 1200—1400 K. Static gauges have been used by Greenberg *et al.*[88] to study the sublimation of $Zn_3P_2(s)$ at 890—1130 K, and by DeLong and

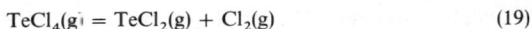

$$TeCl_4(g) = TeCl_2(g) + Cl_2(g) \tag{19}$$

Rosenberger[89] for reaction (19) at 670—1170 K. Topor[90] reports the vapour pressures and enthalpies of vaporization of the liquid chlorides, bromides, and iodides of Na, K, Rb, and Cs, measured by the quasistatic Rodebush–Dixon method, and derives equilibrium constants for the dimerization reactions at 1300 K and compares them with theoretical calculations.

Another study of several alkali-metal halides, by Ewing and Stern,[91] contains a valuable theoretical and experimental investigation of the change-over from molecular to hydrodynamic flow from a Knudsen cell as the pressure becomes higher. Observed rates of vapour loss are shown to be greater in the hydrodynamic and transitional regions than those predicted by the Hertz–Knudsen theory for molecular flow; but vapour pressures are nevertheless still found to be

[84] (*a*) J. Berkowitz, *J. Chem. Phys.*, 1972, **56**, 2766; (*b*) J. Berkowitz and J. L. Dehmer, *ibid.*, 1972, **57**, 3194; (*c*) J. J. Dehmer, J. Berkowitz, and L. C. Cusachs, *ibid.*, 1973, **58**, 5681; (*d*) J. Berkowitz, J. L. Dehmer, and T. E. H. Walker, *ibid.*, 1973, **59**, 3645; (*e*) J. L. Dehmer, J. Berkowitz, L. C. Cusachs, and H. S. Aldrich, *ibid.*, 1974, **61**, 594; (*f*) J. Berkowitz, *ibid.*, p. 407.

[85] G. R. B. Elliott, C. C. Herrick, J. F. Lemons, and P. C. Nordine, *High Temp. Sci.*, 1969, **1**, 58.

[86] G. De Maria and V. Piacente, *J. Chem. Thermodynamics*, 1974, **6**, 1.

[87] D. Battat, M. M. Faktor, I. Garrett, and R. H. Moss, *J.C.S. Faraday I*, 1974, **70**, 2267, 2280, 2293, 2302.

[88] J. H. Greenberg, V. B. Lazarev, S. E. Kozlov, and V. J. Shevchenko, *J. Chem. Thermodynamics*, 1974, **6**, 1005.

[89] M. C. DeLong and F. Rosenberger, *J. Chem. Thermodynamics*, 1974, **6**, 877.

[90] L. Topor, *J. Chem. Thermodynamics*, 1972, **4**, 739.

[91] C. T. Ewing and K. H. Stern, *J. Phys. Chem.*, 1974, **78**, 1998.

calculable from flow rates in these higher pressure regions, and the results agree well with those derived from standard thermodynamic procedures. The ratio of actual flow/(hypothetical) Knudsen flow, in the transition region, depends only on the mean free path and is independent of the diameter of the orifice. A detailed comparison is given of the new data with those recorded in the JANAF tables. The gap of several hundred degrees in the experimental data on which the latter were based was one reason for undertaking this investigation.

If effusion measurements alone are being relied upon for vapour-pressure values, it is important to know how dissociation in the vapour affects the results. Knox and Wyatt[92] reformulate this problem, taking a proper account of the steady-state condition within the Knudsen cell. Haschke,[93] however, finds that equilibrium pressures of $EuBr_3$ are not attained in Knudsen cells at 502—623 K and resorts instead to spectrophotometry to determine the vapour concentration.

Several systems have been examined by the classical Knudsen technique. Nagai *et al.*[94] find that equation (20) applies for the reaction (21) at 1823—1983 K,

$$\Delta G_T^\ominus / kJ\, mol^{-1} = (761.70 \pm 10.46) - 0.2439\ T/K \tag{20}$$

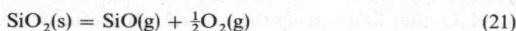

$$SiO_2(s) = SiO(g) + \tfrac{1}{2}O_2(g) \tag{21}$$

making allowance for the dissociation of O_2, and obtain the enthalpy of formation of SiO(g) at 298 K as $(-116.7 \pm 14.6)\,kJ\,mol^{-1}$ and a vaporization coefficient of about 0.02 for $SiO_2(s)$ in the experimental temperature range. The related

$$Si(s) + SiO_2(s) = 2SiO(g) \tag{22}$$

equilibrium (22) has, according to Kubaschewski and Chart,[95] an equilibrium SiO vapour pressure given by equation (23).

$$\log_{10}(P/N\, m^{-2}) = 13.613 - 1.785 \times 10^4\ K/T \tag{23}$$

For the oxides As_2O_3 (arsenolite), Sb_2O_3 (valentinite), and SeO_2, Behrens and co-workers[96] find the vapour pressures to satisfy equation (24) at 367—429 K,

$$\log_{10}[P(As_4O_6)/N\, m^{-2}] = (14.91 \pm 0.32) - (6067 \pm 125)\ K/T \tag{24}$$

equation (25) at 627—732 K (in these cases the vapour species being the doubled

$$\log_{10}[P(Sb_4O_6)/N\, m^{-2}] - (14.39_5 \pm 0.30) - (10\,066 \pm 203)\ K/T \tag{25}$$

formula as shown), and equation (26) at 374—427 K. In all cases, useful ranges

$$\log_{10}[P(SeO_2,s)/N\, m^{-2}] = (14.547 \pm 0.450) - (5785 \pm 180)K/T \tag{26}$$

of thermodynamic data are given. Biefeld and Eick[97] also used the Knudsen method to investigate the sublimation of ZnF_2 from 901 to 1125 K, but collected

[92] J. H. Knox and P. A. H. Wyatt, *J.C.S. Faraday I*, 1973, **69**, 1961.
[93] J. M. Haschke, *J. Chem. Thermodynamics*, 1973, **5**, 283.
[94] S. Nagai, K. Niwa, M. Shimmei, and T. Yokokawa, *J.C.S. Faraday I*, 1973, **69**, 1628.
[95] O. Kubaschewski and T. G. Chart, *J. Chem. Thermodynamics*, 1974, **6**, 467.
[96] R. G. B. Behrens and G. M. Rosenblatt, *J. Chem. Thermodynamics*, 1972, **4**, 175; ibid., 1973, **5**, 173; R. G. Behrens, R. S. Lemons, and G. M. Rosenblatt, *ibid.*, 1974, **6**, 457.
[97] R. M. Biefeld and H. A. Eick, *J. Chem. Thermodynamics*, 1973, **5**, 353.

the sublimate on a target. They deduced equation (27) along with other derived thermodynamic information.

$$\log_{10}\left[P(\text{ZnF}_2)/\text{N m}^{-2}\right] = (13.443 \pm 0.071) - (13\,185 \pm 72)\,\text{K}/T \qquad (27)$$

Metal Sulphates.—Richardson[1] reviews the thermodynamic aspects of many of the pure and mixed solid and liquid inorganic salts of industrial importance. Sulphates are selected here for special mention, partly because they are not treated extensively in Richardson's book, but mainly because their investigation illustrates some of the problems encountered in interpreting vapour-pressure measurements when several possibilities seem open for the choice of vapour species.

Intuitively, most chemists would probably suppose that a solid or liquid sulphate on heating simply dissociates to the oxide plus SO_3, which itself dissociates further into SO_2 and O_2 to an extent that is dependent upon the temperature. The extent of dissociation of SO_3 might also depend upon the time of residence of the vapour in the cell and the accessibility of catalytic surfaces in effusion or transpiration techniques, and detailed kinetic information is obviously necessary to settle that point. To complete the story, the fate of the metal oxide has also to be considered. At moderate temperatures at least, the oxides of most transition and alkaline-earth metals are expected to form a new solid phase, while those of the alkali metals might be expected to vaporize and then perhaps to dissociate further into the free metals and oxygen.

Not all recent investigators would find this description compatible with their experimental findings, though it does seem to accord well with work on $Fe_2(SO_4)_3$, $CuSO_4$, $Al_2(SO_4)_3$, $CaSO_4$, and $MgSO_4$. Halstead and Laxton[98] have used both static and dynamic methods to obtain vapour pressures in the $Fe_2(SO_4)_3$–Fe_2O_3–SO_3 system and have examined the solid residues from their cells by X-ray diffraction and X-ray fluorescence spectroscopy, always finding only $Fe_2(SO_4)_3$ and α-Fe_2O_3. Traces of γ-Fe_2O_3 (probably present as a thin film) were nevertheless detectable in samples in which only a small degree ($<2\%$) of sulphate decomposition had occurred. Further, pressures given by such relatively oxide-free ($<2\%$) $Fe_2(SO_4)_3$ were similar to those obtained using γ-Fe_2O_3 as a starting material, while pressures given by more decomposed ($>4\%$) sulphate samples were similar to those given by partially sulphated α-Fe_2O_3. On this basis, Halstead and Laxton explain discrepancies in the literature and summarize their vapour-pressure values in the equations (28) for

$$\log_{10}\left[P(\text{SO}_3)/\text{N m}^{-2}\right] = (14.18 \pm 1.39) - (0.95 \pm 0.11) \times 10^4\,\text{K}/T \qquad (28)$$

the α-Fe_2O_3–$Fe_2(SO_4)_3$ system and (29) for the γ-Fe_2O_3–$Fe_2(SO_4)_3$ system.

$$\log_{10}\left[P(\text{SO}_3)/\text{N m}^{-2}\right] = (13.17 \pm 1.10) - (0.91 \pm 0.09) \times 10^4\,\text{K}/T \qquad (29)$$

Their dynamic technique resembles that of Dewing and Richardson:[99] pre-heated and equilibrated mixtures of SO_3, SO_2, and O_2, diluted with N_2, were

[98] W. D. Halstead and J. W. Laxton, *J.C.S. Faraday I*, 1974, **70**, 807.
[99] E. W. Dewing and F. D. Richardson, *Trans. Faraday Soc.*, 1959, **55**, 611.

passed over samples in the cell and the temperature at which no weight change occurred was recorded. In their study of $CaSO_4$ and $MgSO_4$, Dewing and Richardson detected the onset of decomposition by a sharp discontinuity in differential thermocouple readings during heating. Supporting the sample in a spiral of Pt–13%Rh thermocouple wire presumably helped to ensure that SO_3 and SO_2 were at equilibrium. Earlier work on the alkaline-earth sulphates illustrates the care with which static and dynamic determinations have to be compared if inconsistencies are to be avoided. In particular, Dewing and Richardson mention objections to the interpretation of Knopf and Staude,[100] whose results were nevertheless taken into account in the $MgSO_4$ entry in the JANAF 1966 supplement (PB 168 370—1).

According to Collins *et al.*,[101] both $CuSO_4$ and $Al_2(SO_4)_3$ decompose initially to give the metal oxides and SO_3, though they obtained a different pattern of results for alunite [presumably $KAl_3(SO_4)_2(OH)_6$] and suggest that the sulphate ion dissociates by at least two different mechanisms. On the other hand, Papazian *et al.*[102a] believe that the primary gaseous products in the decomposition of $Al_2(SO_4)_3$ and $Hf(SO_4)_2$ are SO and O_2, and that SO_2 and SO_3 are subsequently formed from these. Their interpretation has since been criticized by Johnson and Gallagher and defended by the authors themselves.[102b] This unexpected view stems mainly from the fact that $SO_3{}^+$ peaks have been hard to detect in the mass spectrometer while SO^+ and $SO_2{}^+$ show up quite clearly. The question then arises as to whether or not the observed SO_2/SO ratios are compatible with breakdown of SO_3 and SO_2 in the mass spectrometer.

The alkali-metal sulphates are known to vaporize congruently, *i.e.* the vapour has the composition M_2SO_4, but a new feature that now emerges is that K_2SO_4, Rb_2SO_4, and Cs_2SO_4 appear to be present largely as the sulphate molecules in the vapour phase.[103] The results for Li_2SO_4 and Na_2SO_4 indicate more extensive dissociation, however. Cubicciotti and Feneshea[103a] found that their Na_2SO_4 vapour-pressure results, determined by a transpiration technique using N_2 as carrier gas, were greater than those calculated from the literature for the decomposition equilibrium (30) and therefore made further measurements in the

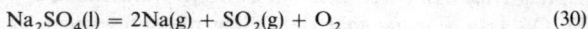

$$Na_2SO_4(l) = 2Na(g) + SO_2(g) + O_2 \qquad (30)$$

presence of SO_2 and O_2 to suppress the dissociation and allow the measurement of the pressure of $Na_2SO_4(g)$ alone. In this way they obtained the equation (31)

$$\log_{10}(P/\text{N m}^{-2}) = (10.964 \pm 0.15) - (15\,540 \pm 380)\,\text{K}/T \qquad (31)$$

for the vapour species Na_2SO_4 over the liquid at 1400—1625 K. Extrapolation of their overall pressures down to lower temperatures produces values an order of magnitude lower than those recorded by an effusion technique by Powell and

[100] H. K. Knopf and H. Staude, *Z. phys. Chem.* (*Leipzig*), 1955, **204**, 265.
[101] L. W. Collins, E. K. Gibson, and W. W. Wendlandt, *Thermochim. Acta*, 1974, **9**, 15.
[102] (a) H. A. Papazian, P. J. Pizzolato, and R. R. Orrell, *Thermochim. Acta*, 1972, **4**, 97; (b) J. W. Johnson, jun., and P. K. Gallagher, *ibid.*, p. 105; H. A. Papazian, P. J. Pizzolato, and R. R. Orrell, *ibid.*, p. 109.
[103] (a) D. Cubicciotti and F. J. Feneshea, *High Temp. Sci.*, 1972, **4**, 32; (b) D. G. Powell and P. A. H. Wyatt, *J. Chem. Soc.* (*A*), 1971, 3614.

Wyatt,[103b] whose results were, however, complicated by the appearance of changes of slope in the $\ln P$ $vs.$ $1/T$ plots for Na_2SO_4 and Li_2SO_4 which have not yet been satisfactorily explained. The presence or absence of the oxides Na_2O and Li_2O in the vapour also raises once again the question of the cracking patterns to be expected in the mass spectrometer.[104]

Sulphates are being studied from a different angle by Rosén and Wittung,[105] who equilibrate the solids with known mixtures of gaseous S_2 and O_2 to deter-

$$5PbO(s) + \tfrac{1}{2}S_2(g) + \tfrac{3}{2}O_2(g) = 4PbO,PbSO_4(s) \tag{32}$$

$$\log_{10}(K/atm^{-2}) = 43\,260\,K/T - 18.51 \tag{33}$$

mine thermodynamic results for the reaction (32), for which they find equation (33) is valid at 973—1073 K.

5 Free-energy Measurements: Electromotive Force

Mass-spectrometric and matrix-isolation techniques, effective as they are at providing detailed information about vapour species, are nevertheless expensive. It is therefore encouraging that similar information can sometimes be arrived at by ingenious variants on conventional physicochemical devices. Ratchford and Rickert have developed an electrochemical Knudsen cell[106] in which the rate of effusion of a species is governed in the steady state by a small electrolysing current and the slope of a graph of the logarithm of the latter against the e.m.f. depends directly upon the number of atoms (x) in such molecular species as S_x and Se_x.

Figure 1 shows the form of cell used for the study of sulphur vapour species between 500 and 800 K, at which temperatures the decomposition of Ag_2S is negligible. Sulphur molecules then only effuse as a result of the passage of an electric current, which is the only measure of the effusion rate here; $i.e.$ no weight-loss measurements are made. An applied positive potential at the Pt electrode next to the Ag_2S pellet releases sulphur at this electrode, and thence into the vapour, and forces silver ions into the solid electrolyte AgI, from which silver is then deposited onto the negative Pt electrode. When the current has settled down to its steady-state value, the Knudsen cell maintains a definite sulphur vapour pressure over the Ag_2S, the fixed activity of which ensures that the prevailing e.m.f., E, of the solid-state cell

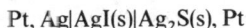

$$Pt, Ag|AgI(s)|Ag_2S(s), Pt$$

measuring the silver activity relative to pure silver, also measures the square root of the S activity, or the $1/2x$ power of the vapour pressure of the species S_x. It then follows that the vapour pressure of S_x is related to its value, $p^0(S_x)$, over pure liquid sulphur by the expression (34), E^0

$$p(S_x) = p^0(S_x)\exp[2x(E - E^0)F/RT]. \tag{34}$$

[104] P. J. Ficalora, O. M. Uy, D. W. Muenow, and J. L. Margrave, *J. Amer. Ceram. Soc.*, 1968, **51**, 574; T. Kosugi, *Kogyo Kagaku Zasshi*, 1970, **73**, 1087.
[105] E. Rosén and L. Wittung, *Acta Chem. Scand.*, 1972, **26**, 2427.
[106] H. Rickert in 'Condensation and Evaporation of Solids', ed. E. Rutner, P. Goldfinger, and J. P. Hirth; Gordon and Breach, London, 1964.

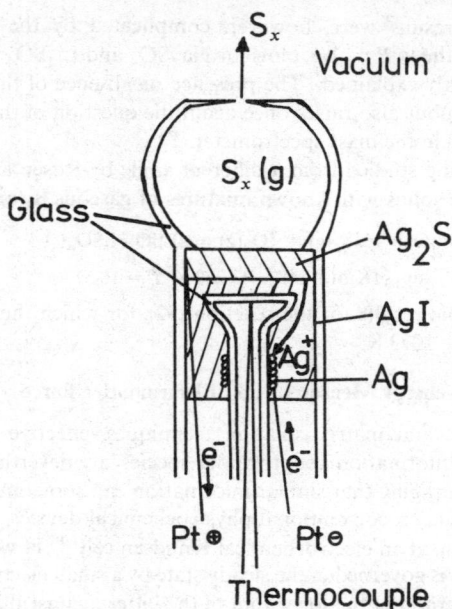

Figure 1   *The electrochemical Knudsen cell*
(Reproduced by permission from 'Condensation and Evaporation of Solids', ed. E. Rutner,
P. Goldfinger, and J. P. Hirth, Gordon and Breach, London, 1964, p. 209)

being the e.m.f. of the cell when Ag_2S is in equilibrium with liquid sulphur. It is
the appearance of x in the exponent that permits its determination, since $p(S_x)$ is
directly proportional to the effusion rate, which is in turn governed here by the
electric current (with x only appearing as a multiplier). Hence the logarithm of
the current is related to E through x, provided that one of the possible S_x species
predominates in the vapour over a certain temperature and pressure range.
Rickert finds that this proves to be the case for both S and Se: the diatomic forms
show up clearly at the lower e.m.f. values for 600—800 K and there are good
indications of higher forms (particularly Se_6) at lower temperatures. The derived
sulphur vapour pressures are in quantitative agreement with the values obtained
by other methods. (For recent information on sulphur and selenium see H. Rau
et al.[107])

This subtle technique obviously has some rather stringent requirements in the
way of solid electrolytes and workable cells, but could perhaps inspire further
exploration along similar lines. Since later developments[108] incorporate mass

[107] H. Rau, T. R. N. Kutty, and J. R. F. Guedes de Carvalho, *J. Chem. Thermodynamics*,
    1973, **5**, 291, 833; H Rau, *ibid.*, 1974, **6**, 525.
[108] D. Detry, J. Drowart, P. Goldfinger, H. Keller, and H. Rickert, *Z. phys. Chem.
    (Frankfurt)*, 1967, **55**, 314; *ibid.*, 1971, **75**, 273; H. Rickert and K. H. Tostmann,
    Werkstoffe und Korrosion, 1970, 965.

spectrometry, it is probably too optimistic to regard the electrochemical Knudsen cell as a serious competitor, though it clearly gives very useful supplementary information in favourable cases.

Galvanic cells continue to be used in more conventional ways at high temperatures, particularly for the study of alloys,[2] in which application molten halides have been popular electrolytes though they have other uses.[109] Thus Nguyen-Duy and Rigaud[110] used the LiCl–KCl eutectic in cells such as

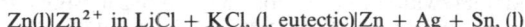

$$Zn(l)|Zn^{2+} \text{ in } LiCl + KCl, (l, \text{ eutectic})|Zn + Ag + Sn, (l)$$

to examine the effects of small additions of Ag (or Cu or Au) on the thermodynamic properties of dilute solutions of Zn in molten Sn in the temperature range 723—923 K. Similar work by Neethling and co-workers on the Na–Cd–In and Na–Pb–In systems[111] employed the slightly more complicated cell

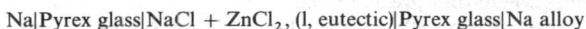

$$Na|Pyrex \text{ glass}|NaCl + ZnCl_2, (l, \text{ eutectic})|Pyrex \text{ glass}|Na \text{ alloy}$$

and required a supplementary determination of the ternary phase diagram, as may frequently become necessary for uncommon ternary systems. Aronson and Lemont[112] have dispensed with the eutectic phase altogether, leaving only the glass as electrolyte:

$$K(l)|K\text{-glass}|K + Tl, (l)$$

The K–Tl alloy composition was varied over the complete range, and values of partial molar enthalpy and entropy, and also of the integral quantities, are given. Curious maxima and minima (like those in the Na system) are believed to indicate a strong chemical interaction between the two metals.

Solid electrolytes[1,2] continue to find applications. Zirconia stabilized with calcium oxide is used in the cell

$$Pt|Co, xCoO + (1 - x)MgO|ZrO_2 + CaO|CoO, Co|Pt$$

with which Rigaud *et al.*[113] have determined the activity of CoO in CoO–MgO mixtures. The derived CoO and MgO activity coefficients show significant positive deviations from ideality. Rezukhina and Kravchenko[114a] used thoria in a similar way for a study of Ta–Co mixtures, and Klinedinst and Stevenson[115] determined the free energy of formation of β-Ga_2O_3 with a cell involving thoria doped with yttria as the solid electrolyte and a flowing $CO + CO_2$ gas mixture as one electrode.

[109] G. Landresse and G. Duyckaerts, *Inorg. Nuclear Chem. Letters*, 1974, **10**, 675.
[110] P. Nguyen-Duy and M. Rigaud, *J. Chem. Thermodynamics*, 1974, **6**, 727, 999.
[111] A. J. Neethling, *J. Chem. Thermodynamics*, 1974, **6**, 707, 1083; H. E. Bartlett, A. J. Neethling, and P. Crowther, *ibid.*, 1970. **2**, 523.
[112] S. Aronson and S. Lemont, *J. Chem. Thermodynamics*, 1973, **5**, 155.
[113] M. Rigaud, G. Giovannetti, and M. Hone, *J. Chem. Thermodynamics*, 1974, **6**, 993.
[114] (a) T. N. Rezukhina and L. I. Kravchenko, *J. Chem. Thermodynamics*, 1972, **4**, 655; (b) T. N. Rezukhina, T. F. Sisoeva, L. I. Holokhonova, and E. G. Ippolitov, *ibid.*, 1974, **6**, 883.
[115] K. A. Klinedinst and D. A. Stevenson, *J. Chem. Thermodynamics*, 1972, **4**, 565.

As a recent example of the use of CaF_2 as a solid electrolyte we cite the paper by Rezukhina et al.[114b] on nine cells of the type

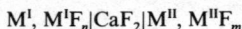

$$M^I, M^IF_n|CaF_2|M^{II}, M^{II}F_m$$

where the metal M^I is baser than M^{II} with respect to fluorine, and M^IF_n and $M^{II}F_m$ are the fluorides in equilibrium with M^I or with M^{II}, respectively. The cell with $M^IF_n = MgF_2$ and $M^{II}F_m = AlF_3$ was used to check the experimental technique, and that with $M^IF_n = CaF_2$ and $M^{II}F_m = MgF_2$ to confirm the absence of electronic conductivity in CaF_2 at very small fluorine activities. The investigation covers the temperature range 710–1120 K and provides values of $\Delta H_f^{\ominus}$ (298.15 K)/kJ mol^{-1} of -1732 (LaF$_3$), -1739 (YF$_3$), -1712 (PrF$_3$), -1649 (ScF$_3$), and -859.4 (MnF$_3$), and of $S^{\ominus}$ (298.15 K)/J K^{-1} mol^{-1} of 99.2, 79.9, 117.2, 92.0, and 87.4, respectively. These results are compared with calorimetric values where they are available.

Finally, some of the cells used by Levitski and Scolis[116] employ CaF_2 and others thoria doped with La_2O_3 or CaO as solid electrolytes for the investigation of the strontium and aluminium tungstates Sr_3WO_6, Sr_2WO_5, $SrWO_4$, $Sr_3Al_2O_6$, and $SrAl_2O_4$ in the range 1100—1400 K. They also give references to the recent Russian literature.

The recent publication of Richardson[1] deals with ionic melt mixtures, reference to which is conveniently included here although techniques other than e.m.f. are also employed. Examples of recent high-temperature thermodynamic mixture studies are the Ce^{III}Cl–alkali-metal chloride and La^{III}Cl–alkali-metal chloride systems by Papatheodorou and co-workers,[117a] bivalent basic oxide–SnO$_2$ systems by Lahiri,[117b] and CaF_2 with alkali fluorides by Kleppa and Hong.[117c] Richardson[1] has also discussed the interaction of gases with ionic melts. Further studies published or in progress include investigations of the solubilities of He, Ar, N_2, O_2, CH_4, H_2, CO, CO_2, and NH_3 in alkali-metal nitrate mixtures by Desimoni, Paniccia and Zambonin[118a] and the high-temperature thermodynamics of solid solutions of H_2 in bcc V, Nb, and Ta by Kleppa et al.[118b] Such reports commonly include useful thermodynamic mixing and solution data.

[116] V. A. Levitski and Y. Y. Scolis, J. Chem. Thermodynamics, 1974, 6, 1181.
[117] (a) G. N. Papatheodorou and O. J. Kleppa, J. Phys. Chem., 1974, 78, 178; G. N. Papatheodorou and T. Østvold, ibid., p. 181; (b) A. K. Lahiri, Trans. Faraday Soc., 1971, 67, 2952; (c) O. J. Kleppa and K. C. Hong, J. Phys. Chem., 1974, 78, 1478.
[118] (a) F. Paniccia and P. G. Zambonin, J.C.S. Faraday I, 1972, 68, 2083; E. Desimoni, F. Paniccia, and P. G. Zambonin, ibid., 1973, 69, 2014; F. Paniccia and P. G. Zambonin, ibid., p. 2019; (b) O. J. Kleppa, P. Dantzer, and M. E. Melnichak, J. Chem. Phys., 1974, 61, 4048.

8 High-pressure Chemistry

By B. CLEAVER

Department of Chemistry, The University, Southampton, SO9 5NH

1 Previous Reviews and General Publications

A new journal entitled *High Temperatures—High Pressures* began in 1969; it contains papers and review articles reporting work either at high temperatures or at high pressures (not necessarily both simultaneously) and also brief reports on conferences and announcements about future meetings. The series *Advances in High-Pressure Research* continues under new Editorship; Volume 4 appeared in June 1974 (five years after Volume 3) and contains articles on the response of solids to shock waves, on X-ray diffraction studies at pressures up to 300 kbar, and on diamond formation at high pressures. A bibliography of high-pressure research[1] has been prepared covering the literature back to 1900; the current literature is surveyed in bimonthly bulletins. Pressure-induced electronic transitions are discussed in a book by Drickamer.[2]

References to review articles on particular aspects of high-pressure chemistry will be given in the appropriate sections below.

2 Technical Advances

High-pressure research has always been experimentally difficult. More rapid progress will result as technical innovations are introduced and are eventually incorporated into commercially available equipment. Some recently described advances are reviewed in this section.

Several improvements to the technique of high-pressure measurement have been described, and the current state of the art has been reviewed.[3]

In diamond anvil cells the frequency shift of the R_1 fluorescence of ruby may be used to indicate the pressure; small pieces of ruby are mixed with the sample.[4] In X-ray diffraction work, a standard substance is mixed with the sample and

[1] 'High Pressure Bibliography' (Published by the High Pressure Data Center, Brigham Young University, Provo, Utah 84602, U.S.A.); 1900–1968 (2 vols.), 1968–1971 (1 vol.), and annual volumes thereafter.

[2] H. G. Drickamer and C. W. Franck, 'Electronic Transitions and the High-Pressure Chemistry and Physics of Solids' (in the series 'Studies in Chemical Physics', ed. A. D. Buckingham), Chapman and Hall, London, 1973; see also H. G. Drickamer, *Chem. in Britain*, 1973, **9**, 353.

[3] C. Y. Liu, K. Ishizaki, J. Paauwe, and I. L. Spain, *High Temps.–High Press.*, 1973, **5**, 359.

[4] S. Block and G. J. Piermarini. *High Temps.–High Press.*, 1973, **5**, 567.

the pressure is deduced from the change in the lattice parameter of the standard. NaCl is the preferred standard,[5] but NaF or LiF may be used if the NaCl diffraction lines overlap with those of the sample or if the sample is appreciably harder than NaCl. If the sample and marker differ in hardness, a pressure intensification effect appears to occur[6] (the pressure being higher in the harder material). Later work[7] showed that this is in part due to non-hydrostatic components in the stress field, notably in opposed anvil apparatus, coupled with the fact that diffraction patterns are then produced only by lattice planes which are parallel with the anvil axes.

Novel high-pressure equipment has been described which permits work at very low temperatures[8] or in high magnetic fields.[9] N.m.r. experiments have been performed using samples in glass tubes at pressures up to 2 kbar; the tubes had been carefully pre-treated with HF to improve their bursting strength and reliability.[10] Vessels transparent to neutrons have been constructed.[11] A pulsed ruby laser has been used to heat to 3000 °C a sample held in a diamond 'opposed anvil' apparatus at 260 kbar.[12] A very efficient form of thermal insulation made from thin metal foil has been described; it can be used to insulate the furnace in internally heated, gas-filled pressure vessels.[13]

The performance of an apparatus employing unusually large Bridgman anvils has been investigated.[14] Using anvils of 78 mm diameter and pyrophyllite gaskets, samples of 5 mm thickness × 10 mm diameter were compressed to over 100 kbar. A large hexahedral press has been built[15] and operated to 100 kbar. The homogeneity of pressure in the sample volume (an isosceles hexahedron) was similar to that normally achieved in tetrahedral or cubic presses (and better than the Bridgman anvil), with the advantage that access to the sample is unobstructed in a mirror plane between the two sets of three pistons. Diffraction experiments can be carried out in this plane, using the Debye–Scherrer method, and the Bragg–Brentano focusing arrangement can be used. The three meridian places at 120° to each other can also be used for neutron diffraction.

The highest static pressures achieved to date were reported by Kawai.[16] The apparatus is described as a split sphere with double-staged pistons. Unlike most multi-piston apparatus, it is simple in design and relatively cheap to construct.

[5] B. Olinger and J. C. Jamieson, *High Temps.–High Press.*, 1970, **2**, 513; D. L. Decker, *J. Appl. Phys.*, 1971, **42**, 3239.
[6] Y. Sato, S. Akimoto, and K. Inone, *High Temps.–High Press.*, 1973, **5**, 289.
[7] Y. Sato, Y. Ida, and S. Akimoto, *High Temps.–High Press.*, 1973, **5**, 679.
[8] J. Wittig, *High Temps.–High Press.*, 1972, **4**, 116; J. S. Schilling, U. F. Klein, and W. B. Holzapfel, *Rev. Sci. Instr.*, 1974, **45**, 1353.
[9] W. B. Holzapfel and D. Severin, *High Temps.–High Press.*, 1969, **1**, 713; G. D. Pitt and D. A. Gunn, *ibid.*, 1970, **2**, 547: 1972, **4**, 353.
[10] H. Yamada, *Rev. Sci. Instr.*, 1974, **45**, 640.
[11] O. Blaschko and G. Ernst, *Rev. Sci. Instr.*, 1974, **45**, 526; D. B. McWhan, D. Bloch, and G. Parisot, *ibid.*, p. 643.
[12] Li-Chung Ming and W. A. Bassett, *Rev. Sci. Instr.*, 1974, **45**, 1115.
[13] P. Malbrunot, P. Meunier, and D. Vidal, *High Temps.–High Press.*, 1969, **1**, 93.
[14] B. Okai and J. Yoshimoto, *High Temps.–High Press.*, 1973, **5**, 675.
[15] M. Contré, *High Temps.–High Press.*, 1969, **1**, 339.
[16] N. Kawai and S. Eudo, *Rev. Sci. Instr.*, 1970, **41**, 1178.

Eight similar steel pistons, each with a separate tip made from carboloy, pack together to form a sphere. The inner surfaces of the tips form a regular octahedron. The assembled pistons are surrounded by a spherical shell made from thick rubber, and the entire device is suspended in oil inside a cylindrical pressure vessel of ~ 30 cm diameter. This vessel is closed by pistons at each end, and these can be driven in by a ram to raise the oil pressure to 3 kbar. Small corner pieces at the outer meeting-points of the pistons ensure that the pistons move synchronously as the oil pressure is raised. Pyrophyllite is used as a gasket material, and conventional tubular graphite heaters can be used to raise the sample temperature to 1500 °C if desired. Pressures in the range 300—500 kbar were originally claimed, but in subsequent experiments[17] vitreous silica was compressed irreversibly to densities comparable with those reached by Al'tshuler in shock-wave experiments. Comparison with the equation of state indicated that pressures in excess of 2 Mbar had been achieved.

In a series of papers,[18] Kumazawa has explored the general design principles of multi-anvil sliding systems (MASS). These employ the principle of massive support, yet offer a relatively large volume compression; as the pressure is raised, the sample volume is reduced by a progressive sliding movement of each anvil against its neighbours. In some versions a controlled *outward* movement of some components is permitted by placing pads of compressible material behind them. Although few of these ideas have yet been put into practice, some interesting possibilities appear to exist.

Spectroscopic Techniques.—The design of pressure cells with windows for spectroscopic studies has been reviewed,[19] and the authors list the maximum pressures attainable with common window materials (sapphire, diamond, germanium, silicon, Irtran). A short-path-length cell suitable for u.v. and visible spectroscopy on corrosive liquids at high temperatures (500 °C, 1 kbar) was described by Lüdemann.[20] In a Raman cell designed for use to 12 kbar,[21] light scattered at 90° from the incident laser beam is collected by a prism and is reflected in a direction parallel to the original beam but offset from it laterally. This ingenious arrangement allows work down to $\Delta v = 20 \text{ cm}^{-1}$ while avoiding the use of cross-bores in the vessel (which would greatly reduce its strength).

3 High-pressure Chromatography

High-speed Liquid Chromatography.—In recent years the technique of liquid–solid chromatography has undergone improvements and is now capable of making separations in times as short as those used in gas chromatography. This has been done by using higher mobile-phase velocities and higher column

[17] N. Kawai, S. Mochizuki, and H. Fujita, *Phys. Letters (A)*, 1971, **34**, 107.
[18] M. Kumazawa, *High Temps.–High Press.*, 1971, 3, 243; M. Kumazawa, K. Masaki, H. Sawamoto, and M. Kato, *ibid.*, 1972, **4**, 293; M. Kumazawa, *ibid.*, 1973, **5**, 599.
[19] J. M. Besson, J. P. Pinceaux, and R. Piotrzowski, *High Temps.–High Press.*, 1974, **6**, 101.
[20] H.-D. Lüdemann and W. A. J. Mahon, *High Temps.–High Press.*, 1969, **1**, 215.
[21] P. Figuiere, M. Ghelfenstein, and H. Szwarc, *High Temps.–High Press.*, 1974, **6**, 61.

efficiencies. The old (1—2 cm diameter) columns packed with particles of 100—150 μm diameter and using gravity feed, have been replaced by narrow (1—3 mm) columns with smaller particles (50 μm), and the liquid phase is driven through by applying a pressure at the column inlet (typically 200—400 bar). Ionic, non-volatile, or thermally unstable solutes (for which gas–liquid chromatography would be unsuitable) can be separated in times of the order of a few minutues. To avoid stagnation of solution trapped in pores, 'porous layer beads' are now being used. These have a solid, inert core of $\sim 40\ \mu$m diameter coated with a 1 μm layer of the active absorbent and rapidly reach equilibrium with the liquid phase. Flow rates of 1—5 cm^3 min^{-1} are used. Column lengths are typically 1 m, and both analytical and preparative-scale columns can be made. The driving pressure is supplied by a gas cylinder (with a bellows to transmit it to the liquid) or by a variable-speed pump, which may incorporate a mixing device to dispense mixed solvents (whose proportions may be changed as elution proceeds). Applications of the technique are being made in the fields of steroids, herbicides, pesticides, antibiotics, dyestuffs, alkaloids, and nucleic acid constituents. The subject has been reviewed by Kirkland.[22]

Supercritical-fluid Chromatography.—In high-speed liquid chromatography, pressure is used simply to boost the flow rate of the eluent; it has virtually no effect on the adsorption equilibrium. The pressure falls linearly from its highest value at the inlet to atmospheric pressure at the outlet. Supercritical-fluid chromatography employs similar pressures, but is based on completely different principles. The eluent is a supercritical gas, at a pressure of up to a few hundred bar, but the pressure *gradient* down the column is very small; most of the pressure drop occurs at the outlet. The solubility of compounds in the eluent is strongly dependent on the pressure, especially in the vicinity of the critical point. Pressure programming is therefore of value; the pressure is held constant at first and then is increased linearly to elute the less soluble components. A good example is given by Bartmann,[23] who demonstrated the separation of n-alkanes from C_5 to C_{22} using supercritical CO_2 at 40 °C. Other potentially useful mobile phases are C_2H_6, C_2H_4, freons, SF_6, and N_2O (all at room temperature). A small amount of a polar substance may be added to the supercritical eluent to increase the solubility (and so reduce the retention times) of substances to be separated. Supercritical-fluid chromatography has been reviewed recently.[24] Like high-speed liquid chromatography, it is an attractive alternative to gas–liquid chromatography which may be used for thermolabile compounds, macromolecules, polymers, biochemicals, and natural products.

[22] 'Modern Practice of Liquid Chromatography', ed. J. J. Kirkland, Wiley, New York, 1971.
[23] D. Bartmann, *Ber. Bunsengesellschaft phys. Chem.*, 1972, **76**, 336; D. Bartmann and G. M. Schneider, *J. Chromatog.*, 1973, **83**, 135.
[24] M. N. Myers and J. C. Giddings, *Progr. Separation and Purification*, 1972, **3**, 133; T. H. Houw and R. E. Jentoft, *J. Chromatog.*, 1972, **68**, 303.

4 Physical Properties of Fluids

Phase Studies.—As part of a wider study on the properties of salts and metals in the supercritical region, Franck. Hensel, Tödheide, and their colleagues have determined the vapour-pressure curves and critical points of NH_4Cl, $BiCl_3$, caesium, and potassium.[25] The possible types of miscibility behaviour which may be found in binary systems at high pressures have been comprehensively described and reviewed by Schneider.[26] Of particular interest in recent years have been the pressure dependence of upper or lower critical solution temperatures (or of closed miscibility loops) and the behaviour of fluid mixtures above the critical temperature of either pure component ('gas–gas systems'). In the former category, a nice example is the study by Peter and Schneider [27] of the systems CHF_3–C_2H_6, CF_4–C_2H_6, and CHF_3–CF_4 at temperatures down to $-150\,°C$ and pressures up to 1700 bar. The ternary system CF_4–CHF_3–C_2H_6 was also studied. In each case the systems show limited miscibility at low temperatures, with an upper critical solution temperature (UCST). The range of immiscibility and the UCST both increase with increasing pressure. The pressure dependence of miscibility has been investigated for some ternary systems, mainly of the type water–salt–organic compound. The addition of the salt often has the same effect on the binary water–organic system as an increase in pressure. A different type of ternary system (water–propan-2-ol–benzene) was recently studied.[28] This forms a single liquid phase at high temperatures and separates into two liquids on cooling. In water-rich mixtures of given composition, the unmixing temperature rises with increasing pressure whereas the opposite is the case in mixtures of low water content. The importance of such studies in the technology of separation and purification of organic compounds is obvious. An interesting and novel application of the temperature-jump technique to the study of the kinetics of phase separation has been reported.[29] Using the system water–KCl–pyridine, a condenser discharge was used to change the temperature from a point in the one-liquid region to one in the two-liquid region, at an applied pressure which could be varied in the range 1—4000 bar. The temperature change was complete in 4 μs, and the onset of turbidity was followed by measuring the optical transmission at a wavelength of 560 nm. Transmission remained constant for 80 μs, then fell rapidly, and finally became constant again after 1 ms. The initial delay

[25] M. Buback and E. U. Franck. *Ber. Bunsengesellschaft phys. Chem.*, 1972, **76**, 350; G. Treiber and K. Tödheide, *ibid.*, 1973, **77**, 1079; H. Renkert, F. Hensel, and E. U. Franck, *ibid.*, 1971, **75**, 507; W. F. Freyland and F. Hensel, *ibid.*, 1972, **76**, 16.
[26] G. M. Schneider, in 'Chemical Thermodynamics', ed. M. L. McGlashan (Specialist Periodical Reports), The Chemical Society, London, Vol. 2, 1975, in preparation; *Adv. Chem. Phys.*, 1970, **17**, 1; 'I.U.P.A.C.; Experimental Thermodynamics', Vol. II, Ch. 16, Butterworths, in the press; *Ber. Bunsengesellschaft phys. Chem.*, 1972, **76**, 325; 'Water—a Comprehensive Treatise', ed. F. Francks, Plenum Press, New York, London, 1973, Ch. 6.
[27] K. Peter and G. M. Schneider, 3rd International Conference on Chemical Thermodynamics, Baden, nr. Vienna, September 1973.
[28] Y. Hirose, P. Engels, and G. M. Schneider, *Chem.-Ing.-Tech.*, 1972, **13**, 857.
[29] A. Jost, personal communication; A. Jost and G. M. Schneider, *J. Phys. Chem.*, in the press; A. Jost, *Ber. Bunsengesellschaft phys. Chem.*, 1974, **78**, 300.

was reduced when light of shorter wavelength was used. Calculation showed that, with $D \sim 10^{-5}$ cm^2 s^{-1}, drops would grow to a size comparable with the wavelength of the light in about 100 μs. The process of phase separation appears, therefore, to be diffusion controlled, with no retardation due to formation of the interface.

The critical behaviour of binary systems continues to attract interest. Systems are classified according to the position of the critical locus in a p–T projection of the equilibrium diagram. In some systems (*e.g.* H_2O–NH_3) the critical locus is a continuous curve joining the critical points of the pure components. In other cases the critical locus is in two separate parts, which apparently do not meet as the pressure is raised. In systems of the 'first kind', the locus runs from the critical point of the less volatile component towards higher temperatures and pressures, and in systems of the 'second kind' it runs to lower temperatures as the pressure is raised, sometimes passing through a temperature minimum. Schneider[30] has listed the examples studied prior to 1970. H_2O–Ar and fifteen systems in which He is one component are systems of the first kind, in which two fluid phases can be formed when the components are mixed at a pressure of a few hundred bar and a temperature above the critical temperature of either component. The systems H_2O–Xe,[31] H_2O–propene,[32] and H_2O–ethylene[32] have recently been studied. The first two are systems of the second kind, with critical curves showing temperature minima at 343 °C and 800 bar and at 329 °C and 1950 bar respectively. In the H_2O–ethylene system, polymerization of the ethylene prevented a direct study of the critical locus; the temperature and pressure were restricted to 300 °C and 900 bar.

Apart from the light that these studies throw on the relationship between intermolecular forces and the form of the phase diagram, they are also of great technological importance; the systems are of interest as solvents for high-temperature chemical and electrochemical processes and are also of some significance in the field of geochemistry.

One study has been reported on the effect of pressure on the reversible sol–gel transformation for solutions of 12-hydroxystearic acid in carbon tetrachloride.[33] The transition temperature rises some 20 °C as the pressure is raised to 2 kbar. The value of (dT/dp) is related to the volume change for formation of cross-links by hydrogen-bonding between adjacent acid molecules, which is negative.

Structure, Dynamics, and Thermodynamics of Fluids.—Accurate measurements of density and permittivity have been made for compressed helium[34] (to 12 kbar) and nitrogen[35] (to 360 bar). The latter results were used to calculate the dielectric virial coefficients of N_2; the work complements earlier studies by the same

[30] G. M. Schneider, *Adv. Chem. Phys.*, 1970, **17**, 1; *Fortschr. chem. Forsch.*, 1970, **13**, 559.
[31] E. U. Franck, H. Lentz, and H. Welsch, *Z. phys. Chem.*, 1974, **93**, 95.
[32] M. Sanchez and H. Lentz, *High Temps.–High Press.*, 1973, **5**, 689.
[33] Y. Taniguchi and K. Suzuki, *J. Phys. Chem.*, 1974, **78**, 759.
[34] A. Dedit, J. Brielles, M. Lallemand, and D. Vidal, *High Temps.–High Press.*, 1974, **6**, 189.
[35] J. F. Ely and G. C. Straty, *J. Chem. Phys.*, 1974, **61**, 1480.

authors on O_2, F_2, and CH_4. A series of papers[36] has appeared reporting systematic $p-V-T$ studies on bromo-alkanes, to 6 kbar and between -70 and $+175\,°C$. Some discussion is given on the relationship between the intermolecular potential and the shapes of the isotherms and isobars. A spectroscopic study has been made of the $\lambda-\mu$ transition in liquid sulphur.[37] The transition temperature falls from 160 °C at atmospheric pressure to 145.5 °C at 840 bar, where the transition line meets the rising freezing-point curve of monoclinic sulphur. The equation of state of liquid nitromethane has been determined, by shock-wave methods, to 100 kbar.[38] This is claimed to be the first such determination for a liquid to this pressure based entirely on experiment. Shock compressions were carried out from a series of initial temperatures between the normal freezing and boiling points. The internal energy–volume–pressure $(U-V-p)$ equation of state was then derived, using the known variation of internal energy with temperature along the atmospheric pressure isobar. The $U-S-V$ equation was also derived by numerical integration. The liquid is metastable ('superpressed') over most of the range studied, the experimental time being *ca.* 1—10 μs. The technique depends on sampling the $U-V-p$ surface by starting at different densities (temperatures) and could be employed for any liquid having a large coefficient of expansion.

An interesting determination of the radial distribution function (RDF) of liquid sodium by X-ray diffraction has been made.[39] Measurements were made at four points along the freezing curve, the highest being at 280 °C and 43 kbar. In this way, larger density changes were obtained than would have been possible by variation of temperature at atmospheric pressure. Measurement of the pressure dependence of the RDF under isothermal conditions would have been desirable, but the method used requires that solid and liquid be present simultaneously; the X-ray pattern due to the solid is then subtracted, to eliminate the effect of background scattering. Over the pressure range studied, the nearest-neighbour distance (first peak in the RDF) decreased by $(12 \pm 2)\%$ and the peak sharpened considerably. The second peak behaved in a qualitatively similar way, but was less well determined because of experimental errors. The RDF's for different pressures could not be brought into coincidence by a simple scaling operation, because of the peak-sharpening effect mentioned.

The study of the pressure dependence of tracer diffusion coefficients has been reviewed by Barton and Speedy.[40] Their paper includes a table listing all systems studied, among which are liquid argon, various organic liquids, and some molten salts.

A diaphragm cell has been described,[41] which is said to be capable of giving self-diffusion coefficients at high pressure, with an accuracy of $\pm 1\%$. The volumes of the two compartments are matched at high pressure using spacer rings. The

[36] G. Jenner and M. Millet, *High Temps.–High Press.*, 1973, **5**, 145; *ibid.*, 1970, **2**, 205; *ibid.*, 1969, **1**, 697.
[37] G. M. Schneider, *Ber. Bunsengesellschaft phys. Chem.*, 1974, **78**, 296.
[38] P. C. Lysne and D. R. Hardesty, *J. Chem. Phys.*, 1973, **59**, 6512.
[39] K. H. Brown and J. D. Barnett, *J. Chem. Phys.*, 1972, **57**, 2009, 2016.
[40] A. F. M. Barton and R. J. Speedy, *High Temps.–High Press.*, 1970, **2**, 587.
[41] M. A. McCool and L. A. Woolf, *High Temps.–High Press.*, 1972, **4**, 85.

compressibility of the liquid must be known, so that appropriate corrections can be applied for the bulk flow which occurs through the diaphragm when the apparatus is pressurized and depressurized.

A study of the temperature and pressure dependence of molecular reorientation in liquid methyl iodide has been reported,[42] covering the ranges 0—90 °C and 0—2.5 kbar. The correlation function for reorientation about an axis perpendicular to the symmetry axis was obtained from the anisotropic component of the Raman v_3 band. The density and viscosity were also measured and a rotational diffusion coefficient, $D_\perp$, was derived. The vibrational relaxation of the molecule was also studied, using the isotropic component of the Raman bands. Finally, the deuterium spin–lattice relaxation time was measured, by n.m.r. This was used to derive a diffusion coefficient, $D_\parallel$, for rotational movement about the C_3 axis. The authors stress the importance of using pressure as an independent variable in investigations of this kind; $D_\perp$ was found to be strongly density-dependent, but $D_\parallel$ not so. In another study[43] of the proton relaxation rate in MeI, and also MeCN, it was shown that at temperatures above room temperature relaxation occurs by spin–rotation interaction. This relaxation mechanism was suppressed when the pressure was increased.

A method for direct measurement of the volume change on mixing for two liquids under pressure was described by Schneider *et al.*[44] It has been used to measure excess volumes for the systems water–acetonitrile and water–3-methyl-pyridine. The results were used to calculate the excess free energy of the mixtures as a function of pressure, using the relationship

$$G^E(p) = G^E(0) + \int_0^p V^E \, dp$$

In an outstanding series of experiments,[45] Franck, Hensel, Tödheide, and their co-workers have investigated the variation of physical properties with density in the critical and supercritical regions for various classes of fluid. The types of fluid studied were metals (Hg, Cs, K), polar liquids (H_2O, HCl), and ionic compounds (NH_4Cl, $BiCl_3$). The object was to study the appearance of typical liquid-like behaviour as the density was increased from gas-like to liquid-like values. For the metals, the properties measured were density, electrical conductivity, optical absorption, and thermoelectric power; for the polar liquids, density, permittivity, electrical conductivity, and i.r. spectra were measured and for the salts, density and conductivity. At the critical point, $BiCl_3$ has a conductivity of only $10^{-2} \, \Omega^{-1} \, cm^{-1}$ and is estimated to be only *ca.* 1 % ionized. At lower

[42] J. H. Campbell, J. F. Fischer, and J. Jonas, *J. Chem. Phys.*, 1974, **61**, 346.
[43] E. U. Franck, H. G. Hertz, and C. Rädle, *Z. phys. Chem. (Frankfurt)*, 1970, **73**, 18.
[44] P. Engels and G. M. Schneider, *Ber. Bunsengesellschaft phys. Chem.*, 1972, **76**, 1239; P. Engels, G. Götze, and G. M. Schneider, 3rd International Conference on Chemical Thermodynamics, Baden, nr. Vienna, September 1973.
[45] E. U. Franck, *Ber. Bunsengesellschaft phys. Chem.*, 1972, **76**, 341; M. Buback and E. U. Franck, *ibid.*, 1973, **77**, 1074; G. Treiber and K. Tödheide, *ibid.*, p. 1079; W. F. Freyland and F. Hensel, *ibid.*, 1972, **76**, 347; R. W. Schmutzler and F. Hensel, *ibid.*, p. 531; F. Hensel, *Phys. Letters (A)*, 1970, **31**, 88.

densities the fluid becomes completely molecular, but at higher densities (pressures) the conductivity rises to over $1\,\Omega^{-1}\,cm^{-1}$. At the lower temperatures employed, the conductivity–pressure plot passed through a maximum. This behaviour indicates that a state of complete ionization is approached at high pressures, with the possibility that 'complex ions' such as $[BiCl_4]^-$ and $[BiCl_2]^-$ are present in the intermediate region. In contrast to $BiCl_3$, NH_4Cl has a high conductivity at its critical point. indicating that it is predominantly ionic under these conditions. Correspondingly, the critical exponent β in the equation

$$(\rho_{liquic} - \rho_{gas}) = \text{const.}\,|T - T_c|^\beta$$

had a different value for the two compounds; it was 0.33 for $BiCl_3$, but 0.50 for NH_4Cl. For the alkali metals and mercury, β was in the range 0.42—0.45.

Water and Aqueous Solutions.—Franck[46] has summarized his work on the density, permittivity, conductivity, viscosity, and vibrational spectrum of pure water and of aqueous electrolyte solutions. Relatively accurate tracer diffusion coefficients ($\pm 1\%$) for THO dissolved in H_2O have been reported,[47] at 25 °C, to 2100 bar. A plot of $D(p)$ against pressure showed a maximum at 900 bar, corresponding roughly to the minimum in the viscosity (at 500 bar at this temperature). Lee and Jonas[48] have reported a number of n.m.r. studies on water and aqueous solutions. The deuteron spin–lattice relaxation time T_1 and the viscosity η were measured for D_2O from 10 to 90 °C and up to 5 kbar. The quantity $(T_1\eta/T)$ was calculated. It was found to be independent of temperature at constant density (in accordance with the Debye theory), but increased significantly with density. This indicates a change in the degree of coupling between rotational and translational motion. Similar studies were reported for water–[$^2H_8$]dioxan and D_2O–dioxan mixtures, and for electrolyte solutions in D_2O. In the latter case, the variation of T_1 with pressure was correlated with the structure-making or -breaking behaviour of the ions.

The solubility in water of benzene (to 1.2 kbar) and toluene (to 3 kbar) have been measured between 25 and 55 °C by a spectrophotometric method.[49] The solubilities of both compounds increased with pressure initially, showing that the volume change on solution, ΔV_s, is negative. For toluene, the solubility passed through a maximum at 1.3 kbar, showing that the volume change is positive above this pressure. Assuming that this high-pressure ΔV_s value corresponds to dissolution of the organic molecule in 'structureless' water, the occurrence of negative ΔV_s values at low pressures indicates that the aromatic molecules have a structure-breaking effect in this region. The solubility of benzene in KNO_3 solution was measured and was found to be enhanced when some K^+ was replaced by Ag^+ at constant ionic strength. This is due to the formation of a

[46] E. U. Franck, *Pure Appl. Chem.*, 1970, **24**, 13.
[47] L. A. Woolf, *J. Chem. Phys.*, 1974, **61**, 1600.
[48] Y. Lee and J. Jonas, *J. Chem. Phys.*, 1972, **57**, 4233; ibid., 1973, **59**, 4845; Y. K. Lee, J. H. Campbell, and J. Jonas, ibid., 1974, **60**, 3537.
[49] R. S. Bradley, M. J. Dew, and D. C. Munro, *High Temps.–High Press.*, 1973, **5**, 169.

charge-transfer complex $[C_6H_6Ag]^+$. The heat and volume change for formation of the complex were deduced from the temperature and pressure dependence of the solubility. The optical absorption corresponding to electronic transitions in various aromatic molecules has been compared in the vapour phase and in solution in water and other solvents.[50] In most solvents a red shift occurs when the solute is transferred from the gas phase to the solvent, and this red shift increases with pressure. The shift is ascribed to the lowering of the energy of the excited state of the solute relative to the ground state, due to interaction with the solvent. When water is the solvent, the red shift is much reduced in magnitude, and the pressure dependence of the shift is also smaller. These differences are explained in terms of water structure, and of the interaction between the water dipole moment and the quadrupole moment of the solute chromophore in the ground and excited states.

Several papers have appeared on the effect of pressure on ion-pair formation in aqueous solutions.[51] Chatterjee *et al.* point out that the value of the ion-pair dissociation constant obtained (and of the corresponding volume change $\Delta \bar{V}$ obtained from the pressure coefficient of this quantity) depends on the technique employed. In a Raman study of $MgSO_4$ solution they obtained $\Delta \bar{V} = -20.3 \pm 1.4$ cm^3 mol^{-1} for the dissociation of ion pairs, whereas a value $\Delta \bar{V} = -7.3$ cm^3 mol^{-1} was reported from conductivity measurements and -7.2 to -8.5 cm^3 mol^{-1} from density measurements. The confusion arises because there are different kinds of ion pair, with two, one, or no solvent molecules held between the ions, and the weighting given to each of these differs from one technique to another. Distèche[52] has written a comprehensive article in which he discusses the effect of pressure on ion-pair formation, ionization equilibria in weak electrolytes, and solubility of sparingly soluble electrolytes, in the context of marine chemistry.

The pressure dependence of electrical conductivity for aqueous electrolyte solutions continues to be studied extensively.[53] Franck, Marshall, and co-workers have reported accurate conductivity values for 0.01 demal KCl solution at round values of temperature and pressure to 800 °C and 12 kbar, which may now be used as standards. The conductivities were derived by averaging the results from eight different sets of experimental measurements, all of which were carried out in the authors' laboratories at Karlsruhe and Oak Ridge respectively. The uncertainty in the smoothed data is $\pm 0.5\%$ at 100 °C and 1000 bar, increasing to 3—4% at 800 °C and 12 kbar.

[50] A. Zipp and W. Kauzmann, *J. Chem. Phys.*, 1973, **59**, 4219.
[51] R. M. Chatterjee, W. A. Adams, and A. R. Davis, *J. Phys. Chem.*, 1974, **78**, 246; F. J. Millero and W. L. Masterson, *ibid.*, p. 1287; F. J. Millero, G. K. Ward, F. K. Lepple, and E. V. Hoff, *ibid.*, p. 1636.
[52] A. Distèche, 'The Effect of Pressure on Dissociation Constants and its Temperature Dependence', in 'The Sea; Volume 5—Marine Chemistry', ed. E. D. Goldberg, Wiley, 1974.
[53] A. S. Quist, W. L. Marshall, E. U. Franck, and W. von Osten, *J. Phys. Chem.*, 1970, **74**, 2241; H. Renkert and E. U. Franck, *Ber. Bunsengesellschaft phys. Chem.*, 1970, **74**, 40; J. U. Hwang, H.-D. Lüdemann, and D. Hartmann, *High Temps.-High Press.*, 1970, **2**, 651.

An interesting spectroscopic study of Cu^{II} chloride complexes in aqueous solutions, to 400 °C and 2 kbar, has been described.[54] The solutions contained cupric perchlorate, cupric chloride, and lithium chloride in various proportions. At low temperatures and in dilute solution the hexa-aquo-complex of Cu^{II} was the predominant species. As the temperature and pressure were increased, mixed chloro–aquo-ions appeared and finally the tetrahedral $[CuCl_4]^{2-}$ complex. Comparable behaviour has previously been found for Co^{II} and Ni^{II} chloride solutions; rising temperature favours the substitution of H_2O as ligand by Cl^-.

Molten Salts.—Relatively few papers have appeared reporting high-pressure work on molten salts. Measurements of isothermal compressibility have been reported for molten alkali nitrates and halides.[55] The pressure dependence of electrical conductivity to 1 kbar was determined for all the molten alkali-metal halides (excluding fluorides), AgCl, and AgBr.[56] The volume ΔV_Λ, defined as $-RT(\partial \ln \Lambda/\partial p)_T$, was found to be zero for lithium halides and increased regularly as the alkali metal or halide ions were changed in the sequence Li^+, Na^+, K^+, Rb^+, Cs^+ or Cl^-, Br^-, I^-. The 'activation energy' for molar conductivity at constant volume, $RT^2(\partial \ln \Lambda/\partial T)_V$, was compared with the corresponding quantity at constant pressure for each salt studied. The ratio E_V/E_p was unity for lithium salts and fell as the cation size was increased, reaching 0.5 for the halides of the heavier alkali metals. Similar results had previously been reported for molten alkali nitrates. No existing theory for molten-salt conductivity was able to account satisfactorily for this behaviour, although some of these theories had been used successfully to account for the temperature dependence of transport processes in molten salts at atmospheric pressure (when the temperature and density are changed simultaneously). This again highlights the advantages to be gained from the use of pressure as an independent variable.

The conductivity of molten nitrates has now been measured at pressures up to 55 kbar.[57] Plots of log(conductivity) against pressure were linear up to *ca.* 10 kbar but showed curvature at higher pressure. An interpretation of this behaviour should await the determination of the densities of the melts at the pressures in question; to date, this information is available only up to 9 kbar. Barton and Speedy[58] have pointed out the advantages of measuring the conductivity and density simultaneously at high pressure and have done this for molten tetrafluoroborates NR_4BF_4 (R = n-butyl, pentyl, hexyl, or heptyl).

Bannard and Treiber[59] have published a very interesting study on the conductivity of molten mercuric iodide, to 4 kbar and 850 °C. This compound is unusual in having a negative temperature coefficient of conductivity at atmospheric

[54] B. Scholz, H.-D. Lüdemann, and E. U. Franck, *Ber. Bunsengesellschaft phys. Chem.*, 1972, **76**, 406.
[55] A. F. M. Barton, G. J. Hills, D. J. Fray, and J. W. Tomlinson, *High Temps.–High Press.*, 1970, **2**, 437.
[56] B. Cleaver, S. I. Smedley, and P. N. Spencer, *J.C.S. Faraday I*, 1972, **68**, 1720.
[57] V. Pilz and K. Tödheide, *Ber. Bunsengesellschaft phys. Chem.*, 1973, **77**, 29.
[58] A. F. M. Barton and R. J. Speedy, *J.C.S. Faraday I*, 1974, **70**, 506.
[59] J. E. Bannard and G. Treiber, *High Temps.–High Press.*, 1973, **5**, 177.

pressure, which has previously been ascribed to displacement of the auto-ionization equilibrium,

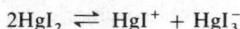

$$2HgI_2 \rightleftharpoons HgI^+ + HgI_3^-$$

to the left with decreasing density. In accordance with this, Bannard and Treiber found that the conductivity increased steeply with pressure (density) at constant temperature. As the pressure was raised, the isobaric temperature coefficient became smaller and eventually changed sign; the isotherms crossed in a narrow pressure range near 2.8 kbar. At the highest pressure reached, the conductivity appeared to have reached a maximum; this represents the balancing of two effects of pressure, namely the increase in degree of ionization and the reduction in ionic mobility. Similar effects were earlier found for $BiCl_3$.[45]

Solvated Electrons.—The effect of pressure on the rates of reactions involving solvated electrons has been studied.[60] In aqueous solution, the rates of reaction of e_{aq}^- with H_3O^+, formamide, acetamide, acetoxime, benzyl alcohol, and 2-chloroethanol were measured, up to 6 kbar. The rate constants increase with pressure, indicating a negative activation volume, but ΔV^+ increased with pressure (*i.e.* became less negative). This was attributed to a reduction with pressure of the cavity volume of e_{aq}^-. The partial molar volume of electrons, $\bar{V}(e_{aq}^-)$, and the cavity volume were estimated to be 7 and 10 cm^3 mol^{-1} respectively at 29 °C and atmospheric pressure, the difference of 3 cm^3 mol^{-1} being due to electrostriction of the water. These values conflict with an earlier estimate of 0 cm^3 mol^{-1} for $\bar{V}(e_{aq}^-)$, which was based on studies of a single reaction. In ethanolic solution, the solvolysis reaction

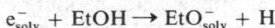

$$e_{solv}^- + EtOH \rightarrow EtO_{solv}^- + H$$

was studied and was found to have an activation volume of -14.4 cm^3 mol^{-1}. However, the bimolecular reaction of e_{solv}^- with scavengers such as nitrobenzene, acetone, and naphthalene had a positive activation volume (7.5, 5.1, and 5.6 cm^3 mol^{-1} respectively). The activation volume for diffusion of solvated electrons was estimated to be $+7$ cm^3 mol^{-1}.

A semi-continuum model has been proposed for the solvated electron in ethanol and methanol to account for the pressure dependence of the absorption spectrum.[61] Increasing pressure leads to a blue shift; this arises mainly from a reduction in the cavity radius and an increase in the energy of the quasi-free electron state. The model takes account of the effect of pressure on density and permittivity. In the model, the first shell of solvent around the electron (4—6 molecules) is regarded as partly oriented dipoles in thermal equilibrium, interacting with the electron by the charge–dipole potential. The remaining solvent is treated as a continuum. When the pressure is changed, the spectral changes arise mainly from changes in the short-range interactions between e^- and the medium; for this reason, recognition of the molecular properties of the solvent in the first co-ordination sphere is essential to the success of the theory.

[60] R. R. Hentz, Farhataziz, and E. M. Hanson, *J. Chem. Phys.*, 1972, **57**, 2959; K. N. Jha and G. R. Freeman, *ibid.*, p. 1408.
[61] D.-F. Feng, K. Feuki, and L. Kevan, *J. Chem. Phys.*, 1972, **57**, 1253.

5 Physical Properties of Solids

Phase Diagrams and Structure.—Papers on the determination of phase diagrams of solids (showing solid–solid and solid–liquid transition lines on a T–p diagram) and on the determination of crystal structures of high-pressure phases form a substantial fraction of the current literature on high-pressure research. Since it is impossible to summarize the results of all this work here, selected examples will be given which illustrate the main uses to which these studies are put.

The techniques used to determine transition lines are visual or spectroscopic observation (using a windowed cell, or a transparent cell made from diamond or sapphire), differential thermal analysis, differential scanning calorimetry (to determine heats of transition), volume measurement (to find the volume change on transition), conductivity measurement, and X-ray diffraction. X-Ray or neutron diffraction is also used to find the crystal class, space group, unit-cell dimensions, and structure of solid phases. The practice of X-ray diffraction at high pressure has been reviewed [62]

A major objective of phase-diagram and structure determination is to study the T–p diagrams of a group of compounds of similar formula but containing different elements and to seek similarities between the diagrams. It is often found that the diagrams of compounds within a group can be brought into coincidence by displacing them relatively along the pressure axis. An empirical rule is that, for simple ionic substances, the same succession of phase changes can be brought about either by increasing the pressure or by changing the cation to one of larger radius. Examples of groups of compounds studied are the alkali-metal nitrates and alkaline-earth carbonates, the spinels AB_2O_4, complex halides of the type ABX_6 (A = alkali metal, B = Group V element, X = halogen), cryolites A_3MF_6 (A = univalent metal, M = tervalent metal),[63] and the univalent metal perchlorates and fluoroborates ($MClO_4$ and MBF_4).[64] In his study on the sodium cryolites, Pistorius found that the phase diagrams of all the compounds were very similar; the I–II phase-transition temperature at atmospheric pressure, and also the unit-cell volume of the monoclinic form, were smooth functions of the crystal radius of the tervalent ion M^{3+}. Measurements were made for M = Al, Fe, V, Ti, and Co, and predictions could then be made on the properties of compounds with M = Mn, Mo, Ta, Nb, Rh, Ru, and Pd. It was also possible to predict that a high-pressure synthesis of rare-earth cryolites Na_3LnF_6 is unlikely to be successful for lanthanide ions bigger than Sc^{3+}, because the unit-cell dimensions would then be such as to make this compound unstable relative to $2NaF + NaLnF_4$. Apart from permitting a prediction to be made of the form of a phase diagram for compounds which have not been studied experimentally, these comparative studies may also allow existing diagrams to be extrapolated into pressure and temperature ranges which are beyond the reach of current experimental techniques. Thus the phase diagrams of germanates at moderate pressures

[62] M. D. Banus, *High Temps.–High Press.*, 1969, **1**, 483.
[63] C. W. F. T. Pistorius, *J. Solid State Chem.*, in the press.
[64] J. B. Clark and C. W. F. T. Pistorius, *Z. phys. Chem.* (*Frankfurt*), 1974, **88**, 242.

resemble those of silicates at much higher pressures, so the germanates can be used as model compounds to predict the behaviour of silicates at pressures such as those encountered in the earth's mantle. A discussion of this point, and of other matters affecting phase relations in geochemical systems, is given in a review by Edgar and Platt.[65]

High-pressure phases are sometimes investigated as examples of 'new' structures, for which no counterpart exists at atmospheric pressure. Conversely some complex compounds, stable under ambient conditions, decompose into simpler compounds, or into elements, when the pressure is raised. Pressure always favours changes which are accompanied by a reduction in volume; although this often corresponds to an increase in co-ordination number in simple lattices (such as elementary lattices), some pressure-induced decompositions involve a *reduction* in co-ordination number around the central atom in a complex group. This is accompanied, of course, by an overall increase in the space-filling efficiency of the atomic arrangement. A nice example of decomposition brought about by application of pressure is provided by the series of chalcogenides Ag_8MX_6 (M = Si, Ge, or Sn; X = S, Se, or Te).[66] These compounds (with the exception of Ag_8GeS_6) are decomposed at high pressure into a mixture of Ag_2X and MX_2.

At pressures above 100 kbar many ionic or semiconducting compounds become metallic. The transition can be detected by measuring the rapid fall in electrical resistance which accompanies it. These transitions have usually been brought about using shock waves (*i.e.* transient high pressures induced by the use of explosives), but one study has been reported in the Mbar range using static techniques. Kawai,[67] using his 'split-sphere' vessel, showed that Fe_2O_3, Cr_2O_3, and TiO_2 all become metallic at pressures between 2 and 3 Mbar; the resistance of the samples fell sharply by between 3 and 6 orders of magnitude as the transition occurred, at room temperature. The transitions were reversed when the pressure was removed. This observation is of great interest in relation to the composition of the core of the earth; this is known to be highly conducting, and has been thought in the past to consist of metals such as iron and nickel. Kawai's result shows that the oxides of these metals are an alternative possibility, since the pressures he reached are approximately equal to those at the centre of the earth. It is an awe-inspiring thought that with his little sphere of less than 30 cm diameter he was able to reproduce the conditions inside that greater globe on which we live.

Two interesting studies have been reported on the behaviour of point defects in solids under pressure.[68] Stoicheiometric TiO is known to have the NaCl structure at atmospheric pressure, but with 14.4% of cation sites and of anion sites vacant. The vacancy concentration falls as the pressure is raised and becomes zero at a pressure given by $p(T - 298) = 90\,000$ kbar deg. The lattice

[65] A. D. Edgar and R. G. Platt, *High Temps.–High Press.*, 1971, **3**, 1; see also Y. Shimizu, Y. Syuno, and S. Akimoto, *ibid.*, 1970, **2**, 113.
[66] C. W. F. T. Pistorius and O. Gorochov, *High Temps.–High Press.*, 1970, **2**, 31.
[67] N. Kawai and S. Mochizuki, *Phys. Letters (A)*, 1971, **36**, 54.
[68] A. Taylor and N. J. Doyle, *High Temps.–High Press.*, 1969, **1**, 679; M. Iqbal and E. H. Baker, *ibid.*, 1973, **5**, 265.

parameter increases from 4.1796 to 4.2062 Å as the vacancy concentration is reduced to zero, but the density rises from 4.97 to 5.69. In the other study, the behaviour of ThO_2 and of ThO_2–YO_2 solid solutions was investigated as a function of oxygen pressure between 10^{-7} and 500 bar, at temperatures from 800 to 1100 °C. The electrical conductivity σ was measured; it was independent of oxygen pressure below 10^{-4} bar and was believed to be ionic in this range. At higher oxygen pressures the conductivity began to increase; σ was then proportional to $p^{\frac{1}{4}}$, and the sample became a p-type semiconductor. In this region the following defect equilibrium is proposed:

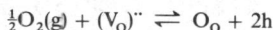

$$\tfrac{1}{2}O_2(g) + (V_O)^{\cdot\cdot} \rightleftharpoons O_O + 2h$$

(where V_O is a vacancy on an oxide site, and h is a hole). Since $[h]^2$ is proportional to $p^{\frac{1}{2}}$ by the law of mass action, and σ is proportional to $[h]$, the fourth-root dependence is explained. At higher oxygen pressures, σ was found to be proportional to $p^{\frac{1}{2}}$. To explain this, the authors propose the following equilibrium involving paired vacancies:

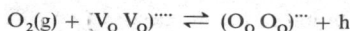

$$O_2(g) + (V_O\,V_O)^{\cdots\cdot} \rightleftharpoons (O_O\,O_O)^{\cdots} + h$$

Transport Properties of Solids.—Baranowski[69] has published an interesting study on the effect of hydrogen pressure on the diffusion of H atoms in β-palladium hydride. Diffusion was followed by recording the resistance of a wire as a function of time following a small pressure change. At pressures up to 500 bar, the composition was in the range $PdH_{0.6}$—$PdH_{0.8}$, and octahedral vacancies provide the sites for H atoms. In one run to 25 kbar, evidence was obtained that tetrahedral interstitial sites were involved. The hydrogen content was relatively high in this case, approaching or even exceeding the stoicheiometric composition (PdH).

Bradley and his colleagues[70] have measured the effect of pressure on conductivity for a number of ionic solids, including $PbCl_2$, AuCN, CuCN, AgCN, CuCl, and the highly conducting KAg_4I_5 and $RbAg_4I_5$. For $PbCl_2$, the enthalpy and volume terms corresponding to defect formation and migration were deduced. It was concluded that anion Frenkel disorder was present, with anion vacancies as the more mobile defect. For AuCN and AgCN, the conductivity increased with pressure and was believed to be partly electronic.

The conductivities of KAg_4I_5 and $RbAg_4I_5$ fell with increasing pressure, with activation volumes of 3.4 and 2.8 cm^3 mol^{-1} respectively. In contrast, the conductivities of Ag_3SBr and Ag_3SI increased with pressure ($\Delta V^{\ddagger} = -1.3$ and -2.3 cm^3 mol^{-1} respectively) at 30 °C.[71]

[69] M. Kuballa and B. Baronowski, *Ber. Bunsengesellschaft phys. Chem.*, 1974, **78**, 335.
[70] R. S. Bradley, D. C. Munro, and P. N. Spencer, *Trans. Faraday Soc.*, 1969, **65**, 1920; R. S. Bradley, 'Ionic Migration in Solids at High Pressures in the Presence of an Electric Field', in 'Atomic Transport in Solids and Liquids', ed. A. Lodding and T. Lagerwall, Verlag der Zeitschrift für Naturforschung, Tübingen, 1971, p. 350; R. S. Bradley, D. C. Munro, and S. I. Ali, *High Temps.–High Press.*, 1969, **1**, 103; see also F. P. Bundy, J. S. Kasper, and M. J. Moore, *ibid.*, 1971, **3**, 303.
[71] H. Hoshino, H. Yanagiya, and M. Shimoji, *J.C.S. Faraday II*, 1974, **70**, 281.

Bundy *et al.*[72] have studied a series of mixed-valence Pt and Pd compounds which show a remarkable increase of electronic conductivity with pressure (6 or 7 orders of magnitude over 140 kbar), the maximum values recorded being in excess of $1 \Omega^{-1} cm^{-1}$ in some cases. The compounds have the following formulae: $[M^{II}(NH_3)_2X_2][M^{IV}(NH_3)_2X_4]$ (M = Pt or Pd; X = halogen), $[Pt^{II}(en)X_2][Pt^{IV}(en)X_4]$, and $[Pt^{II}(EtNH_2)_4][Pt^{IV}(EtNH_2)_4X_2]X_4,4H_2O$. In each case, the M^{II} ion has square-planar and the M^{IV} ion octahedral co-ordination. The structure involves linear chains of M^{II} and M^{IV}, alternately, with X ions in the chains but placed nearer to M^{IV} than M^{II}:

$$\cdots M^{II} \cdots X-M^{IV}-X \cdots M^{II} \cdots X-M^{IV}-X \cdots$$

The remaining ligands form squares, whose planes are perpendicular to the linear chains. The direction of highest conductivity is along the chains, and the compounds show absorption bands which are attributed to transitions between the d_{z^2} orbitals of M^{II} and M^{IV}. These bands shift to lower energy as the pressure rises, and the $M^{II} \cdots X$ and $M^{IV}-X$ distances become closer to each other (though they still remain different at the highest pressure used). These changes indicate increases in delocalization of the d_{z^2} electrons on M^{II}, which is responsible for the steep increase in conductivity observed.

Vibrational Spectroscopy.—A study has been made[73] of the Raman spectrum of solid benzene in two crystalline modifications (I and II). Several advantages arise from the use of pressure as a variable. Pressure tends to reduce non-bonded (intermolecular) atomic separations more than bonded (intramolecular) ones. The lattice-mode frequencies are therefore much more pressure-dependent than the internal-mode frequencies, which facilitates assignments. Also, pressure may remove accidental degeneracies. In this case, a complete assignment of the spectra of I and II was made. The subject of vibrational spectroscopy at high pressure was reviewed in *Annual Reports* for 1972.

Drickamer[74] has made an extensive study of the effect of pressure (to 125 kbar) on the O—H, N—H, and C—H stretching frequencies of 15 hydrogen-bonded solids, mainly phenols. The C—H frequencies increased at all pressures, and the bands broadened slightly. A greater broadening was observed for the hydrogen-bond frequencies; these decreased with pressure initially and sometimes increased at higher pressures. No systematic difference was found between the behaviour of inter- and intra-molecular hydrogen bonds. Substituent effects were discussed; the larger the initial red shift due to the substituent at atmospheric pressure, the greater was the initial red shift with pressure.

6 Inorganic Reactions

Synthesis of Inorganic Compounds.—Several chapters in a recent book[75] survey the techniques of high-pressure synthesis and list some reactions that have been

[72] L. V. Interrante, K. W. Brownall, and F. P. Bundy, *Inorg. Chem.*, 1974, **13**, 1158.
[73] W. D. Ellenson and M. Nicol, *J. Chem. Phys.*, 1974, **61**, 1380.
[74] S. H. Moon and H. G. Drickamer, *J. Chem. Phys.*, 1974, **61**, 48.
[75] 'Preparative Methods in Solid State Chemistry', ed. P. Hagenmuller, Academic Press, New York–London, 1972.

carried out. The following general principles give a guide to the types of syntheses likely to be favoured by pressure:

(i) Pressure generally increases the coupling between d-orbitals on neighbouring metal atoms.

(ii) Pressure favours high co-ordination number in simple compounds and so tends to stabilize high oxidation states of metals (requiring a high co-ordination number of anionic ligands).

(iii) Pressure inhibits the formation of distorted structures, or displacement of ions from the centre of an octahedral site (ferroelectric displacement). These two effects normally operate *against* the introduction of d-electrons by substitution of B′ cations for B cations in ABO_3 structures, where B′ has d-electrons but B does not. Pressure accordingly favours such substitutions.

(iv) Pressure tends to broaden the range of composition attainable in insertion-type bronzes such as Na_xWO_3, because it destabilizes distorted structures which would normally compete.

(v) The range of composition of a non-stoicheiometric compound may be extended to include the stoicheiometric composition, *e.g.* by permitting a very high oxygen activity to be used (*e.g.* $CaMnO_3$ can be made from CaO + MnO in the presence of CrO_3 as a source of oxygen. The conditions required are 20 kbar, 500 °C, 2 h).

(vi) Pressure raises the internal energy of solids, thus tending to stabilize structures in which metal ion–metal ion repulsion is significant.

Pressure may be used in different ways to effect syntheses. Sometimes the 'synthesis' is just a phase transformation to a high-pressure form of a substance stable at atmospheric pressure (*e.g.* diamond synthesis). The compound may be formed at high pressure by chemical reaction; the product may then be cooled and removed from the apparatus and may be metastable at atmospheric pressure. Pressure may be used to increase the *rate* of a reaction which is thermodynamically favoured at atmospheric pressure but which proceeds very slowly.

Sometimes oxygen 'buffers' are required. These consist of a mixture of a metal and its oxide, or of two oxides. The buffer is separated from the reaction mixture by a disc of Pt or BN, with a small hole to transmit the gas. CrO_3, ZrO_2, MnO_2, and PtO_2 may be used as sources of oxygen.

Addition of a small amount of water often improves the yield and crystallinity of the product. In the hydrothermal method the product is formed in super-critical water, usually with the addition of a solubilizing agent such as NaOH. Some novel complex halides have been prepared using HCl, HBr, and HI under hydrothermal conditions.[76] By applying a suitable thermal gradient, large crystals can be grown from fragmented 'nutrient'.

Some syntheses have been carried out using shock-wave conditions.[75] Pressures up to 10 Mbar and temperatures up to 10 000 K can be reached for times of the order 10 μs. Compounds which have been made in this way include TiC, Zn_2SiO_4, CrSe, CrTe, SnS, SnSe, SnTe, K_2PtX_4, and $K_2PtX_2Y_2$ (X and Y being

[76] A. Rabenau, H. Rau, and G. Rosenstein, *Z. anorg. Chem.*, 1970, **374**, 43; *Monatsh.*, 1971, **102**, 1425.

halogens; the last two compounds were made from KX and PtX_2 or PtY_2 respectively). A feature of shock-wave conditions is that the severe distortion causes multiplication of the dislocation density to *ca.* 10^{10}—10^{12} cm^{-2}. This enhances the reaction rate, so reasonable yields can be obtained in spite of the very short reaction time. The high-temperature zone following the shock front also favours high reaction rates, although it can cause thermal decomposition of the initial product.

In the reference previously cited,[75] a review is given of recent progress in the synthesis of compounds ABX_3 (X = O, halogen, or S), and $(AX)_nABX_3$.

Diamond Synthesis.—Synthetic diamonds now account for *ca.* 40% of the world's supply of industrial-quality material. The subject has been reviewed by Wentorf.[77] Diamond exists in nature as a cubic form with ABCABC... stacking, and also as a much rarer hexagonal form with ABAB... stacking. The hexagonal form has been synthesized from graphite and has also been found in meteorites, where it is thought to have been generated during impact.

The growth of gem-quality diamonds of size approaching 1 carat (*ca.* 5 mm diameter) has been described,[78] and the conditions necessary for growth of larger stones are discussed. The diamonds were grown from solutions of carbon in molten Fe or Fe–Ni or Fe–Al mixtures at 57 kbar and 1690—1830 K. The nutrient was diamond grit, augmented by graphite (which was partly converted into diamond *in situ*). A temperature gradient was maintained between the nutrient and the seed, and growth took place at a rate of 1—2.5 mg h^{-1}. Times of up to one week were required for the growth of 1 carat gems. Some of the problems encountered were: dissolution of the seed before the steady state had been reached, spontaneous nucleation and growth of further crystals, incorporation of veils of metal and of other impurities in the growing crystal, and formation of graphite. Graphite can appear even when diamond is thermodynamically stable, because of kinetic factors. The diamonds produced were sometimes coloured, owing to the inclusion of foreign atoms. Nitrogen is a common impurity (as it is in natural diamonds) and imparts a yellow colour; however, it is atomically dispersed in synthetic diamonds, whereas in nature it forms platelets in the (100) planes. Boron may also be incorporated; in the absence of nitrogen, this imparts a blue colour.

It is believed that gem-quality crystals could be grown more rapidly by using as solvent a refractory metal with a melting point nearer to that of diamond (and which does not form stable carbides in the region of the eutectic temperature). Such a possibility is at present beyond the range of experimental techniques.

The growth of gem-quality diamonds in the laboratory is said to be uneconomic (a factor of 10 has been mentioned in the popular press). The main reason for continuing the work is to gain an understanding of the growth process and of

[77] R. H. Wentorf jun., 'Diamond Formation at High Pressure', in *Adv. High Pressure Res.*, 1974, 4; F. P. Bundy, H. M. Strong, and R. H. Wentorf, 'Chemistry and Physics of Carbon', ed. P. L. Walker, Marcel Dekker, New York, 1972, Vol. 10.
[78] H. M. Strong and R. M. Chrenko, *J. Phys. Chem.*, 1971, **75**, 1838; see also R. H. Wentorf, *ibid.*, p. 1833.

the role of impurities. These affect the growth process and also the mechanical, optical, and electronic properties of the product. The control of impurity content is a major aim at the present time.

A polycrystalline form of diamond known as carbonado (or, in a purer form, as ballas) is found in Brazil and West Africa. Since it is not so subject to fracture as single-crystal diamond, having no natural cleavage planes, it is preferred for certain applications in diamond tools. Carbonado is rather rare, and attempts have been made to prepare it in the laboratory, with some success.[79] The technique is to prepare the polycrystalline material directly from graphite or to make it by sintering diamond powder alone or in the presence of a binder. A feature of natural carbonado is that the crystals are held together by carbon–carbon bonds, but in some of the synthetic material this function is performed by the binder, which is less satisfactory. Wentorf considers that the processes by which carbonado and ballas are formed in nature are not understood.[77] Hall points out that the sintering can be carried out in the region of diamond stability (*e.g.* 85 kbar, 2440 K, 3 min) to give a white product, or in the region of graphite stability (*e.g.* 65 kbar, 2500 K, 20 s) to give a black product which still has satisfactory mechanical properties but which contains some non-diamond carbon. The sintered bodies could be produced in any desired simple shape, up to 8 mm in length.

A spherulitic form of graphite has been produced by carbonizing anthracene at a pressure of 1—2 kbar and subsequently graphitizing.[80] A mesophase is produced in the initial process, and the authors discuss the reasons why the small, spherical droplets do not coalesce. In a previous paper the authors describe a study of the effect of pressure on the rate of polymerization of anthracene (which precedes the carbonization reaction). The rate of disappearance of authracene is first-order and is accelerated by pressure ($\Delta V^{\ddagger} = -17 \text{ cm}^3 \text{ mol}^{-1}$). Paramagnetic species are thought to be involved in the early stages of the reaction; they lower the excitation energy of anthracene to the triplet state, which precedes formation of anthryl radicals, which in turn bring about polymerization.

Kinetics of Inorganic Reactions.—Several studies have been reported on the effect of pressure on ligand-substitution reactions in transition-metal complexes. Tong and Swaddle[81] studied the exchange of H_2O between $[Ir(NH_3)_5H_2O]^{3+}$ and water, using $H_2^{18}O$. They measured the rate constant to 4 kbar and found $\Delta V^{\ddagger} = -3.2 \text{ cm}^3 \text{ mol}^{-1}$. The authors had previously obtained similar values for the same reaction using the Rh^{3+} and Cr^{3+} complexes, but for the Co^{3+} compound $\Delta V^{\ddagger}$ was $+1.2 \text{ cm}^3 \text{ mol}^{-1}$. This indicates that the Co^{3+} complex exchanges water by a dissociative mechanism, whereas the other three do so by an associative mechanism. V^{3+} and Mo^{3+} hexa-aquo-complexes and Ru^{3+} amines also exchange ligands by associative mechanisms. Co^{3+} is the smallest

[79] H. T. Hall, *Science*, 1970, **169**, 868; L. F. Vereshchagin, A. A. Semerchan, V. P. Modenov, T. T. Bocharova, and M. E. Dmitriev, *Soviet Phys. Doklady*, 1971, **15**, 1065.
[80] P. W. Whang, F. Dachille, and P. L. Walker, *High Temps.–High Press.*, 1974, **6**, 127, 137; S. Hirano, F. Dachille, and P. L. Walker, *ibid.*, 1973, **5**, 207.
[81] S. B. Tong and T. W. Swaddle, *Inorg. Chem.*, 1974, **13**, 1538.

cation in this group, which may explain its anomalous behaviour. The authors stress that Co^{3+} octahedral complexes should not be used as model compounds representing all M^{3+} complexes.

Jost[29] has used a temperature-jump method to study the process

$$Fe^{3+} + SCN^- \rightarrow FeSCN^{2+}$$

A condenser discharge was used to produce the temperature jump, and the pressure could be varied up to 4 kbar. Optical read-out was employed. The formation of complexes by Ni^{2+}, Co^{2+}, Cu^{2+}, and Zn^{2+} with murexide was also studied. In all cases, the volume change for reaction and the activation volume were both positive.

Caldin and co-workers[82] have made extensive use of a laser temperature-jump cell to study reaction rates at pressures up to 3 kbar. Either optical or conducti-metric read-out were available; in the first case, two windows at right-angles to the laser beam were used. The range of relaxation times which could be studied conveniently was 500 μs—2 s.

The rate of formation of complexes between Ni^{2+} or Co^{2+} and NH_3 or pyri-ridine-2-azodimethylaniline (PADA) was measured as a function of pressure up to 2 kbar. The equilibrium constants for complex formation were also deter-mined at atmospheric pressure by an optical method, and corresponding values at high pressure were found from the amplitude of the relaxation trace (using optical read-out). For each metal ion–ligand system, the enthalpy, entropy, and volume of activation and the standard enthalpy, entropy, and volume changes for reaction were obtained. The activation volumes were very similar for all four reactions and were between 5 and 8 $cm^3\,mol^{-1}$. The authors suggest that formation of the transition state involves the same process in all four cases and that this is the stretching of an M^{2+}—water bond by between 20 and 50%. The standard volume change for reaction showed wide variations between the systems studied; the extreme values recorded were -9 and $+6\,cm^3\,mol^{-1}$ (both for Co^{2+} reactions). A similar study was made using an anionic ligand, the glycinate ion, with Co^{2+}, Ni^{2+}, Cu^{2+}, and Zn^{2+}. ΔV^+ was again positive and was between 7 and 12 $cm^3\,mol^{-1}$. Since charged ligands are now involved, these figures may not be compared directly with those from the previous study without making allowance for the volume change for formation of an outer-sphere complex. When this was done, volumes of activation for breaking of the metal ion–water bond in the range 4—9 $cm^3\,mol^{-1}$ were obtained; these are satisfyingly close to the values for Co^{2+} and Ni^{2+} reactions with NH_3 and PADA, indicating that the rate-determining step is the same in all cases.

The reaction of Co^{2+}, Ni^{2+}, $Cu2^+$, and Zn^{2+} with PADA was also studied in glycerol solution.[82] In contrast to the aqueous reactions, these substitutions all proceeded at the same rate within a factor of 4. The rate constants were smaller by two or three orders of magnitude than those calculated on the basis of dif-fusion control. For Zn^{2+} the activation enthalpy and volume were similar to

[82] E. F. Caldin and M. W. Grant, *J.C.S. Faraday I*, 1973, **69**, 1648, and references cited therein.

those for diffusion control. The low rate constant is attributed to steric factors during the substitution of glycerol by the ligand. The diffusion of a departing glycerol molecule into the bulk solvent is considered to be the main determinant of the activation parameters. For the Co^{2+} reaction, the breaking of the Co^{2+}—glycerol bond is also a significant factor.

Hasinoff[83] has reported measurements of the effect of pressure on the rates of ligand-substitution reactions of biological interest, using the laser temperature-jump apparatus described above. The first reaction studied was that of iodide ion with cobalamin (Vitamin B_{12}). The activation volumes for formation and dissociation of the iodide complex were respectively 6 and 12 $cm^3 mol^{-1}$. After making allowance for formation of an outer-sphere complex, the volume change for the ligand-substitution reaction was $2.3 \pm 0.8 cm^3 mol^{-1}$. This is sufficiently close to the values reported for the other aqueous ligand substitutions described above to suggest that stretching of the Co—H_2O bond is again the rate-determining step. {The relatively low value, 1.2 $cm^3 mol^{-1}$, for H_2O exchange in the $[Co(NH_3)_5H_2O]^{3+}$ complex was mentioned earlier in this section.[81]} There was some evidence that the substitution reaction in cobalamin was preceded by a very fast intramolecular process of unknown nature. If this is substantiated, due allowance should be made for the volume change in this process when calculating ΔV^+ for the Co—H_2O bond-stretching process.

The other reactions studied by Hasinoff were those of oxygen and carbon monoxide with haemoglobin (Hb) and myoglobin (Mb). The laser was used to photodissociate the complex, and the recombination was followed spectro-photometrically. The Hb + CO reaction showed two exponential relaxations, with time constants of 0.1 and 5 s. They are thought to correspond to the 'oxy' and 'deoxy' conformations of Hb, respectively. The activation volumes were positive for the Mb + O_2 and Hb + O_2 reactions (8 and 5 $cm^3 mol^{-1}$ respectively) but negative for the reactions of CO ($-9 cm^3 mol^{-1}$ for Mb, -3 and $-21 cm^3 mol^{-1}$ for the fast and slow reactions of Hb). A detailed discussion is given of the many factors contributing to the ΔV^+ values, in terms of a mechan-istic model due to Perutz. Binding of the ligand and solvation of parts of the Hb molecule during subsequent conformational changes cause negative contri-butions to ΔV^+. ΔV^+ is expected to be larger for the 'oxy' (fast) than for the 'deoxy' (slow) reaction, because fewer salt bridges are broken in the former case. The positive ΔV^+ values for the O_2 reactions are not easily explained; possibly they correspond to a different fit of the ligand into the heme pocket, resulting in different displacements of the surrounding protein. It is evident that experi-mental determinations of ΔV^+ for reactions involving such complex molecules do not lead directly to clear conclusions about the mechanism. However, they do provide some check on the validity of mechanistic proposals based on other evidence.

[83] B. B. Hasinoff, *Canad. J. Chem.*, 1974, **52**, 910; *Biochemistry*, 1974, **13**, 3111.
[84] W. B. Holzapfel, *High Temps.–High Press.*, 1970, **2**, 241; see also ref. 2.

7 Spectroscopic and Photochemical Studies

Mössbauer spectroscopy at high pressures has been reviewed by Holzapfel.[84] The conventional arrangement is used, except that either the absorber or the source (mixed with powdered boron and epoxy resin) is held in a Bridgman anvil. High-pressure experiments have been reported using the nuclei ^{57}Fe, ^{67}Zn, ^{119}Sn, ^{125}Te, ^{151}Eu, and ^{197}Au, and extension to other isotopes is expected to follow rapidly. The reversible transformation of high-spin Fe^{3+} into high-spin Fe^{2+} by pressure has been observed in a variety of compounds. The pressure at which the transformation occurs depends partly on the properties of the ligand that supplies the electron. Studies of hyperfine interactions have led to a better understanding of the changes induced by pressure in the electronic arrangement or band structure.

The effect of pressure on the stability of the toluene–iodine charge-transfer complex in n-hexane has been studied spectrophotometrically to 2 kbar.[85] Values of ΔG, ΔH, ΔS, and ΔV for complex formation were obtained. ΔV was $-6 cm^3 mol^{-1}$ at 1 atm, rising to $-4 cm^3 mol^{-1}$ at 2 kbar. The frequency of the absorption maximum showed a red shift with rising pressure or falling temperature, in line with previous studies.

Offen[86] has reported two photochemical studies on aromatic hydrocarbons in solid matrices at 77 K. The transfer of energy between excited pyrene and perylene was examined in a poly(methyl methacrylate) matrix. The critical transfer distance, at which the rates of intermolecular energy transfer and intramolecular decay are equal, was reduced from 44 Å at atmospheric pressure to 35 Å at 30 kbar. This was due mainly to a reduction in the fluorescence lifetime of the donor. Other significant factors were an increase in the refractive index of the medium and a decrease in the spectral overlap between pyrene emission and perylene absorption. The applicability of the Förster dipole–dipole mechanism of energy transfer is confirmed at high pressure. The phosphorescence spectrum and triplet lifetime of $[^2H_8]$naphthalene were measured to 35 kbar in five different solid matrices. Compression resulted in spectral broadening by a factor of ca. 2, because of increased solvent inhomogeneity and enhanced coupling with the lattice. A pressure-induced red shift was observed in the triplet energy, varying from 0 to 560 cm^{-1} at 30 kbar depending on the matrix. The phosphorescence lifetime was ca. 20 s in each matrix at 77 K, and decreased with pressure.

A theoretical paper on the effect of pressure on molecular electronic spectra and the rates of electronic relaxation processes has been published by Lin.[87] It was predicted that the pressure-induced frequency shift for different vibronic bands is the same; that the pressure dependence of frequency shifts is parabolic; and that the rates of non-radiative decays depend exponentially on the pressure and are also related to the frequency shift of the corresponding electronic transition.

[85] Oh Cheun Kwun and H. Lentz, *Z. phys. Chem.*, in the press. The paper contains references to other work on charge-transfer complexes.
[86] P. C. Johnson and H. W. Offen, *J. Chem. Phys.*, 1972, **57**, 1473; R. A. Beardslee and H. W. Offen, *ibid.*, 1973, **59**, 4633.
[87] S. H. Lin, *J. Chem. Phys.*, 1973, **59**, 4458.

Some comparison of these predictions is made with experimental data for aromatic hydrocarbons.

8 Organic Reactions

Organic Synthesis.—Although most studies involving the use of high pressure in organic chemistry are made principally with the aim of elucidating the mechanism of reaction and the nature of the transition state, the understanding which has come from such studies can be used in the preparative field. If the rate of a desired reaction is increased by pressure, the product can be obtained in a shorter time or at a lower temperature (avoiding thermal decomposition), often in higher yield, by carrying out the process under pressure. Shortening of the reaction time is an important factor if the reactants are sterically hindered.

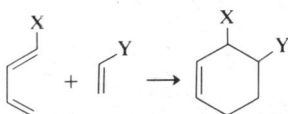

$$(a)\ X = CHO,\ Ac,\ or\ CO_2R\ ;\ Y = NR_2$$
$$(b)\ X = NR_2\ ;\ Y = CHO,\ Ac,\ or\ CO_2R$$

Scheme 1

In one research,[88] a series of addition reactions (Scheme 1) were carried out at room temperature and at pressures between 8 and 20 kbar. The reactants were contained in a beryllium–copper bellows and were allowed to react for times between 10 min and 26 h. Twenty-four different reactions were tried. In all but five cases, an enhancement of the rate was observed, and yields in the region of 70% were obtained.

Kinetics and Mechanism.—Le Noble [89] has published a very extensive review discussing the effect of pressure on the rates of chemical reactions. He lists the factors influencing the sign and magnitude of the activation volume and outlines methods for making numerical estimates of this quantity. A classified tabulation is given containing details of 456 reactions whose rates have been measured as a function of pressure. Other reviews[90] have appeared more recently; that by Neuman considers free-radical initiator decompositions (homolytic scission reactions), while the article by McCabe and Eckert reports on the kinetics of cycloaddition reactions (Diels–Alder reactions), which have received much attention in recent years.

[2 + 2] Cycloadditions are known to occur by a two-step mechanism, the concerted mechanism being forbidden by orbital symmetry rules. However,

[88] W. G. Dauben and A. P. Kozikowski, *J. Amer. Chem. Soc.*, 1974, **96**, 3664.
[89] W. J. Le Noble, *Progr. Phys. Org. Chem.*, 1968, **5**, 208.
[90] G. Kohnstamm, *Progr. Reaction Kinetics*, 1970, **5**, 335: C. A. Eckert, *Ann. Rev. Phys. Chem.*, 1972, **23**, 239; R. C. Neuman, *Accounts Chem. Res.*, 1972, **5**, 381; J. R. McCabe and C. A. Eckert, *ibid.*, 1974, **7**, 251.

[4 + 2] additions can occur either by a two-step mechanism (*i.e.* one bond being formed first, followed by the other, the first step being rate-determining) or by a concerted mechanism in which both new bonds are formed simultaneously (see Scheme 2). These alternatives can often be distinguished by high-pressure studies, the activation volume for the concerted mechanism being roughly twice that for the two-step route. Ignoring electrostriction and solvation effects, the

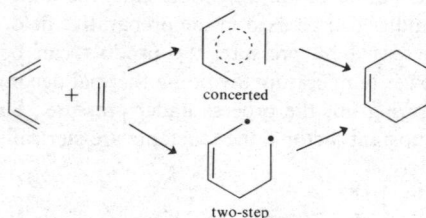

concerted

two-step

Scheme 2

expected values are -20 to $-30\,\mathrm{cm^3\,mol^{-1}}$ and -10 to $-15\,\mathrm{cm^3\,mol^{-1}}$ respectively (based on previous experience, using the rules for estimating $\Delta V^{\ddagger}$ to be found, for example, in ref. 89). For a wide range of [4 + 2] reactions, $\Delta V^{\ddagger}$ is in fact found to be between -30 and $-45\,\mathrm{cm^3\,mol^{-1}}$, confirming the concerted mechanism. Sometimes $\Delta V^{\ddagger}$ was more negative than ΔV for the total reaction. These secondary effects are indicative of interactions in the transition state which make it even more compact than the final product. This effect is particularly marked when there is an electron-donating group in the 1-position of the diene (*e.g.* Me or OMe). Such substituents also lower the activation energy significantly.

When $\Delta V^{\ddagger}$ is known, the partial molar volume $\bar{V}_{\mathrm{m}}$ of the transition state can be calculated. If the actual volume $\bar{V}_{\mathrm{m}}^{0}$ of the transition state can be estimated, the dipole moment can be found using an equation due to J. G. Kirkwood:

$$\bar{V}_{\mathrm{m}} = \bar{V}_{\mathrm{m}}^{\circ} - (\mu^2/r^3)\partial/\partial p[(\epsilon - 1)/(2\epsilon + 1)]$$

Again, 1-substituents such as Me and OMe have the biggest effect on polarity. However, changes in *solvent* polarity have little effect, and the reaction is considered to be 'non-polar'.

In some cases, a [2 + 2] reaction and the [2 + 4] reaction occur simultaneously, and pressure affects the two rate constants differently. In this situation, product analysis is essential. The dimerization of chloroprene is such a case; for the [2 + 2] reaction $\Delta V^{\ddagger} = -22\,\mathrm{cm^3\,mol^{-1}}$, and for the [2 + 4] reaction $\Delta V^{\ddagger} = -30\,\mathrm{cm^3\,mol^{-1}}$. The volume changes for reaction are -27 and $-32\,\mathrm{cm^3\,mol^{-1}}$ respectively. These figures show clearly that only one bond is fully formed in the transition state of the [2 + 2] reaction.

If the dienophile has a carbonyl group conjugated with the double bond, $AlCl_3$ (or other Lewis acids) will complex with this and withdraw electrons from the double bond. This promotes the rate of attack by the diene. In spite of the high polarity of the adduct, which has led to the suggestion that the reaction is two-step, high-pressure studies indicate that a concerted reaction takes place here also; for the addition of 2,3-dimethylbutadiene to n-butyl acrylate, $\Delta V^{\pm}$ is -29 cm^3 mol^{-1} for the uncatalysed reaction and -26 cm^3 mol^{-1} for the catalysed process. One recent study[91] of an addition in which two competing reactions occur shows that great care must be exercised before one can draw the conclusion that the reaction with the greater pressure coefficient is the one involving formation of the larger number of bonds. The reaction (Scheme 3) is the addi-

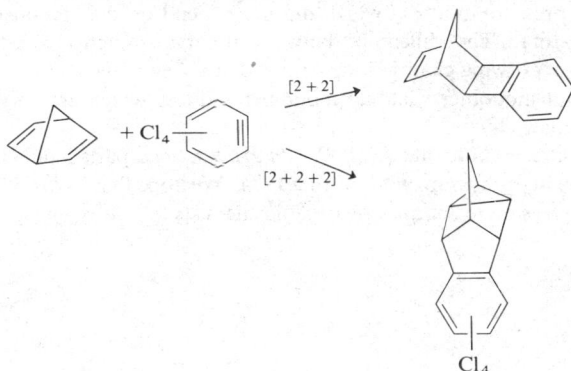

Scheme 3

tion of tetrachlorobenzyne to norbornadiene. Previous experience suggests that the $[2 + 2 + 2]$ reaction would have a more negative $\Delta V^{\pm}$ than the $[2 + 2]$ reaction. However, the reverse was the case. The authors suggest that the reason for this unusual behaviour is that a highly polar (zwitterionic) transition state (1) is formed. This is consistent with an earlier observation that the product ratio

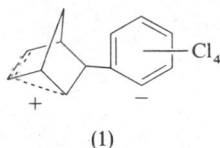

(1)

is influenced by solvent polarity. As noted above, this feature is absent in the majority of Diels–Alder reactions.

Polymerization Reactions.—Weale[92] has recently reviewed the influence of pressure on polymerization reactions. Apart from kinetic effects, other physical effects of pressure, such as freezing and phase separation, are discussed. The other topics covered are radical polymerization, radical addition and chain transfer,

[91] W. J. Le Noble and R. Mukhtar, *J. Amer. Chem. Soc.*, 1974, **96**, 6191.
[92] K. E. Weale, in 'Reactivity, Mechanism, and Structure in Polymer Chemistry', ed. A. D. Jenkins and A. Ledwith, Wiley, Chichester, 1974, Ch. 6.

the effect of pressure on dissociation of radical initiators, and ionic polymerization.

In another study,[93] the radical copolymerization of acenaphthalene (ACN) with methyl methacrylate (MMA) and with maleic anhydride (MA) was investigated, in various solvents, at pressures at up to 4 kbar. The effect of pressure on the relative rate of copolymerization was measured. For the ACN–MMA system, the pressure dependence of the relative rate depends on the ratio of monomers initially present, falling as the proportion of ACN is increased. For ACN–MA mixtures an alternating copolymer was found, and the relative rate varied with pressure in a way which did not depend on the composition of the original mixture. The difference between the two systems is attributed to termination of growing chains in the ACN–MMA system by degradative transfer reactions with monomer, yielding monomer radicals which are very inefficient at initiating new chains.

In conclusion, we note that *J.C.S. Perkin Transactions*, parts I and II, contained no reports on high-pressure work during 1974. We hope that 1975 will see greater use of high-pressure techniques by organic chemists.

[93] M. N. Romani and K. E. Weale, *Brit. Polymer J.*, 1973, **5**, 389.

PART II
INORGANIC CHEMISTRY

9 Introduction

By D. W. A. SHARP

Department of Chemistry, University of Glasgow, Glasgow G12 8QQ

The general form of the 1974 *Annual Reports* (Vol. 71) on Inorganic Chemistry follows that of Volume 70 with the coverage of series of related complexes of transition elements and of properties that relate mainly to the ligand in a separate section rather than separately under each element. In *Annual Reports* specific note is taken of the existence of the *Specialist Periodical Reports* series, and therefore we do not attempt to provide a comprehensive coverage but rather to point out those aspects of inorganic chemistry which the Reporter feels are of particular significance. In particular, we do not cover physical measurements on inorganic compounds unless these lead to significant advances in our understanding of the structure or the bonding of the compound.

Because of the necessity to keep down the length of articles in this year's Report there has been a more conscious attempt to reduce the repetition of material between the various sections of the Report. This could mean that readers will need to consult sections other than those that they would normally use but we hope that in so doing they will find yet other information of use. Of necessity we generally exclude references to full reports when preliminary communications have already been mentioned in a previous volume of *Annual Reports*. We have cut down drastically on making references to review articles although a limited number of more general reviews on inorganic topics are listed in this introduction.

Once again it is difficult to pick out areas of development for special mention but very significant advances continue to be made in the chemistry of the interaction of olefins and polyolefins with metals, and in the incorporation of metals and other heteroatoms into the cage structures of boranes and particularly of carbaboranes. One of the most interesting isolated observations is that Na^- may be present in the product of the reaction between sodium metal and a cyclic polyaminoether (p. 172). It tends to be forgotten that addition of an electron to a gaseous atom is an exothermic process for most elements. Equally interesting is evidence for NaF_2 and $NaCl_2$ in the gas phase (p. 244).

Volume XV of *Inorganic Syntheses* is devoted principally to transition-metal compounds (including olefin, dinitrogen, triphenylphosphine complexes) although there are sections on phosphorus compounds and germanium hydride derivatives.[1] Volumes of Gmelin published during the year include coverage of oxygen (System number 3), transuranium elements (7), boron (13), carbon and organoiron derivatives (14), silicon (15), organonickel derivatives (17), lanthanides (39), tin (46), and silver (61). Amongst the more important books which have become available in 1974 should be mentioned volumes of thermodynamic data on inorganic chalcogenides,[2] selenium and tellurium,[3] the actinides,[4] rare-earth intermetallics,[5] phosphine, arsine, and stibine complexes,[6] iron–sulphur proteins,[7] organotransition-metal chemistry,[8] and catalysis by metal complexes.[9]

Amonst the more important general articles and reviews that have been published in 1974 are a revision of the M^{4+} effective ionic radii,[10] reviews of the chemistry of the metallic elements in the ionosphere and mesosphere,[11] hydrogen bonding in solids,[12] fluoride complexes in aqueous solution,[13] ABX_3 compounds,[14] the marcasite structure,[15] u.v. and X-ray photoelectron spectroscopy,[16] the effects of high pressure on the chemistry and spectra of inorganic compounds,[17] metalloboranes,[18] metalloboroxanes,[19] various aspects of silicon chemistry,[20] metalloenzymes,[21] the relaxation of excited states of

[1] G. W. Parshall (Editor), *Inorg. Synth.*, 1974, XV.
[2] K. C. Mills, 'Thermodynamic Data for Inorganic Sulphides, Selenides, and Tellurides', Butterworths, London, 1973.
[3] A. A. Kudryavtsev, 'Chemistry and Technology of Selenium and Tellurium', Colletts, London, 1974.
[4] A. J. Freeman and J. B. Darby, 'The Actinides', Vols. 1 and 2, Academic Press, New York, 1974.
[5] W. E. Wallace, 'Rare Earth Intermetallics', Academic Press, New York, 1973.
[6] C. A. McAuliffe, 'Transition Metal Complexes of Phosphorus, Arsenic, and Antimony Ligands', MacMillan, London, 1973.
[7] W. Lovenborg, 'Iron–Sulphur Proteins', Vols. 1 and 2, Academic Press, New York, 1974.
[8] R. F. Heck, 'Organotransition Metal Chemistry', Academic Press, New York, 1974; P. W. Jolly and G. Wilke, 'The Organic Chemistry of Nickel', Academic Press, New York, 1974.
[9] M. M. Taqui Khan and A. E. Martell, 'Homogeneous Catalysis by Metal Complexes', Vols. I and II, Academic Press, New York, 1974; P. N. Rylander, 'Organic Syntheses with Noble Metal Catalysts', Academic Press, New York, 1974.
[10] O. Knop and J. S. Carlow, *Canad. J. Chem.*, 1974, **52**, 2175.
[11] T. L. Brown, *Chem. Rev.*, 1973, **73**, 645.
[12] A. Novak, *Structure and Bonding*, 1974, **18**, 177.
[13] G. Hefter, *Co-ordination Chem. Rev.*, 1974, **12**, 221.
[14] J. F. Ackerman, G. M. Cole, and S. L. Holt, *Inorg. Chim. Acta*, 1974, **8**, 323.
[15] A. Kjekshus and T. Rakke, *Structure and Bonding*, 1974, **19**, 85.
[16] W. L. Jolly, *Co-ordination Chem. Rev.*, 1974, **13**, 47; R. L. de Kock and D. R. Lloyd, *Adv. Inorg. Chem. Radiochem.*, 1974, **16**, 66.
[17] E. Sinn, *Co-ordination Chem. Rev.*, 1974, **12**, 185; H. G. Drickamer, *Angew. Chem. Internat. Edn.*, 1974, **13**, 39.
[18] N. N. Greenwood and I. M. Ward, *Chem. Soc. Rev.*, 1974, **3**, 231.
[19] S. K. Mehrotra, G. Srivastava, and R. C. Mehrotra, *J. Organometallic Chem.*, 1974, **73**, 277.
[20] H. Bürger, R. Eujen, G. Fritz, F. Höfler, and E. Hengge, *Topics in Current Chem.*, 1974, Vols. 50 and 51.
[21] U. Weser, R. R. Crichton, M. Llinás, and F. L. Siegel, *Structure and Bonding*, 1973, 17.

transition-metal complexes,[22] transition-metal complexes in cancer chemo-therapy,[23] ligand effects on the kinetics of substitution and redox reactions,[24] acid-catalysed reactions of transition-metal complexes,[25] perchlorate complexes,[26] $(CF_3)_2N$ compounds,[27] fluorinated peroxides,[28] fluorosulphates,[29] metal nitro-syls,[30] transition-metal oxime derivatives,[31] tripodal amine complexes,[32] metalloprotein redox reactions,[33] vitamin B_{12} systems,[34] and cyanide phosphine complexes.[35]

[22] M. K. De Armond, *Accounts Chem. Res.*, 1974, **7**, 309.
[23] M. J. Cleare, *Co-ordination Chem. Rev.*, 1974, **12**, 349.
[24] V. Gutmann and R. Schmid, *Co-ordination Chem. Rev.*, 1974, **12**, 263.
[25] P. J. Staples, *Co-ordination Chem. Rev.*, 1973, **11**, 277.
[26] L. Johansson, *Co-ordination Chem. Rev.*, 1974, **12**, 241.
[27] H. G. Ang and Y. C. Syn, *Adv. Inorg. Chem. Radiochem.*, 1974, **16**, 1.
[28] R. A. De Marco and J. M. Shreeve, *Adv. Inorg. Chem. Radiochem.*, 1974, **16**, 110.
[29] A. W. Jache, *Adv. Inorg. Chem. Radiochem.*, 1974, **16**, 177.
[30] J. H. Enemark and R. D. Feltham, *Co-ordination Chem. Rev.*, 1974, **13**, 539.
[31] A. Chakravorty, *Co-ordination Chem. Rev.*, 1974, **13**, 1.
[32] S. G. Zipp, A. P. Zipp, and S. K. Madan, *Co-ordination Chem. Rev.*, 1974, **14**, 29.
[33] L. E. Bennett, *Progr. Inorg. Chem.*, 1973, **18**, 1.
[34] D. G. Brown, *Progr. Inorg. Chem.*, 1973, **18**, 177.
[35] P. Rigo and A. Turco, *Co-ordination Chem. Rev.*, 1974, **13**, 133.

10   The Typical Elements

By G. W. FRASER

Department of Pure and Applied Chemistry, University of Strathclyde, Glasgow, G1 1XL

D. MILLINGTON

Department of Chemistry, University of Glasgow, Glasgow, G12 8QQ

M. G. H. WALLBRIDGE

Department of Molecular Sciences, University of Warwick, Coventry CV4 7AL

PART I: Groups I—III

1 Group I

Interest in the complexation of the Group I elements with polyethers and other complex organic compounds continues. A salt containing the sodium anion Na$^-$ has been reported from the reaction of sodium metal and (1) in $C_2H_5NH_2$. Two sodium atoms are present in the compound; a sodium cation is trapped in the 'crypt' and a sodium species, which is outside the 'crypt', is a large distance from the other atoms (555 pm from the amine nitrogen and 516 pm from the closest oxygen), which suggests that it is negatively charged.[1]

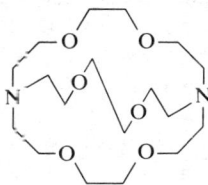

(1)

The reactions of the alkali-metal thiocyanates and halides with a number of 'crown' ethers yield mostly 1 : 1 compounds.[2a,b] Competitive crystallization of the Na or K complex depends upon the relative concentrations and also upon the nature of the 'Crown'.[2c] The reactivity of KF as a fluorinating agent is greatly increased by formation of a 'crown' complex.[2d] Sodium chloride and

[1] J. L. Dye, J. M. Ceraso, M. T. Lok, B. L. Barnett, and F. J. Tehan, *J. Amer. Chem. Soc.*, 1974, **96**, 608, 7203.
[2] (a) J. Petranek and O. Ryba, *Coll. Czech. Chem. Comm.*, 1974, **39**, 2033; (b) D. J. Sam and H. E. Simmons, *J. Amer. Chem. Soc.*, 1974, **96**, 2252; (c) N. S. Poonia, *ibid.*, p. 1012; (d) C. L. Liotta and H. P. Harris, *ibid.*, p. 2250.

other alkali-metal salts are precipitated from aqueous solution with rac-p,p'-diamino-2,3-diphenylbutane (DPB).[3a] The structures of these complexes, Na(DPB)$_3$X (X = Cl, NO$_3$, and CN) show that the DPB molecules act as bridging groups and that the sodium ions are six-co-ordinate.[3b] PhLi reacts with NN'-diphenylethylenediamine in dioxan to form the dilithium salt Li$_2$(PhNCH$_2$CH$_2$NPh); the corresponding sodium and potassium salts are formed from naphthyl-sodium or -potassium.[4a] Li, Na, and K react with (2) in

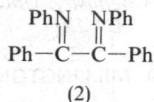

$$\begin{array}{cc} PhN & NPh \\ \| & \| \\ Ph-C & -C-Ph \end{array}$$

(2)

dioxan, ether, or THF (solv) to form a number of compounds of formula M$_2$(L),nsolv.[4b] A series of glycol salts of formula M(OCH$_2$CH$_2$OH) and M$_2$(OCH$_2$CH$_2$O) (M = Groups I and II elements) has been synthesized.[5]

Interest in organolithium compounds has again led to many publications and only a few of more inorganic interest can be reported here. Crotyl-lithium can be synthesized from BuLi and cis- and trans-crotyltrimethyltin; isomerization occurs in the presence of metallic lithium, and this has been explained as occurring via a metal-displacement process.[6] The crystal structure of cyclohexyl-lithium shows that it is hexameric, the molecule containing six triangular lithium atom faces which have two very short (240 pm) and one long (297 pm) Li–Li distances.[7] The exchange between LiR and LiAlR$_4$ (R = Me, Et, and CH$_2$SiMe$_3$) has been studied by ^{1}H and [7]Li n.m.r. spectroscopy, the rate-determining step being the dissociation of the RLi tetramer.[8a] In the reaction between MeLi and LiBr, which was investigated using [1]H n.m.r. spectroscopy, the species Li$_4$Me$_3$Br was observed at low temperature, thus supporting the earlier [7]Li n.m.r. studies.[8b]

The enthalpies of formation of 12 organolithium compounds have been determined and the dissociation energies for the C—Li bond derived.[9] Organolithiums react with peroxides to produce a range of products, EtLi and Bu^{t}OOBu^t yield Bu^{t}OEt, Bu^{t}OLi, C$_4$H$_{10}$, C$_2$H$_4$, and C$_2$H$_6$; the mechanism of production of these compounds is discussed.[10]

Theoretical calculations have been carried out on allyl-lithium[11a] and its bisdimethyl ether complex,[11b] while those on MeLi, Me$_2$Be, and Me$_3$B[11c] lead

[3] (a) N. P. Marullo, J. F. Allen, G. T. Cochran, and R. A. Lloyd, Inorg. Chem., 1974, 13, 115; (b) L. A. Duvall and D. P. Miller, ibid., p. 120.

[4] (a) H. O. Frohlich, M. Freitag, M. Kammerer, and R. Stockmann, Z. Chem., 1974, 14, 412; (b) D. Walther, ibid., p. 285.

[5] G. Gattow and J. Berg, Z. anorg. Chem., 1974, 407, 319.

[6] D. Seyferth and T. F. Jula, J. Organometallic Chem., 1974, 66, 195.

[7] R. Zerger, W. Rhine, and G. Stucky, J. Amer. Chem. Soc., 1974, 96, 6048.

[8] (a) R. L. Kieft and T. L. Brown, J. Organometallic Chem., 1974, 77, 289; (b) R. L. Kieft, D. P. Novak, and T. L. Brown, ibid., p. 299.

[9] T. Holm, J. Organometallic Chem., 1974, 77, 27.

[10] W. A. Nugent, F. Bertini, and J. K. Kochi, J. Amer. Chem. Soc., 1974, 96, 4945.

[11] (a) E. R. Tidwell and B. R. Russell, J. Organometallic Chem., 1974, 80, 175; (b) J. F. Sebastian, J. R. Grunwell, and B. Hsu, ibid., 1974, 78, C1; (c) N. J. Fitzpatrick, Inorg. Nuclear Chem. Letters, 1974, 10, 263.

to predicted M—C bond lengths which agree closely with the experimental values. Li_2Sb can be prepared from LiN_3 and metallic antimony[12a] and is stable to 1300 K and Na_2K, Na_2Cs, K_2Cs, and K_7Cs_6 can be isolated from the appropriate Na–K–Cs system.[12b] The crystal structure of potassium dithioformate has been analysed; the potassium is eight-co-ordinate and there are sheets of dithioformate ions.[13]

Thermal decompositions of solid CsO_2 and Cs_2O_2 have been investigated over the temperature ranges 600—720 K and 600—770 K, respectively.[14] The reaction paths were determined as $2CsO_2(s) = 2Cs_2O_2(s) + O_2(g)$ and $2Cs_2O_2(s) = 2Cs_2O(s) + O_2(g)$.

2 Group II

Beryllium.—Several papers on organoberyllium compounds have appeared, including a group on alkoxide derivatives in which a large number of compounds have been synthesized by a variety of methods and their structures rationalized.[15a] In contrast to the ready reduction of aldehydes and ketones by dialkylberylliums, only a few azomethines and no alkyl cyanides are reduced. However, RCN reacts with methylberyllium hydride in trimethylamine to produce $[Me(RCH=N)-Be, NMe_3]_2$, from which the NMe_3 can be replaced by pyridine without cleaving the Be_2N_2 ring.[15b] $BeEt_2$ reacts rapidly with triallylaluminium to produce $(CH_2=CHCH_2)_2Be$, which oligomerizes under the reaction conditions used.[16] The structures of two more cyclopentadienylberyllium compounds, C_5H_5BeBr and $C_5H_5BeC\equiv CH$, have been studied by electron diffraction. Both compounds contain symmetrically π-bonded C_5H_5 rings.[17]

Hydridoberyllium chloride has been identified as an intermediate in the reaction of AlH_3 with $BeCl_2$ in Et_2O and of $BeCl_2$ with BeH_2. Molecular weight and i.r. spectral measurements are in agreement with structure (3), in which both hydrogens are bridging.[18]

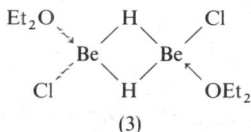

(3)

$BeCl_2$ reacts with MCl (M = NH_4, K, or Rb) at 670 K in the mole ratios 2:1 to produce the new diberyllates MBe_2Cl_5, confirmed by X-ray, i.r., and

[12] (a) R. Gerardin and J. Aubry, *Compt. rend.*, 1974, **278**, C, 1097; (b) A. Simon and G. Ebbinghaus, *Z. Naturforsch.*, 1974, **29b**, 616.
[13] R. Engler, G. Kiel, and G. Gattow, *Z. anorg. Chem.*, 1974, **404**, 71.
[14] S. P. Berardinelli and D. L. Kraus, *Inorg. Chem.*, 1974, **13**, 189.
[15] (a) R. A. Andersen and G. E. Coates, *J.C.S. Dalton*, 1974, 1171, 1440, 1729; (b) G. E. Coates and D. L. Smith, *ibid.*, p. 1737.
[16] G. Wiegand and K. H. Thiele, *Z. anorg. Chem.*, 1974, **405**, 101.
[17] A. Haaland and D. P. Novak, *Acta Chem. Scand.*, 1974, **A28**, 153.
[18] E. C. Ashby, P. Claudy, and R. D. Schwartz, *Inorg. Chem.*, 1974, **13**, 192.

Raman spectroscopy.[19a] The reaction of $BeCl_2$ with ClCN at 273 K yields the complex $BeCl_2,2ClCN$. Some of the physical and spectroscopic properties of the compound are discussed and are compared with those of the corresponding acetonitrile complex $BeCl_2,2MeCN$.[19b] A number of fluoroberyllates have been synthesized[20] and ion-exchange evidence for the $Be_3(OH)_3^{3+}$ ion has been presented, in agreement with the earlier potentiometric work.[21] The stability constants of beryllium complexes with glycine and alanine (HA) have been obtained and the species $Be(HA)^{2+}$, $Be_3(OH)_3(HA)_2^{3+}$, $Be_3(OH)_3A^{2+}$, and $Be_2(OH)(HA)_2^{3+}$ identified.[22] The i.r. spectra of a series of compounds $M(NH_2)_2$ and $M(NH)$ (M includes Be, Mg, Ca, Sr, and Ba) have been analysed and the valence force constants of the NH bond shown to increase with decreasing radius and increasing charge of the cation.[23]

Magnesium, Calcium, Strontium, and Barium.—Further details of very reactive magnesium have been published, including many of its physical properties. Its reactivity can be indicated by the conversion of C_6H_5Br into phenylmagnesium bromide at 195 K in a few minutes.[24] Interest continues in the organo Group II element derivatives, particularly those of magnesium and calcium. The structures of $(C_5H_5)_2Mg$ and $(C_5H_5)_2Ca$ have been analysed, the former by electron diffraction[25a] and the latter by X-ray crystallography.[25b] A detailed investigation of the formation of R_2Mg in hydrocarbon solvents in the absence of organic base has resulted in over 60 Grignard reagents being synthesized.[26] RMgF (R = Me, Bu, or Ph) can be prepared by the reaction of R_2Mg with a suitable mild fluorinating agent, e.g. BF_3,OEt_2 or SiF_4.[27] The structure of the dimer, $(EtMgBr,OPr_2^i)_2$, has been determined; the magnesium is four-co-ordinate and bridging occurs through the bromine atoms.[28]

A number of organocalcium compounds have been synthesized by metallation reactions using $(C_6H_5)_3CCaCl^{29a}$ or Ph_2Ca^{29b} and the appropriate arene in THF, and also by the reaction of metallic calcium and the corresponding mercury derivative.[29b]

[19] (a) J. MacCordick, *Chem. Ber.*, 1974, **107**, 1066; (b) *Compt. rend.*, 1974, **278**, C, 1177.

[20] T. K. Ghosh, *Current Sci.*, 1974, **43**, 407 (*Chem. Abs.*, 1974, **81**, 98700).

[21] M. K. Cooper, D. E. J. Garman, and D. W. Yaniuk, *J.C.S. Dalton*, 1974, 1282.

[22] G. Duc, F. Bertin, and G. Thomas-David, *Bull. Soc. chim. France*, 1974, 793.

[23] G. Linde and R. Juza, *Z. anorg. Chem.*, 1974, **409**, 199.

[24] R. D. Rieke and S. E. Bales, *J. Amer. Chem. Soc.*, 1974, **96**, 1775.

[25] (a) A. Haaland, J. Lusztyk, D. P. Novak, J. Brunvoll, and K. B. Starowieyski, *J.C.S. Chem. Comm.*, 1974, 54; (b) R. Zerger and G. Stucky, *J. Organometallic Chem.*, 1974, **80**, 7.

[26] W. N. Smith, *J. Organometallic Chem.*, 1974, **64**, 25.

[27] E. C. Ashby and J. Nackashi, *J. Organometallic Chem.*, 1974, **72**, 11.

[28] A. L. Spek, P. Voorbergen, G. Schat, C. Blomberg, and F. Bickelhaupt, *J. Organometallic Chem.*, 1974, **77**, 147.

[29] (a) K. A. Allan, B. G. Gowenlock, and W. E. Lindsell, *J. Organometallic Chem.*, 1974, **65**, 1; (b) I. E. Paleeva, N. I. Sheverdina, and K. E. Kocheshkov, *Zhur. obshchei Khim.*, 1974, **44**, 1133, 1135.

Two papers have appeared on the fixation of CO_2 using the system $TiCl_4$–Mg–THF; in the presence of H_2, magnesium formate results.[30] The reaction of magnesium with fused NaOH at 670 K forms NaH and MgO together with some Na_2O and H_2[31a] while K_2O and MgO at 1070 K produce K_6MgO_4.[31b] As might be expected, the i.r. and Raman spectra of the $MgCl_4^{2-}$ ion are consistent with it having a tetrahedral structure.[32] Some detailed i.r. studies of Ca, Sr, and Ba hydroxyapatites used D, ^{18}O, ^{44}Ca, and ^{48}Ca isotopes to help in the spectral analysis.[33] Calcium complexes with N-benzoylphenylhydroxylamine and N-cinnamoylhydroxylamine were synthesized and their i.r. spectra and electrical conductivity were measured and compared with those of similar species.[34a] Calcium and magnesium complexes of partially deprotonated H_4SiO_4 have been investigated at 288 K and the data explained by postulating a series of equilibria involving $[MH_3SiO_4]^+$, $M[H_3SiO_4]_2$, and MH_2SiO_4.[34b]

3 Group III

The distribution of publications in the various subject areas has followed the trend established in previous years, namely boron being the dominant element with the boron hydrides, carbaboranes, and their metallo-derivatives being featured. Several novel extensions have been reported, notably in the metallo-borane and -carbaborane systems where the polyhedral subrogation type of reaction has been used to prepare other 13-vertex bimetallo- (and one trimetallo-) carbaboranes,[35a] which supplement the first such example containing the atoms $FeCoC_2B_9$ in the cage reported last year, and also the first 14-vertex polyhedron made up from $Co_2C_2B_{10}$ atoms.[35b] In another set of novel cage-expansion reactions leading to $[Me_2C_2B_7H_7]Pt[PEt_3]_2$ the product has been found to possess a *nido*-10-atom polyhedron, C_2B_7Pt, arranged in the unexpected form of a near bicapped square anti-prism.[36]

Notable progress has also been made in developing direct methods of inserting metal atoms into the smaller carbaboranes,[37] and a series of such compounds,

[30] B. Jezowska-Trzebiatowska and P. Sobota, *J. Organometallic Chem.*, 1974, **76**, 43; 1974, **80**, C27.
[31] (a) G. A. Vorobev and V. L. Kubasov, *Zhur. neorg. Khim.*, 1974, **19**, 339; (b) J. C. Bardin, M. Avallet, and M. Cassou, *Compt. rend.*, 1974, **278**, C, 709.
[32] J. E. Davies, *J. Inorg. Nuclear Chem.*, 1974, **36**, 1711.
[33] B. O. Fowler, *Inorg. Chem.*, 1974, **13**, 194, 207
[34] (a) A. T. Pilipenko, L. L. Shevchenko, R. I. Sukhomlin, V. L. Ryzhenko, and M. S. Ostrovskaya, *Zhur. obshchei Khim.*, 1974, **44**, 997; (b) P. N. Santschi and P. W. Schindler, *J.C.S. Dalton*, 1974, 181.
[35] (a) D. F. Dustin and M. F. Hawthorne, *J. Amer. Chem. Soc.*, 1974, **96**, 3462; (b) W. J. Evans and M. F. Hawthorne, *J.C.S. Chem. Comm.*, 1974, 38.
[36] M. Green, J. L. Spencer, F. G. A. Stone, and A. J. Welch, *J.C.S. Chem. Comm.*, 1974, 571.
[37] V. R. Miller, L. G. Sneddon, D. C. Beer, and R. N. Grimes, *J. Amer. Chem. Soc.*, 1974, **96**, 3090.

derived from $C_2B_{n-2}H_n$ ($n = 5, 6, 7, 11$, or 12), which contain iron, cobalt, or nickel as $(CO)_3FeC_2B_3H_5$ and $(\eta-C_5H_5)_2Co_2C_2B_3H_5$ for example, have been prepared. Novel developments in the metallo-borane area have led to the isolation of compounds such as $B_4H_8Fe(CO)_3$, and $[Bu_4N]^+[\mu-Fe(CO)_4-B_7H_{12}]^-$,[38b] which contain $Fe(CO)_3$ and $Fe(CO)_4$ groups respectively bonded into borane fragments, and $(CO)_3MnB_8H_{12}$,[39c] where the B_8H_{12} moiety acts essentially as a terdentate ligand towards the manganese atom. Further reports this year have shown that the I-tetragonal form of boron first reported in 1943 as B_{50} [or $(B_{12})_4B_2$] is now believed to be a boron-rich tetragonal boron carbide $B_{50}C_2$ [or $(B_{12})_4B_2C_2$],[39a] and the existence in the solid phase of a linear metaborate anion, $O-B-O^-$, has finally been verified.[39b]

The capability of the carbonyl oxygen atom in various metal carbonyls to act as a donor centre towards Group III halides has been demonstrated,[40,41] and the preparation of the potentially useful tris(trimethylsilyl)aluminium, $(Me_3SiCH_2)_3Al$, has been achieved as the pure compound.[42] A more detailed investigation of the behaviour of the thallous ion in donor solvents and with ligands of the edta-type has been undertaken using ^{205}Tl and ^{13}C n.m.r. techniques respectively.[43,44]

Boron.—While the publications on this element are dominated by the boranes and carbaboranes, the properties of boron–nitrogen compounds, the boron halides, and heterocycles containing boron also make up a reasonable proportion of the publications for this year.

Boron Hydrides. In this section the boron hydrides are discussed in order of increasing number of boron atoms, with the exception that studies relating to the borohydride and related ions are discussed last.

Further theoretical studies on diborane have been carried out with particular reference to the dimerization of two borane fragments as $2BH_3 \rightleftharpoons B_2H_6$. Two different pathways have been proposed, one involving a transition state with a single BHB bridge, the other involving two symmetrically equivalent BHB bridges. The theoretical energies calculated for these two pathways indicate that the symmetric (C_{2h}) double-bridge system is energetically favoured. In the transition state it is proposed that two equivalent unsymmetrical bridges are formed with a B—B distance of 3.0 Å, and these become the symmetric bent

[38] (a) N. N. Greenwood, C. G. Savory, R. N. Grimes, L. G. Sneddon, A. Davison, and S. S. Wreford, *J.C.S. Chem. Comm.*, 1974, 718; (b) O. Hollander, W. R. Clayton, and S. G. Shore, *ibid.*, p. 604; (c) J. C. Calabrese, M. B. Fischer, D. F. Gaines, and J. W. Lott, *J. Amer. Chem. Soc.*, 1974, **96**, 6318.
[39] (a) G. Will and K. Ploog, *Nature*, 1974, **251**, 406; (b) C. Calvo and R. Faggiani, *J.C.S. Chem. Comm.*, 1974, 714.
[40] J. S. Kristoff and D. F. Shriver, *Inorg. Chem.*, 1974, **13**, 499.
[41] G. Schmid and V. Batzel, *J. Organometallic Chem.*, 1974, **81**, 321.
[42] J. Z. Nyathi, J. M. Ressner, and J. D. Smith, *J. Organometallic Chem.*, 1974, **70**, 35.
[43] J. J. Dechter and J. I. Zink, *J.C.S. Chem. Comm.*, 1974, 96;
[44] O. W. Howarth, P. Moore, and N. Winterton, *J.C.S. Chem. Comm.*, 1974, 664.

B—H—B bonds as the B—B distance closes to *ca.* 2.1 Å.[45a] The question of the nature of active species in reactions of diborane has been studied experimentally using a discharge flow reactor and a time-of-flight mass spectrometer with diborane and oxygen as reactants under conditions which do not produce an explosion. When diborane is present in excess a chain reaction ensues, initiated by $O + B_2H_6 \rightarrow BH_3O + BH_3$ and propagated by $BH_3 + O \rightarrow OH + BH_2$; $OH + B_2H_6 \rightarrow H_2O + BH_3 + BH_2$; when O atoms are in excess the $O + B_2H_6$ reaction still occurs but O atoms are now removed also by $BH_3O + O \rightarrow BH_3 + O_2$; $2BH_3 \rightarrow B_2H_6$ etc. There is little doubt that $O(^3P)$ is the reactive species and not $O_2(^1\Delta_g)$ since the rate constant is invariant when nitrogen carrier is added to eliminate $O_2(^1\Delta_g)$. The species H_2O, BH_2O_2, and possibly BH_2O have been identified as products or intermediates in the reaction and although OH was not identified here its presence is highly probable since it has been previously observed in the flash photolysis of B_2H_6–O_2 mixtures.[45b]

An *ab initio* calculation on the borane adduct H_3BCO has quantified the amount of charge transferred ($\sim 2e$) which consists of transfer from the lone-pair orbital on the C atom to a $p\sigma$-orbital on boron. An unexpected feature is the apparently large back donation of the π-type from the LEMO of BH_3 to the π^* orbital of the CO ligand.[46a] The ease with which the B—C bond is cleaved is demonstrated in the reaction with trimethylamine when, although a 1 : 1 adduct formulated as $Me_3N \rightarrow C(O) \rightarrow BH_3$ is formed first at 153 K, dissociation occurs above 228 K, and H_3BNMe_3 and CO are the only products at room temperature. The reaction is of interest since the 1 : 1 adduct may be considered as an analogue of the still unknown carbamic acid, and extension to the use of primary or secondary amines yields boranocarbamate salts of the type $R^1R^2NH_2^+$ $[R^1R^2NC(O) \rightarrow BH_3]^-$ ($R^1, R^2 = H$ or Me).[46b] The triborane carbonyl, B_3H_7CO, undergoes spontaneous decomposition at 273 K to yield the bis(carbonyl)-diborane(4), $B_2H_4(CO)_2$, m.p. 163–173 K, and an *X*-ray study on a sample cooled to 123 K shows the molecule has an ethane-like structure with a centre of inversion, as might be expected since the $H_2(C)B$—$B(C)H_2$ 14-electron skeleton is iso-electronic with H_3C—CH_3.[46c] All four nitrogen atoms in hexamethylenetetramine act as donor centres towards borane, and all the four possible adducts have been isolated. Heating the adducts causes migration of the hydrogen atoms and derivatives of the type $(Me_2N)BH$ and $(Me_2NBH_2)_2$ are obtained. In contrast to other amine boranes, the corresponding cations were not easily obtained and only unstable cations such as $[(CH_2)_6N_4BH_2 \cdot C_5H_5N]^+$ could be identified.[46d] There is only limited information available on interactions between diborane and organometallic compounds, and it has now been shown that phenols are produced when the product from an aryl halide, lithium (or potassium, or calcium) and diborane is treated with hydrogen peroxide.[46e]

[45] (a) D. A. Dixon, I. M. Pepperberg, and W. N. Lipscomb, *J. Amer. Chem. Soc.*, 1974, **96**, 1325; (b) C. W. Hand and L. K. Derr, *Inorg. Chem.*, 1974, **13**, 339.

[46] (a) S. Kato, H. Fujimoto, S. Yamabe, and K. Fukui, *J. Amer. Chem. Soc.*, 1974, **96**, 2024; (b) J. C. Carter, A. L. Moye, and G. W. Luther, *ibid.*, p. 3071; (c) J. Rathke and R. Schaeffer, *Inorg. Chem.*, 1974, **13**, 760; (d) M. D. Riley and N. E. Miller, *ibid.*, p. 707; (e) G. M. Pickles and F. G. Thorpe, *J. Organometallic Chem.*, 1974, **76**, C23.

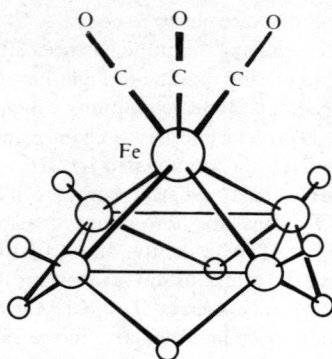

(4)

A new type of small ferraborane, $B_4H_8Fe(CO)_3$, (4), has been prepared as an orange liquid from the action of iron pentacarbonyl on pentaborane(9) in a hot–cold reactor at 493—293 K.[38a] It is also a product from the reaction mixture $Fe_3(CO)_{12}$–B_4H_{10}, and spectral properties indicate the replacement of a BH by an $Fe(CO)_3$ group, making the compound similar to the carbaborane, $(\pi\text{-}C_2B_3H_7)Fe(CO)_3$, where an apical BH group in $C_2B_4H_8$ is also replaced by an $Fe(CO)_3$ group. It is also relevant to compare this product to the $B_4H_8Co(C_5H_5)$ compound, reported last year,[47a] where a basal BH group in pentaborane(9) is replaced by the $Co(C_5H_5)$ group. A different type of metal–boron bonding is found in the known $\mu\text{-}Fe(CO)_4B_6H_{10}$ compound where the iron atom bridges two basal boron atoms, and this compound undergoes insertion reactions by the established route of reaction with hydride ion, to yield the borane anion, $\mu\text{-}Fe(CO)_4B_6H_9^-$, followed by borane (as diborane) when $[\mu\text{-}Fe(CO)_4B_7H_{12}]^-$ and also $\mu\text{-}Fe(CO)_4B_7H_{11}$ are obtained. The anion $[\mu\text{-}Fe(CO)_4B_7H_{12}]^-$ has both $Fe(CO)_4$ and BH_3 groups inserted as bridging groups into the base of a B_6 pentagonal pyramid.[47b] The insertion of an $Fe(CO)_4$ fragment is not exclusive and heavier transition metals, Rh(acac), RhCl, and IrCl behave similarly, but the products are less stable and, although not characterized structurally, are presumably similar to the olefin presursor compounds.[47c]

A further variation is provided by the product of the reaction between $Ir(CO)Cl(PMe_3)_2$ and B_5H_9 (or 1- or 2-BrB_5H_8), namely $[Ir(CO)X_2(PMe_2)]B_5H_8$ (where X_2 = ClH or Br_2) where a B(basal)—Ir bond is formed as shown by the crystal structure of the dibromo-product (5).[47d] The exacting requirements of

[47] (a) D. Millington, J. M. Winfield, and M. G. H. Wallbridge, Ann. Reports (A), 1973, 70, 279; (b) O. Hollander, W. R. Clayton, and S. G. Shore, J.C.S. Chem. Comm., 1974, 604; (c) A. Davison, D. D. Traficante, and S. S. Wreford, J. Amer. Chem. Soc., 1974, 96, 2802; (d) M. R. Churchill, J. J. Hackbarth, A. Davison, D. D. Traficante, and S. S. Wreford, ibid., p. 4041.

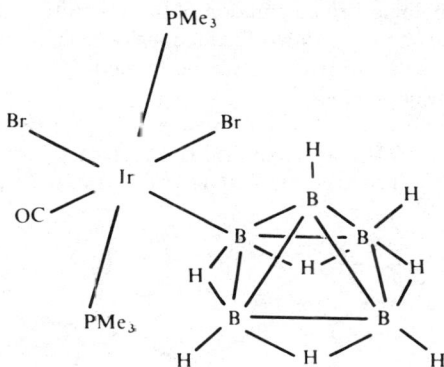

(5)

the reaction are illustrated by the fact that there is no reaction if the *trans*-Ir(CO)Cl(PPh$_3$)$_2$ compound is used. Other studies on pentaborane(9) have shown that insertion reactions, which may be considered as analogous to carbene insertion reactions, can be achieved by the action of BMe$_3$ (assisted by GaMe$_3$) which yields 2-MeB$_6$H$_9$, while the action of H$_2$BCl,OEt$_2$ on the [Me$_3$MB$_5$H$_7$]$^-$ anion (M = Si or Ge) affords the first apically substituted hexaborane(10), namely 1-Me$_3$MB$_6$H$_9$.[48a] The crystal structure of the adduct B$_5$H$_9$,2PMe$_3$ shows the two phosphorus atoms to be located on base and apical boron

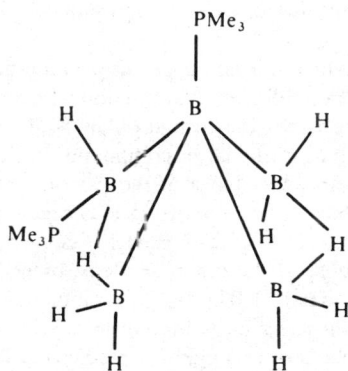

(6)

[48] (a) D. F. Gaines, S. Hildebrandt, and J. Ulman, *Inorg. Chem.*, 1974, **13**, 1217; (b) A. V. Fratini, G. W. Sullivan, M. L. Denniston, R. K. Hertz, and S. G. Shore, *J. Amer. Chem. Soc.*, 1974, **96**, 3013; (c) A. E. Burg, *Inorg. Chem.*, 1974, **13**, 1010.

atoms. The topological representation of the structure is shown in (6), and not unexpectedly the molecule shows fluxional behaviour in solution.[48b] Trifluoro-silylpentaboranes, $2\text{-}SiF_3B_5H_8$, have been prepared by treating B_5H_9 with silicon tetrafluoride at 195 K.[48c]

The main product of the reaction between $Mn(CO)_5Br$ and KB_9H_{14} in THF is $(CO)_3MnB_9H_{12}$,THF, as mentioned last year. A by-product of this reaction has now been identified as $(CO)_3MnB_8H_{13}$ (7) and, as the crystal structure shows,

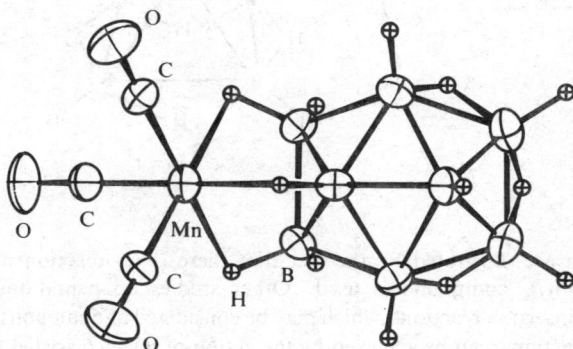

(7)

this is a new type of metalloborane with the B_8H_{13} fragment acting as a terdentate ligand through three B—H—Mn bridge bonds.[38c] The only other reported example of a MB_8 system is $(PEt_3)_2PtB_8H_{12}$, and while this has been suggested on the basis of spectral evidence to possess a platinum atom bonded directly to boron atoms in a substituted B_9 cage, it may be that both compounds possess similar structures.

Several new heteroatom boranes, and some metallo-derivatives, have been characterized. It is well established that reaction of decaborane(14) with ammonium polysulphide leads to the formation of $B_9H_{12}S^-$ and $B_9H_{11}S$. As reported last year, pyrolysis of the latter compound produces the $closo\text{-}B_9H_9S$, and it has now been established that a further product of this pyrolysis is the thiaborane $(B_9H_8S)_2$, which consists of two B_9H_8S cages linked by a two-centre B—B bond. The structure of the $2,2'\text{-}(1\text{-}B_9H_8S)_2$ isomer has been determined from X-ray data, and those of the other isolated isomers, 2,6'- and 6,6'-, were inferred from [11]B n.m.r. data.[49a] The $B_9H_{12}S^-$ ion, and related species, have been deprotonated *in situ* using butyl-lithium (*e.g.* Scheme 1). The structures proposed for the products are as might be predicted. Thus (A) is similar to $[Ni(B_{10}H_{12})_2]^{2-}$ except that B–S–metal interactions replace B–B–metal bonds, and (B) contains a B_9S fragment, similar to the B_{10} fragment in $B_{10}H_{14}$, with the

[49] (a) W. R. Pretzer and R. W. Rudolph, *J.C.S. Chem. Comm.*, 1974, 629; (b) A. R. Siedle, D. McDowell, and L. J. Todd, *Inorg. Chem.*, 1974, **13**, 2735; (c) L. J. Guggenberger, *J. Organometallic Chem.*, 1974, **81**, 271.

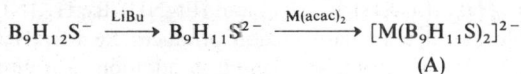

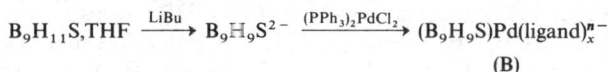

$$B_9H_{12}S^- \xrightarrow{\text{LiBu}} B_9H_{11}S^{2-} \xrightarrow{\text{M(acac)}_2} [M(B_9H_{11}S)_2]^{2-}$$

$$(A)$$

$$B_9H_{14}^- \xrightarrow{\text{LiBu}} B_9H_{14}^{2-} \xrightarrow{\text{PhBCl}_2} \text{6-PhB}_{10}H_{13}$$

$$B_9H_{11}S,\text{THF} \xrightarrow{\text{LiBu}} B_9H_9S^{2-} \xrightarrow{\text{(PPh}_3)_2\text{PdCl}_2} (B_9H_9S)\text{Pd(ligand)}_x^{n-}$$

$$(B)$$

(ligand $= PPh_3$, $x = 2$, $n = 0$; phenanthroline, $x = 1$, $n = 0$; $C_2S_2(CN)_2$, $x = 1$, $n = 2$)

Scheme 1

metal atom situated centrally above the cage on a mirror plane of the B_9S cage.[49b] The crystal structure of a related complex $(PPh_3)_3Au^+B_9H_{12}S^-$ confirms that the B_9S cage is an open icosahedral fragment with the S atom in the 6-position on the periphery of the decaborane polyhedron. However, in this case there is apparently no interaction of the nearly trigonal cation with the B_9S cage.[49c]

While a variety of phospha- and arsa-carboranes and metallo-compounds are known, it is only this year that the corresponding borane compounds have been isolated. When decaborane is deprotonated with a strong base in the presence of a reducing agent and arsenic trichloride, various arsaboranes are produced (Scheme 2).[50a] These compounds, and related phospha-systems, such

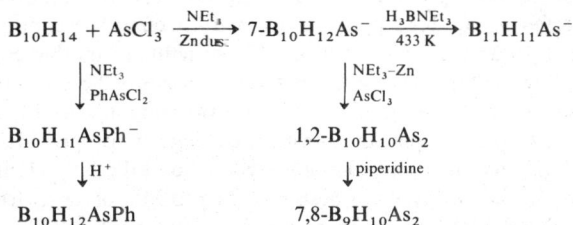

$$B_{10}H_{14} + AsCl_3 \xrightarrow[\text{Zn dust}]{\text{NEt}_4} \text{7-}B_{10}H_{12}As^- \xrightarrow[\text{433 K}]{\text{H}_3\text{BNEt}_3} B_{11}H_{11}As^-$$

$$\downarrow \begin{array}{l}\text{NEt}_3\\\text{PhAsCl}_2\end{array} \qquad\qquad \downarrow \begin{array}{l}\text{NEt}_3\text{-Zn}\\\text{AsCl}_3\end{array}$$

$$B_{10}H_{11}AsPh^- \qquad\qquad 1,2\text{-}B_{10}H_{10}As_2$$

$$\downarrow \text{H}^+ \qquad\qquad\qquad \downarrow \text{piperidine}$$

$$B_{10}H_{12}AsPh \qquad\qquad 7,8\text{-}B_9H_{10}As_2$$

Scheme 2

as $7\text{-}B_{10}H_{12}As^-$, $7,8\text{-}B_9H_{10}As_2{}^-$, $B_{10}H_{12}P^-$, and $7,8\text{-}B_9H_{10}CHP^-$, interact with the organometallic derivatives $[(C_5H_5)Fe(CO)_2(\text{cyclohexene})]^+$ and $[(C_7H_7)Mo(CO)_3]^+$ to form compounds such as $(C_5H_5)Fe(CO)_2B_{10}H_{12}As$ and $(C_7H_7)Mo(CO)_2B_{10}H_{12}P$. In each case it is suggested that the heteroatom in the borane cage acts as a donor centre towards the metal atom.[50b]

Several higher boranes have been produced by oxidative coupling reactions, thus the addition of $Fe(NO_3)_3,9H_2O$ to $(Et_3NH)B_{10}H_{10}$ in aqueous solution

[50] (a) J. L. Little, S. S. Pao, and K. K. Sugathan, *Inorg. Chem.*, 1974, **13**, 1752; (b) T. Yamamoto and L. J. Todd, *J. Organometallic Chem.*, 1974, **67**, 75.

yields $(Et_3NH)_3B_{20}H_{19}$, $(Et_3NH)_2B_{20}H_{18}$, and $(Et_3NH)_3B_{20}H_{18}(NO)$, but the use of iron(III) chloride in a similar reaction produces $1,6,8\text{-}B_{10}H_7Cl_3{}^{2-}$ and $1,6$ (or $2,4$)-$B_{10}H_8Cl_2$.[51a] If nitriles are present in addition to iron(III) chloride then coupling reactions do occur, and the anions produced, *e.g.* $B_{20}H_{18}{}^{2-}$, $B_{24}H_{23}{}^{3-}$, $B_{24}H_{22}Cl^{3-}$, are dependent upon the reaction conditions.[51b] Electrochemical oxidation may also be used to transform ions such as $B_{10}H_9I^{2-}$ and $B_{10}H_9L^-$ ($L = NH_3$, NMe_3, or SMe_2) in acetonitrile into higher borane anions including $B_{20}H_{17}L_2{}^-$, $B_{20}H_{18}L_2{}^{2-}$, and $B_{20}H_{16}L_2$.[51c] Even higher species may be obtained from the low-temperature decomposition of $(H_3O)_2B_{12}H_{12}$, and in this case the products include $B_{48}H_{45}{}^{5-}$, $B_{24}H_{23}{}^{3-}$, and $B_{24}H_{22}OH^{3-}$.[51d] Localized MO diagrams have been given for some higher boranes, $B_{16}H_{20}$, $B_{18}H_{22}$, and $B_{20}H_{18}{}^{2-}$, and it has been found that the localized MO's derived for the smaller boranes may be transferred to the larger molecules. For example, the three-centre orbitals derived for $B_{16}H_{20}$ are closely related to those for $B_{10}H_{14}$ and B_8H_{12}, with two boron atoms being shared to make up the B_{16} fragment.[51e]

Studies on the $BH_4{}^-$ ion have shown that the value of $J(^{11}B-H)$ decreases as *n* decreases in $LiBH_nD_{4-n}$, and provides the first report of an isotropic shift in a ^{11}B n.m.r. spectrum.[52a] The hindered rotational levels of the $BH_4{}^-$ ion in high-temperature phases of $NaBH_4$ and KBH_4 have been investigated, and the large differences between the rotational heat capacities of the ion in the two compounds suggest differences in the crystal structure, with the $BH_4{}^-$ ion being in an octa-hedral and tetrahedral symmetry in $NaBH_4$ and KBH_4, respectively.[52b] New $Cu(Me_5dien)(NCBH_3)_2$ [where $Me_5dien = Me_2N(CH_2)_2N(Me)(CH_2)_2NMe_2$][53a] has a distorted square-base pyramid geometry around the copper atom with the $NCBH_3$ ligand in the axial position. The structure of another copper deriva-tive $(PPh_3)_2Cu(NCBH_3)_2$ is different in that it consists of dimeric units with each copper atom being near tetrahedral and with two bridging $Cu\text{-}N\text{-}C\text{-}B(H)_2\text{-}H\text{-}Cu$ groups making up a non-planar 10-membered ring. It also provides an unusual example of only one $B-H$ bond being used to bond the $NCBH_3$ ligand to the metal atom.[53b] An improved synthesis giving up to 94% of the hydroxytrihydro-borate anion, BH_3OH^-, has been devised by acidifying the $BH_4{}^-$ ion in a flow reactor and quenching the reaction with hydroxide ions.[53c]

While it is known that the $B_3H_8{}^-$ ion can act as a bidentate ligand towards metals, an interesting extension has shown that it may also function as a terdentate

[51] (a) Z. B. Curtis, C. Young, R. Dickerson, K. K. Lai, and A. Kaczmarczyk, *Inorg. Chem.*, 1974, **13**, 1760; (b) A. H. Norman and A. Kaczmarczyk, *ibid.*, p. 2316; (c) A. P. Schmitt and R. L. Middaugh, *ibid.*, p. 163; (d) R. Bechtold and A. Kaczmarczyk, *J. Amer. Chem. Soc.*, 1974, **96**, 5954; (e) D. A. Dixon, D. A. Kleier, T. A. Halgren, and W. N. Lipscomb, *ibid.*, p. 2293.

[52] (a) B. D. James, B. E. Smith, and R. H. Newman, *J.C.S. Chem. Comm.*, 1974, 294; (b) D. Smith, *J. Chem. Phys.*, 1974, **60**, 958.

[53] (a) B. G. Segal and S. J. Lippard, *Inorg. Chem.*, 1974, **13**, 822; (b) K. M. Melmed, T. Li, J. J. Mayerle, and S. J. Lippard, *J. Amer. Chem. Soc.*, 1974, **96**, 69; (c) J. W. Reed, H. H. Ho, and W. L. Jolly, *ibid.*, p. 1248.

ligand, thus:

$$\text{Mn (or Re) (CO)}_5\text{Br} + \text{Me}_4\text{NB}_3\text{H}_8 \xrightarrow[\text{293 K}]{\text{CH}_2\text{Cl}_2} \text{(CO)}_4\text{MnB}_3\text{H}_8$$
$$\text{(A)}$$

$$\text{(CO)}_4\text{MnB}_3\text{H}_8 \underset{\text{CO}}{\overset{\text{u.v.}}{\rightleftarrows}} \text{(CO)}_3\text{MnB}_3\text{H}_8$$
$$\text{(B)}$$

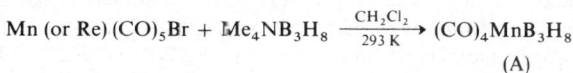

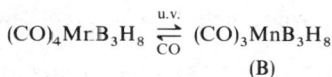

Compound (A), which is a yellow volatile liquid, contains the B_3H_8 group bonded by two $B-H-Mn$ bridge bonds, but in (B), which may be formed reversibly from (A), the position vacated by the removal of the CO molecule is taken up by a further $B-H-Mn$ bond, making the B_3H_8 group terdentate.[54]

Carbaboranes and Metallo-carbaboranes. These compounds are discussed generally in terms of increasing number of atoms in the cage system, but a strict order is not always possible since papers often refer to the preparation of different size cage systems by one general method.

It was reported last year and in more detail recently[55] that cobalt could be inserted into the smaller carbaboranes, *nido*-2,3-$B_4C_2H_8$, *closo*-1,6-$B_4C_2H_6$, and *closo*-$B_5C_2H_7$. A typical reaction involved treating the carbaborane with a Na naphthenide–$CoCl_2$–NaC_5H_5 mixture followed by air oxidation to yield products formally containing the anions $B_3C_2H_5^{4-}$ and $B_4C_2H_6^{2-}$. These, and related compounds, have now been prepared by the direct insertion of a metal atom (iron, cobalt, or nickel) into the polyhedral cage of $C_2B_{n-2}H_n$ (where $n = 5, 6, 7, 11,$ or 12), and it is reasonable to expect that the method would also succeed for the $n = 8, 9,$ or 10 systems. The procedure consists of heating the carbaborane with transition-metal compounds such as $Fe(CO)_5$, $(C_5H_5)Co(CO)_2$, and $(PPh_3)_2Ni(C_2H_4)$,[37] *e.g.*

$$1,5\text{-}B_3C_2H_5 \xrightarrow{Fe(CO)_5} (CO)_3Fe(B_3C_2H_5) + (CO)_6Fe_2(B_3C_2H_5)$$

$$\xrightarrow{(C_5H_5)Co(CO)_2} (C_5H_5)Co(B_3C_2H_5) + (C_5H_5)_2Co(B_3C_2H_5)$$

$$2,4\text{-}B_5C_2H_7 \xrightarrow{Fe(CO)_5} 1,2,4\text{-}(CO)_3Fe(B_4C_2H_6) + 3,1,7\text{-}(CO)_3Fe(B_5C_2H_7)$$

$$\xrightarrow{(PPh_3)_2Ni(C_2H_4)} (PPh_3)_2Ni(B_5C_2H_7)$$

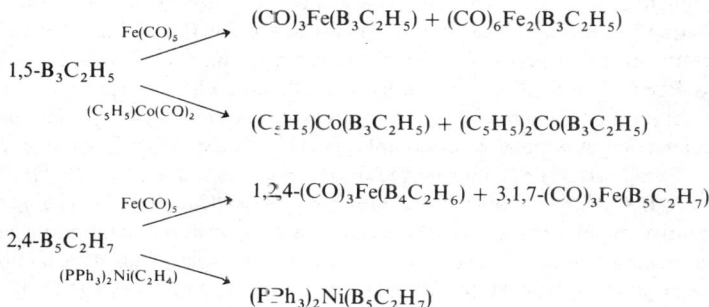

While this method is simple and direct it does have some limitations in that it is likely that intermediates and some isomers will decompose at the elevated

[54] D. F. Gaines and S. J. Hildebrandt, *J. Amer. Chem. Soc.*, 1974, **96**, 5574.
[55] R. N. Grimes, D. C. Beer, L. G. Sneddon, V. R. Miller, and R. Weiss, *Inorg. Chem.*, 1974, **13**, 1138.

reaction temperatures, and the position of substitution cannot easily be controlled. The properties of the $nido$-$C_2Me_2B_4H_5^-$ anion have been clarified, and the reaction of this anion with halogen compounds (ICl and Br_2) leads to 2- or 3-halogen substituted carbaboranes $C_2Me_2B_4H_5X$ (X = Cl, Br, or I).[56a] Bis(trifluoromethyl)phosphine derivatives of $closo$-2,4-$C_2B_5H_7$ have been prepared:[56b]

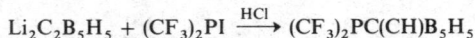

$$Li_2C_2B_5H_5 + (CF_3)_2PI \xrightarrow{HCl} (CF_3)_2PC(CH)B_5H_5$$

The molecular structure of the $closo$-CB_5H_7 carbaborane has been determined from microwave data as a distorted octahedron of CB_5 atoms with one face containing three long B—B bonds; it is in, or above, this face that the extra hydrogen atom is located.[57a] An improved synthesis of $nido$-dicarbaoctaborane-(10), $B_6C_2H_{10}$, results from the use of a hot–cold reactor in the reaction of $closo$-$C_2B_3H_5$ with diborane.[57b]

Two new carbaboranes with a C_2B_8 cage have been prepared. The reduction of 5,6-$B_8C_2H_{12}$ with sodium amalgam in ethanol yields $nido$-6,9-$B_8C_2H_{14}$, which is proposed to be isostructural with the isoelectronic $B_{10}H_{14}^{2-}$ anion.[58a] The same initial carbaborane, 5,6-$C_2B_8H_{12}$, which may be obtained in reasonable yield by the oxidation of the 7,8-$B_9C_2H_{12}^-$ anion with aqueous iron(III) chloride, undergoes dehydrogenation at 513 K to yield the $closo$-carbaborane 1,2-B_8C_2-H_{10}.[58b]

The structure of the nine-atom cage metallocarbaborane anion $[(C_5H_5)Co-(B_7CH_8)]^-$ has been determined from X-ray data, and consists of a Co^{III} atom sandwiched between a π-C_5H_5 ring and the $B_7CH_8^{3-}$ group. The CoB_7C cage is an approximate tricapped trigonal prism in which one carbon and two boron atoms are in the low-co-ordinate cap positions, and is very similar to the structure proposed when the compound was first reported in 1972. The structure also follows the general rule that the carbon atoms in such compounds occupy low-co-ordinate positions.[59] This rule is also obeyed in the structure of the $nido$-10,10-$(Et_3P)_2$-2,8-Me_2-10,2,8-Pt-$B_7C_2H_7$ (8) where the PtC_2B_7 skeleton, whose geometry approximates to a bicapped square antiprism, but the separation of the atoms Pt—C(2) and B(7)—B(9), which are 2.83 and 2.52 Å, respectively, is too great for bonding interactions. The parent carbaborane 1,6-$C_2B_7H_9$ affords a similar compound, and both complexes exhibit fluxional behaviour in solution. The $nido$-structure is unexpected since the related species $(C_5H_5)CoB_7C_2H_9$ and $[Co(B_7C_2H_9)_2]^-$ possess the predicted $closo$-structures.[36] The general preparative method of the direct insertion of a d^{10} transition metal into a carbaborane cage has been extended to prepare an example of an η-carbadiboraallyl complex $[Ni(B_7C_2H_9Me_2)(PEt_3)_2]$ from the reaction of $Ni(PEt_3)_4$ with

[56] (a) C. G. Savory and M. G. H. Wallbridge, *J.C.S. Dalton*, 1974, 880; (b) L. Maya and A. B. Burg, *Inorg. Chem.*, 1974, **13**, 1522.

[57] (a) G. L. McKown, B. P. Don, R. A. Beaudet, P. J. Vergamini, and L. H. Jones, *J.C.S. Chem. Comm.*, 1974, 765; (b) T. J. Reilly and A. B. Burg, *Inorg. Chem.*, 1974, **13**, 1250.

[58] (a) B. Štíbr, J. Plešek, and S. Heřmánek, *Coll. Czech. Chem. Comm.*, 1974, **39**, 1805; (b) J. Plešek and S. Heřmánek, *ibid.*, p. 821.

[59] K. P. Callahan, C. E. Strouse, A. L. Sims, and M. F. Hawthorne, *Inorg. Chem.*, 1974, **13**, 1393.

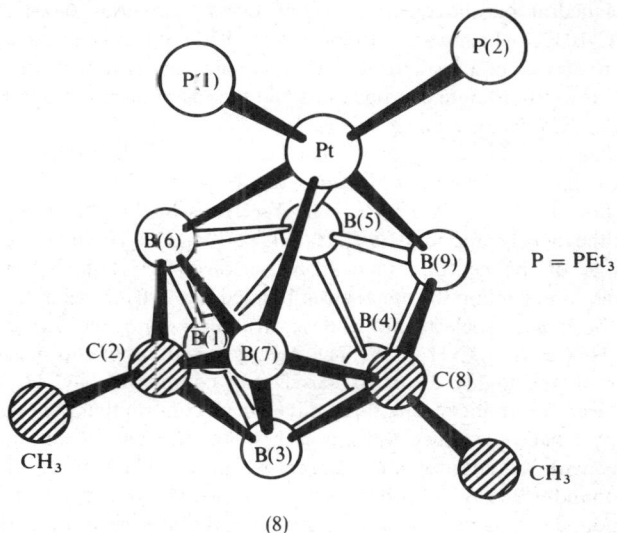

(8)

the *arachno*-carbaborane $1,3-B_7C_2H_{11}Me_2$. The *nido*-metallocarbaborane (9) is isoelectronic and nearly isostructural with $B_{10}H_{14}$, and it is suggested that the carbaborane fragment acts as a $1,2,3-B_2C$ ligand group in both this compound and the similar platinum compounds, *e.g.* $[Pt(B_7C_2H_{11})(PEt_3)_2]$, and behaves therefore like a π-allyl group in the corresponding derivatives $[ML_2(C_3H_5)]^+$ (M = Ni or Pt).[50] The first confirmed metal–metal bond in

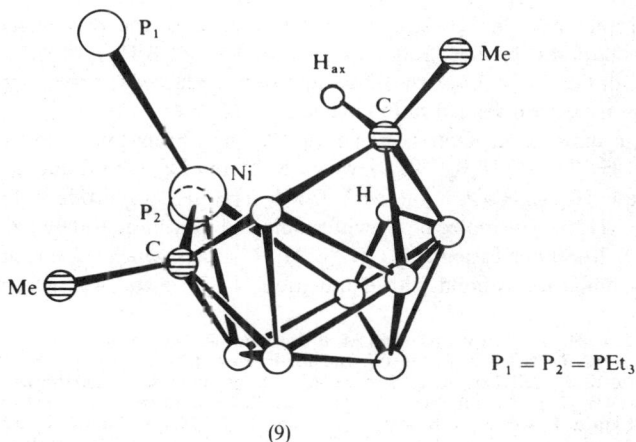

(9)

[60] M. Green, J. Howard, J. L. Spencer, and F. G. A. Stone, *J.C.S. Chem. Comm.*, 1974, 153.

a polyhedral borane or carbaborane occurs in $closo$-2,6-$(\eta$-$C_5H_5)_2$-2,6-Co_2-1,10-$C_2B_6H_8$, which was first reported in 1970. X-Ray data show the $Co_2C_2B_6$ cage to consist of a distorted bicapped square antiprism with the carbon atoms in the low-co-ordinate cap positions and the metal atoms in the tropical planes, the Co—Co distance being 2.49 Å.[61]

Instead of the insertion of d^{10} metal atoms into carbaboranes it has also been shown that an icosohedral cage can be constructed by the addition of $[Ni^{IV}(\eta$-$C_5H_5)]^{3+}$ or $[Co^{III}(\eta$-$C_5H_5)]^{2+}$ vertices to the known $B_{10}CH_{11}^{3-}$ ion. Both the nickel compound, $(\eta$-$C_5H_5)Ni^{IV}(\eta$-7-$B_{10}CH_{11})$, which is the first known member of the neutral monocarbon series $(\eta$-$C_5H_5)Ni^{IV}(B_nCH_{n+1})$, and the cobalt anion contain the metal atom bonded into a B_4C face of the carbaborane, and both are isoelectronic and isostructural with the two-carbon species $(\eta$-$C_5H_5)Co^{III}(\eta$-$B_9C_2H_{11})$.[62a] The precise positions of the metal and carbon atoms in the highly distorted icosahedral compound $(\eta$-$C_5H_5)_2Co_2C_2B_8H_{10}$ have been determined from X-ray data, which confirm that the two cobalt atoms occupy adjacent vertices and are each bonded to two boron and two carbon atoms, with the latter in non-adjacent positions. The Co—Co distance in this compound is 2.39 Å.[62b] It has also proved possible to prepare an isomer of this compound having the $(\eta$-$C_5H_5)Co$ groups in the $para$-positions of the icosahedral cage by heating the 11-vertex system $(\eta$-$C_5H_5)CoC_2B_8H_{10}$ to 623 K under vacuum.[62c] When the polyhedral expansion reaction of $C_2B_8H_{10}$ is carried out at 193 K instead of room temperature, in addition to isomers of $(\eta$-$C_5H_5)_2$-$Co_2C_2B_8H_{10}$ being produced, a further bimetallic anion, $[(\eta$-$C_5H_5)CoC_2B_8H_{10}$-$CoC_2B_8H_{10}]^-$, is obtained, which contains two different carbaborane units as shown in (10), with one cobalt atom of the $Co_2C_2B_8H_{10}$ unit being shared by another $C_2B_8H_{10}$ terminal unit.[62d] Some empirical rules which may be applied to thermal rearrangements of the non-icosahedral cobaltacarbaboranes $(\eta$-$C_5H_5)CoC_2B_nH_{n+2}$ $(n = 6, 7, 8,$ or $10)$ have been summarized.[62e] The isotropic shifts in ^{11}B and ^{13}C n.m.r. spectra of the paramagnetic metallo-carbaboranes $(\eta$-$C_5H_5)M(C_2B_nH_{n+2})$ and $M(C_2B_nH_{n+2})_2$ $(M = Cr^{III}, Fe^{III},$ $Ni^{III},$ or $Co^{III}; n = 6, 7, 8,$ or 9) have also been evaluated to provide some information on electron delocalization in such systems.[62f]

An unusual geometry shown in (11) has been found for the molecule $(\eta$-$C_5H_5)Co(\eta$-$C_5H_4B_9C_2H_{11})$, which is found as a side-product in the reaction of the $[B_{10}C_2H_{12}]^{2-}$ ion with $CoCl_2$–NaC_5H_5 to produce $(\eta$-$C_5H_5)Co(\eta$-$B_{10}C_2H_{12})$. The molecule is zwitterionic and may be considered as arising from a cobaltocenium cation and a $[B_{10}C_2H_{12}]^-$ anion, which become linked through a carbon–carbon bond with elimination of hydrogen. Apart from the overall

[61] E. L. Hoel, C. E. Strouse, and M. F. Hawthorne, *Inorg. Chem.*, 1974, **13**, 1388.
[62] (a) R. R. Rietz, D. F. Dustin, and M. F. Hawthorne, *Inorg. Chem.*, 1974, **13**, 1580;
(b) K. P. Callahan, C. E. Strouse, A. L. Sims, and M. F. Hawthorne, *ibid.*, p. 1397;
(c) W. J. Evans and M. F. Hawthorne, *J. Amer. Chem. Soc.*, 1974, **96**, 301; (d) G. Evrard, J. A. Ricci, I. Bernal, W. J. Evans, D. F. Dustin, and M. F. Hawthorne, *J.C.S. Chem. Comm.*, 1974, 234; (e) D. F. Dustin, W. J. Evans, C. J. Jones, R. J. Wiersema, H. Gong, S. Chan, and M. F. Hawthorne, *J. Amer. Chem. Soc.*, 1974, **96**, 3085; (f) R. J. Wiersema and M. F. Hawthorne, *ibid.*, p. 761.

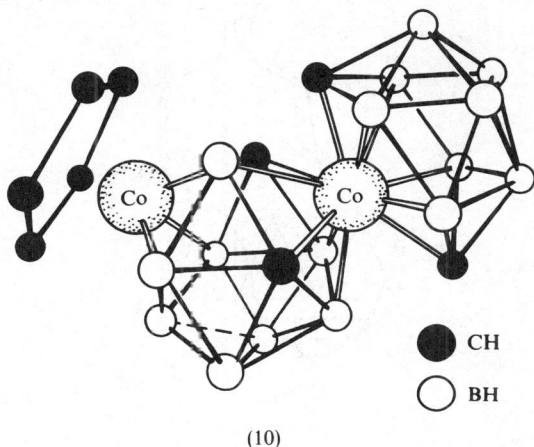

(10)

geometry, the point of interest lies in the position of the extra hydrogen atom, which lies asymmetrically above the open C_2B_3 pentagonal face, and although this atom interacts strongly with one of the boron atoms (B—H distance 1.33 Å) it also bonds, albeit more weakly, with the remaining four atoms in the face.[63]

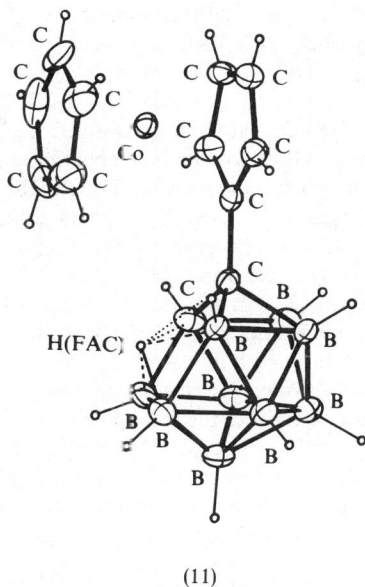

(11)

[63] M. R. Churchill and B. G. Deboer, *J. Amer. Chem. Soc.*, 1974, **96**, 6310.

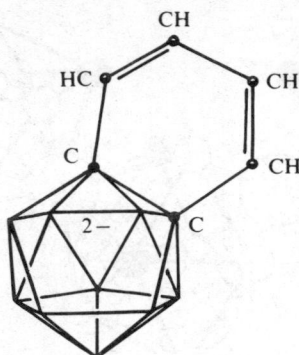

(12)

Various other types of metallocarbaborane compounds have been discovered. The benzodicarbollide ion (12) may be obtained from the (1,4-dihydrobenzo)-carbaborane by bromination and deprotonation reactions, and this dianion reacts with manganese, cobalt, and nickel compounds by established procedures to yield derivatives such as $\{[B_9H_9C_2(C_4H_4)]Mn(CO)_3\}^-$ and $\{[B_9H_9C_2-(C_4H_4)]_2Co\}^-$.[64a] Several reported complexes containing metal–carbaborane σ-bonds include cis- or trans-$(Et_3P)_2PtH(\sigma$-carbaborane)[64b] and the three-co-ordinate rhodium compounds $(PPh_3)_2Rh(\sigma$-carbaborane)[64c] (where for example σ-carbaborane = 2-R-1,2- and 7-R-1,7-$B_{10}C_2H_{10}$; R = H, Me, or Ph). They have been prepared by treating the lithio-derivative $LiRB_{10}C_2H_{10}$ with trans-$(Et_3P)_2PtHCl$ or $(PPh_3)_3RhCl$. Similar derivatives, such as $(PPh_3)_2Rh(H)-(B_9C_2H_{11})$, are effective catalysts for the isomerization and hydrogenation of hex-1-ene,[64d] while these compounds and $(PPh_3)_3RuHCl$ catalyse the H–D exchange in the B—H bonds of $1,2-B_{10}C_2H_{12}$ and other carbaboranes and boranes.[64e] The crystal structure of a previously isolated platinum compound, $(Pr_3P)_2Pt(2-Ph-1,2-B_{10}C_2H_{10})$, confirms the presence of a Pt–carbaborane σ-bond with a Pt—C distance of 2.13 Å, but also reveals a novel feature in that, while one phosphine ligand is bonded normally to the metal through the phosphorus atom, the other is bonded through both the phosphorus atom and the α-carbon atom of the n-propyl side-group. Unfortunately, the hydrogen atoms in the alkyl side-chains were not detected in the refinement of the structure.[64f]

The syntheses of nido- and closo-arsacarbaboranes have been achieved by reaction of alkyl (or aryl) arsenic(III) halides with the thallium(I) derivatives of the $B_9C_2H_{11}{}^{2-}$ ion. If an alkali metal is used as the cation then polymeric products are formed. It is proposed that the closo-compound has an icosahedral

[64] (a) D. S. Matteson and R. E. Grunzinger, Inorg. Chem., 1974, 13, 671; (b) S. Bresadola, B. Longato, and F. Morandini, J.C.S. Chem. Comm., 1974, 510; (c) S. Bresadola and B. Longato, Inorg. Chem., 1974, 13, 539; (d) T. E. Paxson and M. F. Hawthorne, J. Amer. Chem. Soc., 1974, 96, 4674; (e) E. L. Hoel and M. F. Hawthorne, ibid., p. 4676; (f) N. Bresciani, M. Calligaris, P. Delise, G. Nardin, and L. Randaccio, ibid., p. 5642.

$$closo\text{-}3\text{-}R\text{-}3\text{-}As\text{-}1,2\text{-}C_2B_9H_{11}$$

$$Tl_2C_2B_9H_{11} \underset{Me_2AsBr}{\overset{RAsX_2 \diagdown Et_2O}{\diagup}} {\diagdown}2\,moles$$

$$nido\text{-}(Me_2As)_2C_2B_9H_{11}$$

(R = Me, X = Br; R = Ph or Bu^n, X = Cl)

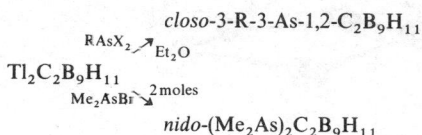

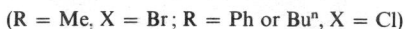

AsC$_2$B$_9$ skeleton, but in contrast to the same compound reported last year, the ^{11}B n.m.r. spectrum is invariant from room temperature down to 173 K, and the molecule does not therefore appear to be fluxional. The *nido*-compound has a 12-vertex structure with one of the Me$_2$As groups terminally bonded to a boron atom.[65a] Two new aza-dicarba-*nido*-boranes have also been prepared by the action of aqueous nitrous acid on the 7,8-C$_2$B$_9$H$_{12}^-$ anion and have been formulated as HNC$_2$B$_8$H$_{10}$ and H$_2$NC$_2$B$_8$H$_{11}$. Both compounds appear to contain an 11-atom, NC$_2$B$_8$, icosahedral fragment with the NH and NH$_2$ groups bonded into an open pentagonal face consisting of C$_2$B$_2$N atoms.[65b]

The replacement of a boron atom by a metal atom in a polyhedral framework has been termed a 'polyhedral subrogation' reaction, and it provides a useful pathway for the synthesis of bimetallo-carbaboranes from the monometallo-compounds. Thus if the general reaction

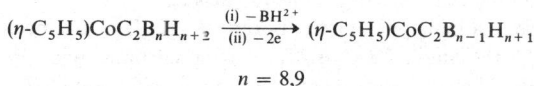

$$(\eta\text{-}C_5H_5)CoC_2B_nH_{n+2} \xrightarrow[\text{(ii)} -2e]{\text{(i)} -BH^{2+}} (\eta\text{-}C_5H_5)CoC_2B_{n-1}H_{n+1}$$

$$n = 8,9$$

is carried out in the presence of excess metal ion then this ion may be inserted into the position previously occupied in the framework by the departing boron atom. Such a reaction was used to prepare the first 13-atom bimetallic carbaborane reported last year, and has now been extended to prepare other such compounds, and one trimetallic 13-atom species. The bimetallic species have been prepared by the reaction:

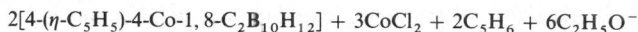

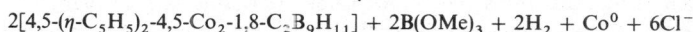

$$2[4\text{-}(\eta\text{-}C_5H_5)\text{-}4\text{-}Co\text{-}1,8\text{-}C_2B_{10}H_{12}] + 3CoCl_2 + 2C_5H_6 + 6C_2H_5O^-$$

$$\downarrow \text{KOH-EtOH}$$

$$2[4,5\text{-}(\eta\text{-}C_5H_5)_2\text{-}4,5\text{-}Co_2\text{-}1,8\text{-}C_2B_9H_{11}] + 2B(OMe)_3 + 2H_2 + Co^0 + 6Cl^-$$

If iron(II) chloride was substituted for cobalt(II) chloride the mixed FeCo compound (13) was obtained as 4,5-(η-C$_5$H$_5$)$_2$-4-Co-5-Fe-1,8-C$_2$B$_9$H$_{11}$. The trimetallic species [{(η-C$_5$H$_5$)CoC$_2$B$_9$H$_{11}$}$_2$Co]$^-$ was obtained in a similar reaction with the cyclopentadiene omitted.[35a]

The more usual type of cage expansion, namely reduction of a carbaborane, or metallocarbaborane, in the presence of a metal halide and sodium cyclo-pentadienide has led to the isolation of an even larger 14-vertex polyhedron,

---

[65] (a) H. D. Smith and M. F. Hawthorne, *Inorg. Chem.*, 1974, **13**, 2312; (b) J. Plešek, B. Štíbr, and S. Heřmánek, *Chem. and Ind.*, 1974, 662.

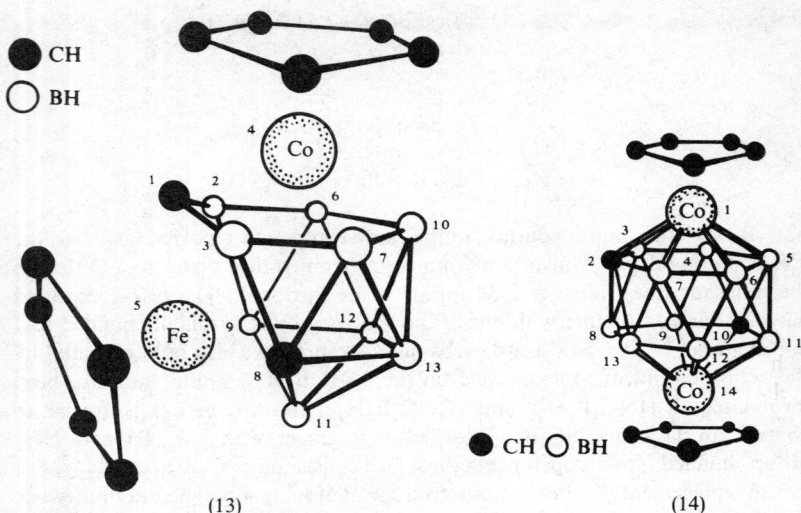

● CH
○ BH

(13)   (14)

$(C_5H_5)Co_2C_2B_{10}H_{12}$ (14), from the 13-vertex compounds 4,1,12- and 4,1,8-$C_5H_5CoC_2B_{10}H_{12}$. As in most of these derivatives, various isomers are possible depending upon the positions of the carbon atoms.[35b]

The synthesis of a new type of carbaborane, $C_4B_{18}H_{22}$, has been achieved by the oxidation of the anion $7,8-C_2B_9H_{12}{}^-$ using chromic acid. By varying the quantity of acid either the neutral carbaborane or the anion $C_4B_{18}H_{23}{}^-$ can be obtained. The structure has been proposed to consist of two linked $C_2B_9H_{11}$ cages on the basis of ^{11}B n.m.r. measurements. The carbaborane reacts with diborane under pressure to yield $1,2-C_2B_{10}H_{12}$ and $(C_2B_{10}H_{11})_2$, m.p. 503 K, and the latter compound appears to be an isomer of the previously prepared derivative which melts at 582 K.[66]

Compounds containing B—C Bonds. The crystal structure of triphenylborane shows that, unlike the corresponding derivatives of gallium and indium, there are no significant intermolecular contacts. While the BC_3 skeleton is planar, the phenyl rings are tilted by *ca.* 30° from the boron valence plane.[67a] The mechanism of stereoisomerism of triarylboranes in solution has been further studied, and the two-ring-flip process proposed earlier has been substantiated by a study of the temperature dependence of the ^1H n.m.r. spectrum of bis-(2,6-xylyl)-1-(3-isopropyl-2,4,6-trimethylphenyl)borane. The presence of the

[66] Z. Janousek, S. Heřmánek, J. Plešek, and B. Štibr, *Coll. Czech. Chem. Comm.*, 1974, **39**, 2363.
[67] (a) F. Zettler, H. D. Hausen, and H. Hess, *J. Organometallic Chem.*, 1974, **72**, 157; (b) J. P. Hummel, D. Gust, and K. Mislow, *J. Amer. Chem. Soc.*, 1974, **96**, 3679; (c) A. K. Holliday, W. Reade, K. R. Seddon, and I. A. Steer, *J. Organometallic Chem.*, 1974, **67**, 1; (d) J. D. Odom, L. W. Hall, S. Riethmiller, and J. R. Durig, *Inorg. Chem.*, 1974, **13**, 170; (e) J. R. Durig, R. O. Carter, and J. D. Odom, *ibid.*, p. 701.

meta-isopropyl group allows the rate of reversal of the helicity to be measured, while the xylyl methyl groups allow a simultaneous determination to be made of those ligand permutations which are independent of the reversal of helicity in an achiral medium.[67b] Evidence which suggests that π-bonding occurs in the B—C bond of trivinylborane,[67c] and that the molecule exists in planar and non-planar conformers in the fluid state,[67d] has been obtained from spectroscopic data, although other similar data on vinyl boron difluoride indicate that the B—C bond distance is very similar to that in other organoboranes, and that in this case the existence of any B—C π-bonding is doubtful.[67e] A comparison of the reaction pathways in the two exchange reactions between (i) Me_3PBMe_3 and excess BMe_3 and (ii) Me_3PAlMe_3 and excess Al_2Me_6, indicates that for the former the dissociation of the complex is rate-determining, but for the latter an exchange intermediate involving the complex and monomeric trimethylalane is likely. This paper also includes lineshape equations for exchange between singlet and first-order doublet sites.[68a] Reaction of the bis(borinato)-cobalt compounds $Co(C_5H_5BR)_2$ (R = Me or Ph) with iron carbonyls yields the dimeric species $[Fe(CO)_2(C_5H_5BR)]_2$, which consists of a doubly CO-bridged structure, as confirmed by an X-ray study.[68b] Pyrolysis of these compounds at 503 K yields the corresponding bis(borinato)iron compounds $Fe(C_5H_5BR)_2$ (15).[68c]

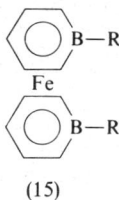

(15)

The liquid tetra-alkylammonium tetra-alkylboride $Et_3(C_6H_{11})N^+BEt_3$-$(C_6H_{11})^-$, $(N_{2226}B_{2226})$, has been used as a solvent for the reactions between tetraethylammonium halides with methyl toluene-*p*-sulphonate to determine the nucleophilic reactivities of the halide ions. The relative rate constants are in the order $Cl^- > Br^- > I^-$, as found in other polar aprotic solvents, but the range of the constants in $N_{2226}B_{2226}$ is much smaller, probably due to the effect of the neighbouring R_4N^+ ions.[69a] The use of the trialkylalkynylborates, and cyanoborates, in organic reactions has been extended to the stereospecific synthesis of allyl amides,[69b] 4-substituted pyridines,[69c] and for the preparation of derivatives of the type $R_2^1BC(R)^1=C(R^2)-PD_2^3$ via reaction with R_2^3PCl

[68] (a) E. Alaluf, K. J. Alford, E. O. Bishop, and J. D. Smith, *J.C.S. Dalton*, 1974, 669; (b) G. Huttner and W. Gartzke, *Chem. Ber.*, 1974, **107**, 3786; (c) G. E. Herberich, H. J. Becker, and G. Greiss, *ibid.*, p. 3780.
[69] (a) W. T. Ford, R. J. Hauri, and S. G. Smith, *J. Amer. Chem. Soc.*, 1974, **96**, 4316; (b) A. Pelter, A. Arase, and M. G. Hutchings, *J.C.S. Chem. Comm.*, 1974, 346; (c) A. Pelter and K. J. Gould. *ibid.* p. 347; (d) P. Binger and R. Koster, *J. Organometallic Chem.*, 1974, **73**, 205; (e) A. Pelter, K. J. Gould, and L. A. P. Kane-Maguire, *J.C.S. Chem. Comm.*, 1974, 1029.

compounds.[69d] The $R_3^1BC{\equiv}CR^2$ anions attack cationic metal complexes in a stereo- and regio-specific manner to yield, e.g. with dienyl-iron complexes of the type $[(C_6H_6OMe(Fe(CO)_3]^+ BF_4{}^-$, the *exo*-5-substituted cyclohexadiene derivatives $R_2^1BC(R^1){=}C(R^2)(C_6H_6OMe)Fe(CO)_3$, which upon hydrolysis and oxidation yield cyclohexenones, and with trimethylamine oxide and oxidation produce diketones.[69e]

Compounds containing B—N *Bonds.* MO calculations on the simple adduct H_3B,NH_3 have shown that the transfer of charge occurs almost exclusively from the highest occupied MO on the NH_3 molecule to the lowest empty MO on BH_3.[70a] The relative effectiveness of the N and P atoms as donor centres in the series of molecules Me_2NPMe_2, $(Me_2N)_2PMe$, and $(Et_2NCH_2)_3P$ has been investigated by n.m.r. spectroscopy, using BH_3 as the acceptor. In the first two, phosphorus acts as the donor but in the last phosphine, where the N and P atoms are separated by a methylene group to inhibit any π-bonding, it is the nitrogen centre which binds to boron.[70b] Exchange reactions involving the adduct Me_3N,BH_2I with the various anions X^- (X = Cl, Br, NCS, NCO, or BH_3CN) have been used as a convenient synthetic route to the compounds Me_3B,BH_2X, but the choice of reagents is important since the related compounds Me_3N,BH_2Y (Y = Cl or Br) and Me_3N,BHZ_2 (Z = Br or I) do not undergo such reactions.[70c] Similar amine cyanoboranes R_2HN,BH_2CN have been synthesized, and the mechanism of the hydrolysis reaction

$$R_2HN,BH_2CN + 2H_2O + 2OH^- \longrightarrow R_2HN + B(OH)_4{}^- + 2H_2 + CN^-$$

has been determined for different amines. It appears to be analogous to the accepted mechanism for the base hydrolysis of ammine cobalt(III) complexes, namely removal of the N-bonded proton by the hydroxide ion followed by a rate-determining decomposition, probably *via* a dissociative pathway involving the substrate conjugate base.[70d] New amine adducts, $Me_{3-n}B(SMe)_nC_5H_5N$, have been obtained directly from the methyl(methylthio)boranes, and the corresponding boronium salts, e.g. $[Me_2B(C_5H_5N)_2]Cl$, are formed when the adducts are dissolved in CH_2Cl_2.

Several studies relate to dialkylaminoboranes. An improved preparation of $B(NR^1R^2)_3$ compounds has been achieved:[71a]

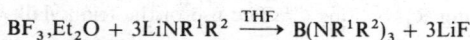

$$BF_3,Et_2O + 3LiNR^1R^2 \xrightarrow{THF} B(NR^1R^2)_3 + 3LiF$$

($R^1 = R^2$ = Me, Et, Pr^i, Ph, or $PhCH_2$; R^1 = Ph, R^2 = Me; *etc.*)

[70] (a) H. Fujimoto, S. Kato, S. Yamabe, and K. Fukui, *J. Chem. Phys.*, 1974, **60**, 572; (b) C. Jouany, J. P. Laurent, and G. Jugie, *J.C.S. Dalton*, 1974, 1510; (c) P. J. Bratt, M. P. Brown, and K. R. Seddon, *ibid.*, p. 2161; (d) C. Weidig, S. S. Uppal, and H. C. Kelly, *Inorg. Chem.*, 1974, **13**, 1763; (e) H. Nöth and U. Schuchardt, *Chem. Ber.*, 1974, **107**, 3104.

[71] (a) W. R. Purdum and E. M. Kaiser, *J. Inorg. Nuclear Chem.*, 1974, **36**, 1465; (b) R. H. Neilson and R. L. Wells, *Inorg. Chem.*, 1974, **13**, 480; (c) H. Nöth and W. Storch, *Chem. Ber.*, 1974, **107**, 1028; (d) H. Nöth, W. Tinhof, and B. Wrackmeyer, *ibid.*, p. 518.

A similar reaction of LiN(alkyl)(SiMe$_3$) with PhBCl$_2$ or PhB(NMe$_2$)Cl has led to the corresponding aminoboranes (Me$_3$Si)(alkyl)NB(X)Ph (alkyl = Me, Bu^t, X = NMe$_2$; alkyl = Me, X = Cl; *etc.*), and these compounds decompose at temperatures ranging from 293—423 K to yield the substituted borazines, *e.g.* (Pr^{i}NBPh)$_3$.[71b] The silicon–nitrogen bond in the related derivatives B[N(Me)-SiMe$_3$]$_3$ may be successively cleaved by Me$_2$BBr, and all members of the series [Me$_2$BNMe]$_n$B[N(Me)SiMe$_3$]$_{3-n}$ (*n* = 0—3) can be isolated.[71c] Evidence for (*p–d*)π- and (*p–p*)π-bonding in the Si—N and B—N bonds respectively has been obtained from ^{14}N and ^{11}B n.m.r. studies.[71d]

The reaction of excess diborane with tris(dimethylamine)alane, Al(NMe$_2$)$_3$, in ether affords the trimeric borane (Me$_2$NBH$_2$)$_3$ in good yields; this method also provides a route to higher borane derivatives *via*:[72a]

$$Me_2NHBH_2NMe_3 + KH \rightarrow K[Me_2NBH_2NMe_2BH_3] + \tfrac{1}{2}H_2$$
$$\downarrow B_2H_6-Et_2O$$
$$HB(NMe_2)_2 + B_2H_5 \xrightarrow{Et_2O} \mu\text{-}[Me_2N]_2B_3H_7 \, [+KBH_4]$$

The same triborane, μ-(Me$_2$N)$_2$B$_3$H$_7$, also occurs as one of several by-products in the reaction of excess diborane with HAl(NMe$_2$)$_2$ and [H$_2$B(NMe$_2$)$_2$]$_2$AlH but the main product is the substituted aluminium hydroborate, H$_2$B(NMe$_2$)$_2$-Al(BH$_4$)$_2$.[72b]

Tetrakis(dimethylamino)diborane(4) reduces vanadium tetrachloride to yield the vanadium(III) derivative VCl$_3$,(NMe$_2$)$_2$CH$_2$ but the bis-dimethylamino)-methane, which possibly acts as a chelating ligand in the complex, is easily removed by the addition of pyridine, when VCl$_3$,(py)$_3$ is formed.[73a] The N-boryl ketimines R$_2$C=N=BR$_2$, which are isoelectronic with allene, are capable of acting as ligand species and react readily with bis(benzonitrile)palladium(II) chloride to yield, for example, (Ph$_2$CNBEt$_2$)$_2$PdCl$_2$. It is believed that it is the C=N group which is the active donor centre, and that the N=B bond is not involved in the metal–ligand bonding.[73b]

Several new compounds involving the polypyrazolylborate anions, H$_n$B(pz)$_{4-n}$ (pz = pyrazol-1-yl, *n* = 0, 1, or 2), with metal ions have been prepared. The crystal structure of one of the platinum compounds reported last year, [HB(pz)$_3$]PtMe-[C$_2$(CF$_3$)$_2$], has confirmed that the co-ordination about the metal atom is nearly a trigonal bipyramid, with the borate anion terdentate, and the acetylene, considered as a unidentate ligand, occupying one of the equatorial positions.[74a] In contrast, the structure of [HB(pz)$_3$]Pt(Me)CO, also determined from X-ray data, shows the ligand to be only bidentate, with the platinum being surrounded by 2N and 2C atoms in a slightly distorted square-planar arrangement.[74b]

[72] P. C. Keller, *J. Amer. Chem. Soc.*, 1974, **96**, 3078; (b) P. C. Keller, *ibid.*, p. 3073.
[73] (a) R. F. Kiesel and E. P. Schram, *Inorg. Chem.*, 1974, **13**, 1313; (b) G. Schmid and L. Weber, *Chem. Ber.*, 1974, **107**, 547.
[74] (a) B. W. Davies and N. C. Payne, *Inorg. Chem.*, 1974, **13**, 1843; (b) P. E. Rush and J. D. Oliver, *J.C.S. Chem. Comm.*, 1974, 996.

It has been established from X-ray data that the B_3N_3 borazine ring in $(Me_3B_3N_3Me_3)Cr(CO)_3$ is π-bonded to the chromium atom, and further derivatives $(R_3^1B_3N_3R_3^2)Cr(CO)_3$ (R^1 = Me, Pr^n, Me, Pr^i, Me when R^2 = H, Me, Pr^n, Me, Pr^i, respectively) have now been prepared.[75a,b] Substituted borazines which

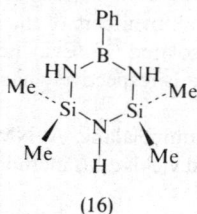

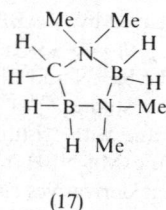

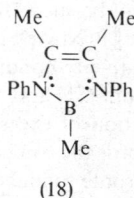

(16)   (17)   (18)

contain a silicon atom in a BSi_2N_3 ring (16) have been obtained from the reaction of $PhBCl_2$ with $(Me_2SiNH)_3$.[75c] The synthesis of the new small ring heterocyclic boron compound (17) has been achieved by the action of Na–K alloy on $(Me_3N)_2BH_2{}^+I^-$. It is a clear liquid, stable at 298 K, but is decomposed on contact with acid or alkali.[76a] The range of diazaborolines (18), which are isoelectronic with the $C_5H_5{}^-$ ion and were reported for the first time last year, has been extended by the reaction of $PhN=C(Me)C(Me)=NPh$ with $MeBBr_2$ to yield an intermediate boronium salt, which is then reduced by Na–Hg amalgam.[76b] Previous attempts to prepare borazarenes have been unsuccessful but now the tetrahydro-borazarene $Me\overline{N(CH_2)_4BH}$ has been dehydrogenated over a

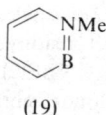

(19)

Pd/Al_2O_3 catalyst to yield the N-methyl derivative (19). However, as other workers have suggested, this compound is very reactive, and it has only been identified by spectroscopic results, but it is apparently not at all like benzene but resembles more a polarized butadiene molecule.[76c]

Compounds containing B—O and B—S Bonds. Condensation of triborylmethide anions $[(R^1O)_2B]_3C^-$ with ketones or aldehydes yields alkene-1,1-diboronic esters, $R_2^2C=C[B(OR^1)_2]_2$, which are potentially useful intermediates in that they react easily, *e.g.* hydrogen peroxide yields carboxylic acids, and a mercury(II) chloride–sodium acetate mixture produces the mercury compound

[75] (a) M. Scotti and H. Werner, *Helv. Chim. Acta*, 1974, **57**, 1234; (b) M. Scotti and H. Werner, *J. Organometallic Chem.*, 1974, **81**, C17; (c) H. Nöth, W. Tinhof, and T. Taezer, *Chem. Ber.*, 1974, **107**, 3113.
[76] (a) B. R. Gragg and G. E. Ryschkewitsch, *J. Amer. Chem. Soc.*, 1974, **96**, 4717; (b) L. Weber and G. Schmid, *Angew. Chem. Internat. Edn.*, 1974, **13**, 467; (c) H. Wille and J. Goubeau, *Chem. Ber.*, 1974, **107**, 110.

$R_2^2C=C(HgCl)_2$.[77a] Several papers have been concerned with heterocycles containing boron and oxygen (or sulphur) atoms in the ring. Triphenylborthiin, $(PhBS)_3$, shows an intense molecular ion in its mass spectrum, suggesting that the B_3S_3 ring system is more stable, probably due to charge delocalization in the ring, than had previously been supposed.[77b] A series of 2-substituted 1,3,2-oxathiaborinans, $\overline{O(CH_2)_3SBX}$ (X = Ph, OR, NHR, NR_2, or SR; R = alkyl),[77c] and boron derivatives of o-hydroxybenzyl alcohol, $\overline{XBO(o\text{-}C_6H_4)CH_2O}$ (X = Ph, NR_2, NHR, OR, or SR; R = alkyl),[77d] have been prepared from $B(NR_2)_3$ and $(PhBO)_3$, respectively. The related cyclic system $\overline{H_2BO(o\text{-}C_6H_3R^1)CH_2N(R^2)}(R^3)$ $(R^1, R^2, R^3$ = alkyl) has also been prepared by the action of diborane on the benzoxazine $\overline{CH_2O(o\text{-}C_6H_3R^1)CH_2N(R^2)}$.[77e]

Several derivatives of the 1,2,5-thiadiborole system (20) have been characterized from reactions using tri-iodoborthiin, $(IBS)_3$, e.g.

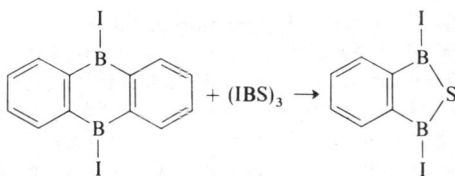

(20)

The sulphur atom in the ring can be substituted by NR groups (R = alkyl or phenyl) by treatment with the corresponding primary amine.[78]

When a mixture of $M_3P_2O_8$ and $Na_2B_4O_7,10H_2O$ (M = Ca or Sr) is heated to 1673 K the crystalline product is of composition $M_{9+y}Na_x(PO_4)_6B_{x+2y}O_2$ $(9 + y + x < 10, x + 2y < 1, x > 0, y > 0)$. Different crystals were examined using X-rays and were found to be structurally related to apatite, with a linear OBO grouping at the centre of symmetry. The B—O distance is 1.25 ± 0.02 Å, and this is the first identification of such a grouping in the solid state.[39b] The structures of sodium diborate, $Na_2O,2B_2O_3$, and triborate, $\alpha\text{-}Na_2O,3B_2O_3$, show that the former contains one non-bridging oxygen atom in borate layers composed of di-pentaborate and triborate groups.[79a] In the latter the borate anion forms two separate interpenetrating infinite frameworks, with each framework containing pentaborate and diborate groups in equal amounts.[79b] In nickel orthoborate, $Ni_3(BO_3)_2$, all the boron atoms are triangularly co-ordinated, but there are two types of nickel atom, both octahedrally surrounded by oxygen atoms and linked by oxygen-sharing to form a three-dimensional network.[79c]

[77] (a) D. S. Matteson and P. B. Tripathy, J. Organometallic Chem., 1974, 69, 53; (b) R. H. Cragg and A. F. Weston, J.C.S. Chem. Comm., 1974, 22; (c) R. H. Cragg and M. Nazery, J.C.S. Dalton, 1974, 1438; (d) ibid., p. 162; (e) R. E. Lyle and D. A. Walsh, J. Organometallic Chem., 1974, 67, 363.
[78] B. Asgarouladi, R. Full, K. J. Schaper, and W. Siebert, Chem. Ber., 1974, 107, 34.
[79] (a) J. Krogh-Moe, Acta Cryst., 1974, B30, 578; (b) J. Krogh-Moe, ibid., p. 747; (c) J. Pardo, M. M. Ripoll, and S. G. Blanco, ibid., p. 37; (d) P. M. Gasperin, ibid., p. 1181.

The new borate $TlNbB_6O_6$ has been synthesized by heating the respective oxides, and it also contains an octahedrally co-ordinated metal atom (niobium), the octahedra being linked by the corners into a zig-zag chain, with each chain joined together by BO_3 triangles sharing one oxygen atom.[79d]

Compounds containing Boron–Halogen Bonds. An examination of the electronic states of B^+ ions produced by electron bombardment of boron halides at 70 eV shows that the percentage of excited-state (^3P) atoms compared with ground-state (^1S) atoms varies in a regular manner from BF_3 (35% ^3P, 65% ^1S) to BI_3 (100% ^1S). There is no apparent rationale for this observation.[80] An *ab initio* SCF study on the boron subhalides B_4F_4 and B_4Cl_4 has been made using a minimum basis set of Slater orbitals. The results show that the bonding MO's involved are of E, T_1, and T_2 symmetry and are composed mainly of F(2p) and Cl(3p) orbitals, and therefore both molecules are stabilized by back-donation into the B_4 tetrahedron, with the back-bonding being greater for B_4F_4 than for B_4Cl_4.[81]

The boron halides can react with transition-metal compounds in a variety of ways. Thus BX_3 (X = Cl, Br, or I) form a series of unstable 1 : 1 adducts with cobalt carbonyl,[40b] but the rather more stable adducts $[(\eta\text{-}C_5H_5)Fe(CO)_2]_2,BX_3$ (X = Cl or Br), $(\eta\text{-}C_5H_5)Ni_3(CO)_2,BF_3$ (formed at 195 K), and $[\eta\text{-}C_5H_5FeCO]_4$, nBX_3 (X = F, n = 1, 2, or 4; X = Cl, n = 1 or 2; X = Br, n = 1 or 2) are known where the bridging carbonyl group is believed to act as the donor centre.[40a] In the case of rhodium compounds direct metal–boron bonding is probable in some adducts (*cf.* p. 262).[82]

Halogen-exchange reactions in adducts derived from boron halides have led to mixed halogeno-adducts between $MeCl_2P,BX_3$ (X = Cl, Br, or I), to form *e.g.* $MeClBrP,BBr_2Cl$ from $MeCl_2P,BBr_3$,[83a] and $MeC(X)YMe,BZ_3$ (X = O or S, Y = O or S; Z = halogen) to form *e.g.* $MeC(O)SMe,BCl_2F$. In the latter case, where the O or S atom can act as the donor centre, it is found that donation always occurs from the carbonyl or thiocarbonyl centre.[83b] The general nature of exchange reactions occurring in adducts of nitrogen donors with boron halides, *e.g.* $o\text{-}MeC_6H_4NMe_2,BCl_3$, has been examined, and unimolecular bond fission of a boron–chlorine bond in the molecular adduct, $DBX_3 \rightleftharpoons DBX_2\cdots X$, is the rate-determining step for amine scrambling. The existence of multiply charged boronium ions in the mixture has also been established from spectral results in the presence of tetraphenylarsonium chloride.[83c] Mixed halide adducts with tetramethylurea (tmu) have also been prepared, *e.g.* tmu,BCl_2F, by mixing solutions of the respective tmu,BX_3 adducts.[83d]

[80] K. Lin, R. J. Cotter, and W. S. Koski, *J. Chem. Phys.*, 1974, **60**, 3412.
[81] J. H. Hall and W. N. Lipscomb, *Inorg. Chem.*, 1974, **13**, 710.
[82] D. D. Lehman and D. F. Shriver, *Inorg. Chem.*, 1974, **13**, 2203.
[83] (a) R. M. Kren, M. A. Mathur, and H. H. Sisler, *Inorg. Chem.*, 1974, **13**, 174; (b) M. J. Bula, J. S. Hartman, and C. V. Raman, *J.C.S. Dalton*, 1974, 725; (c) J. R. Blackborow, M. N. S. Hill, and S. Kumar, *ibid.*, p. 411; (d) J. S. Hartman and G. J. Schrobilgen, *Inorg. Chem.*, 1974, **13**, 874.

A chloro-substituted boro-adamantane $(BCl)_6(CH)_4$ [the methyl derivative $(BMe)_6(CH)_4$ was reported last year] has been obtained in low yield from the thermal decomposition of 1,2-bis(dichloroboryl)ethane, $Cl_2BCH_2CH_2BCl_2$, at 673—773 K. It is a solid which is very easily hydrolysed in air.[84] A novel vapour pump for the synthesis of B_2Cl_4 from BCl_3 has been described.[85]

Metal Borides and Elemental Boron. Four different forms of elemental boron have been reported, namely α- and β-rhombohedral boron, both of which contain B_{12} icosahedra, and I- and II-tetragonal boron, the structures of which have not been definitely established. Doubts concerning the existence of the I-tetragonal form have been expressed recently by one of the workers who first reported this form in 1943, and a report this year substantiates those doubts by showing that I-tetragonal boron is not $B_{50}[(B_{12})_4B_2]$ but is a boron-rich tetragonal boron carbide $B_{50}C_2$, which has the lattice form of the I-tetragonal boron. The crystal structures of both this compound and a nitride, $B_{50}N_2$, prepared by the pyrolysis of $BBr_3-CH_4-H_2$ or $BBr_3-N_2-H_2$ at 1473 and 1673 K, respectively, show that 48 B atoms are located in four B_{12} icosahedra and the remaining B, C (or N) atoms take positions in the tetragonal icosahedral framework.[39a]

Aluminium.—A general review has considered the structure and behaviour of aluminate ions in solution.[86a] The ^{27}Al n.m.r. spectra of the aluminium perchlorate complexes $AlL_6(ClO_4)_3$ [L = $P(O)(OMe)_3$; $MeP(O)(OMe)_2$] in nitromethane solution show sharp heptets, indicating octahedral co-ordination around the metal atom. The addition of water gives extra signals ascribed to the mixed octahedral solvates $AlL_n(H_2O)^{3+}_{6-n}$. When hexamethylphosphoramide (HMPA) is used as the ligand a sharp quintet in the ^{27}Al n.m.r. spectrum of $AlL_4(ClO_4)_3$ indicates tetrahedral co-ordination at the aluminium, but in this case addition of water affords mixed octahedral solvates, $Al(HMPA)_n(H_2O)^{3+}_{6-n}$, and no mixed tetrahedral solvates can be detected.[86b]

X-Ray emission spectra have been obtained for a series of aluminium compounds, including the metal itself, using the aluminium $K\alpha$ and $K\beta_{1.3}$ X-rays. The relative intensities of the two peaks can be related to bond length and the degree of ionic character, while the relative intensities of the components of any one peak are dependent upon the degree of atomic orbital participation in particular molecular orbitals. In the cases considered here for Al—F, Al—Cl, and Al—O bonds the fine structure of the $K\beta_{1.3}$ peak has been correlated with the occurrence of σ- and π-bonding between the metal and the ligand. Thus in Na_3AlF_6, AlF_3, and tris(acetylacetonato)aluminium, π-bonding appears to be present, while in α-alumina, topaz, kyanite, and microcline (aluminosilicate minerals) there appears to be little or no π-bonding.[87]

[84] M. S. Reason, A. G. Briggs, J. D. Lee, and A. G. Massey, *J. Organometallic Chem.*, 1974, **77**, C9.
[85] J. P. Brennan, *Inorg. Chem.*, 1974, **13**, 490.
[86] (a) N. I. Eremin, Yu. A. Volokhov, and V. E. Mironov, *Russ. Chem. Rev.*, 1974, **43**, 92; (b) J. J. Delpuech, M. R. Khaddar, A. Regny, and P. Rubini, *J.C.S. Chem. Comm.*, 1974, 154.
[87] C. J. Nicholls and D. S. Urch, *J.C.S. Dalton*, 1974, 901.

Tricyclic pyrazolyl derivatives of aluminium and gallium, $[N_2C_3H_3,MR_2]_2$ (M = Al or Ga; R = H, D, Me, Et, or Cl), have been synthesized from pyrazole and the appropriate aluminium or gallium precursor such as Me_6Al_2 and Me_3N,GaD_3. The compounds are dimeric in solution and in all cases a boat conformation for the six-membered ring of N_4M_2 (M = Al or Ga) atoms is proposed.[88] The existence of Sn—M (M = Al, Ga, In, or Tl) bonding in the known compounds $Li^+[Me_3SnMMe_3]^-$, and the new compounds $Li^+[Me_3-Sn)_nTlMe_{4-n}]$ (n = 0—4), has been established from $J(\underline{Sn}MC\underline{H})$.[89]

Compounds containing Al—C Bonds. An X-ray study of $K(HAlMe_3)$, which is obtained by pyrolysis of $K(H_3SiAlMe_3)$, shows that it contains a tetrahedral $[HAlMe_3]^-$ anion.[90a] The general phenomenon of exchange of bridge and terminal alkyl groups in Al_2R_6 derivatives has been studied by a total lineshape analysis of ^{13}C n.m.r. spectra for Al_2Et_6 and $Al_2Pr^n_6$ in toluene and cyclopentane solvents. The mechanism proposed involves an intramolecular process for both compounds in cyclopentane, but it may become intermolecular in toluene. The interconversion becomes faster as the chain length of R increases.[90b] The observation that some MR_3 (M = Al or Ga) compounds occur as monomers when the R group contains alkene groups, *e.g.* $Al[(CH_2)_2C(H)=CH_2]_3$ and $M[(CH_2)_3C(H)=CH_2]_3$, has been rationalized in terms of intramolecular π-interactions between the olefinic bond and the metal atom on the basis of the ^1H n.m.r. spectra of these compounds.[90c] A determination of ΔH_f^0 values for several aluminium alkyls and related compounds, notably the primary alkyl aluminium halides, has been made from the heats of redistribution, and the values have been compared to show that many of the existing literature values on these compounds, especially those obtained from heat of combustion measurements, require revision.[91a] A redetermination of the structure of $(Me_2AlCl)_2$ in the gas phase by electron diffraction confirms the D_{2h} symmetry of the molecule with bridging chlorine atoms, and shows that the terminal Al—C bonds (1.93 Å) are shorter than those in Al_2Me_6 (1.96 Å), while the Al—Cl bridge bonds (2.30 Å) are significantly longer than those in Al_2Cl_6 (2.25 Å).[91b]

The $Me_3SiCH_2^-$ group is known to be effective for stabilizing a range of transition-metal alkyls, and the aluminium compound $Al(CH_2SiMe_3)_3$, which may prove to be an effective alkylating agent as preliminary studies indicated last year, has now been prepared. It is obtained as a colourless inflammable liquid from the action of $(Me_3SiCH_2)_2Hg$ on aluminium foil, and appears to be a mixture of monomer and dimer in benzene solution.[41]

The thermal decomposition of the adducts Et_2AlX,NH_2R (X = Et, Cl, Br, or I; R = Me or Bu^i) has been shown to yield the $(EtAlXNHR)_2$ compounds containing

[88] A. Arduini and A. Storr, *J.C.S. Dalton*, 1974, 503.

[89] A. T. Weibel and J. P. Oliver, *J. Organometallic Chem.*, 1974, **74**, 155.

[90] (a) G. Hencken and E. Weiss, *J. Organometallic Chem.*, 1974, **73**, 35; (b) O. Yamamoto, K. Hayamizu, and M. Yanagisawa, *ibid.*, p. 17; (c) T. W. Dolzine and J. P. Oliver, *J. Amer. Chem. Soc.*, 1974, **96**, 1737.

[91] (a) M. B. Smith, *J. Organometallic Chem.*, 1974, **76**, 171; (b) K. Brendhaugen, A. Haaland, and D. P. Novak, *Acta Chem. Scand.*, 1974, **A28**, 45.

an $(Al-N)_2$ four-membered non-planar ring, as confirmed by an X-ray structure determination of the *cis*-isomer of $[Et(Br)AlNHBu^t]_2$.[92a,b] A similar $(Al-N)_2$ ring is present in the dimeric derivative $[Me_2AlNCMe_2]_2$ prepared by heating $K[Al_2Me_6SCN]$ to 393 K.[92c] Elimination of alkane also occurs in the reaction of $X_2P(O)N(H)Me$ with Al_2Et_6 to form compounds $X_2P(O)N(Me)AlEt_2$ (X = OR or NR_2; R = alkyl) containing an $Al-N$ bond. In these compounds an equilibrium exists between monomers and dimers as:

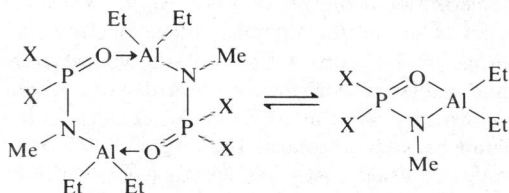

On reaction with aldehydes (*e.g.* PhCHO) the corresponding *N*-arylidene methylamine, $PhC(H)=NMe$, is obtained[92d] together with the aluminium compound $Et_2AlOP(X_2)O$.

The use of aluminium alkyls in organic reactions has been extended to show that ethers are cleaved photochemically in the presence of aluminium alkyls, and benzylic alcohols (R^3OH) are alkylated under similar conditions:[93a]

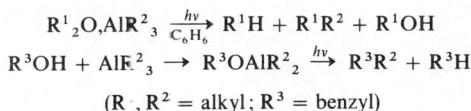

$$R^1_2O,AlR^2_3 \xrightarrow[C_6H_6]{h\nu} R^1H + R^1R^2 + R^1OH$$

$$R^3OH + AlR^2_3 \rightarrow R^3OAlR^2_2 \xrightarrow{h\nu} R^3R^2 + R^3H$$

$$(R^, R^2 = \text{alkyl}; R^3 = \text{benzyl})$$

N-Benzylanilines, $PhNHCH_2Ph$, rearrange on irradiation in the presence of $Et_2Al_2Cl_4$ to yield the *meta*-substituted product, m-$PhCH_2C_6H_4NH_2$, whereas in the absence of the aluminium compound it is the *o*- and *p*-isomers which are obtained.[93b] A series of papers has considered the methylation of alcohols $(ROH \rightarrow RMe)$,[93c] ketones $(R_2CO \rightarrow R_2CMe_2)$,[93d] and carboxylic acids $(RCO_2H \rightarrow RCMe_3)$ by trimethylalane at 373—473 K. Methyl-group transfer also occurs when Al_2Me_6 is heated at 408 K with lead sulphide to give $PbMe_4$, metallic lead, and $(Me_2Al)_2S$,[94a] the reaction being very similar to that involving lead oxide, PbO, reported in 1972. The insertion of carbon disulphide into the

[92] (a) R. E. Bowen and K. Gosling, *J.C.S. Dalton*, 1974, 964; (b) R. E. Bowen and K. Gosling, *ibid.*, p. 1961; (c) S. K. Seale and J. L. Atwood, *J. Organometallic Chem.*, 1974, **73**, 27; (d) K. Urata, K. Itoh and Y. Ishii, *ibid.*, 1974, **76**, 203.

[93] (a) J. Furukawa, K. Omura, O. Yamamoto, and K. Ishikawa, *J.C.S. Chem. Comm.*, 1974, 77; (b) J. Furukawa, K. Omura, and S. Sawada, *ibid.*, p. 78; (c) D. W. Harney, A. Meisters, and T. Mole, *Austral. J. Chem.*, 1974, **27**, 1639; (d) A. Meisters and T. Mole, *ibid.*, p. 1655; (e) *ibid.*, p. 1655.

[94] (a) M. Boleslawski, S. Pasynkiewicz, and A. Kunicki, *J. Organometallic Chem.*, 1974, **73**, 193; (b) K. Wakatsuki, Y. Takeda, and T. Tanaka, *Inorg. Nuclear Chem. Letters*, 1974, **10**, 383.

Al—N bond in Me_2AlNPh_2, to yield $Me_2AlS_2CNPh_2$, occurs when the reactants are refluxed together in hexane.[94b]

Compounds containing Al—O and Al—Halogen Bonds. γ-Alumina is known to be an effective hydrogenation catalyst only above *ca.* 773 K, but its activity may be increased by certain treatments such as the introduction of iron impurity centres. Another simple means of achieving higher activity now reported is mild alkali treatment using sodium acetate (or hydroxide), after which hydrogenation and isomerization of alkenes occurs at 373 K. The alkali is believed to generate two types of site on the alumina surface, one of which is capable of dissociating hydrogen and is probably an 'iron' site, while the other behaves as a Lewis-acid site and is able to absorb the unsaturated hydrocarbon.[95] Aluminium alkoxides are known to possess unusual properties, such as their variation of degree of association between the solid and liquid phases. The nature of $Al(OPr^i)_3$ has been reinvestigated using ^{27}Al n.m.r. spectroscopy, which indicates that at 353 K there are two types of aluminium centres in a 3 : 1 ratio which contain the metal atom tetrahedrally and octahedrally co-ordinated respectively, the two centres being linked by bridging OPr^i groups.[96]

The crystal structure of the adduct between tetramethylcyclobutadiene, C_4Me_4, and aluminium trichloride, which was suggested to contain an Al—C

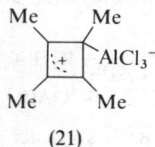

Me   Me

$AlCl_3^-$

Me   Me

(21)

σ-bond last year from n.m.r. evidence, has now been confirmed as (21). This compound is believed to be an intermediate in the trimerization of but-2-yne to hexamethyl Dewar benzene.[97a] The ^{35}Cl n.q.r. transitions in a series of tetrachloraluminate compounds have been studied to diagnose interactions between the $AlCl_4^-$ anions and different cations. In $NaAlCl_4$, $Te_4(AlCl_4)_2$, $ICl_2(AlCl_4)$, $Bi_5(AlCl_4)_3$, and $GaAlCl_4$ the transitions lie between 10.6 and 11.3 MHz and are indicative of ionic compounds containing discrete $AlCl_4^-$ ions. In $Co(AlCl_4)_2$ and $Hg_3(AlCl_4)_2$ both the average frequency and range of transitions are increased, reflecting anion–cation interaction. The existence of high anions $(Al_2Cl_7^-, Al_3Cl_{10}^-$ *etc.*) in compounds may also be verified using this method and in contrast to $Te_4^{2+}(AlCl_4^-)_2$ there is weak halogen bridging interaction

[95] P. A. Sermon, G. C. Bond, and G. Webb, *J.C.S. Chem. Comm.*, 1974, 417.
[96] J. W. Akitt and R. H. Duncan, *J. Magn. Resonance*, 1974, **15**, 162.
[97] (a) C. Kruger, P. J. Roberts, Y. H. Tsay, and J. B. Koster, *J. Organometallic Chem.*, 1974, **78**, 69; (b) D. J. Merryman, P. A. Edwards, J. D. Corbett, and R. E. McCarley, *Inorg. Chem.*, 1974, **13**, 1471; (c) E. R. Alton, R. G. Montemayor, and R. W. Parry, *ibid.*, p. 2267.

between the Te and Al atoms in Te(Al$_2$Cl$_7$).[97b] Phosphorus trifluoride, which shows only weak donor properties towards the Group III elements, will form a 1 : 1 adduct with aluminium chloride, F$_3$P,AlCl$_3$, which rearranges slowly at 298 K to produce PCl$_3$ and AlF$_3$. No reaction is observed between PCl$_3$ and Al$_2$Cl$_6$.[97c]

Compounds of Gallium and Indium.—Cyclic ylide compounds of gallium, indium, and thallium have been prepared by the action of Me$_3$P=CH$_2$ on the MeMCl (M = Ga, In, or Tl) compounds. The ylides [Me$_2$MCH$_2$PMe$_2$(CH$_2$)]$_n$

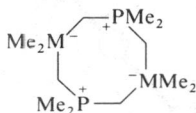

(22)

can be isolated as eight-membered ring species (22) when $n = 2$, containing alternate onium and metallate units.[98]

A convenient preparation of GaBu$^n_3$, Bu$^n_2$GaCl, and Bu^{n}GaCl$_2$ has been devised using the mixture of GaCl$_3$–LiBu^n in appropriate molar ratios. The trialkyl compound is monomeric in benzene solution, in contrast to the halogeno species, which are dimeric.[99a] The [115]In and halogen n.q.r. studies on Me$_2$InX (X = Br or I) indicate that these compounds possess a structure with linear InMe$_2$ groups equatorially surrounded by a square-planar arrangement of halogens. The corresponding chloro-compound has a distorted form of this structure.[99b] The 1 : 1 adduct Ph$_3$In,SO$_2$, previously formulated as PhSO$_2$InPh$_2$, is now believed to be O—S—O→InPh$_3$.[99c]

Intermediates in the reaction of InBr with RBr (R = alkyl) to produce RInBr$_2$ (with some In$_2$Br$_4$ as side-product when R = Pr^n or Bu^n) have been identified as In$_5$Br$_7$ (which reacts further with RBr to produce RInBr$_2$ and In$_2$Br$_4$) and a solid of composition In$^+$[InBr$_3$Me]$^-$ when R = Me.[100a] A known gallium iodide GaI$_{1.5}$ has been formulated as 2Ga$^+$[Ga$_2$I$_6$]$^{2-}$, with a Ga—Ga bond in the anion. The new halide 2Ga$^+$Ga$_2$Br$_6^{2-}$ has been prepared but the corresponding chloro-compound could not be formed.[100b] The e.s.r. spectra of adducts formed between the free-radical base 2,2,6,6-tetramethylpiperidine

[98] H. Schmidbaur and H. J. Fuller, *Chem. Ber.*, 1974, **107**, 3674.
[99] (a) R. A. Kovar, G. Loaris, H. Derr, and J. O. Callaway, *Inorg. Chem.*, 1974, **13**, 1476; (b) D. B. Patterson and A. Carnevale, *ibid.*, p. 1479; (c) A. T. T. Hsieh and G. B. Deacon, *J. Organometallic Chem.*, 1974, **70**, 39.
[100] (a) L. G. Waterworth and I. J. Worrall, *J. Organometallic Chem.*, 1974, **81**, 23; (b) M. Wilkinson and I. J. Worrall, *Inorg. Nuclear Chem. Letters*, 1974, **10**, 747; (c) C. Hambly and J. B. Raynor, *J.C.S. Dalton*, 1974, 604; (d) J. G. Contreras and D. G. Tuck, *ibid.*, p. 1249.

nitroxide and $GaCl_3$ or Ga_2Cl_6, *e.g.* $Me_2C(CH_2)_3CMe_2NO \rightarrow Ga(Cl_2)ClGaCl_3$, show that the principal values of the g-tensor are close to those of the free base itself.[100c] The reaction between InX and Hacac yields $In(acac)_3$ and $InX_2(acac)$, but the dihalogeno-compound could not be isolated pure unless complexed by N-donors as $[InX_2(acac)L]EtOH$ (L = 2,2'-bipyridyl or 1,10-phenanthroline) and $[InX_2(acac)L_2]EtOH$ (L = C_5H_5N or C_5D_5N), when it forms crystalline solids. The structure of these compounds is believed to involve a six-co-ordinate In^{III} ion, $InX_2O_2N_2$.[100d]

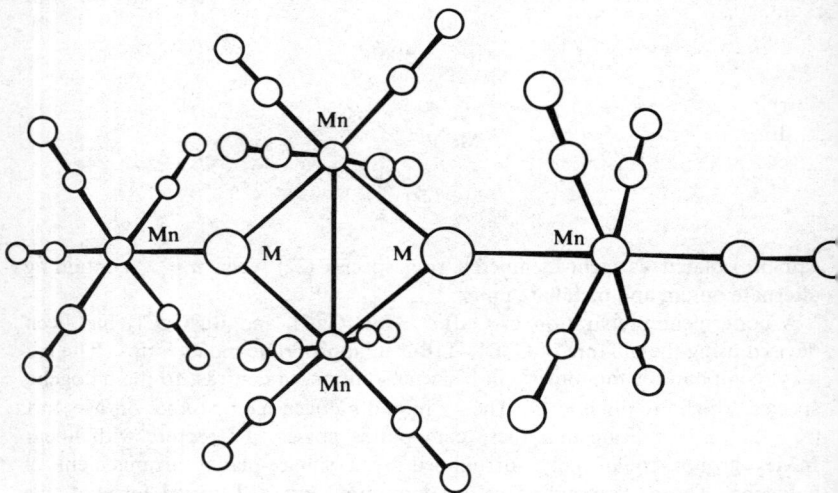

(23) M = Ga or In

The crystal structures of the isomorphous compounds $Mn_2(CO)_8$ [μ-MMn-$(CO)_5$]$_2$ (M = Ga or In) (23) show them to contain a planar Mn_2M_2 ring, the Mn—Mn distances of 3.05 Å (Ga) and 3.22 Å (In) being consistent with the presence of a Mn—Mn bond.[101]

Compounds of Thallium.—The dependence of the ^{205}Tl n.m.r. shifts on solvent and anion has been studied using water, formamide, dimethylformamide, methanol, pyridine, pyrrolidine, and n-butylamine. The shifts vary over a wide range (1900 p.p.m.), and the region of the shift corresponds to the donor atom, making this a useful diagnostic technique.[42a] Relaxation of the ^{205}Tl^I n.m.r. signal in aqueous solution appears to be dependent only upon the amount of oxygen present, and although no Tl^I–O$_2$ complex can be detected, the oxygen probably penetrates the hydration sphere, causing relaxation of the Tl nucleus by electron–nuclear dipole–dipole interactions.[102a] The use of ^{13}C n.m.r.

[101] H. Preut and H. J. Haupt, *Chem. Ber.*, 1974, **107**, 2860.
[102] (a) S. O. Chan and L. W. Reeves, *J. Amer. Chem. Soc.*, 1974, **96**, 404; (b) B. Falcinella, P. D. Felgate, and G. S. Laurence, *J.C.S. Dalton*, 1974, 1367.

techniques has been applied to the study of complexes of Tl^I; $[Tl^I(edta)OH]^{4-}$ is formed from $[Tl^I(edta)]^{3-}$ at high pH. This is the first example of a metal with charge $1+$ forming such complexes, and this technique appears to offer distinct possibilities in the study of the solution behaviour of such complexes.[42b] The standard reduction potentials for the reactions $Tl^{3+} + e = Tl^{2+}$ and $Tl^{2+} + e = Tl^+$ have been determined as $+0.33$ and $+2.22$ V, respectively, and the thermal exchange reaction between Tl^I and Tl^{III} appears to be a two-electron transfer, and not to involve Tl^{II} ions.[102b]

The crystal structure of thallium(I) fluoride shows that there are two independent thallium atoms in the lattice, each being surrounded by octahedra of F^- ions with three pairs of F^- ions at distances of 2.25—2.62 Å, 2.79 Å, and 3.07—3.90 Å.[103a] Trimethylsilylmethylthallium(III) halides $(Me_3SiCH_2)_2TlX$ (X = Cl or Br) have been prepared, and unlike the polymeric dialkyl thallium(III) halides, are dimeric, with two halogen atoms bridging the two metal atoms.[103b] Both π-cycloheptatrienylthallium(III) dichloride,[103c] and dicyclo-octatetraenethallium-(III), $(C_8H_8)_2Tl_2$,[103d] have been prepared; the former is a reddish-brown solid prepared by the action of a C_7H_3–NEt_3 mixture on $(C_5H_6N)_2TlCl_5$, and the latter is obtained when $C_8H_8Tl^{III}Cl$ is treated with sodium naphthenide, with the two thallium atoms in the product being in the $1+$ and $3+$ oxidation states.

Thallium(I) uranates of composition Tl_2UO_4 and $Tl_2U_2O_7$ have been identified in reaction products from the action of UO_3 or U_3O_8 on Tl_2CO_3 or Tl_2O_3. The $Tl_2U_2O_7$ was the more stable phase and was formed from the various other phases on hydrolytic treatment.[104a] Several transition-metal carbonyl derivatives of thallium, including $Tl[Co(CO)_4]$ and $Tl[(C_5H_5)M(CO)_3]$ (M = Cr, Mo, or W), have been prepared and their relative rates of disproportionation depend on the attached group in the order $Co(CO)_4 \gg (C_5H_5)M(CO)_3$ (W > Mo > Cr) > $Co(CO)_3PPh_3 \approx Mn(CO)_5 \gg (C_5H_5)Fe(CO)_2$. The corresponding Tl^{III} derivatives, except $Tl[(C_5H_5)Fe(CO)_2]_3$, have also been isolated.[104b]

PART II: Groups IV and V

1 Group IV

Carbon.—The intercalate of aluminium trichloride in graphite acts as a milder catalyst for Friedel–Crafts alkylation reactions than pure $AlCl_3$;[1a] moreover

[103] (a) N. W. Alcock and H. D. B. Jenkins, *J.C.S. Dalton*, 1974, 1907; (b) S. Numata, H. Kurosawa, and R. Okawara, *J. Organometallic Chem.*, 1974, **70**, C21; (c) N. Kumar and R. K. Sharma, *Chem. and Ind.*, 1974, 773; (d) *ibid.*, p. 261.

[104] (a) A. S. Giridharan, M. R. Udupa, and G. Aravamudan, *Z. anorg. Chem.*, 1974, **403**, 211; (b) J. M. Burlitch and T. W. Theyson, *J.C.S. Dalton*, 1974, 828.

[1] (a) J. M. Lalancette, M. J. Fournier-Breault, and R. Thiffault, *Canad. J. Chem.*, 1974, **52**, 589; (b) T. Tominaga, T. Sakai, and T. Kimura, *Chem. Letters*, 1974, 853; (c) D. Hohlwein and W. Metz, *Z. Krist.*, 1974, **139**, 279; (d) A. G. Freeman, *J.C.S. Chem. Comm.*, 1974, 746; (e) L. B. Ebert, R. A. Huggins, and J. I. Brauman, *Carbon*, 1974, **12**, 199; (f) L. B. Ebert, R. A. Huggins, and J. I. Brauman, *J.C.S. Chem. Comm.*, 1974, 924.

the extent of polysubstitution of the organic substrate is much reduced, and the selective nature of such a catalyst presents interesting possibilities for this type of reaction. Mössbauer spectra obtained from the intercalation compounds of graphite–$AlCl_3$–$FeCl_3$ have shown that when the chlorides are present in a ratio of 0.3 : 0.5 (Fe : Al) the Fe^{III} exists predominantly in the high-spin state.[1b] X-Ray diffraction studies on graphite–$FeCl_3$ intercalates have led to the proposal of a new structural model involving a disturbance in the sequence of the layers of graphite and $FeCl_3$.[1c] A doubly filled graphite–$FeCl_3$–N_2O_5 intercalate has been reported in which the graphite–iron(III) chloride layers are interleaved with graphite–N_2O_5 layers.[1d] The intercalation of chromium trioxide in graphite takes place in the presence of glacial acetic acid as a solvent and not, as reported previously, by a direct reaction, which instead yields a mixture of lower chromium oxides and unreacted graphite.[1e] An interesting wide-line ^{19}F n.m.r. study on the SbF_5–graphite intercalate suggests that there is little interaction between the graphite and the inserted SbF_5 molecules, which continue to display liquid-like behaviour well below the freezing point of the pure fluoride.[1f] An examination of the oxidation products of natural diamonds has produced no evidence for the presence of SO and SO_2, which were predicted by one of the theories of diamond formation involving a reduction of CO_2 with FeS.[2a] The production of artificial diamonds continues to be a subject of considerable interest[2b] and one report in particular suggests that the extreme local conditions found at the tip of a cutting tool may provide suitable sites for diamond nucleation.[2c]

A considerable amount of synthetic work on carbenes has been reported and, although the majority of this is in the area of organic chemistry, a few specific cases are mentioned briefly here. Alkoxycarbonylcarbenes[3a] (A) and 1,2,2-trifluoroethylidene[3b] (B) react readily with tetra- and tri-alkylsilanes to give α-C—H and β-C—H insertion products of the carbene with (A) and exclusive Si—H insertion with (B); addition of $:CH_2$ and $:CCl_2$ to an enamine ester gives rise to cyclopropane derivatives, which on hydrolysis yield a γ-keto-ester.[3c] The electronic spectra of ring-halogenated difluorophenyl carbenium ions are particularly sensitive to substituent effects, and the observed wavelength shifts were rationalized in terms of earlier theoretical calculations.[3d]

A flash photolysis–e.s.r. kinetic study on the addition of $Me_3Si\cdot$ to C_2H_4 confirms that ethylene is an efficient scavenger for silyl radicals and, as the first direct investigation of its type, this may be considered a prototype for the

[2] (a) R. E. Langford, C. E. Melton, and A. A. Giardini, Nature, 1974, 249, 647; (b) D. Millington, J. M. Winfield, and M. G. H. Wallbridge, Ann. Reports (A), 1973, 70, 315; (c) R. Komanduri and M. C. Shaw, Nature, 1974, 249, 582.
[3] (a) W. Ando, K. Konishi, and T. Migita, J. Organometallic Chem., 1974, 67, C7; (b) R. N. Haszeldine, A. E. Tipping, and R. O'B. Watts, J.C.S. Perkin I, 1974, 2391; (c) H. Bieräugel, J. M. Akkerman, J. C. Lapierre Armande, and U. K. Pandit, Tetrahedron Letters, 1974, 2817; (d) Y. K. Mo, R. E. Linder, G. Barth, E. Bunnenberg, and C. Djerassi, J. Amer. Chem. Soc., 1974, 96, 4569.

chain-carrying steps in the free-radical polymerization of olefins.[4a] The platinum-catalysed addition of H_2SiCl_2 to acetylenes yields mainly *trans*-adducts although addition to terminal acetylenes gives the novel bis-(*trans*-dialkenyl)dichlorosilanes.[4b] Photoelectron data on H_3Si-substituted acetylenes have been interpreted without invoking silicon *d*-orbital participation[4c] (*cf.* similar data on Me_4Si[4d]).

A theoretical study on the carbon–silicon double bond has shown there are unusually large *d*-orbital contributions to the bonding in $R_2^1Si{=}CR_2^2$ ($R^1 = H$ or F; $R^2 = H$ or F)(A). The high reactivity of such compounds has been ascribed to the polar nature of the Si—C bond.[4e] The reaction of (A) ($R^1 = Me$, Et, or Ph; $R^2 = H$) with aldehydes and ketones results in either olefin or silyl enol ether formation,[4f] the latter being favoured by aliphatic carbonyls. The reactivity of $Me_2Si{=}CH_2$ towards a variety of substrates[4g] suggests that while polarity changes have little effect on the reaction rate, increased steric effects cause it to be markedly decreased.

Tetracyanomethane has a trigonal structure related to that of cubic SiF_4;[5a] the short $N{\cdot}{\cdot}{\cdot}C$ intermolecular distances are taken as evidence of donor–acceptor interactions. An interesting charge-transfer complexation process occurs between tetracyanoethylene (TCNE) and permethylpolysilanes[5b] although the latter possess neither π-bonds nor lone pairs; e.s.r. and absorption spectra have been interpreted in terms of σ-electron donation from the Group IVB catenate.[5c] Similar complexes have also been reported with TCNE and distannanes in which cleavage of the Sn—Sn bond takes place to give $R_3Sn(TCNE)$ ($R = Me$ or Bu).[5d] E.s.r. studies on the radicals $BrCN^-$ and Br_2CN,[6] obtained from ^{60}Co irradiation of cyanogen bromide, suggest linear structures in contrast to the $F\dot{C}O$ and $Cl\dot{C}O$ radicals, which are strongly bent. A vibrational analysis[7a] of $(NO_2)_3CH$ has shown that the symmetry of the molecules is below C_3 in the liquid state, while a similar study[7b] on $(NO_2)_6C_2$ has established a D_{3h} configuration. Aryl isocyanates react with Me_3SiCN to give ring compounds of the type $Me_3SiN\dot{C}{-}N(R){-}C(O){-}N(R){-}\dot{C}(O)$ ($R = Ph$, $p\text{-}ClC_6H_4$, or α-naphthyl); tosyl isocyanate, however, gave only a 1 : 1 adduct, even when used in excess.[8]

[4] (a) K. Y. Choo and P. P. Gaspar, *J. Amer. Chem. Soc.*, 1974, **96**, 1284; (b) R. A. Benkeser and D. F. Ehler, *J. Organometallic Chem.*, 1974, **69**, 193; (c) W. Ensslin, H. Bock, and G. Becker, *J. Amer. Chem. Soc.*, 1974, **96**, 2757; (d) D. Millington, J. M. Winfield, and M. G. H. Wallbridge, *Ann. Reports (A)*, 1973, **70**, 316; (e) R. Dramauer and D. R. Williams, *J. Organometallic Chem.*, 1974, **66**, 241; (f) R. D. Bush, G. M. Golino, D. N. Roark, and L. H. Sommer, *ibid.*, p. 29; (g) R. D. Bush, G. M. Golino, G. D. Homer, and L. H. Sommer, *ibid.*, 1974, **80**, 37.

[5] (a) D. Britton, *Acta Cryst.*, 1974, **B30**, 1818; (b) V. F. Traven and R. West, *J. Amer. Chem. Soc.*, 1973, **95**, 6824; (c) H. Sakurai, M. Kira, and T. Uchida, *ibid.*, p. 6826; (d) A. B. Cornwell, P. G. Harrison, and J. A. Richards, *J. Organometallic Chem.*, 1974, **67**, C43.

[6] S. P. Mishra, G. W. Neilson, and M. C. R. Symons, *J.C.S. Faraday II*, 1974, **70**, 1280.

[7] (a) M. I. Dakhis and V. A. Shlyapochnikov, *J. Mol. Structure*, 1974, **21**, 305; (b) A. Loewenschuss, N. Yellin, and A. Gabai, *Spectrochim. Acta*, 1974, **30A**, 371.

[8] I. Ojima, S. Inaba, and Y. Nagai, *J.C.S. Chem. Comm.*, 1974, 826.

Two reports have appeared on the conversion of CO and NO into CO_2 and NO_2 *via* iridium[9a] and rhodium[9b] dinitrosyl complexes; the '20-electron' dinitrosyls postulated as intermediates in these reactions would be more appropriately regarded as 18-electron NN-bonded *cis*-dinitrogen dioxide complexes. Significant differences have been observed in the reactivities of $Bu^tO\cdot$ and $Bu^t_fO\cdot$;[10] the perfluoro-radical adds to double bonds rather than abstracting an allylic hydrogen (Scheme 1) while preferential proton abstraction takes place with trialkyl phosphites and trialkylboranes rather than an exclusive attack at the metal centre.

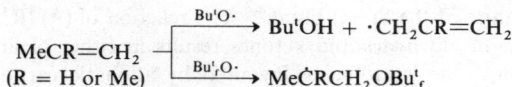

$$MeCR{=}CH_2 \quad \overset{Bu^tO\cdot}{\underset{Bu^t_fO\cdot}{\Bigg[}} \quad \begin{array}{l} \longrightarrow Bu^tOH + \cdot CH_2CR{=}CH_2 \\ \longrightarrow Me\dot{C}RCH_2OBu^t_f \end{array}$$

$(R = H \text{ or } Me)$

Scheme 1

A simple synthesis of COFX (X = Cl or Br) from CFX_3 and SO_3 has been reported;[11] the product contains only $CFCl_3$ as an impurity. New methods of synthesizing poly(carbon monofluoride), $(CF)_n$[12a] and C_4F have been reported;[12b] the former has high potential as a solid lubricant and as a cathode material in high-energy batteries. X-Ray powder patterns of $(CF)_n$ suggest an intercalation arrangement with either F_2 or F_2^- between the graphite layers.[12c] The substitution of C_6F_6 with trichloro-2-thienyl-lithium (A), Bu^nLi (B), and Bu^tLi (C) gives in all cases a 1,4-bis-derivative though, in addition, a tetrakis-compound was found with (A) and (B), and mono- and tris-compounds with (B) only.[13a] C_6F_6 reacts with $O^+[AsF_6]^-$ to give $[C_6F_6]^+$ $[AsF_6]^-$ in which the $[C_6F_6]^+$ ion is either disordered in its lattice placement or possesses three-fold axial symmetry.[13b] Perfluoromethylenecyclopropane,[14a] $(CF_2)_2C{=}CF_2$, has been obtained by the reaction sequence shown in Scheme 2. Although the olefin is extremely reactive towards nucleophiles, it surprisingly fails to undergo thermal [2 + 2] cycloaddition reactions.[14b]

$$F_2C{=}CClCClF_2 + F_3C{-}CF\overset{O}{\diagup}CF_2 \longrightarrow \underset{F_2}{\overset{F_2}{\triangleright}}\!\!\overset{Cl}{\underset{}{\diagdown}}CF_2Cl \xrightarrow{\text{Zn–dioxan}} \underset{F_2}{\overset{F_2}{\triangleright}}{=}CF_2$$

Scheme 2

[9] (a) B. L. Haymore and J. A. Ibers, *J. Amer. Chem. Soc.*, 1974, **96**, 3325; (b) S. Bhaduri, B. F. G. Johnson, C. J. Savory, J. A. Segal, and R. H. Walter, *J.C.S. Chem. Comm.*, 1974, 809.

[10] A. G. Davies, R. W. Dennis, B. P. Roberts, and R. C. Dobbie, *J.C.S. Chem. Comm.*, 1974, 468.

[11] G. Siegemund, *Angew. Chem. Internat. Edn.*, 1973, **12**, 918.

[12] R. J. Lagow, R. B. Badachhape, J. L. Wood, and J. L. Margrave, (a) *J. Amer. Chem. Soc.*, 1974, **96**, 2628; (b) *J.C.S. Dalton*, 1974, 1268; (c) V. K. Mahajan, R. B. Badachhape, and J. L. Margrave, *Inorg. Nuclear Chem. Letters*, 1974, **10**, 1103.

[13] (a) S. S. Dua, R. D. Howells, and H. Gilman, *J. Fluorine Chem.*, 1974, **4**, 381; (b) T. J. Richardson and N. Bartlett, *J.C.S. Chem. Comm.*, 1974, 427.

[14] (a) B. E. Smart, *J. Amer. Chem. Soc.*, 1974, **96**, 927; (b) *ibid.*, p. 929; (c) J. B. Levy and R. C. Kennedy, *ibid.*, p. 4791; (d) J. D. Park, T. S. Croft, and R. W. Anderson, *J. Organometallic Chem.*, 1974, **64**, 19.

A kinetic study of the thermal reaction of fluorine and perfluorocyclobutane mixtures has shown that the products C_nF_{2n+2} ($n = 1$—4) may be explained on the basis of the decomposition of thermally excited radicals.[14c] Grignard reagents react with perfluorocyclobutene to give vinylic monoalkyl products in the case of magnesium bromides and vinylic dialkyl products with magnesium chlorides.[14d]

Silicon.—The electronic ground state of $^{31}SiH_2$ is a singlet although in the addition to butadiene it is present as 80% triplet and 20% singlet.[15a] *Ab initio* calculations on H_3SiCl including the Si and Cl d-orbitals have shown that a charge transfer, which also involves the hydrogen atoms, takes place towards silicon.[15b] Cl_3SiH reacts with tetrafluoropropyne to give a mixture of the 1 : 1 adducts;[16a] with $CF_2{=}CHCl$ it gives $CHF_2CH_2SiCl_3$ (after CCl conversion to CH) using an excess of the silane and $ClCF_2CH_2SiCl_3$ with an excess of olefin.[16b] The latter products are rationalized by the addition of $Cl_3Si\cdot$ at the CF_2 group in competition with olefin reduction and radical attack at the CH_2 group.

$(C_6F_5)_2SiMeH$ reacts with RM (R = Me or Bu^n, M = Li; R = Me or Et, M = MgX) to give C_6F_5M and R_2SiMeH;[17] however, removal of the second C_6F_5 group is more difficult, and this may be an indication of the importance of steric hindrance in the cleavage of Si—C bonds. Two novel silanes, $HSi(NCO)_3$ and $HSi(NCS)_3$, have been prepared from a reaction of trihalogenosilanes with AgX (X = NCO or NCS);[18] the partially substituted derivatives prepared from exchange reactions have been characterized spectroscopically.

An interesting synthesis of allylsilanes has been reported from the $Cr(CO)_6$-photocatalysed addition of a silane containing an Si—H bond to a 1,3-diene.[19a] Allylsilanes undergo anionic cycloaddition reactions with $(PhCH)_2N^-Li^+$ to yield silyl-substituted pyrrolidines.[19b] Three papers have appeared on 1,2-anionic rearrangements of organo-silanes and -germanes; these involve silicon migration both from oxygen to carbon[20a] and from sulphur to carbon.[20b] The former process takes place stereospecifically with inversion at the benzyl carbon.[20c] ^{1}H n.m.r. studies on tri-[21a] and tetra-arylsilanes[21b] have shown that the barrier to the interconversion of the stereoisomers is low, and in the latter case, contrary to an earlier report, precludes isolation of the various isomers. A number of reports have appeared on the synthesis of optically active silanes;

[15] (a) O. F. Zeck, Y. Y. Su, G. P. Gennaro, and Y.-N. Tang, *J. Amer. Chem. Soc.*, 1974, **96**, 5967; (b) J. M. Howell and J. R. Van Wazer, *ibid.*, p. 3064.
[16] (a) R. N. Haszeldine, C. R. Pool and A. E. Tipping, *J.C.S. Perkin I*, 1974, 2293; (b) W. I. Bevan, R. N. Haszeldine, J. Middleton, and A. E. Tipping, *J.C.S. Dalton*, 1974, 2305.
[17] D. Sethi, R. D. Howells, and H. Gilman, *J. Organometallic Chem.*, 1974, **69**, 377.
[18] F. Höfler, G. Jägerhuber, and W. Veigl, *Monatsh.*, 1974, **105**, 539.
[19] (a) M. S. Wrighton and M. A. Schroeder, *J. Amer. Chem. Soc.*, 1974, **96**, 6235; (b) E. Popowski, *Z. Chem.*, 1974, **14**, 360.
[20] (a) A. Wright and R. West, *J. Amer. Chem. Soc.*, 1974, **96**, 3214; (b) *ibid.*, p. 3222; (c) *ibid.*, p. 3227.
[21] (a) R. J. Boettcher, D. Gust, and K. Mislow, *J. Amer. Chem. Soc.*, 1974, **96**, 7157; (b) *ibid.*, p. 7158; (c) R. J. P. Corriu and G. F. Lanneau, *J. Organometallic Chem.*, 1974, **64**, 63; *ibid.*, 1974, **67**, 243; (d) R. J. P. Corriu, G. F. Lanneau, and M. Leard, *ibid.*, p. 79.

in particular, monofunctional silanes[21c] have been prepared from the reaction of Grignard and organolithium reagents with bifunctional silanes, the latter being obtained from (−)-menthol and phenyl-α-naphthylchlorosilane.[21d]

A considerable amount of work has been published over the past year on cycloalkane rings which incorporate a Group IV atom. The formation of a benzosilacyclopentene has been reported from the pyrolysis of $Me_3SiC(N_2)Ph$ and is thought to proceed *via* a silacyclopropane[22a] intermediate (Scheme 3).

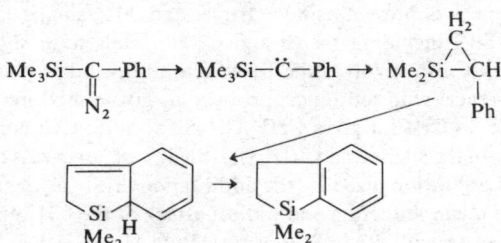

Scheme 3

Silenes react with cyclohexene to give silacyclopropane derivatives[22b] which, interestingly, on further photolysis rearrange to a 3-silylcyclohexene. $(Me_3Si)_2$-CBrLi reacts with Me_2MCl_2 (M = Si, Ge, or Sn) to give highly substituted 1,3-disila-, 1,3-digerma-, and 1,3-distanna-cyclobutane;[22c] the tin derivative (1) is the first known compound of this type.

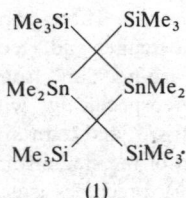

(1)

The pyrolysis of 1,1-disubstituted silacyclobutanes is thought to proceed *via* an intermediate containing a silicon–nitrogen double bond, $PhN{=}SiR_2$ (R = Me or Ph).[22d] The latter has been definitely established by a trapping reaction with benzophenone. Oxametallacyclopentanes[22e] (Si, Ge, and Sn)

²² (a) W. Ando, A. Sekiguchi, T. Hagiwara, and T. Migita, *J.C.S. Chem. Comm.*, 1974, 372; (b) M. Ishikawa and M. Kumada, *J. Organometallic Chem.*, 1974, **81**, C3; (c) D. Seyferth and J. L. Lefferts, *J. Amer. Chem. Soc.*, 1974, **96**, 6237; (d) C. M. Golino, R. D. Bush, and L. H. Sommer, *ibid.*, p. 614; (e) M. Massol, J. Barrau, J. Satgé, and B. Bouyssieres, *J. Organometallic Chem.*, 1974, **80**, 47; (f) J. V. Swisher and H.-H. Chen, *ibid.*, 1974, **69**, 83; (g) G. Manuel, G. Cauquy, and P. Mazerolles, *Synth. Inorg. Metal-org. Chem.*, 1974, **4**, 143; (h) J. W. Connolly, *J. Organometallic Chem.*, 1974, **64**, 343; (i) J. Nagy, E. Gergö, K. A. Andrianov, L. M. Volkova, and N. V. Delazari, *ibid.*, 1974, **67**, 19; (j) R. J. Ouellette, *J. Amer. Chem. Soc.*, 1974, **96**, 2421; (k) J. H. Burk and W. A. Kriner, *J. Organometallic Chem.*, 1974, **63**, C1.

have been obtained from the cyclization of γ-hydroxy-metal hydrides; the latter are prepared from the reduction of the corresponding γ-acetoxychloro- or γ-hydroxychloro-compounds. The cyclization of $CH_2\!=\!CH(CH_2)_nSiMe_2H$ ($n = 0$—6) has been investigated using a chloroplatinic acid catalyst.[22f] Ring closure was not observed for $n = 0$ or 1 and with $n = 2$ only 1,1-dimethyl-1-silacyclopentane was formed; with $n = 3$ both cyclopentane and cyclohexane derivatives were obtained; a reaction mechanism has been proposed to account for the formation of the various products. Dihalogenocarbenes react with 1-sila- and 1-germa-cyclopent-3-enes to give 6,6-dihalogeno-3-sila- and -3-germa-bicyclo[3,1,0]hexanes.[22g] The photolysis of 2,4,4-trimethyl-4-sila-3-methylene-1,5-hexadiene yields a silabicyclo[2,1,1]hexane derivative;[22h] however, under similar reaction conditions related butadienes only undergo *cis–trans* isomerization while the analogous heptadiene is unreactive. Dipole moment calculations on 1-oxa-2,6-disilacyclohexane derivatives indicate that the compounds exist in the chair conformation with the substituents on the silicon atoms in *trans* positions.[22i] A conformational analysis on silacyclohexanes by force-field methods also predicts a chair arrangement though the ring is flatter than cyclohexane about the Si atom.[22j] The controlled hydrolysis of [ClMeSi-CH$_2$]$_3$ gives rise to the unusual silicon cage compound (2) in high yield.[22k]

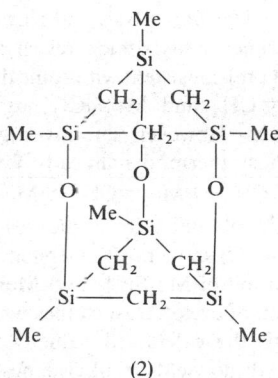

(2)

t-Butylmethyldichlorosilane undergoes a condensation reaction with Na–K alloy in THF to give the four-membered ring compounds (Bu^{t}MeSi)$_4$; remarkably the compound is inert to oxygen at room temperature and is not attacked by H_2SO_4 or HCl–AlCl$_3$.[23a] Silanes Si$_n$H$_{2n+2}$ ($n = 2$—5) react with Bu^{n}Li to give a variety of mono- and di-substituted polysilanes[23b] in which substitution takes place at the terminal silicon atoms.

[23] (a) M. Biernbaum and R. West, *J Organometallic Chem.*, 1974, **77**, C13; (b) F. Fehér and R. Freund, *Inorg. Nuclear Chem. Letters*, 1974, **10**, 569.

The thermal decomposition of silicon di-imide in vacuum results in very high purity amorphous Si_3N_4 powders;[24a] studies have also been made of the hot-pressing behaviour of Si_3N_4[24b] and some of its physical properties.[24c]

The first cyclotrisilazanes containing three different substituents on the ring nitrogen atoms[25a] have been prepared by the selective aminolysis of $(ClMe_2Si)_2$-NMe with primary amines followed by ring closure with LiBu and Me_2SiCl_2.

The donor behaviour of octamethylcyclotetrasilazane has been investigated with certain transition metals[25b] (Ti, V, and Cr); co-ordination is through two of the four available nitrogen atoms in the Si–N ring. A number of interesting cyclo-silazanes and -siloxazanes have been reported which contain N—N bonds within the ring. In particular the $Si_3N_2O_2$ ring system[25c] (3) is obtained from $(LiNMe)_2$ and $[Me_2Si(Cl)O]_2SiMe_2$ while the tetrasilatetra-azacyclo-octane derivative[25d] (4) is prepared from $(HMeN)_2$ and $(ClMe_2Si)_2$. The five-membered

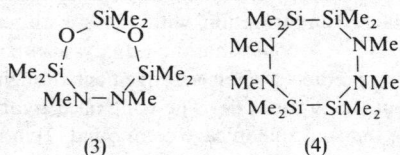

(3)   (4)

heterocycles Si_4Ph_8NR (R = Me or Et) are prepared from the reaction of $(SiPh_2)_4$ with iodine followed by aminolysis and ring closure.[26] X-Ray studies on bis(dioxan)potassium(trimethylsilyl)amide reveal an ionic structure with a very short Si—N bond (164 pm) consistent with some degree of $(p–d)\pi$-bonding.[27]

The reaction of $Me_3P=CH_2$ and Me_2SiCl_2 has been reported as giving disilacyclobutane derivatives. However, a re-investigation has shown that on prolonged reaction the most thermodynamically favoured product is a six-membered ring $Me_2\overline{Si}—CH_2—PMe_2=CH—SiMe_2—\overline{C}=PMe_3$.[28a] Cyclo-disilaphospha(III)diazanes[28b] obtained from the reaction of $(ClMe_2Si)_2$ and $RP(NHMe)_2$ (R = Bu^t or MeO) show no P inversion up to 453 K.

The hydrolytic condensation of Me_2SiCl_2 and $MePhSiCl_2$ in either benzene or ether yields a distribution of cyclotetrasiloxanes which in the case of benzene is in close agreement with the calculated values.[29a] The hydrogenolysis of disiloxanes with sodium hydride yields trialkylsilanes and trialkylsilanoates.[29b]

24 (a) K. S. Mazdiyasni and C. M. Cooke, J. Amer. Ceram. Soc., 1973, 56, 628; (b) G. R. Terwilliger and F. F. Lange, ibid., 1974, 57, 25; F. F. Lange, ibid., p. 84; (c) W. A. Fate, ibid., p. 49; S. D. Hartline, R. C. Bradt, D. W. Richardson, and M. L. Torti, ibid., p. 190.
25 (a) U. Wannagat and D. Labuhn, Monatsh., 1974, 105, 209; (b) J. Hughes and G. R. Willey, J. Amer. Chem. Soc., 1973, 95, 8758; (c) U. Wannagat and M. Schlingmann, Z. anorg. Chem., 1974, 406, 7; (d) F. Höfler and D. Wolfer, ibid., p. 19.
26 E. Hengge and D. Wolfer, J. Organometallic Chem., 1974, 66, 413.
27 A. M. Domingos and G. M. Sheldrick, Acta Cryst., 1974, B30, 517.
28 (a) W. Malisch and H. Schmidbaur, Angew. Chem., Internat. Edn., 1974, 13, 540; (b) O. J. Scherer, W. Glässel, and R. Thalacker, J. Organometallic Chem., 1974, 70, 61.
29 (a) A. G. Kuznetsov, S. A. Golubtsov, N. P. Telegina, V. I. Ivanov, and G. G. Pchelintseva, J. Gen. Chem. (U.S.S.R.), 1973, 43, 300; (b) E. A. Batyaev, N. P. Kharitonov, and B. A. Bolotov, ibid., p. 1281; (c) J. E. Griffiths, Spectrochim. Acta, 1974, 30A, 945.

The vibrational spectra of hexachlorodisiloxane[29c] confirm that the Si—O—Si skeleton is bent and suggest that the molecule has C_{2v} symmetry. Synthetic $K_4[Si_8O_{18}]$[30a] is the first example of a silicate in which the $[SiO_4]$ tetrahedra form double $[Si_4O_{10}]$ chains, which are connected to give single layers of $[Si_8O_{18}]$ units. The structure of $Si_5O[PO_4]_6$ consists[30b] of $[SiO_6]$ octahedra and $[Si_2O_7]$ units linked by $[PO_4]$ tetrahedra. Tetra-(2-thienyl)silane has an orientationally disordered structure in which each thienyl ring has two conformations with respect to a particular Si—C bond.[31]

A simple synthesis of Si_2F_5H has been reported[32] from the fluorination of Si_2BrF_4H with SbF_3; unlike previous methods the compound is produced in high yields and is easily separated from the reaction by-products. The dissociation parameters for various tetrafluorosilane–amine adducts have been considered in terms of steric and hydrogen-bonding effects.[33a] The data suggest that no adduct formation is possible at room temperature with bulky tertiary amines unless a large excess of SiF_4 is present. Further work on the self-dehydrofluorination of the above adducts has shown that by changing the reaction conditions polymers of the type SiF_2NR[33b] (R = Me, Bu, or Ph) may be obtained instead of the aminofluorosilanes and hexafluorodisilazanes reported earlier.[33c]

Germanium.—The photolysis of $GeH_{4-n}X_n$ (X = Cl or Br; n = 0–4) in argon and carbon monoxide matrices has led to the characterization of a variety of free-radical germanium species.[34] Although the vibrational spectra of the four solid phases of GeH_4 have proved to be complex it has been possible to interpret the data in terms of a limited number of site symmetries.[35] An investigation of the reaction involving recoiling germanium atoms with GeH_4 and Ge_2H_6 has shown that GeH_2 is an important intermediate in the process which gives rise to the next higher homologue.[36] The stretching frequency of the Ge—H bond in triorganogermanes[37] is influenced by $(p–d)\pi$-bonding with the other substituents; it has also been shown that germanium is less capable of a $(p–d)\pi$-interaction than silicon.

Two interesting radical reactions have been reported[38] involving migration of phenyl from carbon to germanium and germanium to carbon. The former concerns the thermal rearrangement of $PhCXY(CH_2)_2GeMe_2H$ to $HCXY-(CH_2)_2GeMePh$ (X = Y = Me; X = H, Y = Ph) and the latter, the reduction of $PhMe_2Ge(CH_2)_4Cl$ to $Ph(CH_2)_4GeMe_2H$ with Bu_3SnH.

30 (a) H. Schweinsberg and F. Liebau, *Acta Cryst.*, 1974, **B30**, 2206; S. Ďurovič, *ibid.*, p. 2214; (b) H. Mayer, *Monatsh.*, 1974, **105**, 46.
31 A. Karipides, A. T. Reed, and R. H. P. Thomas, *Acta Cryst.*, 1974, **B30**, 1372.
32 J. F. Bald, K. G. Sharp, and A. G. Macdiarmid, *J. Fluorine Chem.*, 1974, **3**, 433.
33 (a) C. J. Porritt, *Chem. and Ind.*, 1974, 415; (b) *ibid.*, p. 574; (c) D. Millington, J. M. Winfield, and M. G. H. Wallbridge, *Ann. Reports (A)*, 1973, **70**, 322.
34 W. A. Guillory, R. J. Isabel, and G. R. Smith, *J. Mol. Structure*, 1974, **19**, 473.
35 N. D. The, J.-M. Gagnon, R. Belzile, and A. Cabana, *Canad. J. Chem.*, 1974, **52**, 327.
36 P. P. Gaspar and J. Frost, *J. Amer. Chem. Soc.*, 1973, **95**, 6567.
37 A. N. Egorochkin, S. Ya. Khorshev, N. S. Ostasheva, J. Satgé, P. Rivière, J. Barrau, and M. Massol, *J. Organometallic Chem.*, 1974, **76**, 29.
38 H. Sakurai, I. Nozue, and A. Hosomi, *J. Organometallic Chem.*, 1974, **80**, 71.

A series of pentafluorophenylgermanium hydrides has been obtained by the reduction of the corresponding bromides with $LiAlH_4$.[39a] These compounds react with Et_3SnNEt_2 to give derivatives containing a Ge—Sn bond. The crystal structures of $(C_6F_5)_4M$ (M = Ge or Sn) consist of discrete molecular units with $\bar{4}$ crystallographic symmetry;[39b] the C_6F_5 groups are planar with a significantly shorter mean C—C bond length (137.9 and 138.5 pm for Ge and Sn respectively) than in C_6F_6 (139.4 pm). $(C_6F_5)_6Ge_2$ reacts with a variety of reagents, including H_2O, HCl, and MeOH in THF, via cleavage of the Ge—Ge bond[39c] while $(C_6F_5)_3GeBr$ undergoes exchange reactions with Et_3GeSH to give $(C_6F_5)_3GeSH$.[39d] Spectroscopic data on pentamethyldigermanyl compounds[40a] indicate that there is little change in the stability of the Ge—Ge bond with different ligands. The digermanes $[PhY_2Ge]_2$ (Y = MeO, MeS, Me_2N, or Et_2P) decompose thermally by an α-elimination process to give new organogermylenes PhGeY.[40b] The latter react by insertion into Ge—M bonds (M = O, S, N, or P) to give digermanes.[40c] The germylgermylenes $Ph(PhX_2Ge)Ge$: (X = Cl or Me) have been obtained from the photolysis of $(PhX_2Ge)_3GePh$.[40d] Optically active alkoxygermanes react with RLi with retention of configuration when R is saturated and with inversion when R is unsaturated;[41a] reaction with germyl-lithium reagents gives the threo- and meso-isomers of digermanes.[41b] $PhCH_2MgCl$ and α-NpPhMeGeH react to give an optically active digermane;[41c] the experimental results suggest the formation of a germylmagnesium intermediate, obtained from a cleavage of the Ge—H bond, with retention of configuration.

$K_2Ge_8O_{17}$ is orthorhombic, with 32 Ge atoms in the unit cell;[42] twenty-four of these are tetrahedrally co-ordinated by oxygen and remarkably the remaining eight are five-co-ordinate in the form of distorted trigonal bipyramids which are joined by edge-sharing. The structure of $PbGeS_3$ consists[43] of $[GeS_4]$ tetrahedra linked to form infinite $(GeS_3)_n^{2n-}$ chains. The lead atom possesses a highly distorted octahedral geometry, being surrounded by six sulphur atoms and a lone pair.

The vibrational spectra of matrix-isolated GeF_2 and $(GeF_2)_2$ suggest[44] that the latter has a centrosymmetric chair configuration with fluorine bridges.

39 (a) M. N. Bochkarev, L. P. Maiorova, S. P. Korneva, L. N. Bochkarev, and N. S. Vyazankin, J. Organometallic Chem., 1974, 73, 229; (b) A. Karipides, C. Forman, R. H. P. Thomas, and A. T. Reed, Inorg. Chem., 1974, 13, 811; (c) M. N. Bochkarev, G. A. Razuvaev, N. S. Vyazankin, and O. Yu. Semenov, J. Organometallic Chem., 1974, 74, C4; (d) M. N. Bochkarev, L. P. Maiorova, N. S. Vyazankin, and G. A. Razuvaev, ibid., 1974, 82, 64.
40 (a) A. J. Andy and J. S. Thayer, J. Organometallic Chem., 1974, 76, 339; (b) P. Rivière, J. Satgé, and D. Soula, ibid., 1974, 72, 329; (c) P. Rivière, J. Satgé, G. Dousse, M. Rivière-Baudet, and C. Couret, ibid., p. 339; (d) P. Rivière, J. Satgé, and D. Soula, ibid., 1974, 63, 167.
41 (a) F. Carré and R. Corriu, J. Organometallic Chem., 1974, 65, 343; (b) ibid., p. 349; (c) ibid., 1974, 73, C49.
42 E. Fáy, H. Völlenkle, and A. Wittmann, Z. Krist., 1974, 138, 439.
43 M. Ribes, J. Olivier-Fourcade, E. Philippot, and M. Maurin, Acta Cryst., 1974, B30, 1391.
44 H. Huber, E. P. Kündig, G. A. Ozin, and V. Voet, Canad. J. Chem., 1974, 52, 95.

MCl_2 (M = Ge or Sn) and Si_2Cl_6 react with R^1Li and R_2^2NLi [$R^1 = (Me_3Si)_2CH$, $R^2 = Me_3Si$] on irradiation to give the stable metal-centred radicals $R^1_3M\cdot$ and $R^1_3Si\cdot$ and $(R^2_2N)_3M\cdot$.[45] It appears that the stability of such radicals is remarkably sensitive to steric effects. The direct reaction of germanium with α,ω-dihalogenoalkanes gives rise to a new class of compounds $X_3Ge(CH_2)_nM$ (X = Cl or Br; M = GeX_3 or X; n = 1—3).[46a] The first known example of a Ge^{IV}–Ge^{II} reduction in a non-aqueous medium has been reported.[46b] $GeCl_4$ is reduced instantaneously by NaH_2PO_2,H_2O in MeOH. Evaporation of the solvent under vacuum causes the precipitation of $Ge(HPO_3)$, which on the basis of i.r. data has been assigned the structure (7).

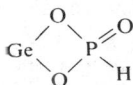

(7)

Tin.—The 1,4-concerted addition of Me_3SnH to singlet excited penta-1,3-dienes[47] is inconsistent with the presence of an allylmethylene configuration said to be favoured in this type of reaction. An improved synthesis of diallyldibromostannane has been reported (in the absence of oxygen) from allyl bromide and tin at room temperature.[48a] The u.v. photoelectron spectra of cyclopropyl-carbinyltrimethyltin and allyltrimethyltin[48b] indicate that the interaction of the cyclopropane orbitals with the C—Sn σ-bond in the former is similar in magnitude to the σ–π interactions in the latter. Dibutyldiallyltin and carboxylic acids react to give compounds of the type $(RCO_2)Bu_2SnOSnBu_2(O_2CR)$ and $(RCO_2)Bu_2SnOSnBu_2(OH)$ (R = H, Me, CH_2Cl, $CHCl_2$, CCl_3, or CF_3); the nature of the products depends on R and the hydrolysing ability of the solvent.[48c] An electron diffraction study on Me_4Sn has shown that the Me groups undergo very nearly free rotation;[49a] the Sn—C bond length (214.4 pm) is about the same as those found in the corresponding methyl hydrides but is longer than the Sn—C distance in the methyl chloride compounds. The e.s.r. spectra of the radicals obtained from a variety of organotin compounds, $R_{4-n}SnCl_n$ (R = Me or Bu; n = 0 or 1), Me_6Sn_2, and Ph_3SnH, have been reported. Particularly interesting results are the low spin density on tin in $Me_5Sn\cdot$, thought to have

[45] J. D. Cotton, C. S. Cundy, D. H. Harris, A. Hudson, M. F. Lappert, and P. W. Lednor, *J.C.S. Chem. Comm.*, 1974, 651.

[46] (a) V. F. Mironov, T. K. Gar, and A. A. Buyakov, *J. Gen. Chem. (U.S.S.R.)*, 1973, **43**, 797; (b) P. S. Poskozim and C. P. Guengerich, *Inorg. Chem.*, 1974, **13**, 241.

[47] M. Bigwood and S. Boué, *J.C.S. Chem. Comm.*, 1974, 529.

[48] (a) V. V. Pozdeev and V. E. Gel'fan, *J. Gen. Chem. (U.S.S.R.)*, 1973, **43**, 1196; (b) R. S. Brown, D. F. Eaton, A. Hosomi, T. G. Traylor, and J. M. Wright, *J. Organometallic Chem.*, 1974, **66**, 249; (c) V. Peruzzo and G. Tagliavini, *ibid.*, p. 437.

[49] (a) M. Nagashima, H. Fujii, and M. Kimura, *Bull. Chem. Soc. Japan*, 1973, **46**, 3708; (b) S. A. Fieldhouse, A. R. Lyons, H. C. Starkie, and M. C. R. Symonds, *J.C.S. Dalton*, 1974, 1966; (c) A. Folaranmi, R. A. N. McLean, and N. Wadibia, *J. Organometallic Chem.*, 1974, **73**, 59.

D_{3h} symmetry, and the pyramidal nature of $R_3Sn\cdot$.[49b] The reaction of R_4Sn ($R = Bu^n$ or Ph) with ICl and IBr has been reported to give good yields of R_3SnBr, R_nSnCl_{4-n} ($n = 3$ or 2);[49c] the mechanism of the cleavage of the Sn—C bond by the interhalogen is discussed. The rate of cleavage of the H_2C—$SnMe_3$ bond in benzyltrimethylstannanes[50a] is not much greater than the rate of cleavage of the Sn—Me bonds; in the case of the m-MeC$_6$H$_4$ and m-MeOC$_6$H$_4$ derivatives[50b] ring-protonation is an important step in the mechanism. Reaction of Me_3SnCH_2I[51] with $NaMo(CO)_3Cp$ and related complexes does not afford $Me_3SnCH_2Mo(CO)_3Cp$ but results in cleavage of an Sn—C bond to give the trimethyl-substituted derivatives. A number of trifluoromethylphenyltin(IV) derivatives, $Bu_nSn(C_6H_4CF_3-m)_{4-n}$ ($n = 1$—3), have been obtained from the reaction of a suitable n-butyltin chloride and an appropriate Grignard reagent;[52] the tin–fluorine coupling is thought to arise from a 'through-space' interaction. Structural assignments have been made[53] to the three trigonal-bipyramidal isomeric forms of R_3SnL_2 ($R = Me$ or Ph; $L_2 = $ bidentate ligand) on the basis of an analysis of the ^{119}Sn quadrupole splitting data; the wide range of splitting values obtained places doubt on structural conclusions made in earlier reports. A number of complexes of the type RClSnL ($R = $ Me, Ph, or n-octyl; $L = $ dianion of a terdentate ligand with ONO and SNO donor atoms) have been synthesized[54] and examined spectroscopically; trigonal-bipyramidal structures have been assigned to the complexes although monomeric and dimeric species cannot be ruled out. Two related compounds, Ph_2SnL [L is the dianion of either 2-(o-hydroxyphenyl)benzothiazoline[55a] or 2-hydroxy-N-(2-hydroxybenzylidene)aniline[55b]], also possess distorted trigonal-bipyramidal structures. SnX_2 ($X = $ Cp, OMe, or SPh) reacts with $(PhS)_2$ to give $SnX_3(SPh)$ and $SnX(SPh)_3$.[56] The structure of 2,2′-bipyridyldichlorodiphenyltin consists[57] of octahedrally co-ordinated tin atoms with cis-chlorine atoms and $trans$-phenyl groups which are not equivalent, the plane of one being at an angle of 79.5° to the other.

A free-radical chain mechanism has been proposed for the reaction of hexa-alkylditins and N-chlorosuccinimide;[58a] the reaction is catalysed by molecular oxygen. $R_3Sn\cdot$ ($R = $ Ph or Bu^n) abstracts halogen atoms from alkyl halides to

50 (a) R. Alexander, M. T. Attar-Bashi, C. Eaborn, and D. R. M. Walton, Tetrahedron, 1974, 30, 899; (b) C. J. Moore, M. L. Bullpitt, and W. Kitching, J. Organometallic Chem., 1974, 64, 93.
51 R. B. King and K. C. Hodges, J. Organometallic Chem., 1974, 65, 77.
52 M. Barnard, P. J. Smith, and R. F. M. White, J. Organometallic Chem., 1974, 77, 189.
53 G. M. Bancroft, V. G. K. Das, T. K. Sham, and M. G. Clark, J.C.S. Chem. Comm., 1974, 236.
54 L. Pellerito, R. Cefalù, A. Silvestri, F. Di Bianca, R. Barbieri, H.-J. Haupt, H. Preut, and F. Huber, J. Organometallic Chem., 1974, 78, 101.
55 (a) H. Preut, H.-J. Haupt, F. Huber, R. Cefalù, and R. Barbieri, Z. anorg. Chem., 1974, 407, 257; (b) ibid., 1974, 410, 88.
56 K. D. Bos, E. J. Bulter, and J. G. Noltes, J. Organometallic Chem., 1974, 67, C13.
57 P. G. Harrison, T. J. King, and J. A. Richards, J.C.S. Dalton, 1974, 1723.
58 (a) P. M. Digiacomo and H. G. Kuivila, J. Organometallic Chem., 1974, 63, 251; (b) H. G. Kuivila and C. C. H. Pian, J.C.S. Chem. Comm., 1974, 369; (c) E. J. Bulten and H. A. Budding, J. Organometallic Chem., 1974, 78, 385; (d) A. Peloso, ibid., 1974, 67, 423.

give alkyl radicals whose lifetimes are dependent upon the concentration and nature of the hydrogen-atom donor.[58b] Steric factors appear to be important in the disproportionation of asymmetric hexa-alkylditins to the symmetric compounds;[58c] the reactivity of organoditins to oxidation seems to be related to the inductive nature of the organic groups.[58d]

The first four-membered cyclostannazanes (8)[59a] have been obtained from the self-condensation of $Bu^t_2Sn(NHR)_2$ (R = Me or $PhCH_2$), prepared from $Bu^t_2Sn(NMe_2)_2$ and RNH_2. An unusual cyclotristannazane (9) has been prepared from the reaction of $(Me_3Sn)_2NP_3N_3F_5$ and trifluoroacetic acid anhydride.[59b] The reaction of Bu^t_3SnPh with NH_3 affords the first primary

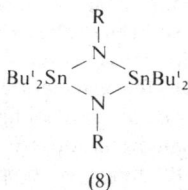

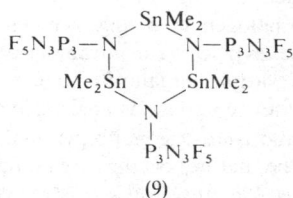

stannylamines.[60] Unlike other stannylamines the Sn—N bond in these compounds is not cleaved by CCl_4. The Sn atom in $Me_3SnN(Me)NO_2$ is co-ordinated by two axial nitrogen atoms and three equatorial methyl groups in a trigonal-bipyramidal arrangement.[61] This is not in agreement with an earlier report and it seems likely that the compound has a different structure in solution.

Stannylated phosphorus ylides of the type $Ph_3P=C(SnMe_3)C(O)R$ have been obtained from the treatment of $Ph_3P=CHC(R)=O$ with BuLi and Me_3SnCl.[62] However, when R = OMe the elimination of Me_3SnOMe from the ylide takes place, giving a keten, $Ph_3P=C=C=O$. The structure of $(Me_2Sn)_3(PO_4)_2.8H_2O$ has one tin atom octahedrally co-ordinated with *trans*-methyl groups and the other two tin atoms in a highly distorted tetrahedral arrangement due to weak co-ordination to two water molecules.[63] The methyldichlorophosphates of tin and lead[64] have been obtained from the reaction of the appropriate methyltin and methyl-lead chloride with $P_2O_3Cl_4$. Spectral data suggest that the tin compounds have polymeric structures with O—P—O bridges.

[59] (a) D. Hänssgen and I. Pohl, *Angew. Chem. Internat. Edn.*, 1974, **13**, 607; (b) H. W. Roesky and H. Wiezer, *Chem. Ber.*, 1974, **107**, 1153.
[60] H.-J. Götze, *Angew. Chem. Internat. Edn.*, 1974, **13**, 88.
[61] A. M. Domingos and G. M. Sheldrick, *J. Organometallic Chem.*, 1974, **69**, 207.
[62] J. Buckle and P. G. Harrison, *J. Organometallic Chem.*, 1974, **77**, C22.
[63] J. P. Ashmore, T. Chivers, K. A. Kerr, and J. H. G. Van Roode, *J.C.S. Chem. Comm.*, 1974, 653.
[64] K. Dehnicke, R. Schmitt, A.-F. Shihada, and J. Pebler, *Z. anorg. Chem.*, 1974, **404**, 249.
[65] (a) D. P. Gaur, G. Srivastava and R. C. Mehrotra, *J. Organometallic Chem.*, 1974, **63**, 221; (b) I. Wakeshima and I. Kijima, *ibid.*, 1974, **76**, 37; (c) D. P. Gaur, G. Srivastava, and R. C. Mehrotra, *ibid*, 1974, **65**, 195; (d) G. Plazzogna, V. Peruzzo, and G. Tagliavini, *ibid.*, 1974, **66**, 57.

Contrary to an earlier report, alkyltin trialkoxides are easily synthesized from alkyltin trichlorides and the appropriate sodium alkoxide.[65a] The related compounds $R^1Sn(OR^2)_2I$ (R^1 = Me, Et, or Bu^n; R^2 = MeC=CHOMe, MeC=CHCO$_2$Et, PhC=CHCOPh, or o-C$_6$H$_4$CO$_2$Et) have been obtained from Sn(OR^2)$_2$ and n-alkyl iodides.[65b] BuSn(OPr^i)$_3$ reacts with various alkanolamines to give cyclic derivatives;[65c] *e.g.* reaction with HOC$_2$H$_4$NHMe gives Pr^{i}OBuSn—NMe—C$_2$H$_4$—O. I.r. data on compounds of the type Ph$_4$Sn$_2$-(O$_2$CMPh$_3$)$_2$ (M = C, Si, or Ge) indicate significant $(d–p)\pi$-bonding between the carbonyl groups and M in the case of the silicon and germanium derivatives.[65d] The crystal structure of Sn$_2$(O$_2$CC$_6$H$_4$NO$_2$-o)$_4$O,THF shows that the compound is polymeric with octahedral Sn^{IV} atoms and pentagonal-pyramidal Sn^{II} atoms bridged by the carboxylate groups.[66a] The rearrangement of Sn(acac)$_2$X$_2$ (X = halogen) involving exchange of the acac groups has been studied kinetically and takes place *via* an intramolecular mechanism.[66b] Spectroscopic data obtained from various organotin mercaptoesters indicate that the glycolate and propionate[67] compounds possess, in addition to the *trans*-octahedral configuration, *cis*- and *trans*-trigonal-bipyramidal forms. The configuration adopted depends on the nature of the organotin compound and its mode of preparation. Bis-(NN'-diethyldithiocarbamato)diphenylstannane[68] possesses two bidentate dithiocarbamato-ligands, resulting in a distorted *cis*-octahedral arrangement about the tin atom.

Sn$_3$F$_8$ has been synthesized[69] from the oxidation of SnF$_2$ in HF; the structure of this compound consists of *trans*-fluorine-bridged Sn^{IV}F$_6$ units linked to polymeric Sn^{II}F chains. Only mononuclear complexes have been obtained from the reaction of alkyltin(IV) cations and fluoride ligands.[70] The stability constants of such complexes are strongly affected by the chain length of the alkyl group. A new class of compounds, trichloro(silylmethyl)stannanes, has been synthesized from chloromethylsilanes and tin dichloride.[71] There continues to be interest in the chemistry of tin halide adducts,[72a] and in particular one report has appeared on the adducts of SnCl$_2$ with nitrogen bases.[72b] The isolation of these compounds is considered to be evidence for the possible participation of the tin d-orbitals in the bonding. The i.r. spectra of K$_2$SnCl$_4$,H$_2$O indicate that both the hydrogen atoms of the water molecule are involved in hydrogen-bonding and that one of these is weak and bifurcated.[73a] Similar studies on KSnCl$_3$,H$_2$O indicate that

[66] (a) P. F. R. Ewings, P. G. Harrison, T. J. King, and A. Morris, *J.C.S. Chem. Comm.*, 1974, 53; (b) R. W. Jones and R. C. Fay, *Inorg. Chem.*, 1973, **12**, 2598.
[67] C. H. Stapfer and R. H. Herber, *J. Organometallic Chem.*, 1974, **66**, 425.
[68] P. F. Lindley and P. Carr, *J. Cryst. Mol. Structure*, 1974, **4**, 173.
[69] M. F. A. Dove, R. King, and T. J. King, *J.C.S. Chem. Comm.*, 1973, 944.
[70] F. Magno, G. Bontempelli, G.-A. Mazzocchin, and G. Pilloni, *J. Organometallic Chem.*, 1974, **67**, 33.
[71] V. F. Mironov, V. I. Shiryaev, V. V. Yankov, A. F. Gladchenko, and A. D. Naumov, *J. Gen. Chem. (U.S.S.R.)*, 1974, **44**, 776.
[72] (a) D. Millington, J. M. Winfield, and M. G. H. Wallbridge, *Ann. Reports (A)*, 1973, **70**, 331; (b) D. L. Perry and R. A. Geanangel, *J. Inorg. Nuclear Chem.*, 1974, **36**, 207.
[73] (a) M. Falk, C.-H. Huang, and O. Knop, *Canad. J. Chem.*, 1974, **52**, 2380; (b) *ibid.*, p. 2928.

the water molecules are crystallographically distinct, asymmetric, and possess weak hydrogen bonds which are highly bent.[73b] The structure of [SnCl$_3$(OEt), EtOH]$_2$ consists of centrosymmetric dimeric units in which the tin atoms are octahedrally co-ordinated by three *cis*-chlorine atoms and bridged by ethoxy-groups.[74]

Lead.—Activated olefins and acetylenes undergo cycloaddition reactions with $\overset{+}{N}{=}N{-}\overset{-}{C}(PbMe_3)CO_2Et$ (A) to give pyrazoles and pyrazolines substituted by the Me$_3$Pb group.[75a] The diazoacetic ester derivative (A) is one of a series of diazo-lead derivatives which have been prepared from the reaction of Me$_3$-PbN(SiMe$_3$)$_2$ and HC(N$_2$)R (R = CO$_2$Et, COMe, COPh, or Me$_3$Pb)[75b] The ^{207}Pb chemical shifts of a number of methyl-lead derivatives appear to be strongly influenced by π-bonding effects. Data for the propenyl compounds indicate complex formation in benzene and pyridine solutions.[76] (*m*-NO$_2$C$_6$H$_4$)$_4$Pb has been synthesized in a symmetrization reaction of (*m*-NO$_2$C$_6$H$_4$)$_2$PbX$_2$ (X = Cl or Br) with hydrazine hydrate in the presence of Na$_2$CO$_3$.[77] The product reacts with H$_2$, X$_2$, or MeCO$_2$H to re-form (*m*-NO$_2$C$_6$H$_4$)$_2$PbX$_2$. The structure of (π-C$_6$H$_6$)Pb(AlCl$_4$)$_2$,C$_6$H$_6$ consists of a chain of [AlCl$_4$]$^-$ tetrahedra bridged by Pb^{II} atoms.[78] The remaining AlCl$_4$$^-$ ion acts as a bidentate ligand forming axial and equatorial Pb—Cl bonds. The lead atom has a distorted trigonal-bipyramidal geometry with one axial co-ordination site occupied by the centre of a benzene ring.

The PbO–Bi$_2$O$_3$ system[79] has been examined at various temperatures and found to contain five different phases: *n*Bi$_2$O$_3$: *m*PbO (*m* : *n* = 6 : 1, 3 : 2, 4 : 5, 3 : 7, or 1 : 3). The structure of [Pb$_4$(OH)$_4$]$_3$(CO$_3$)(ClO$_4$)$_{10}$,6H$_2$O contains discrete [Pb$_4$(OH)$_4$]$^{4+}$ units in which the four lead atoms form a distorted tetrahedron with the OH groups situated on its outside faces.[80] The strong interest in the synthetic importance of lead tetra-acetate stressed last year[81a] has been maintained. Reports have appeared on the synthetic[81b] and stereochemical[81c] consequences of the oxidation of alcohols and on the preparation of bis(aryloxy)diarylplumbanes.[81d] Two important mechanistic studies have also been reported. The first concerns the oxidation of MeOH in which the rate-determining step involves the cleavage of a C—H bond. Contrary to an earlier

[74] M. Webster and P. H. Collins, *Inorg. Chim. Acta*, 1974, **9**, 157.
[75] (*a*) R. Grüning and J. Lorberth, *J. Organometallic Chem.*, 1974, **69**, 213; (*b*) ibid., 1974, **78**, 221.
[76] M. J. Cooper, A. K. Holliday, P. H. Makin, R. J. Puddephatt, and P. J. Smith, *J. Organometallic Chem.*, 1974, **65**, 377.
[77] E. Kunze and F. Huber, *J. Organometallic Chem.*, 1974, **63**, 287.
[78] A. G. Gash, P. F. Rodesiler, and E. L. Amma, *Inorg. Chem.*, 1974, **13**, 2429.
[79] J. C. Boivin and G. Tridot, *Compt. rend.*, 1974, **278**, *C*, 865.
[80] S.-H. Hong and A. Olin, *Acta Cryst.*, 1973, **B29**, 2242.
[81] (*a*) D. Millington, J. M. Winfield, and M. G. H. Wallbridge, *Ann. Reports (A)*, 1973, **70**, 332; (*b*) J. Ehrenfreund, M. P. Zink, and H. R. Wolf, *Helv. Chim. Acta*, 1974, **57**, 1098, 1116; (*c*) M. Lj. Mihailović, J. Bošnjak, and Ž. Čeković, ibid., p. 1015; (*d*) O. P. Syutkina, E. M. Panov, and K. A. Kocheshkov, *J. Gen. Chem. (U.S.S.R.)*, 1973, **43**, 1313; (*e*) Y. Pocker and B. C. Davis, *J.C.S. Chem. Comm.*, 1974, 803; (*f*) A. L. J. Beckwith, R. T. Cross, and G. E. Gream, *Austral. J. Chem.*, 1974, **27**, 1673; ibid., p. 1693.

report, the 'product', HCHO, is further oxidized by the methanol hemiacetal to methyl formate.[81e] The second[81f] involves the decarboxylation of tertiary carboxylic acids, which is thought to involve an organolead intermediate which decomposes *via* an $S_N 1$ displacement or a cyclic *cis*-elimination process.

The structure of plagionite, $Pb_5Sb_8S_{17}$,[82] consists of two lead atoms co-ordinated by six and seven S atoms in distorted octahedral arrangements and a third, eight-co-ordinate lead atom with a square-antiprismatic geometry.

An organolead intermediate has been obtained during the reaction of $PbF_2(OAc)_2$ and pregnenolone;[83] although such compounds have been previously postulated in reactions of Pb^{IV} with carbon–carbon double bonds this is the first report of the actual isolation of this type of intermediate.

2 Group V

Nitrogen.—*Ab initio* calculations on N_3^+, N_4, and N_6 have shown that in the cyclic form N_3^+ possesses a secondary minimum and that the stability of N_6 relative to N_4 may be attributed to the presence of a delocalized π-system in the former.[84] N_5H_5 consists of linear N_3^- ions and staggered $N_2H_5^+$ ions. The system is held together by NH···N bonds which are weak and probably bifurcated.[85] Pure di-imine has been obtained as a yellow solid from the thermolysis of alkali-metal tosylhydrazides.[86a] Spectroscopic evidence shows that the molecule has a *trans* configuration. However, i.r. studies on labelled μ-di-imine complexes of the type $^mN_2X_2[M(CO)_5]_2$ (M = W[86b] or Cr;[86c] X = H or D; $m = 14, 15$) suggest that the di-imine is in the *cis*-form. A crystal structure determination[86d] of $Me_3SiN=NSiMe_3$ reveals a remarkably short N—N bond (117 pm) and an unusually long Si—N bond (181 pm). Certain geometrical aspects of the structure result from the stereochemical activity of the nitrogen lone pair. The thermolysis of bis(trimethylsilyl)di-imine[86e] leads *inter alia* to the formation of the tris(trimethylsilyl)hydrazyl radical, which reacts with hydrogen donors to form the corresponding hydrazine.[86f] Mechanistic studies on the deuteriated di-imine have shown that the various products are formed from either radical chain reactions or a disproportionation and dimerization reaction of the starting material.[86g] *Ab initio* calculations on PH_4NH_2 have shown that the favoured conformation is the one in which the nitrogen lone pair lies in the equatorial plane of the phosphorane moiety.[87] ^{13}C n.m.r. studies on bicyclic hydrazines indicate that the double inversion process at the two nitrogen

[82] S.-A. Cho and B. J. Wuensch, *Z. Krist.*, 1974, **139**, 351.
[83] M. Ephritikhine and J. Levisalles, *J.C.S. Chem. Comm.*, 1974, 429.
[84] J. S. Wright, *J. Amer. Chem. Soc.*, 1974, **96**, 4753.
[85] G. Chilgen, J. Etienne, S. Jaulmes, and P. Laruelle, *Acta Cryst.*, 1974, **B30**, 2229.
[86] (a) N. Wiberg, G. Fischer, and H. Bachhuber, *Chem. Ber.*, 1974, **107**, 1456; (b) D. Sellman, A. Brandl, and R. Endell, *Angew. Chem. Internat. Edn.*, 1973, **12**, 1019; (c) *ibid.*, p. 1019; (d) M. Veith and H. Bärnighausen, *Acta Cryst.*, 1974, **B30**, 1806; (e) M. Wiberg and W. Uhlenbrock, *J. Organometallic Chem.*, 1974, **70**, 239; (f) *ibid.*, p. 249; (g) N. Wiberg, W. Uhlenbrock, and W. Baumeister, *ibid.*, p. 259.
[87] J. M. Mowell, *Chem. Phys. Letters*, 1974, **25**, 51.

atoms is *trans* $\rightleftharpoons$ *trans* rather than *cis* $\rightleftharpoons$ *cis*.[88a] Variable-temperature n.m.r. data obtained on N-substituted hydrazines indicate that steric hindrance between eclipsed substituents in the transition state is of considerable importance in determining the size of the energy barriers involved in the nitrogen inversion process.[88b] Similar arguments have also been put forward to explain the conformational fluctuations of tetrazines.[88c]

Tertiary amines may be dealkylated by an unusually mild reaction involving treatment of the amine with $AgNO_2$ in DMF to yield N-dealkyl-N-nitrosoamines.[89] A detailed spectroscopic study has been carried out on the complexes formed by perfluoro-organo-bromides and -iodides with nitrogen bases;[90a] in particular, the i.r. frequency shifts suggest that NMe_3 forms stronger complexes than NEt_3.[90b] The barriers to rotation about the amide bonds in N-acetylformimide have been measured for the first time using n.m.r. spectroscopy.[91] The favoured conformation is that with C_{3h} symmetry and requires rotation about all three amide bonds.

Hexafluoroacetone azine. $(CF_3)_2CN_2$, reacts with phosphites to yield phosphoranes[92a] and with isobutylene to give a 'criss-cross' cycloaddition product.[92b] Contrary to an earlier report the adduct is a zwitterionic azomethinimine derivative and not a Diels–Alder adduct;[92c] the latter, itself, forms 1 : 1 adducts with nucleophiles[92d] and undergoes [3 + 2] cycloaddition reactions.[92e] Sulphonylisocyanates react with $(CF_3)_2CN_2$ to give the ring system (10),[93] which on treatment with H_2O and Et_2NH undergoes substitution of X.

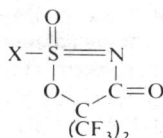

$$X - \overset{\displaystyle O}{\underset{\displaystyle O}{\overset{\displaystyle \|}{\underset{\displaystyle |}{S}}}} = N$$

(in ring) C=O, C, $(CF_3)_2$

(10) X = C, Cl, or CF_3

The amounts of NO and NO_2 present in the stratosphere have been measured using various spectroscopic techniques.[94a] It is thought that an increase in the

[88] (a) Y. Nomura, N. Masai, and Y. Takeuchi, *J.C.S. Chem. Comm.*, 1974, 288; (b) V. J. Baker, A. R. Katritzky, J.-P. Majoral, S. F. Nelson, and P. J. Hintz, *J.C.S. Chem. Comm.*, 1974, 823; (c) R. A. Y. Jones, A. R. Katritzky, A. R. Martin, D. L. Ostercamp, A. C. Richards, and J. M. Sullivan, *J.C.S. Perkin II*, 1974, 948.
[89] L. Bernardi and G. Bosisio, *J.C.S. Chem. Comm.*, 1974, 690.
[90] (a) M. F. Cheetham, I. J. McNaught, and A. D. E. Pullin, *Austral. J. Chem.*, 1974, **27**, 973; ibid., p. 987; (b) I. J. McNaught and A. D. E. Pullin, ibid., p. 1009.
[91] E. A. Noe and M. Raban, *J. Amer. Chem. Soc.*, 1974, **96**, 1598.
[92] (a) K. Burger, W. Thenn, J. Fehn, A. Gieren, and P. Narayanan, *Chem. Ber.*, 1974, **107**, 1526; (b) K. Burger, W. Thenn, and A. Gieren, *Angew. Chem. Internat. Edn.*, 1974, **13**, 474; (c) K. Burger, W. Thenn, A. Gieren, and P. Narayanan, ibid., p. 475; (d) K. Burger, W. Thenn, H. Schickaneder, and H. Peuker, ibid., p. 476; (e) K. Burger, W. Thenn, R. Rauh, and H. Schickaneder, ibid., p. 477.
[93] H. Steinbeisser, R. Mews, and O. Glemser, *Z. anorg. Chem.*, 1974, **406**, 299.
[94] (a) M. Ackerman, D. Frimout, C. Muller, D. Nevejans, J. C. Fontanella, A. Girard, L. Gramont, and N. Louisnard, *Canad. J. Chem.*, 1974, **52**, 1532; H. I. Schiff, ibid., p. 1536; (b) S. C. Wofsy and M. B. McElroy, ibid., p. 1582.

concentration of these gases could lead to a significant reduction in the amount of atmospheric ozone.[94b] Kinetic studies on the reaction of NO^+ with $MeOH$[95] indicate that the nitroxide ion participates in a series of clustering reactions until four molecules of methanol have been added. A new, highly active, low-temperature catalyst, obtained from the coprecipitation of SnO_2–CuO gels, for the reduction of NO by CO_2 has been reported.[96] Hypofluorite isomers NOF and $ONOF$ of FNO and O_2NF have been isolated (see p. 236).

Reaction of NF_3 with P_4S_3 and P_4S_{10} gives SPF_3 and PF_3 at 601—633 K and $(NPF_2)_n$ ($n = 3$—9) at 453—488 K.[97] N_2H_5F is structurally similar to N_5H_5 mentioned earlier[85] in that it consists of $N_2H_5^+$ ions and F^- ions linked by NH$\cdots$N and NH$\cdots$F bonds;[98] each fluoride ion is connected to four different $N_2H_5^+$ ions. The structures of NI_3,C_5H_5N[99] and $NI_3,I_2,(CH_2)_6N_4$[100] have been reported. The former contains NI_4 tetrahedra which are connected by corners to form infinite chains while the latter is linked by three intermolecular N—I bonds and an I_2 bridge.

Phosphorus.—*Ab initio* calculations[101] on H_2NPH_2 suggest that the barrier to inversion at phosphorus is strongly dependent on the electronegativity of the NH_2 group. Further, the planar geometry of the nitrogen atom is related to an inductive electron release from the PH_2 to the NH_2 moiety and remarkably does not necessarily involve $(p$–$d)\pi$-bonding. Silylphosphine[102a] reacts easily with a variety of amines to give silylamines; the reaction represents a useful alternative route to the formation of Si—N bonds. X_3SiPH_2 have been synthesized in the reaction of $BrSiF_3$ and $LiAl(PH_2)_4$ for $X = F$[102b] and from $SiCl_4$ and $MeSiHClPH_2$ for $X = Cl$.[102c] The metallation of the P—H bond in $HPMe_2$ and various silylphosphines has been accomplished with $LiBu$[102d] and $LiPHMe$[102e] respectively. The products from the last reaction form derivatives of the type $LiAl(PHSiH_3)_4$ and $LiAl(PHSiH_2Me)_4$ on treatment with $AlCl_3$. Several reports have appeared on the preparation and characterization of various phosphine–borane adducts.[103a] In particular, a normal co-ordinate analysis of the spectroscopic data obtained from PH_3,BX_3 (X = Cl, Br, or I) is consistent with an HPH angle of 105—106°.[103b] A rubidium phosphide,

[95] D. L. Turner and L. I. Bone, *J. Phys. Chem.*, 1974, **78**, 501.

[96] M. J. Fuller and M. E. Warwick, *J.C.S. Chem. Comm.*, 1974, 57.

[97] A. Tasaka and O. Glemser, *Z. anorg. Chem.*, 1974, **409**, 163.

[98] L. Golič and F. Lazarini, *Monatsh.*, 1974, **105**, 735.

[99] H. Hartl and D. Ullrich, *Z. anorg. Chem.*, 1974, **409**, 228.

[100] H. Pritzkow, *Z. anorg. Chem.*, 1974, **409**, 237.

[101] I. G. Csizmadia, A. H. Cowley, M. W. Taylor, and S. Wolfe, *J.C.S. Chem. Comm.*, 1974, 432.

[102] (a) C. Glidewell, *Inorg. Nuclear Chem. Letters*, 1974, **10**, 39; (b) G. Fritz, H. Schäfer, R. Demuth, and J. Grobe, *Z. anorg. Chem.*, 1974, **407**, 287; (c) G. Fritz and H. Schäfer, *ibid.*, p. 295; (d) *ibid.*, 1974, **406**, 169; (e) G. Fritz, H. Schäfer, and W. Hölderich, *ibid.*, 1974, **407**, 266.

[103] (a) B. Rapp and J. E. Drake, *Inorg. Chem.*, 1973, **12**, 2868; R. M. Kren, M. A. Mathur, and H. H. Sisler, *ibid.*, 1974, **13**, 174; (b) J. E. Drake, J. L. Hencher, and B. Rapp, *J.C.S. Dalton*, 1974, 595.

Rb_4P_6, has been obtained by treating rubidium with red phosphorus.[104] The compound contains planar P_6 rings which form hexagonal-bipyramidal Rb_2P_6 units.

A number of reports have appeared on the preparation of alkenyl-phosphines. These are of particular interest as the presence of a carbon–carbon double bond confers a high synthetic potential on such compounds. Vinyldifluorophosphine has been obtained from PF_2Br and $Hg(CH=CH_2)_2$.[105a] The phosphines $PPh_n[(CH_2)_mCH=CH_2]_{3-n}$ ($n = 2$—0; $m = 2$ or 3)[105b] have been prepared by the reaction of an appropriate Grignard reagent with PCl_3. The first 4-mono-substituted phospha- and arsa-benzenes have been obtained from the reaction of a 1,4-dihydrostannin with PBr_3 and $AsCl_3$ respectively.[106] Alkylenebi-phosphines react with silyl azides to give a new series of silylated bisiminophos-phoranes and the cyclic silane (11).[107]

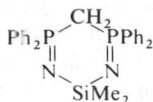

$$Ph_2P \overset{\displaystyle CH_2}{\underset{N \diagdown_{SiMe_2}\diagup N}{\overset{\|}{}\overset{\|}{}}} PPh_2$$

(11)

A useful route to compounds containing the $P(X)-CH_2-Si$ ($X = O$ or S) skeleton has been reported from the reaction of R_3SiCl ($R = Me$, Et, Ph, or naphthyl) with $LiCH_2P(X)Ph_2$.[108] When $X = S$ and $R = Me$ the substitution of two R_3Si groups was observed. Scrambling reactions between $(Me_3Si)_3P$ and Me_3MCl ($M = Ge$ or Sn) give rise to all ten possible mixed products,[109a] while $Me_3SiM(CF_3)_2$ ($M = P$ or As) have been reported from exchange reactions of Me_3SiMMe_2 and $(CF_3)_2MH$.[109b] Me_3MCl ($M = Si$, Ge, or Sn) reacts with magnesium and $Bu^t_nPCl_{3-n}$ ($n = 2$—0) to form $Bu^t_nP(Me_3M)_{3-n}$[109c] in high yield.

Two interesting n.m.r. studies on trifluoromethylphosphoranes have been reported. The first[110a] examines the fluxional behaviour of H_2PF_3 and $CF_3(H)PF_3$ in which intramolecular molecular exchange takes place between the fluorine atoms. The second[110b] concerns the preference for an axial or

[104] W. Schmettow, A. Lipka, and H. G. von Schnering, *Angew. Chem. Internat. Edn.*, 1974, **13**, 345.
[105] (a) E. L. Lines and L. F. Centofanti, *Inorg. Chem.*, 1974, **13**, 1517; (b) P. W. Clark, J. L. S. Curtis, P. E. Garrou, and G. E. Hartwell, *Canad. J. Chem.*, 1974, **52**, 1714.
[106] G. Märkl and F. Kneidl, *Angew. Chem. Internat. Edn.*, 1973, **12**, 931.
[107] R. Appel and I. Ruppert, *Z. anorg. Chem.* 1974, **406**, 131.
[108] G. Schott and K. Golz, *Z. anorg. Chem.*, 1974, **404**, 204.
[109] (a) H. Schumann, H. J. Kroth, and L. Rösch, *Z. Naturforsch.*, 1974, **29b**, 608; (b) J. E. Byrne and C. R. Russ, *J. Inorg. Nuclear Chem.*, 1974, **36**, 35; (c) H. Schumann and L. Rösch, *Chem. Ber.*, 1974, **107**, 854.
[110] (a) J. W. Gilje, R. W. Braun, and A. H. Cowley, *J.C.S. Chem. Comm.*, 1974, 15; (b) R. G. Cavell, D. D. Poulin, K. I. The, and A. J. Tomlinson, *ibid.*, p. 19; (c) H. Schmidbaur, W. Buchner, and F. H. Köhler, *J. Amer. Chem. Soc.*, 1974, **96**, 6208.

equatorial placement of the substituents in a number of trifluoromethylphosphoranes. An apicophilicity series is suggested in which the position of a substituent is related to the inductive rather than the mesomeric or electronegativity effects of a particular group. A similar study has been published[110c] on the exchange between MeOH and $Me_3P=CH_2$ in which an intermolecular scrambling of the methoxy-groups and the methyl protons takes place. An unusual cyclobutane derivative $[PhHC=\overset{|}{C}-\overset{|}{C}=NPH]_2$ has been obtained[111] from the phosphorane $Ph_3P=C=C=NPh$ and PhCHO. The phenyl groups in (2,2-diethoxyvinylidene)triphenylphosphorane are orientated in a 'paddle wheel' arrangement;[112] significantly the P—C(vinylidene) bond (168.2 pm) is longer than the sum of the covalent radii. The structure of $Ph_3PNS(O_2)PhMe$ is consistent with a delocalized electronic structure in the PNS system with a significant degree of d-orbital participation.[113] A number of reports have appeared on the effects of constraining five-co-ordinate phosphorus atoms within various cyclic systems. $(RO)_3P$ (R = Me or Et) reacts with tetramethyl-1,2-dioxetan to give a cyclic phosphorane of the type $(RO)_3P$—O—CMe_2—CMe_2—O;[114a] kinetic studies suggest that the reaction is either homolytic or concerted in nature. The highly constrained phosphorane (12) is remarkably stable, surviving at 473 K for several hours.[114b] This has been accounted for in terms of orbital symmetry rules which, thermally, only allow axial–axial and equatorial–equatorial fragmentations, and in this case these would lead to thermodynamically unfavoured products. The related

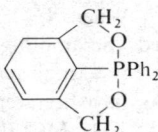

(12)

spirophosphoranes have also been the subject of intensive investigation. Certain spirophosphoranes containing a P—H bond display a new type of tautomeric equilibrium between the three- and five-co-ordinate forms.[115a] E.s.r. studies on a variety of biphenylenephosphoranyl radicals indicate that the molecules have a tetrahedral geometry[115b] and that the odd electron is delocalized on the aromatic rings and the d-orbitals of the phosphorus atom.[115c] N.m.r. and X-ray data on spirocyclic oxyphosphoranes are consistent with square-pyramidal geometries. These are favoured when relatively electronegative atoms are

[111] H.-J. Bestmann and G. Schmid, Angew. Chem. Internat. Edn., 1974, 13, 473.
[112] H. Burzlaff, U. Voll, and H.-J. Bestmann, Chem. Ber., 1974, 107, 1949.
[113] A. F. Cameron, N. J. Hair, and D. G. Morris, Acta Cryst., 1974, B30, 221.
[114] P. D. Bartlett, A. L. Baumstark, M. E. Landis, and C. L. Lerman, J. Amer. Chem. Soc., 1974, 96, 5268; (b) D. Hellwinkel and W. Krapp, Angew. Chem. Internat. Edn., 1974, 13, 542.
[115] (a) R. Burgada and C. Laurenço, J. Organometallic Chem., 1974, 66, 255; (b) R. Rothuis, J. J. H. M. Font-Freide, and H. M. Buck, Rec. Trav. chim., 1973, 92, 1308; (c) R. Rothuis, J. J. H. M. Font-Freide, J. M. F. van Dijk, and H. M. Buck, ibid., 1974, 93, 128; (d) R. R. Holmes, J. Amer. Chem. Soc., 1974, 96, 4143.

attached to phosphorus,[115d] and, although the C_{4v} structure is stabilized by reducing ring strain and a better electronic balance, the trigonal-bipyramidal form is favoured as the electronegativity of the attached atoms decreases.

Trifluoromethylalkylaminophosphoranes, $(CF_3)_3P(NMe_2)_2$ and $(CF_3)_{4-n}$-$PCl_n(NMe_2)$ ($n = 1$ or 2),[116a] have been obtained from trifluoromethylchlorophosphoranes and $HNMe_2$. The axial CF_3 groups display an unusually low, but characteristic, $^2J(31P-^{31}P)$ (50 Hz) value. However, the halogen atoms occupy the axial positions preferentially even when they have a lower formal electronegativity than CF_3. The P—N torsional barriers in $Me_2NPCl_n(CF_3)_{2-n}$ ($n = 0$—2) are thought to arise mainly from steric and lone pair–lone pair repulsion effects rather than $(p-d)\pi$-bonding.[116b] Aminolysis reactions of iminotriorganylphosphoranes have shown that cleavage of the phosphorus–carbon bond is strongly related to the bond polarity and the electronegativity of the substituents.[117] Chloroamination of bis(diphenylphosphino)amines containing the P—N(R)—P skeleton gives rise to a phosphonium salt which, when R = H or hydrocarbon, rearranges with migration of the R group to a nitrogen atom bonded to only one phosphorus.[118a] $[Cl_2(S)P]_2NMe$ reacts with $HNMe_2$ to give mono-, non-geminal bis-, and tetrakis-derivatives, and also a new ring compound $Me_2N(S)P—NMe—P(S)(NMe_2)—S$;[118b] attempts to synthesize a geminal bis(dimethylamino)-compound have been unsuccessful. $Ph_2MeP{=}NSiMe_3$ (A) and acid anhydrides react as expected *via* cleavage of the Si—N bond; however, treatment of (A) with phenyl isocyanate, isothiocyanate, or CS_2 results in a Wittig addition–elimination reaction.[119] This was explained in terms of the strong affinity of the Me_3Si group for the anionic oxygen atom preventing elimination of Ph_2MePO. Crystal structure determinations of $P_4(NMe)_6,MeI$[120a] and $P_4S_4(NMe)_6$[120b] have shown that the P—N bonds to the four-bonded phosphorus atom are shorter than the others while the NPN bond angles are larger. These changes are consistent with greater d-orbital participation of the unique P atom in π-bonding. The birdcage compound $P_2(NMe)_6$ reacts with PhN_3 to give $PhNP(NMe)_6PNPh$[120c] and not, as was expected from an earlier report, the mono-substituted product. The basicity and polarity of various phosphoramides appear to increase as the methyl substituents are replaced by tetramethylene groups and consequently $NN'N''$-tris(tetramethylene)phosphoramide was found to have a higher donor number than any other known solvent.[120d] A 1 : 1 complex is formed between

[116] (a) D. D. Poulin and R. G. Cavell, *Inorg. Chem.*, 1974, **13**, 2324; (b) A. H. Cowley, M. J. S. Dewar, J. W. Gilje, D. W. Goodman, and J. R. Schweiger, *J.C.S. Chem. Comm.*, 1974, 340.

[117] B. Ross and K.-P. Reetz, *Chem. Ber.*, 1974, **107**, 2720.

[118] D. F. Clemens and W. E. Perkinson, *Inorg. Chem.*, 1974, **13**, 333; (b) G. Bulloch, R. Keat, and N. H. Tennent, *J.C.S. Dalton*, 1974, 2329.

[119] K. Itoh, M. Okamura, and Y. Ishii, *J. Organometallic Chem.*, 1974, **65**, 327.

[120] (a) G. W. Hunt and A. W. Cordes, *Inorg. Chem.*, 1974, **13**, 1688; (b) G. W. Hunt and A. W. Cordes, *Inorg. Nuclear Chem. Letters*, 1974, **10**, 637; (c) M. Bermann and J. R. Van Wazer, *ibid.*, p. 737; (d) Y. Ozari and J. Jagur-Grodzinski, *J.C.S. Chem. Comm.*, 1974, 295; (e) P. Bruno, M. Caselli, and D. Monica, *Inorg. Chim. Acta*, 1974, **10**, 121.

HMPA and I_2 which conductiometric measurements suggest contains the HMPAI$^+$ and I^- ions.[120e] $(NMe_2)_2PCl$ reacts with $AlCl_3$ to give a 1 : 1 adduct with the remarkable ionic structure $[(Me_2N)_2P]^+[AlCl_4]^-$;[121a] this contains a two-co-ordinate phosphorus cation and displays evidence of $(p-p)\pi$-bonding between the P and N atoms. The di-iminophosphorane $R_2NP(=NR)_2$ (R = $SiMe_3$) has been obtained from the reaction of $R_2NP=NR$ with RN_3;[121b] i.r. data suggest that a polar resonance structure, $RN-\overset{+}{P}[\overset{-}{\cdots}NR]_2$, contributes significantly to the bonding in this compound. Heptamethyldisilazane reacts with the heterocycle $ClP-[NMe]_2-PCl-[NMe]_2$ to give the new bicyclic

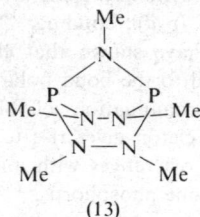

$$
\begin{array}{c}
\text{Me} \\
| \\
\text{N} \\
\diagup \quad \diagdown \\
\text{P} \qquad \text{P} \\
\text{Me}\diagdown\diagup\text{N}\!-\!\text{N}\diagdown\diagup\text{Me} \\
\text{N}\!-\!\text{N} \\
\diagup \qquad \diagdown \\
\text{Me} \qquad\quad \text{Me}
\end{array}
$$

(13)

P—N ring compound (13).[122a] Cyclodiphosphazanes of the type $Cl(X)$-$P-NR-P(X)Cl-NBu^t$ (X = lone pair; R = Me, Et, or Bu^t; X = O or S; R = Me) have been obtained from a reaction of $[Cl_2P(X)]_2NR$ and Bu^tNH_2.[122b] Interestingly only one of the two possible isomers is obtained in these heterocycles and, although the reason for this is not clear, it is probably related to a balance of electronic and steric factors. The related diazaphosphetidine di-imides $[Me_3C(RN=)P-NR]_2$ (R = Me or Et), which do form cis- and trans-isomers, have been obtained by the elimination of Me_3SiCl from $RN=PCMe_3$-$ClNRSiMe_3$.[122c] The thermal decomposition of cyclodiphosphazanes results in the formation of various oligomeric and polymeric phosphazenes.[122d]

Phosphazene chemistry continues to be an area subject to intensive investigation[123a] although the majority of the reports over the past year have been concerned with the cyclic derivatives. A novel phosphazene ring, $(Me_2NHPN)_3$, in which one of the substituents at phosphorus is hydrogen, has been synthesized by the reaction of $(Me_2N)_2PCl$ and NH_3,[123b] while a useful route to perfluoroalkylphosphazenes[123c] has been reported from the reaction of $R^f_2PCl_3$ (R^f = perfluoroalkyl) and NH_3. A number of phenylpiperidinocyclotriphosphazenes[123d]

[121] (a) M. G. Thomas, R. W. Kopp, C. W. Schultz, and R. W. Parry, J. Amer. Chem. Soc., 1974, 96, 2646; (b) E. Niecke and W. Flick, Angew. Chem. Internat. Edn., 1974, 13, 134.
[122] (a) H. Nöth and R. Ullman, Chem. Ber., 1974, 107, 1019; (b) G. Bulloch and R. Keat, J.C.S. Dalton, 1974, 2010; (c) O. J. Scherer, P. Klusmann, and N. Kuhn, Chem. Ber., 1974, 107, 552; (d) H.-G. Horn, Z. anorg. Chem., 1974, 406, 119.
[123] (a) D. Millington, J. M. Winfield, and M. G. H. Wallbridge, Ann. Reports (A), 1973, 70, 338; (b) A. Schmidpeter and H. Rossknecht, Chem. Ber., 1974, 107, 3146; (c) V. N. Prons, M. P. Grinblat, and A. L. Klebanskii, J. Gen. Chem. (U.S.S.R.), 1973, 43, 695; (d) S. Das, R. A. Shaw, and B. C. Smith, J.C.S. Dalton, 1974, 1610; (e) A. T. Fields and C. W. Allen, J. Inorg. Nuclear Chem., 1974, 36, 1929.
[124] (a) P. L. Markila and J. Trotter, Canad. J. Chem., 1974, 52, 2197; (b) H. P. Calhoun and J. Trotter, J.C.S. Dalton, 1974, 377.

have been obtained from a Friedel–Crafts reaction of the corresponding chloro-compounds in benzene; chlorine replacement only takes place at the non-geminally substituted phosphorus atoms. Silylamino-substituted cyclotri-phosphazenes have been prepared from the reaction of hexamethyldisilazane with $P_3N_3Cl_6$ and related amino-substituted compounds.[123e] The 1 : 1 adduct $N_3P_3Me_6,I_2$ contains a non-planar ring in which the iodine molecule is weakly bonded to a nitrogen atom;[124a] the two unique P—N bonds are longer (164 pm) than the other four (159.8 pm). $N_4P_4Me_9Cr(CO)_5I$[124b] consists of $[Cr(CO)_5I]^-$ anions and $[N_4P_4Me_9]^+$ cations in which the PN ring adopts a distorted tub conformation with a methyl group attached to one of the ring nitrogen atoms. $N_4P_4(NMe_2)_8$ reacts with SbF_3 to yield a series of non-geminal dimethylamino-fluorotetraphosphazenes $N_4P_4F_n(NMe_2)_{8-n}$ ($n = 1$—4);[125a] this is a convenient route to the less highly substituted fluoro-derivatives and in particular permits the isolation of stable heptakis(dimethylamino)-compound. $N_3P_3(NMe_2)Cl_5$ has been partially fluorinated with KSO_2F to give a series of compounds $N_3P_3F_nCl_{5-n}NMe_2$ ($n = 1$—4);[125b] in this case geminal substitution takes place with preferential attack at the PCl_2 group. The crystal structure of 1-*trans*, 3-*cis*, 5-*trans*, 7-tetrakis(dimethylamino)-1,3,5,7-tetrafluorocyclotetraphosphaz-ene[125c] reveals a distorted saddle ring conformation in which the four phosphorus atoms are very nearly planar. The ring shape in this, and in other cyclotetra-phosphazenes, is thought to be determined by a balance between ring π-bonding and steric interactions between the substituents. Cyclophosphazenes containing a phosphazenyl side-chain have been prepared[126a] from pentakis(aryloxy)chloro-cyclotriphosphazenes by the substitution of the P—Cl bond with NH_3 and PCl_5 to give the P—N=PCl_3 group. Further reaction with alcohols affords the dialkoxyhydroxyphosphaza-derivatives. A crystal structure determination of a related compound $N_3P_3Cl_4Ph(NPPh_3)$ reveals two equal exocyclic P—N bonds (158 pm) which are shorter than the adjacent endocyclic P—N bonds (162.5 pm) and indicates an extensive delocalization of the bonding electrons.[126b] A number of pentakisaryloxycyclotriphosphazenes[127a] have been obtained from the reaction of $N_3P_3Cl_6$ with $NaOC_6H_4X$ (X = H, p-Me, p-OMe, m-Me, or m-OMe). Clathrates have been reported with tris-(2,3-naphthalenedioxy)-cyclotriphosphazene and benzene[127b] and tris-(1,8-naphthalenedioxy)cyclotri-phosphazene and p-xylene.[127c] In the former, channel inclusion adducts are formed in which the benzene molecules are free to tumble, while X-ray studies on the latter show that the p-xylene molecules are physically trapped in a fixed posi-tion within the channels. $N_6P_6(OMe)_{12}$ is centrosymmetric with a double-tub

[125] (a) D. Millington and D. B. Sowerby, *J.C.S. Dalton*, 1974, 1070; (b) B. Green, *ibid.*, p. 113; (c) M. J. Begley, D. Millington, T. J. King, and D. B. Sowerby, *ibid.*, p. 1162.
[126] (a) A. A. Volodin, V. V. Kireev, V. V. Korshak, and A. A. Fomin, *J. Gen. Chem. (U.S.S.R.)*, 1973, **43**, 2198; (b) M. Biddlestone, G. J. Bullen, P. E. Dann, and R. A. Shaw, *J.C.S. Chem. Comm.*, 1974, 56.
[127] (a) I. B. Telkova, V. V. Kireev. V. V. Korshak, A. A. Volodin, and A. A. Fomin, *J. Gen. Chem. (U.S.S.R.)*, 1973, **43**, 1247; (b) H. R. Allcock and M. T. Stein, *J. Amer. Chem. Soc.*, 1974, **96**, 49; (c) H. R. Allcock, M. T. Stein, and E. C. Bissell, *ibid.*, p. 4795; (d) M. W. Dougill and N. L. Paddock, *J.C.S. Dalton*, 1974, 1022.

ring conformation in which all the P—N bonds are equal (156.7 pm);[127d] non-bonded interactions and the participation of the phosphorus d-orbitals in the bonding are two important factors in determining the ring shape adopted by the molecule. Two sulphur-containing phosphazenes have been reported: the first, a six-membered ring, $Cl_2\overline{P=N-SO_2-N=PCl_2-NMe}$,[128a] has been obtained from the thermolysis of $ClSO_2N=PCl_2-N=PCl_2-NMeSiMe_3$ while the second is an eight-membered ring $(NPCl_2)_3NSOCl$[128b] prepared from the reaction of $[Cl_3P=N-PCl_2=N-PCl_3]^+[PCl_6]^-$ and $SO_2(NH_2)_2$. $N_5P_5Cl_{10}$ reacts with KSO_2F to give a series of mixed chloride–fluorides, $N_5P_5Cl_{10-n}F_n$ ($n = 1$—9);[129a] substitution takes place geminally and the rates of the first two fluorination steps have been related to the type of π-bonding expected in a N_5P_5 ring. [1]H n.m.r. studies[129b] on $N_3P_3F_{6-n}Ph_n$ ($n = 1$—4) suggest that the phenyl group is subject to a strong conjugative electron withdrawal by the phosphazene ring. The crystal structures have been reported for two non-geminally substituted cyclotriphosphazenes $N_3P_3F_3X_3$ (X = Cl or Br)[129c] in which the substituents have a *cis* orientation. When X = Cl the ring has a flattened boat conformation whereas when X = Br a slight chair shape is obtained. However, in both cases the deviations from planarity are attributed to crystal packing effects.

Microwave studies[130a] on P_2H_4 have shown that the stable conformer is the *gauche*-form while photoelectron spectra of various diphosphines[130b] have indicated that the *trans*-conformer is favoured by increasing the P—P bond length and substituent electronegativity. Low-temperature n.m.r. studies on $(Bu^t_2P)_2$ suggest that the *gauche*-isomer[130c] is more stable than the *trans*-form although molecular models indicate that steric crowding is smallest for the latter.

The first silylated diphosphine $(PhMe_3SiP)_2$ has been obtained[130d] from $(KPhP)_2$ and Me_3SiCl; spectroscopic data suggest that it exists in the *trans*-conformation. P_2F_4 reacts with C_2H_4 and $F_2PCH_2CH=CH_2$ to give 1,2-bis(difluorophosphino)ethane[130e] and $F_2PCH_2CHPF_2CH_2PF_2$.[130f] Stereochemical and substituent effects in $(PR)_5$ (R = Me[131a] or CF_3[131b]) strongly influence the value of $^1J(^{31}P-^{31}P)$. Although eclipsing of the lone pairs on coupled nuclei would favour large negative values of $^1J(^{31}P-^{31}P)$ it has not been possible to relate this to other lone pair orientations. Photoelectron data on

[128] (a) H. W. Roesky and W. Grosse-Böwing, *Z. anorg. Chem.*, 1974, **406**, 260; (b) C. Voswijk and J. C. van de Grampel, *Rec. Trav. chim.*, 1974, **93**, 120.

[129] (a) N. L. Paddock and J. Serreqi, *Canad. J. Chem.*, 1974, **52**, 2546; (b) C. W. Allen and A. J. White, *Inorg. Chem.*, 1974, **13**, 1220; (c) P. Clare, T. J. King, and D. B. Sowerby, *J.C.S. Dalton*, 1974, 2071.

[130] (a) J. R. Durig, L. A. Carreira, and J. D. Odom, *J. Amer. Chem. Soc.*, 1974, **96**, 2688; (b) A. H. Cowley, M. J. S. Dewar, D. W. Goodman, and M. C. Padolina, *ibid.*, p. 2650; (c) S. Aime, R. K. Harris, and E. M. McVicker, *J.C.S. Chem. Comm.*, 1974, 426; (d) M. Baudler, M. Hallab, A. Zarkadas, and E. Tolls, *Chem. Ber.*, 1973, **106**, 3962; (e) K. W. Morse and J. G. Morse, *J. Amer. Chem. Soc.*, 1973, **95**, 8469; (f) E. R. Falardeau, K. W. Morse, and J. G. Morse, *Inorg. Chem.*, 1974, **13**, 2333.

[131] (a) J. P. Albrand, D. Gagnaire, and J. B. Robert, *J. Amer. Chem. Soc.*, 1973, **95**, 6498; (b) J. P. Albrand and J. B. Robert, *J.C.S. Chem. Comm.*, 1974, 644; (c) A. H. Cowley, M. J. S. Dewar, D. W. Goodman, and M. C. Padolina, *J. Amer. Chem. Soc.*, 1974, **96**, 3666; (d) L. R. Smith and J. L. Mills, *J.C.S. Chem. Comm.*, 1974, 808; (e) T. C. Wallace, R. West, and A. H. Cowley, *Inorg. Chem.*, 1974, **13**, 182.

polyphosphines indicate that $(p-c)\pi$-bonding is relatively unimportant in such compounds,[131c] while [31]P n.m.r. data on cyclopolyphosphines have been used as a simple means of determining ring size.[131d] N.m.r. coupling constants obtained from various polyphosphine anion radicals[131e] suggested that the unpaired electron is localized mainly on the carbon–carbon double bond.

Microwave studies on $MeOPF_2$ have shown that there is a low barrier to rotation of the methyl group which is tilted away from the fluorine atoms.[132] X-Ray studies on two phospholes $(C_6H_4O)_2PR$ reveal that when $R = Me$[133a] the geometry about phosphorus is a rectangular pyramid whereas when $R = F$[133b] the geometry lies between the trigonal-bipyramidal and square-pyramidal forms. A simple route to $Me_2PH(O)$ has been described from the reaction of Me_2PCl and H_2O.[134a] A number of optically active phosphines have been synthesized by the reduction of racemic phosphine oxides with chiral non-racemic aluminium hydride derivatives.[134b] A stereospecific synthesis of optically active phosphine oxides has also been reported from the reaction of Grignard reagents with 1,3,2-dioxaphosphorinan-2-ones.[134c] A six-co-ordinate phosphorus compound has been prepared in which, for the first time, three different ligands have been successively attached to phosphorus.[135] Intramolecular cyclization reactions of various phosphoramidates have led to the formation of diazaphospholanes,[136a] oxazaphospholanes, and phosphorinans[136b] in good yields. $RECl_2$ and R_2ECl ($E = C$, Si, Sn, Al, or P; $R = Me$, Et, or Ph) react with alkali phosphides $Ph(M)P(CH_2)_nP(M)Ph$ ($M = Li$, Na, or K; $n = 2$ or 3) to form respectively diphospholans and diphosphorinans.[136c] The first compound with a $>P(O)NC$ structure has been synthesized as shown in Scheme 4; the compound is very reactive and readily isomerizes to the cyano-derivative.[136d]

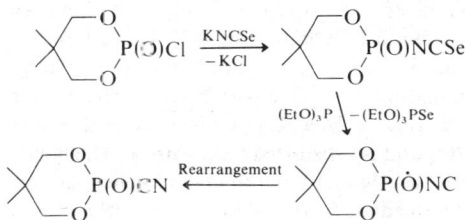

Scheme 4

[132] E. G. Codding, C. E. Jones, and R. H. Schwendeman, *Inorg. Chem.*, 1974, **13**, 178.
[133] (a) H. Wunderlich, *Acta Cryst.*, 1974, **B30**, 939; (b) H. Wunderlich and D. Mootz, *ibid.*, p. 935.
[134] (a) H.-J. Kleiner, *Annalen*, 1974, 751; (b) E. Cernia, G. M. Giongo, F. Marcati, W. Marconi, and N. Palladino, *Inorg. Chim. Acta*, 1974, **11**, 195; (c) D. B. Cooper, T. D. Inch, and G. J. Lewis, *J.C.S. Perkin I*, 1974, 1043.
[135] M. Koenig, A. Munoz, D. Houalla, and R. Wolf, *J.C.S. Chem. Comm.*, 1974, 182.
[136] (a) P. Savignac and M. Dreux, *J. Organometallic Chem.*, 1974, **66**, 81; (b) P. Savignac, J. Chenault, and M. Dreux, *ibid.*, p. 63; (c) K. Issleib and W. Böttcher, *Z. anorg. Chem.*, 1974, **406**, 178; (d) W. J. Stec, A. Konopka, and B. Uznański, *J.C.S. Chem. Comm.*, 1974, 923; (e) W. G. Bentrude, W. A. Khan, M. Murakami, and H.-W. Tan, *J. Amer. Chem. Soc.*, 1974, **96**, 5566.

The first report has appeared on the stereochemistry of free-radical displacements at tervalent phosphorus.[136e] A *cis–trans* mixture of the dioxaphosphorinans (14a and b) reacts easily with $Me_2N\cdot$ to give the corresponding phosphoramidites (14c). The reactions are highly stereoselective and result in almost complete configurational inversion at phosphorus.

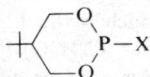

$$(14) \ a; \ X = Bu^t$$
$$b; \ X = PhCH_2$$
$$c; \ X = Me_2N$$

Nucleophilic substitution of *trans*-1-chloro-2,2,3,4,4-pentamethylphosphetan 1-oxide by alcohols does not, as expected, occur with retention of configuration but produces both *cis*- and *trans*-isomers.[137a] Evidence for the existence of monomeric metaphosphorimidates, postulated as reactive intermediates in phosphorus chemistry, has been obtained from the photolysis of *cis*- and *trans*-1-azido-2,2,3,4,4-pentamethylphosphetan 1-oxide.[137b] The first example of the cyclization of phosphoranyl radicals to oxaphosphetans or oxaphospholans has been reported;[137c] the products have an unpaired electron centred on an exocyclic carbon atom.

A variety of cyclic phosphates have been obtained from condensation reactions of orthophosphoric acid[138a] and carbodi-imides. The products were identified from an analysis of the n.m.r. spectra and a number of unusual ring and chain structures proposed. The thermal dehydration of 1-hydroxyethylidenediphosphonic acid gives rise to a compound containing the $[C_4H_8O_{12}P_4]^{4-}$ anion, which possesses a $\overline{POCPOC}$ ring.[138b] This adopts a chain conformation with the phosphorus and carbon atoms almost coplanar. The first crystal structure determination of a five-membered cyclic acylphosphate has been reported.[138c] The ring in $MeC(O)\!-\!\overline{CMe\!-\!C(O)\!-\!O\!-\!PMe(O)\!-\!O}$ is almost planar and the PO_4 unit is a distorted tetrahedron. The bond-angle deformations necessary to form a trigonal-bipyramidal oxyphosphorane (which might be obtained as an intermediate by the addition of a nucleophile) are small and must, in part, account for the remarkable reactivity of this compound towards water, alcohols, and phenols. A refined crystal structure determination[139a] of H_3PO_4, $\frac{1}{2}H_2O$ has shown that the $O(H_2O)\!-\!H\cdots O$(phosphate) hydrogen bonds are

[137] (*a*) J. Emsley, T. B. Middleton, and J. K. Williams, *J.C.S. Dalton*, 1974, 633; (*b*) J. Wiseman and F. H. Westheimer, *J. Amer. Chem. Soc.*, 1974, **96**, 4262; (*c*) A. G. Davies, M. J. Parrott, and B. P. Roberts, *J.C.S. Chem. Comm.*, 1974, 27.
[138] (*a*) T. Glonek, J. R. Van Wazer, R. A. Kleps, and T. C. Myers, *Inorg. Chem.*, 1974, **13**, 2337; (*b*) A. J. Collins, G. W. Fraser, P. G. Perkins, and D. R. Russell, *J.C.S. Dalton*, 1974, 960; (*c*) G. D. Smith, C. N. Caughlan, F. Ramirez, S. L. Glaser, and P. Stern, *J. Amer. Chem. Soc.*, 1974, **96**, 2698.
[139] (*a*) B. Dickens, E. Prince, L. W. Schroeder, and T. H. Jordan, *Acta Cryst.*, 1974, **B30**, 1470; (*b*) M. Catti and G. Ferraris, *ibid.*, p. 1; (*c*) J. L. Galigné, J. Durand, and L. Cot, *ibid.*, p. 697.

markedly bent, with the positions of the hydrogen atoms (H_2O) affected by
$H \cdots H$(phosphate) repulsions. X-Ray studies[139b] on NaH_2PO_4 reveal the
presence of a short asymmetric hydrogen bond of the type $O-H \cdots O$ in which
the $O \cdots O$ distance is 250.0 pm. The structure of $LiKPO_3F,H_2O$ consists of
chains of $[PO_3F]^{2-}$ and $[LiO_4]$ tetrahedra[139c] which are linked by the sharing
of the three oxygen atoms of the $[PO_3F]^{2-}$ ion with three different $[LiO_4]$ units.
   The interest in the electronic structure of PF_5 and related compounds has
continued.[140a] LCAO–MO–SCF calculations on PF_5 and PF_4H have shown
that charge feedback from fluorine takes place into the phosphorus d-orbitals[140b]
and may be described in terms of $(p-d)\pi$-bonding. γ-Irradiation of NH_4PF_6 and
KPF_6 gives rise to the $[PF_5]^-$ ion and not as reported previously the $\cdot PF_4$
radical.[140c] Vibrational studies[140d] on the polymorphism of KPF_6 indicate that
the room-temperature form is cubic with disordered $(PF_6)^-$ ions. A crystal
structure determination on the PF_5–pyridine adduct shows that the molecule
adopts an octahedral geometry with the pyridine ring staggered at 40.8° to one
F_3PN plane.[141a] The related adduct (15) has been obtained from a reaction of
RPF_4 (R = F, Me, Et, or Ph) and 8-trimethylsiloxyquinolines. The molecule
contains the first known example of an intramolecular $N \rightarrow P$ co-ordinate bond
(191.1 pm), causing the geometry about phosphorus to be almost perfectly
octahedral.[141b] An interesting 1 : 1 complex has been reported[141c] between
PF_3 and $AlCl_3$ which probably has an ethane-type structure in which the
phosphorus lone pair is used to form a σ-bond.

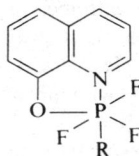

(15)

   Perfluoropinacolyltrifluorophosphorane, $RF_2P-O-C(CF_3)_2-C(CF_3)_2-O$
(R = F) (A), has been obtained from a reaction of PF_5 with the corresponding
bis(trimethylsilyl)amino-derivative $[R = (Me_3Si)_2N]$ (B),[142] which itself is
prepared from the reaction of $(Me_3Si)_2NPF_2$ and $(CF_3)_2CO$. Two interesting
points arise from the synthesis of this type of compound: firstly (A) could not be
obtained from a reaction of PF_3 and hexafluoroacetone and secondly (B)
eliminates Me_3SiF above 423 K to give a phosphetidine $LF-N(R)-LF-NR$
(R = $SiMe_3$; L = $O-(CF_3)_2C-C(CF_3)_2-O-FP<$). The Berry pseudorotation process in F_4PR (R = Me_2N, Cl, or Me) has been examined using

140 (a) D. Millington, J. M. Winfield, and M. G. H. Wallbridge, *Ann. Reports (A)*, 1973,
   70, 344; (b) J. M. Howell, J. R. Van Wazer, and A. R. Rossi, *Inorg. Chem.*, 1974, **13**,
   1747; (c) S. P. Mishra and M. C. R. Symons, *J.C.S. Chem. Comm.*, 1974, 279; (d)
   A. M. Heyns and C. W. F. T. Pistorius, *Spectrochim. Acta*, 1974, **30A**, 99.
141 (a) W. S. Sheldrick, *J.C.S. Dalton*, 1974, 1402; (b) K.-P. John, R. Schmutzler, and
   W. S. Sheldrick, *ibid.*, p. 1841; (c) E. R. Alton, R. G. Montemayor, and R. W. Parry,
   Inorg. Chem., 1974, **13**, 2267.
142 J. A. Gibson and G.-V. Röschenthaler, *J.C.S. Chem. Comm.*, 1974, 694.

^{31}P n.m.r. spectroscopy. The rate of rotation is shown to increase according to the series $Me_2N < H < Cl < Me,F$.[143a] ^{1}H n.m.r. data on 2-isopropylphenyl-bis-(4,4'-dimethyl-2,2'-biphenylene)phosphorane suggest that pseudorotation does not occur by the Berry process but requires a mechanism involving a square-pyramidal intermediate.[143b] CF_3MI_2 (M = P or As) reacts with secondary amines HNR_2 (R = alkyl) to give good yields of $CF_3M(NR_2)_2$, which on treatment with acids undergo a reaction involving cleavage of the P—N bond.[144] POI_3 has been obtained for the first time from the reaction of an alkyl phosphoro-di-iodite with iodine; the product readily decomposes to HI and H_3PO_4 on exposure to moisture.[145]

Arsenic.—$NaAsHR^1$ (R^1 = Bu or Ph) reacts with $Cl(CH_2)_3NH_2$ to give $R^1As(H)(CH_2)_3NH_2$;[146] further reaction with ketones, R^2R^3CO, generates perhydro-1,3-azarsenines, $R^1As—(CH_2)_3—NH—CR^2R^3$ (R^2 = H or Me; R^3 = H). Spectroscopic data on Me_3As,BH_3 have been interpreted in terms of C_{3v} molecular symmetry;[147a] microwave studies have established the As—B distance as 203.5 pm, which is close to the sum of the covalent radii of the atoms and appears to be the first such value to be reported. 1 : 1 Adducts of the type Me_3M,BX_3 (X = Br or I) are indefinitely stable for M = As at room temperature in an inert atmosphere;[147b] however, when M = Sb slow decomposition to Me_3SbX takes place.

Monocyclopentadienylarsines, $CpAsX_2$,[148] in which Cp groups are σ-bonded, have been obtained from the reaction of CpLi or $CpSiMe_3$ with AsX_3 (X = F, Cl, or Br); the compounds are extremely air- and moisture-sensitive and are thermally unstable. Me_2AsNMe_2 reacts with 1,2-diols to give esters of the type $Me_2AsOCR_2CR_2OAsMe_2$ and $Me_2AsOCR_2CR_2OH$ (R = H or Me)[149a] while the related reaction of $(Me_2N)_3As$ with alcohols and thiols gives esters of general formula $As(XR)_3$ (X = O or S; R = alkyl);[149b] a stepwise replacement process of the amino-groups affords the mixed compounds $(Me_2N)_{3-n}As(OR)_n$ (n = 1 or 2). Me_2AsNMe_2 also reacts with oximes to form a new class of compound, O-(dimethylarsino)oximes.[149c] Various phenylalkylarsines $Me_{3-n}AsPh_n$ (n = 1 or 2), $EtAsPh_2$, $(CH_2)_n(AsPh_2)_2$ (n = 1 or 2), and $C(CH_2AsPh_2)_4$ react with HI to give cleavage of the As—Ph bonds to form the corresponding arsenic iodides; in particular, i.r. data on $EtAsI_2$ indicate the presence of both *trans*- and *gauche*-

[143] (a) M. Eisenhut, H. L. Mitchell, D. D. Traficante, R. J. Kaufman, J. M. Deutch, and G. M. Whitesides, *J. Amer. Chem. Soc.*, 1974, **96**, 5385; (b) M. Eisenhut, W. M. Bunting, and G. M. Whitesides, *ibid.*, p. 5398.
[144] O. Adler and F. Kober, *J. Organometallic Chem.*, 1974, **72**, 351.
[145] V. G. Kostina, N. G. Feshchenko, and A. V. Kirsanov, *J. Gen. Chem. (U.S.S.R.)*, 1973, **43**, 207.
[146] A. Tzschach and P. Franke, *J. Organometallic Chem.*, 1974, **81**, 187.
[147] (a) J. R. Durig, B. A. Hudgens, and J. D. Odom, *Inorg. Chem.*, 1974, **13**, 2306; (b) M. L. Denniston and D. R. Martin, *J. Inorg. Nuclear Chem.*, 1974, **36**, 2175.
[148] P. Jutzi and M. Kahn, *Chem. Ber.*, 1974, **107**, 1228.
[149] (a) F. Kober and W. J. Rühl, *Z. anorg. Chem.*, 1974, **406**, 52; (b) *ibid.*, 1974, **403**, 56; (c) J. Kaufmann and F. Kober, *J. Organometallic Chem.*, 1974, **71**, 49.

isomers.[150] The iodides undergo a cyclization reaction in the mass spectrometer to give cycloarsines, *e.g.* $(AsR)_3$ (R = Me or Et). The first all-*cis* organocyclotriarsane[151] has been obtained by the reaction sequence shown in Scheme 5.

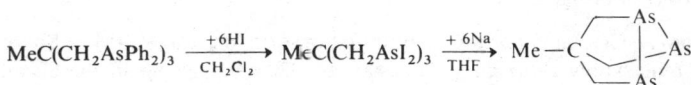

$$MeC(CH_2AsPh_2)_3 \xrightarrow[CH_2Cl_2]{+6HI} MeC(CH_2AsI_2)_3 \xrightarrow[THF]{+6Na} Me-C\begin{array}{c} As \\ \diagdown \\ As \diagup \end{array}As$$

Scheme 5

Electron-impact studies have been reported on $NAs(OR)_2$ (R = Me or Et)[152a] and $As_2O(OMe)_6NR$ (R = Pr or Bu).[152b] In the former case the arsenic–nitrogen analogues of cyclophosphazenes with six-, eight-, ten-, and twelve-membered rings have been identified in the fragmentation process, while the latter affords four-membered $\overline{AsNAsO}$ ring compounds. The alkylarsoxanes of the general formula $(RAsO)_n$ (R = Me, Et, Pr^n, or Bu^n) exist as cyclic trimers except when R = Me, when a tetramer is thought to be formed;[153] these compounds have the oxygen atom in the bridging positions and are thought to possess a limited degree of electron delocalization. The alkyl-halide-promoted rearrangement of R_3AsO (R = Et^{154a} or Pr^{i154b}), in contrast to the reaction with the corresponding sulphides, leads to the formation of triethylhydroxyarsonium trihalide, $[Et_3AsOH]X_3$, as the major product. Guérinite, $Ca_5(HAsO_4)_2(AsO_4)_2,9H_2O$,[155] contains twenty-five molecules in a unit cell in which one calcium atom and five water molecules are disordered in large cavities; the structure is built up from Ca polyhedra and $[AsO_4]$ tetrahedra which have common edges, vertices, and faces. The first tris(dialkylselenocarbamate) complexes of As, Sb, and Bi have been prepared and examined spectroscopically;[156] the shifts observed in the electronic spectra suggest that the metal–ligand π-bonding is less pronounced in the seleno-derivatives than in the corresponding thio-compounds.

Dialkylaminodifluoroarsines have been prepared[157] both from the direct aminolysis of AsF_3 and from a transamination reaction of Me_2NAsF_2 (A) with a secondary amine; (A) also reacts with primary amines to yield polymeric aminoarsines. Ion–molecule reactions of SF_4 and AsF_3 produce the secondary $[AsF_4]^-$ ion while in pure AsF_3 the secondary $[AsF_5]^-$ and $[AsF_6]^-$ ions are

[150] J. Ellermann, H. Schössner, A. Haag, and H. Schödel, *J. Organometallic Chem.*, 1974, **65**, 33.
[151] J. Ellermann and H. Schössner, *Angew. Chem. Internat. Edn.*, 1974, **13**, 601.
[152] (a) H. Preiss and D. Hass, *Z. anorg. Chem.*, 1974, **404**, 190; (b) ibid., p. 199.
[153] M. Durand and J.-P. Laurent, *J. Organometallic Chem.*, 1974, **77**, 225.
[154] (a) Yu. F. Gatilov and V. A. Perov, *J. Gen. Chem. (U.S.S.R.)*, 1973, **43**, 1126; (b) ibid., p. 1129.
[155] M. Catti and G. Ferraris, *Acta Cryst.*, 1974, **B30**, 1789.
[156] C. J. Shea and A. Hain, *Inorg. Chem.*, 1973, **12**, 3013.
[157] F. Kober and O. Adler, *J. Fluorine Chem.*, 1974, **4**, 73.

obtained.[158] Crystal structure determinations[159] of $M_2(As_2F_{10}O),H_2O$ (M = Rb or K) have shown that there is no hydrogen-bonding between the water molecules and the $[As_2F_{10}O]^{2-}$ anion in which the two arsenic atoms are linked by an oxygen bridge.

Antimony.—Crystal structure determinations[160a] on Ca_2M (M = Sb or Bi) reveal remarkably short Ca—Ca distances (330 and 334 pm in the Sb and Bi compounds respectively); the Group V atom is only co-ordinated to calcium atoms (a related compound Li_2Sb has also been reported, cf. p. 175). The antimony atoms in Ca_5Sb_3 are in two environments, being surrounded by either eight or nine calcium atoms;[160b] the Sb—Ca bond distances (average 330 pm) are indicative of some charge transfer to the Sb atom and confer a partial ionic character on the bond; Ca_5Bi_3 is isotypic.

A number of organoantimony aldehydes and ketones have been synthesized from dialkylaminostibines and enol acetates;[161a] it was found that increasing the degree of steric hindrance at the α-carbon or the antimony atom increased the amount of O-isomer formed.[161b] Organoantimony compounds of the type R_3SbL (R = Me or Ph; L = terdentate dianion containing ONO or SNO donor atoms) have been prepared and examined spectroscopically; u.v. and i.r. data suggest that the L ligand is planar, although n.m.r. spectra indicate that at room temperature the molecules are stereochemically non-rigid.[162] [121]Sb Mössbauer studies on Ph_4SbX and Ph_3SbX_2 (X = electronegative group) indicate trigonal-bipyramidal geometries in which the X group occupies an axial position.[163] A certain amount of confusion exists in the literature concerning the exact geometry of Ph_2SbCl_3 (A), either as the monohydrate or in the anhydrous form; however, a careful X-ray analysis[164] on anhydrous (A) has revealed a dimeric structure held together by chlorine bridges. $SbCl_5$ reacts with NaOEt to give a dimeric monosubstituted derivative $(SbCl_4OEt)_2$,[165] which on further attempted substitution does not give a readily identifiable product. An X-ray study[166] on the cyclic stibocan $ClŠb—S—(CH_2)_2—S—(CH_2)_2—Š$ has shown that the eight-membered ring has a deformed boat conformation and contains a re-entrant sulphur atom which gives rise to a 1,5-intra-annular Sb—S interaction (Sb—S distance 286 pm).

[158] T. C. Rhyne and J. G. Dillard, *Inorg. Chem.*, 1974, **13**, 322.
[159] W. Haase, *Acta Cryst.*, 1974, **B30**, 1722.
[160] (a) B. Eisenmann and H. Schäfer, *Z. Naturforsch.*, 1974, **29b**, 13; (b) M. Martinez-Ripoll, A. Haase, and G. Brauer, *Acta Cryst.*, 1974, **B30**, 1083, 2004.
[161] (a) V. L. Foss, N. M. Semenenko, N. M. Sorokin, and I. F. Lutsenko, *J. Organometallic Chem.*, 1974, **78**, 107; (b) V. L. Foss, N. M. Sorokin, I. M. Avrutov, and I. F. Lutsenko, *ibid.*, p. 115.
[162] F. Di Bianca, E. Rivarola, A. L. Spek, H. A. Meinema, and J. G. Noltes, *J. Organometallic Chem.*, 1974, **63**, 293.
[163] J. N. R. Ruddick, J. R. Sams, and J. C. Scott, *Inorg. Chem.*, 1974, **13**, 1503.
[164] J. Bordner, G. O. Doak, and J. R. Peters, *Inorg. Chem.*, 1974, **13**, 6763.
[165] R.-A. Laber and A. Schmidt, *Z. anorg. Chem.*, 1974, **405**, 71.
[166] M. Dräger, and R. Engler, *Z. anorg. Chem.*, 1974, **405**, 183.

The antimony atom in $SbO(H_2PO_4),H_2O$ has a trigonal-bipyramidal geometry with a lone pair in an equatorial position.[167a] The crystal is built up from corner-sharing $[PO_4]$ tetrahedra and $[SbO_4]$ polyhedra which form infinite layers with the composition $SbO(H_2PO_4)$. $K_3Sb_5O_{14}$ and $K_2Sb_4O_{11}$ both have structures in which the antimony is octahedrally surrounded by oxygen atoms;[167b] the octahedra are linked by edge- and corner-sharing to form b- and c-axis tunnels in which the K^+ ions are located. The structures of $Rb_2(Sb_2F_{10}O)$[168a] and $Cs_3(Sb_3F_{12}O_3)$[168b] have been reported. The former is similar to the arsenic analogue mentioned earlier[159] in that they both contain the $[M_2F_{10}O]^{2-}$ (M = As or Sb) anion but in this case the symmetry is lower, being m (C_s). The latter contains the $[Sb_3F_{12}O_3]^{3-}$ ion, which exists as a six-membered Sb_3O_3 ring with a boat conformation. A new antimony fluoride, $Sb_{11}F_{43}$, has been shown[169a] to contain the $[SbF_6]^-$ anion and a polymeric chain cation $[Sb_6F_{13}]_n^{5+}$; by considering only the shorter Sb^{III}—F bonds the cation can be separated into $[SbF_2]^+$ and $[Sb_2F_5]^+$ units linked by longer bridging bonds. ^{19}F N.m.r. studies on solutions of SbF_3 in water and organic solvents[169b] have shown that exchange of fluoride ion takes place with the Sb^{III} species and that the $[SbF_4]^-$ is formed in aqueous solution. E.s.r. studies on a variety of $[SbX_4]^-$ and $[Sb_2X_9]^{3-}$ anions (X = F, Cl, Br, or I) have shown that only in $[SbF_4]^-$, $[SbCl_4]^-$, and $[Sb_2F_9]^{3-}$ are the 5s electrons stereochemically active.[169c] A structural determination[170a] of K_2SbCl_5 has shown that the $[SbCl_5]^{2-}$ ion has a distorted square-pyramidal geometry in which the axial Sb—Cl bonds (238.5 pm) are shorter than the mean basal Sb—Cl bonds (average 263.9 pm); the antimony lone pair seems to have little stereochemical effect as the Sb—Sb distance is comparatively short (393.2 pm). The structure of β-$Cs_3M_2Cl_9$ (M = Sb or Bi) consists[170b] of close-packed Cs and Cl ions with the M atoms in the octahedral holes; in both cases there are two types of MCl_3 molecule present in the structure. A 1:1 adduct, $SbCl_3,S_2C_4H_8$, has been prepared from a reaction of $SbCl_3$ with 2-mercaptoethanol in benzene;[171] an X-ray structure determination has shown that the Sb—S distances are 313.5 and 306.5 pm. The structure of quinolinium hexabromoantimonate consists[172] of quinolinium, Br_3^-, and $[SbBr_6]^-$ ions; the latter has a slightly distorted octahedral geometry and retains its centre of symmetry.

[167] (a) C. Särnstrand, *Acta Chem. Scand.*, 1974, **A28**, 275; (b) H. Y.-P. Hong, *Acta Cryst.*, 1974, **B30**, 945.
[168] (a) W. Haase, *Acta Cryst.*, 1974, **B30**, 2508; (b) *ibid.*, p. 2465.
[169] (a) A. J. Edwards and D. R. Slim, *J.C.S. Chem. Comm.*, 1974, 178; (b) Yu. A. Buslaev and V. V. Peshkov, *Russ. J. Inorg. Chem.*, 1973, **18**, 803; (c) J. G. Ballard, T. Birchall, J. B. Milne, and W. D. Moffett, *Canad. J. Chem.*, 1974, **52**, 2375.
[170] (a) R. K. Wismer and R. A. Jacobson, *Inorg. Chem.*, 1974, **13**, 1678; (b) K. Kihara and T. Sudo, *Acta Cryst.*, 1974, **B30**, 1088.
[171] G. Kiel and R. Engler, *Chem. Ber.*, 1974, **107**, 3444.
[172] S. L. Lawton, E. R. McAfee, J. E. Benson, and R. A. Jacobson, *Inorg. Chem.*, 1973, **12**, 2939.

236   *G. W. Fraser, D. Millington, and M. G. H. Wallbridge*

Bismuth.—Low-temperature electron diffraction studies[173] on 'amorphous' bismuth films have shown that as the thickness of the film decreases the spatial co-ordination of the bismuth atoms becomes quite sharply defined. The hydration of the Bi^+ ion has been studied *via* gas-phase clustering reactions.[174] A comparison with lead-ion hydrates reveals evidence of chemical bonding in the $Bi^+(H_2O)_n$ ($n \leq 6$) clusters. Ba_2Bi is isostructural with Sr_2Sb, each bismuth atom being surrounded by nine barium atoms.[175] Two polymorphs of bismuth orthoborate, $BiBO_3$, have been obtained[176] by cooling a melt of H_3BO_3 and Bi_2O_3. The compound rapidly decomposes at 873 K to $3Bi_2O_3,B_2O_3$ and $3Bi_2O_3,5B_2O_3$. A number of differing reports are to be found in the literature on the thermal decomposition of $Bi_2(SO_4)_3$. A detailed study[177] reported this year shows that the decomposition may be represented by:

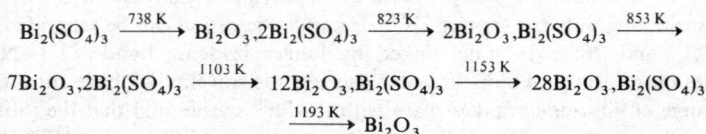

$$Bi_2(SO_4)_3 \xrightarrow{738\,K} Bi_2O_3,2Bi_2(SO_4)_3 \xrightarrow{823\,K} 2Bi_2O_3,Bi_2(SO_4)_3 \xrightarrow{853\,K}$$

$$7Bi_2O_3,2Bi_2(SO_4)_3 \xrightarrow{1103\,K} 12Bi_2O_3,Bi_2(SO_4)_3 \xrightarrow{1153\,K} 28Bi_2O_3,Bi_2(SO_4)_3$$

$$\xrightarrow{1193\,K} Bi_2O_3$$

$Bi_2Cu_3S_4Cl$[178] contains bismuth polyhedra linked by common edges to form $[(Bi_2S_4)_\infty]^{2-}$ and $[(BiS_2Cl)_\infty]^{2-}$ chains which are joined by chlorine atoms. The controlled oxidation of the $Bi_2O_3-TeO_2$ system has led to the identification of two new bismuth tellurates, $Bi_6Te_2O_{15}$ and Bi_2TeO_6.[179] In the latter compound the bismuth atoms are pyramidal while the geometry about tellurium is a distorted octahedron.

PART III: Groups VI—VIII

1 Group VI

Oxygen.—The electron affinity of O^{2-} has been computed[1] from interaction potential functions applied to the oxides MO (M = Be, Mg, Ca, Sr, or Ba). The value (728.7 kJ mol^{-1}) obtained agrees well with those derived from other methods.

Reactions of photolytically generated fluorine atoms with NO and NO_2 produce, in addition to FNO and FNO_2, the new hypofluorite tautomers NOF and ONOF.[2a] Both molecules have been studied spectroscopically using

[173] Yu. F. Komnik, L. A. Yatsuk, A. A. Motornaya, M. L. Bolotina, and B. I. Belevtsev, *Soviet Phys. Cryst.*, 1974, **18**, 791.
[174] I. N. Tang and A. W. Castleman, *J. Chem. Phys.*, 1974, **60**, 3981.
[175] M. Martinez-Ripoll, A. Haase, and G. Brauer, *Acta Cryst.*, 1974, **B30**, 2003.
[176] M. J. Pottier, *Bull. Soc. chim. belges*, 1974, **83**, 235.
[177] R. Matsuzaki, A. Sofue, H. Masumizu, and Y. Saeki, *Chem. Letters*, 1974, 737.
[178] J. Lewis and V. Kupičík, *Acta Cryst.*, 1974, **B30**, 848.
[179] B. Frit and M. Jaymes, *Bull. Soc. chim. France*, 1974, 402.
[1] K. P. Thakur, *J. Inorg. Nuclear Chem.*, 1974, **36**, 2171.
[2] (a) R. R. Smardzewski and W. B. Fox, *J.C.S. Chem. Comm.*, 1974, 241; (b) *J. Amer. Chem. Soc.*, 1974, **96**, 304; *J. Chem. Phys.*, 1974, **60**, 2980.

matrix-isolation techniques and their i.r. frequencies have been assigned.[2b] O_3 reacts with a number of covalent hypohalites; with $ClOClO_3$, $ClOSO_2F$, $BrONO_2$, and $BrOClO_3$ oxidative oxygenation of the terminal halogen occurs, giving $O_2ClOClO_3$, O_2ClOSO_2F, O_2BrONO_2, and $O_2BrOClO_3$ respectively. Under similar conditions $ClONO_2$ yields $NO_2{}^+ClO_4{}^-$.[3]

CF_3OONO_2 has been prepared by the reactions of CF_3OOH and N_2O_5 and of CF_3OOF and N_2O_4 and its i.r. and Raman spectra have been assigned. It decomposes slowly at $400\,K$ into COF_2, NO_2, and O_2.[4a] The reactions of CF_3OOF with SO_2, CO, and SF_4 have been examined but no CF_3OO compounds were obtained; the products contained CF_3OSO_2, $CF_3OC(O)$, and CF_3OSF_4 groups, indicating that O—O rather than O—F bond cleavage had occurred.[4b]

Sulphur.—As in previous years, a large amount of work on this element has appeared, and the selection made is arranged as in last year's *Annual Reports*.

Compounds containing S—S, S—C, or S—O Bonds. Two large sulphur rings S_{18} and S_{20} have been synthesized from H_2S_8 (or H_2S_{10}) and $S_{10}Cl_2$ and their structures analysed by X-ray crystallography.[5] S_8O (mentioned last year) has been conveniently prepared by the oxidation of S_8 with CF_3CO_3H in the molar ratio $1:1$,[6a] and its i.r. spectrum has been assigned.[6b] A number of six-, seven-, and eight-membered heterocycles have been formed by the reactions of hydrazine-1,2-dicarboxylic esters with sulphanyl chlorides containing two SCl, one SSCl and one SCl, and two SSCl groups respectively. An example of these reactions is:

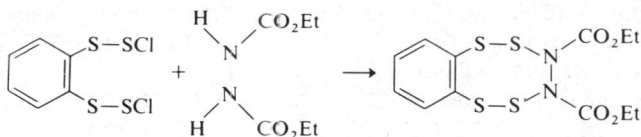

Some of the compounds react with $SbCl_5$ to form crystalline $1:1$ adducts.[7] CF_3SSCl is formed from CF_3SH and SCl_2. It undergoes a number of reactions in which the S—S bond is retained, *e.g.* $R_2NH \rightarrow R_2NSSCF_3$ and $RLi \rightarrow RSSCF_3$, in contrast to those of S_2Cl_2, in which both S—Cl and S—S bond cleavage occurs.[8]

Two reactions in which sulphonium salts are the unexpected products have been reported. Me_2S reacts with WCl_6 or WCl_5 to produce $[Me_3S^+]_2[WCl_6]^{2-}$,[9a]

[3] C. J. Schack and K. O. Christe, *Inorg. Chem.*, 1974, **13**, 2378.
[4] (a) F. A. Hohorst and D. D. DesMarteau, *Inorg. Chem.*, 1974, **13**, 715; (b) R. A. De Marco and W. B. Fox, *Inorg. Nuclear Chem. Letters*, 1974, **10**, 965.
[5] M. Schmidt, E. Wilhelm, T. Debaerdemaeker, E. Hellner, and A. Kutoglu, *Z. anorg. Chem.*, 1974, **405**, 153.
[6] (a) R. Steudal and J. Latte, *Angew. Chem. Internat. Edn.*, 1974, **13**, 603; (b) R. Steudal and M. Rebsch, *J. Mol. Spectroscopy*, 1974, **51**, 334.
[7] K.-H. Linke and R. Binczok, *Chem. Ber.*, 1974, **107**, 771.
[8] N. R. Zack and J. M. Shreeve, *Inorg. Nuclear Chem. Letters*, 1974, **10**, 619.
[9] (a) P. N. Boorman, T. Chivers, and K. N. Mahandev, *J.C.S. Chem. Comm.*, 1974, 502; (b) Y. Hara and M. Matsuda, *ibid.*. p. 919.

and $MeOS(O)Cl$ and Me_2SO in the presence of $SbCl_5$ form $[MeOMe_2S]^+[SbCl_6]^-$ at 195 K and $[MeSCH_2Me_2S]^+[SbCl_6]^-$ at 273 K. The latter reaction can be extended to produce a range of compounds of the type $[RMe_2S]^+[SbCl_6]^-$.[9b]

The thermodynamic parameters for the formation of a number of substituted aniline–SO_2 adducts have been calculated. The values of ΔH, ΔS, and ΔG for these adducts agree closely with those of similar complexes published earlier.[10] A series of Me_2SO and other sulphoxide complexes in which bonding can occur *via* the O or S atoms has been studied by ESCA. The relative differences between S $2p_{\frac{3}{2}}$ and O $1s$ ionization potentials mean that oxygen-bonded and sulphur-bonded complexes can be identified by this method.[11]

A number of optically active compounds $R^1{}_2NS(O)SR^2$ (*e.g.* $R^1 = Me$, $R^2 = Pr^n$) have been prepared from $(R^1{}_2N)_2SO$. $(-)$-$Me_2NS(O)Bu^t$ can be converted into the corresponding $(+)$-compound by reaction with $HgCl_2$ and then with Bu^tOH and K_2CO_3.[12] The sulphonyl isocyanates RSO_2NCO ($R = Me$, Ph, p-MeC_6H_4) can be synthesized from RSO_2NH_2 and $ClSO_2NCO$ *via* the intermediate $RSO_2NHCONHSO_2Cl$.[13]

Halogenosulphate chemistry continues to attract interest. $Ca(SO_3Cl)_2$, $Sr(SO_3Cl)_2$,$2HSO_3Cl$, and $Ba(SO_3Cl)_2$,$nHSO_3Cl$ ($n = 1, 2$, or 3) were produced from HSO_3Cl and the corresponding chloride and were characterized by X-ray diffraction and i.r. spectroscopy.[14a] The salts $M(S_2O_6F)_2$ ($M = Ba$ or Ca) were prepared from $M(SO_3F)_2$ and SO_3 in HSO_3F.[14b] $CHCl_3$ and $CHCl{=}CCl_2$ react with SO_3F_2 at 195 K to produce the new fluorosulphates $CHCl_2OSO_2F$ and $CHClFCCl_2OSO_2F$, which were characterized by i.r. and ^{19}F n.m.r. spectroscopy.[15a] $(CF_3)_2C(CN)O^-Na^+$ and $(CF_3)_2C(N_3)O^-Na^+$ [prepared from $(CF_3)_2CO$] react with $S_2O_5Cl_2$ or S_2O_5FCl to form $(CF_3)_2C(CN)OSO_2X$ and $(CF_3)_2C(N_3)OSO_2X$ ($X = F$ or Cl).[15b]

A large number of esters of $CF_3S(O)OH$ and $C_4F_9^tS(O)OH$ have been synthesized, including $XS(O)Cl$, $XS(O)SiMe_3$, and $XS(O)NCO$ ($X = CF_3$ or $C_4F_9^t$).[15c]

Compounds containing S—N *or* S—*Halogen Bonds.* This area of chemistry continues to provide a wealth of compounds of varying structures. $S_{11}NH$ has been synthesized from S_4N_4 and N_2H_4;[16] it can be considered as an S_{12} ring in which one of the S atoms is replaced by an NH group (similar to S_7NH and S_8). $S_3N_5PF_2$ was reported last year[17a] but the structure postulated on the basis of ^{19}F and

[10] A. P. Zipp, *J. Inorg. Nuclear Chem.*, 1974, **36**, 1399.
[11] C. C. Su and J. W. Faller, *Inorg. Chem.*, 1974, **13**, 1734.
[12] M. Mikolajczyk and J. Drabowicz, *J.C.S. Chem. Comm.*, 1974, 775.
[13] R. Appel and M. Montenarh, *Chem. Ber.*, 1974, **107**, 706.
[14] (*a*) G. Mairesse, P. Barbier, and J. Huebal, *Bull. Soc. chim. France*, 1974, 1297; (*b*) P. Bernard and P. Vast, *Compt. rend.*, 1974, **278**, C, 255.
[15] (*a*) L. F. R. Cafferata and J. E. Sicre, *Inorg. Chem.*, 1974, **13**, 242; (*b*) T. M. Churchill and M. Lustig, *J. Inorg. Nuclear Chem.*, 1974, **36**, 1426; (*c*) H. W. Roesky and S. Tutkunkardes, *Chem. Ber.*, 1974, **107**, 508.
[16] H. Garcia-Fernandez and H. G. Heal, *Compt. rend.*, 1974, **278**, C, 517.
[17] (*a*) H. W. Roesky and O. Peterson, *Angew. Chem. Internat. Edn.*, 1973, **12**, 415; (*b*) J. Weiss, I. Ruppert, and R. Appel, *Z. anorg. Chem.*, 1974, **406**, 329; (*c*) R. Appel, I. Ruppert, R. Milker, and V. Bastian, *Chem. Ber.*, 1974, **107**, 380.

[31]P n.m.r. evidence has been shown to be incorrect. An X-ray crystallographic determination shows that the molecule and the related $PhFPS_3N_5$ have structure (1) ($R^1 = R^2 = F$; $R^1 = F$, $R^2 = Ph$) in which the P atoms are tetrahedral

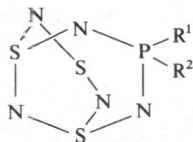

(1)

rather than five-co-ordinate.[17b] Thermal decomposition of (1) produces S_4N_4 and $SN_2^{\overline{\cdot}}$ radical anion in high yield, the reaction being monitored by e.s.r and by ^{1}H, ^{19}F, and [31]P n.m.r. spectroscopy.[17c] Two more S_7NH derivatives have been characterized; $Ph_2P(S)Cl$ reacts with S_7N^- to produce $Ph_2P(S)NS_7$,[18a] and S_7NHgPh is formed from S_7NH and $PhHgO_2CMe$.[18b] The mercury derivative can be used to prepare other S_7N compounds, e.g. S_7NSnMe_3 from Me_3SnCl. From the deep-blue solutions of S_7NH in $(Me_2N)_3PO$, the salt $[Bu^n_4N]^+[NS_4]^-$ has been isolated.[19] The vibrational spectra suggest that the anion adopts the branched chain structure $[SN(S)SS]^-$.

The structure of $S_6N_4(S_2O_6Cl)_2$ (synthesized from S_3N_2Cl and HSO_3Cl) has been determined by X-ray crystallography. The $[S_6N_4]^{2+}$ ion consists of two S_3N_2 rings linked by two S—S bonds as shown in (2). The S—S distance of 303 pm is much less than the usual van der Waals separation (~ 370 pm) and indicates some covalent contribution.[20]

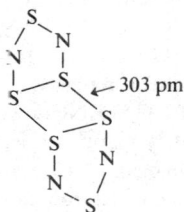

(2)

The sulphur di-imides $Me_3SnN=S=NSnMe_3$ and $Me_3SnN=S=NSiMe_3$ have been prepared from S_4N_4.[21a] Reaction of the former compound with

[18] (a) J. M. Kanamueller, *J. Inorg. Nuclear Chem.*, 1974, **36**, 3855; (b) H. G. Heal and R. J. Ramsay, *ibid.*, p. 950.

[19] T. Chivers and I. Drummond, *Inorg. Chem.*, 1974, **13**, 1222.

[20] A. J. Banister, H. G. Clarke, I. Rayment, and H. M. M. Shearer, *Inorg. Nuclear Chem. Letters*, 1974, **10**, 647.

[21] (a) H. W. Roesky and H. Wiezer, *Chem. Ber.*, 1974, **107**, 3186; (b) *Angew. Chem. Internat. Edn.*, 1974, **13**, 146; (c) A. Golloch and M. Kuss, *Z. Naturforsch.*, 1974, **29b**, 320; (d) H. W. Roesky and W. Scharer, *Chem. Ber.*, 1974, **107**, 3451; (e) W. Lidy and W. Sundemeyer, *Z. Naturforsch.*, 1974, **29b**, 276.

MeSiCl$_3$ produces MeSi(NSN)$_3$SiMe, the structure of which is postulated as (3) on i.r. and ^{1}H n.m.r. spectral evidence.[21b] Some reactions of N$_3$S$_3$Cl$_3$ in which di-imides are produced have been prepared: *e.g.* S$_7$NH forms S$_4$N$_2$ and C$_6$F$_5$SNHSiMe$_3$ gives C$_6$F$_5$SN=S=NSiMe$_3$.[21c] Me$_3$SiNR$_2$ (R = Me or Et) reacts with S(NSO)$_2$ to form R$_2$NSN=S=O and Me$_3$SiN=S=O.[21d]

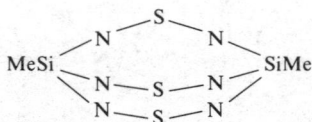

(3)

XN(SiMe$_3$)$_2$ (X = Cl, Br, or I) and SOCl$_2$ give Me$_3$SiON=S=O, which was characterized by its i.r. and mass spectra.[21e] Several S$_3$N$_4$ compounds, including R^{1}R^2$_2$P=N—S$_3$N$_3$ (R^1 = Me, R^2 = Ph), have been synthesized from S$_4$N$_4$ and iminophosphoranes and characterized by a variety of spectroscopic methods.[22a] The structures of both (Ph$_2$C)$_2$S$_3$N$_4$[22b] and Ph$_3$PS$_3$N$_4$[22c] have been determined by *X*-ray crystallography. In the former compound, the S$_3$N$_4$ entity is a conjugated chain with a Ph$_2$C group at each end whereas in the latter Ph$_3$P=N is bonded to the sulphur atom in an S$_3$N$_3$ ring. CF$_3$C(O)S$_3$N$_3$ and S(NSNSiMe$_3$)$_2$ are prepared from (Me$_3$SiN)$_2$S and CF$_3$C(O)Cl and SCl$_2$ respectively.[23] S(NSNSiMe$_3$)$_2$ and similar compounds were also products of one of the reactions described above. [22a]

The reactions of (NSO)$_3$F$_3$ with secondary and primary amines have been described; the products include (NSO)$_3$F$_2$NR$_2$, (NSO)$_3$F(NR$_2$)$_2$ (R = Me or Et), and (NSO)$_3$F$_2$NHR (R = Me, Et, or Ph). Treatment of the last compound with [Ph$_4$P]$^+$Cl$^-$ and BuLi yields [Ph$_4$P]$^+$[(NSO)$_3$F$_2$NR]$^-$ and (NSO)$_3$F$_2$NRLi respectively.[24a] When (NSOF)$_2$NPCl$_2$ reacts with NH$_3$, Me$_3$SiNHR, or Me$_3$SiNR$_2$, one of the Cl atoms is replaced by an NH$_2$, NHR, or NR$_2$ group.[24b]

A number of reactions of CF$_3$SO$_2$NCO with organic compounds have been reported; with P$_2$S$_5$ and PCl$_5$ it forms CF$_3$SO$_2$NCS and CF$_3$SO$_2$NCCl$_2$.[25a] and with (CF$_3$)$_2$CN$_2$ (4) is formed.[25b] (Me$_3$SiN)$_2$S reacts with ClSO$_2$NCO and HN(SO$_2$Cl)$_2$ to form the new heterocycles (5a) and (5b) respectively. The molecules were characterized by their i.r. and mass spectra.[25c] Several salts of CF$_3$SO$_2$NH$_2$ have been prepared and the silver compounds have been used as intermediates in the preparation of other CF$_3$SO$_2$ derivatives; *e.g.* CF$_3$SCl and

[22] (*a*) I. Ruppert, V. Bastion, and R. Appel, *Chem. Ber.*, 1974, **107**, 3426; (*b*) E. M. Holt, S. L. Holt, and K. J. Watson, *J.C.S. Dalton*, 1974, 1357; (*c*) E. M. Holt and S. L. Holt, *ibid.*, p. 1990.

[23] W. Lidy, W. Sundemeyer, and W. Verbeck, *Z. anorg. Chem.*, 1974, **406**, 228.

[24] (*a*) H. Wagner, R. Mews, T. P. Lin, and O. Glemser, *Chem. Ber.*, 1974, **107**, 584; (*b*) W. Heider, U. Klingebiel, T. P. Lin, and O. Glemser, *ibid.*, p. 592.

[25] (*a*) E. Behrend and A. Haas, *J. Fluorine Chem.*, 1974, **4**, 83; (*b*) H. Steinbeisser, R. Mews, and O. Glemser, *Z. anorg. Chem.*, 1974, **406**, 299; (*c*) H. W. Roesky and B. Kuhtz, *Chem. Ber.*, 1974, **107**, 1.

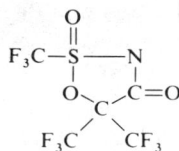

(4)

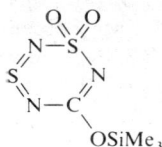

(5a)

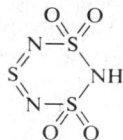

(5b)

$COCl_2$ yield $CF_3SO_2N(SCF_3)_2$ and CF_3SO_2NCO respectively.[26] The structure of $SO_2N_2H_2Ag_2$ has been studied by i.r. and broad-line ^1H n.m.r. spectroscopy

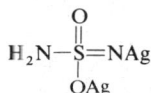

(6)

and is formulated as (6).[27] $SO_2(NSOF_2)_2$ can be prepared from $F_2OSN\cdot SO_2\cdot$ $NPCl_3$ [from $Hg(NSOF_2)_2$ and $ClSO_2NPCl_3$] and SOF_4; $ClSO_2NPCl_3$ and SOF_4 form FSO_2NSOF_2.[28] NSF_3 and C_3F_6 produce $NSF_2CF(CF_3)_2$ [which isomerizes above 333 K to $F_2S{=}NCF(CF_3)_2$]. Further reaction with C_3F_6 yields $(CF_3)_2CFSF{=}NCF(CF_3)_2$ and SF_4 forms *cis*- and *trans*-$(CF_3)_2CFSF_4N{=}$ SF_2.[29a] $SF_2(NSiMe_3)_2$ and OSF_4 produce $N{\equiv}SF_2{-}N{=}SF_2{=}O$, which hydrolyses to $HN{=}SF_2{=}NSO_2F$, identified as its $[Ph_4As]^+$ salt.[29b]

The importance of the 3*d*-orbitals in the bonding of sulphur compounds is still of interest. Core-electron bonding energies for the atoms in NSF_3, SF_5Cl, and S_2Cl_2 have been measured; the results indicate that the bonding in many sulphur compounds can be explained using only *s* and *p* valence orbitals.[30]

Radical addition of C_3F_6 to $ClNSO_2$ yields both $CF_3CFClCF_2NSOF_2$ and $CF_3CF(NSOF_2)CF_2Cl$, whereas $ClN(SO_2F)_2$ gives exclusively $CF_3CFClCF_2{-}N(SO_2F)_2$. $(CF_3)_2CN_2$ reacts with $ClNSF_2$, $BrNSF_2$, $ClNSOF_2$, and $ClN(SO_2F)_2$ by insertion into the N—halogen bond to produce $(CF_3)_2CClNSF_2$, $(CF_3)_2CBrNSF_2$, $(CF_3)_2CClNSOF_2$, and $(CF_3)_2CClN(SO_2F)_2$ respectively.[31] Some of the reactions of $(CF_3S)_xNH_{3-x}$ (x = 1, 2, or 3) have been described; $(CF_3S)_2NH$ reacts with SCl_2 and S_2Cl_2 to yield $[(CF_3S)_2N]_2S$ and $[(CF_3S)_2NS]_2$ respectively. CF_3-substituted organic compounds are obtained by irradiation of hydrocarbons in the presence of $(CF_3S)_3N$. The pyrolysis of $(CF_3S)_3N$ yields $CF_3SN{=}S{=}NCF_3$ and its reaction with Ph_3P produces $Ph_3P{=}NSCF_3$.[32]

[26] E. Behrend and A. Haas, *J. Fluorine Chem.*, 1974, **4**, 99.

[27] E. Nachbaur, A. Popitsch, and P. Burkert, *Monatsh.*, 1974, **105**, 822.

[28] C. Jäckh and W. Sundemeyer, *Angew. Chem. Internat. Edn.*, 1974, **13**, 401.

[29] (a) A. F. Clifford and J. S. Harman, *J.C.S. Dalton*, 1974, 571; (b) O. Glemser and R. Hofer, *Z. Naturforsch.*, 1974, **29b**, 120.

[30] W. L. Jolly, M. S. Lazarus, and O. Glemser, *Z. anorg. Chem.*, 1974, **406**, 209.

[31] J. Varung, R. Mews, and O. Glemser, *Chem. Ber.*, 1974, **107**, 2468.

[32] A. Haas, J. Helmbrecht, and E. Wittke, *Z. anorg. Chem.*, 1974, **406**, 185.

Bu^{t}NHSiMe$_3$ and SCl$_2$ form Bu^t(Me$_3$Si)NSCl, which reacts with Bu^{t}NH$_2$ to give Bu^t(Me$_3$Si)NSNHBu^t.[33a] [Me$_3$Si(Me)N]$_2$S, [the product of SCl$_2$ and Me$_3$Si(Me)NLi] reacts with (ClSiMe$_2$)$_2$ to produce the novel heterocycle (7a) whereas [ClMe$_2$Si(Me)N]$_2$S forms (7b).[33b]

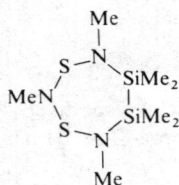   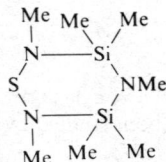

(7a)   (7b)

A series of S—N compounds of formula XC$_6$H$_4$S(O)N=CR$_2$ has been synthesized.[34] (Me$_2$N)$_2$S and [Ph$_2$P(S)]$_2$S produce Ph$_2$P(S)SSNMe$_2$; a number of other P—S compounds are reported, including (8).[35]

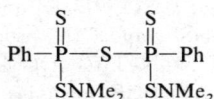

(8)

The i.r. spectra of ClS≡N and BrS≡N (prepared from [S$_4$N$_3$]$^+$Cl$^-$ or [S$_4$N$_3$]$^+$Br$^-$) have been measured in an argon matrix at 15 K. Analyses of the spectra show that both molecules are bent, with ∠XSN = 118°.[36]

The use of CS in the syntheses of S-halides has been described. Reaction with Cl$_2$ and Br$_2$ yields Cl$_3$CSCl and Br$_3$CSBr while HCl gives HClC=S, which on warming produces (HClCS)$_3$.[37] CF$_3$SMe is oxidized at 273 K by *m*-ClC$_6$H$_4$CO$_3$H to CF$_3$S(O)CH$_3$ and by ClF at 298 K to *trans*-CF$_3$SF$_4$Me, in contrast to (CF$_3$)$_2$S which forms a mixture of *cis*- and *trans*-(CF$_3$)$_2$SF$_4$.[38]

SF$_6$ has been used as a fluorinating agent in the temperature range 298—773 K and pressure range 1—4000 atm. MgO, NiO, and SiO$_2$ are converted into MgF$_2$, NiF$_2$, and SiF$_4$, SO$_2$F$_2$ being the other product.[39] S$_2$I$_2$ and SOI$_2$ can be prepared from HI and S$_2$Cl$_2$ and SOCl$_2$ at 195 K; the dark-brown solids are stable below 240 K and in the absence of water.[40]

[33] (a) O. J. Scherer and G. Wolmershauser, *Z. Naturforsch.*, 1974, **29b**, 277; (b) *ibid.*, p. 444.
[34] F. A. Davies, A. J. Friedman, and E. W. Kluger, *J. Amer. Chem. Soc.*, 1974, **96**, 5000.
[35] E. Fluck, G. Gonzalez, and H. Binder, *Z. anorg. Chem.*, 1974, **406**, 161.
[36] S. C. Peake and A. J. Downs, *J.C.S. Dalton*, 1974, 859.
[37] K. J. Klabunde, C. M. White, and H. F. Efner, *Inorg. Chem.*, 1974, **13**, 1778.
[38] S. L. Yu, D. T. Sauer, and J. M. Shreeve, *Inorg. Chem.*, 1974, **13**, 484.
[39] A. P. Hagen, D. J. Jones, and S. R. Rutteman, *J. Inorg. Nuclear Chem.*, 1974, **36**, 1217.
[40] D. K. Padma, *Indian J. Chem.*, 1974, **12**, 417.

Selenium and Tellurium.—The crystal structures of α-TeI and β-TeI have been determined and in both cases are more complex than might be expected. α-TeI consists of discrete Te_4I_4 units (9a) whereas β-TeI crystallizes in zig-zag chains

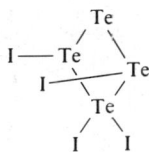

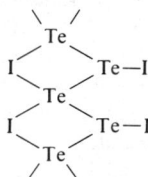

(9a)   (9b)

(9b).[41] In $BaTeS_3,2H_2O$ and $(NH_4)_2TeS_3$ [prepared from $Te_2O_3(OH)NO_3$ using $Ba(OH)_2$ and NH_3 respectively], the Te atoms have a distorted trigonal-bipyramidal environment.[42]

The indium selenides In_5Se_6 and In_2Se can be synthesized from the elements using zone-melting or fusion techniques.[43]

$Se(SCMe_2CO_2H)_2$ can be synthesized from $HSCMe_2CO_2H$ and H_2SeO_3 or H_2SeO_4 but the isomeric compounds $Se[S(CH_2)_3CO_2H]_2$ and $Se(SCHMe-CH_2CO_2H)_2$ could not be obtained pure.[44] The irradiation of some organoselenium compounds with 350 nm light has been reported. $(PhCH_2)_2Se_2$ produces PhCHO and Se in the presence of O_2 and $(PhCH_2)_2Se$ in the absence of O_2. $(PhCH_2Se)_2CSe$ under the same conditions yields $(PhCH_2)_2Se_2$.[45]

Alkylammonium selenocyanates react with SO_2Cl_2 or Br_2 to give $[R_4N]^+[SeX_2CN]^-$ (R = Pr^n, X = Cl or Br; R = Me, X = Cl). Further reactions with SO_2Cl_2 or Br_2 produce $[R_4N]^+[SeX_3]^-$. The i.r. and Raman spectra are consistent with the latter compounds containing $[Se_2X_6]^{2-}$ in the solid state and $[SeX_3]^-$ in solution.[46] A number of methylseleno-silanes and -germanes, including $MeSeSiH_3$, $(MeSe)_2SiH_2$, and $MeSeGeH_3$, have been synthesized from $LiAl(SeMe)_4$ or germyl halides. They have been fully characterized by cleavage reactions and by ^1H n.m.r. and vibrational spectroscopy.[47]

$MeHgNO_3$ and H_2Se form $[Se(HgMe)_3]^+[NO_3]^-$ whereas MeHgBr reacts with H_2Se and H_2Te, to produce $Se(HgMe)_2$ and $Te(HgMe)_2$ respectively. The i.r. and Raman spectra indicate that formation of $[Se(HgMe)_3]^+$ weakens the Se—Hg bonds.[48] The reactive compounds $[Ph_2SeCHR^1R^2]^+[BF_4]^-$, ($R^1$ = H, R^2 = H or Me) have been synthesized from Ph_2Se, $AgBF_4$, and R^1R^2CH halides and have been used in organic synthesis to produce oxirans.[49]

[41] R. Kniep, D. Mootz, and A. Rabeneau, *Angew. Chem. Internat. Edn.*, 1974, **13**, 403.
[42] H. Gerl, B. Eisenmann, P. Roth, and H. Schäfer, *Z. anorg. Chem.*, 1974, **407**, 135.
[43] B. Celustka and S. Popovic, *J. Phys. and Chem. Solids*, 1974, **35**, 287.
[44] E. R. Clark and A. J. Collett, *J. Inorg. Nuclear Chem.*, 1974, **36**, 3860.
[45] W. Stanley, M. R. Van de Mark and P. L. Kunler, *J.C.S. Chem. Comm.*, 1974, 700.
[46] K. J. Wynne and J. Golen, *Inorg. Chem.*, 1974, **13**, 184.
[47] G. K. Barker, J. E. Drake, and R. T. Hennings, *J.C.S. Dalton*, 1974, 450.
[48] D. Breitinger and W. Morell, *Inorg. Nuclear Chem. Letters*, 1974, **10**, 409.
[49] W. Dumont, P. Bayet, and A. Krief, *Angew. Chem. Internat. Edn.*, 1974, **13**, 274.

A number of papers on organo-selenium and -tellurium halides have been published this year. $TeCl_4$–$AlCl_3$ mixtures are useful for synthesizing organo-tellurium compounds: e.g. C_6H_6 forms Ph_2Te_2 (from reduction of $PhTeCl_3$), Ph_2TeCl_2, and Ph_3TeCl. Extension of the reaction to other aromatic tellurium derivatives appears feasible.[50] The reactions of Ph_2SeX_2 (X = Cl or Br) with amines and aminosilanes have been described; with NH_3, $MeNH_2$, and Me_2NH reduction occurs and Ph_2Se is formed, whereas Ph_2SeCl_2 and Me_3SiNMe_2 produce $Ph_2Se(NMe_2)Cl$, and $(Me_3Si)_2NH$ yields $[Ph_2Se=N=SePh_2]^+Cl^-$.[51] Diaryltellurium diacetates can be prepared from the corresponding dichloride and $MeCO_2^-$ Ag^+ and by the oxidation of the diaryltelluride with lead tetra-acetate.[52a] $[ArTeI_2]^-$ and $[ArTeIBr]^-$ salts (Ar = aryl) are obtained from the appropriate ArTeI and $[Ph_3PMe]^+X^-$ (X = Br or I). The selenium compound $[Ph_3PMe]^+$ $[PhSeBr_2]^-$ can be prepared similarly from $PhSeBr$.[52b]

The i.r. spectra of $(CH_2NH)_2SeO$ and $(MeNH)_2SeO$ have been analysed and interpreted to show that the molecules have strong hydrogen-bonding between the $Se=O$ and NH groups.[53]

High-purity H_2SeO_4 has been prepared by ion exchange from a solution of K_2SeO_4 and $MnSeO_4$.[54] SeO_3 forms complexes with $SeCl_4$, $TeCl_4$, PCl_5, Ph_3CCl, and S_4N_3Cl. Conductivity and i.r. spectral studies suggest that the complexes are ionic, probably containing the $[SeO_3Cl]^-$ ion.[55]

Interest in the chemistry of TeF_6 derivatives continues. Hydrolysis produces the acids $F_xTe(OH)_{6-x}$ (x = 1—4)[56a] and alcohols yield the mono- and di-substituted derivatives TeF_5OR and $TeF_4(OR)_2$ (R = Me, Et, etc.).[56b] CsF and RbF react with TeF_6 to form 1 : 1 and 2 : 1 complexes respectively. The vibrational spectra of these compounds are consistent with them containing the $[TeF_7]^-$ and $[TeF_8]^{2-}$ ions.[56c]

The use of ClF as a fluorinating agent in tellurium chemistry has been described. $TeClF_5$ can be conveniently synthesized from TeF_4, $TeCl_4$, or TeO_2;[57a] $(C_2F_5)_2Te_2$ forms $C_2F_5TeF_3$ and $trans$-$C_2F_5TeClF_4$ whereas $(C_2F_5)_2Te$ gives $(C_2F_5)_2TeF_2$, $trans$-$C_2F_5TeClF_4$, and $trans$-$(C_2F_5)_2TeF_4$. The Te^{IV} derivatives $C_2F_5TeF_3$ and $(C_2F_5)_2TeF_2$ form 1 : 1 complexes with CsF and SbF_5.[57b]

After many attempts $SeOF_4$ has been prepared[58] by the pyrolysis of $Na^+[SeOF_5]^-$. It can be trapped at 77 K and is stable below 173 K, but above

50 W. H. H. Gunther, J. Nepywoda, and J. Y. C. Chu, J. Organometallic Chem., 1974, 74, 79.
51 V. Horn and R. Paetzold, Z. anorg. Chem., 1974, 404, 213.
52 (a) B. C. Pant, J. Organometallic Chem., 1974, 65, 51; (b) N. Petragnani, L. Torres, K. J. Wynne, and D. J. Williams, ibid., 1974, 76, 241.
53 G. Hopf and R. Paetzold, Z. anorg. Chem., 1974, 403, 137.
54 B. Blanka, Chem. Zvesti, 1974, 28, 298 (Chem. Abs., 1974, 81, 85 313).
55 R. C. Paul, R. D. Sharma, and K. C. Malhotra, Indian J. Chem., 1974, 12, 320.
56 (a) G. W. Fraser and G. D. Meikle, J.C.S. Chem. Comm., 1974, 624; (b) G. W. Fraser and J. B. Millar, J.C.S. Dalton, 1974, 2029; (c) H. Selig, S. Sarig, and S. Abramowitz, Inorg. Chem., 1974, 13, 1508.
57 (a) C. Lau and J. Passmore, Inorg. Chem., 1974, 13, 2278; (b) C. D. Desjardins, C. Lau, and J. Passmore, Inorg. Nuclear Chem. Letters, 1974, 10, 151.
58 K. Seppelt, Z. anorg. Chem., 1974, 406, 287; Angew Chem. Internat. Edn., 1974, 13, 91, 92.

this temperature it dimerizes, the dimer $Se_2O_2F_8$ being confirmed by mass spectroscopy. In contrast, pyrolysis of $LiOTeF_5$ yields only $Te_2O_2F_8$, no $TeOF_4$ being detected. The ^{19}F n.m.r. spectra of the dimers are consistent with structure (10) (M = Se or Te).[58]

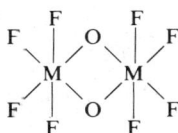

$$\begin{array}{c} F \quad \underset{|}{\overset{F}{\diagup}} \quad O \quad \underset{|}{\overset{F}{\diagup}} F \\ F \diagdown \overset{|}{M} \diagup \diagdown \overset{|}{M} \diagdown F \\ F \quad \overset{|}{O} \quad F \\ F \qquad F \end{array}$$

(10)

The crystal structure of $[Ph_4As]^+[TeCl_4OH]^-,H_2O$ shows that it contains infinite chains of distorted square-pyramidal $TeOCl_4$ units.[59]

2 Group VII

The standard enthalpy of formation of F^- in aqueous solution has been determined from measurements on the heats of dilution of HF solutions. The value obtained ($-333\ kJ\ mol^{-1}$) agrees well with those obtained by other methods.[60] Emission spectra show the presence of NaF_2 and $NaCl_2$ (containing X_2^- ions) resulting from the reaction of Na and X_2 in the gas phase.[60a] CF_3OClO_3 has been synthesized[61a] from CF_3I and $ClOClO_3$ and identified by its mass, ^{19}F n.m.r., and vibrational spectra.[61b] The molecule is relatively stable in the presence of CsF: only 30% decomposition into COF_2 and $FClO_3$ occurred after 18 h at 373 K. Cl_2O_7 reacts with a range of primary, secondary, and tertiary alcohols to produce the corresponding alkyl perchlorates $ROClO_3$; ethylene glycol and other diols form disubstituted compounds of the type $O_3ClOCH_2CH_2OClO_3$. The use of these compounds as intermediates in organic synthesis has been demonstrated.[62a] The reactions of Cl_2O_7 with amines have been reported by two groups of workers;[62b] secondary amines give R_2NClO_3 and primary amines $RNHClO_3$. The latter compounds are acidic and the salts $M^+[RNClO_3]^-$ (M = Na, K, or Ag) have been prepared and characterized. $ClOClO_3$ reacts with CsBr to form $Cs^+[Br(OClO_3)_2]^-$, in contrast to CsI which was previously reported to form $Cs^+[I(OClO_3)_4]^-$.[63]

A new oxide, Br_2O_3, has been prepared by thermal decomposition of Br_2O_4; the vibrational spectrum was measured but it was impossible to distinguish between the possible structures OBrOBrO and $BrOBrO_2$.[64] The structure of

[59] P. H. Collins and M. Webster, *J.C.S. Dalton*, 1974, 1545.
[60] V. P. Vasilev and E. V. Kozlovskii, *Zhur. neorg. Khim.*, 1974, **19**, 267.
[60a] D. O. Ham and H. W. Chang, *Chem. Phys. Letters*, 1974, **24**, 579.
[61] (a) C. J. Schack, D. Pilipovich, and K. O. Christe, *Inorg. Nuclear Chem. Letters*, 1974, **10**, 449; (b) C. J. Schack and K. O. Christe, *Inorg. Chem.*, 1974, **13**, 2374.
[62] (a) K. Baum and C. D. Beard. *J. Amer. Chem. Soc.*, 1974, **96**, 3233; (b) ibid., p. 3237; D. Baumgarten, E. Hilth, J. Jander, and J. N. Meussdoerffer, *Z. anorg. Chem.*, 1974, **405**, 77.
[63] K. O. Christe and C. J. Schack, *Inorg. Chem.*, 1974, **13**, 1452.
[64] J. L. Pascal, A. C. Pavia, J. Potier, and A. Potier, *Compt. rend.*, 1974, **279**, *C*, 43.

$(IO)_2SO_4$ has been analysed by X-ray crystallography[65] and the molecule shown to consist of infinite parallel chains (11).

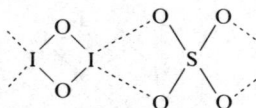

(11)

A series of compounds Ba_2MIO_6 (M = Li, Na, K, or Ag) and Sr_2NaIO_6 has been synthesized; their vibrational spectra have been measured and the force constants calculated.[66]

Adducts of the type $ClOF_3,MF_5$ (M = P, As, V, Sb, Ta, Nb, or Bi) have been synthesized and their structures formulated as $[ClOF_2]^+[MF_6]^-$ on the basis of the X-ray powder diffraction patterns and their vibrational spectra.[67] IO_2F_3, originally thought to be monomeric and to exist as two isomers, has been shown by gas-phase molecular weight measurements[68a] and mass spectroscopy[68b] to be polymeric, probably *via* oxygen bridges. It is a Lewis acid, forming $[IO_2F_4]^-$ and $(IO_2F_3,SbF_5)_n$ with HF and SbF_5 respectively. Calculations show that in monomeric IO_2F_3 the C_{2v} isomer (equatorial O atoms in a trigonal bipyramid) is energetically favoured over the C_s isomer.[68c]

IF_5 and its derivatives have attracted interest; its crystal structure at 193 K has been determined[69] and its reaction with Me_3SiOMe shown to yield the series of compounds $IF_{5-n}(OMe)_n$ (n = 1—4).[70a] CF_3I, C_6F_5I, and $C_6F_4I_2$ can be fluorinated with ClF_3 to produce CF_3IF_4,[70b] $C_6F_5IF_4$, and $C_6F_4(IF_4)_2$[70c] respectively. Further reactions of the fluorobenzene derivatives with ClF_3 produce viscous liquids, the mass spectra of which suggest that attack on the aromatic ring has occurred. IF_7 has been shown to be a weak Lewis acid; NOF and CsF form $NO^+[IF_8]^-$ and $Cs^+[IF_8]^-$, characterized by Raman spectra and X-ray powder data.[71]

[65] S. Furnseth, K. Selte, H. Hope, A. Kjekshus, and B. Klewe, *Acta Chem. Scand.*, 1974, **A28**, 71.

[66] J. T. W. Dehair, A. F. Corsmit, and G. Blasse, *J. Inorg. Nuclear Chem.*, 1974, **36**, 313.

[67] R. Bougon, T. Buithuy, A. Cadet, P. Charpin, and R. Rousson, *Inorg. Chem.*, 1974, **13**, 690.

[68] (a) I. R. Beattie and G. J. Van Schalkwyk, *Inorg. Nuclear Chem. Letters*, 1974, **10**, 343; (b) A. Engelbrecht, O. Mayr, G. Ziller, and E. Schandara, *Monatsh.*, 1974, **105**, 796; (c) B. M. Rode, *ibid.*, p. 807.

[69] R. D. Burbank and G. R. Jones, *Inorg. Chem.*, 1974, **13**, 1071.

[70] (a) G. Oates, J. M. Winfield, and O. R. Chambers, *J.C.S. Dalton*, 1974, 1381; (b) G. Oates and J. M. Winfield, *ibid.*, p. 119; (c) J. A. Berry, G. Oates, and J. M. Winfield, *ibid.*, p. 509.

[71] C. J. Adams, *Inorg. Nuclear Chem. Letters*, 1974, **10**, 831.

The synthesis of pure BrCl has been described[72a] and the far-i.r. spectra of ICl and IBr have been measured.[72b] Interhalogen fluorosulphates IX_2SO_3F and I_2XSO_3F (X = Br or Cl) have been synthesized by the oxidative addition of the appropriate X_2 or IX to ISO_3F. The vibrational spectra of the molecules are consistent with their formulation as $[IX_2]^+[SO_3F]^-$ and $[I_2X]^+[SO_3F]^-$.[73a] An n.q.r. spectroscopic study of the complex $2ICl,SbCl_5$ has shown that its structure is $[I_2Cl]^+[SbCl_6]^-$.[73b] The X-ray crystal structures of $I_2^+[Sb_2F_{11}]^-$[74a] and $I_2,P_3N_3Me_6$[74b] have been investigated. In the former molecule the I—I distance (256 pm) is shorter whereas in the latter compound it is larger (282 pm) than that found in molecular I_2 (256 pm).

I_2 complexes with $POCl_3$, $SOCl_2$, and $SeOCl_2$ have been prepared and the 'blue shift' which occurs in the visible iodine band on complex formation correlates linearly with the standard enthalpy of complex formation and with the M–O force constant of the donor molecule.[75]

3 Group VIII

Interest this year has centred on structural and spectroscopic studies of the complexes of xenon and, to a lesser extent, krypton fluorides. The complexes KrF_2,MF_5 (M = As, Sb, or Pt) and $2KrF_2,MF_5$ (M = As or Sb) are ionic materials containing the $[KrF]^+$ and $[Kr_2F_3]^+$ ions respectively.[76]

The first compound containing a xenon–nitrogen bond has been reported.[77] $FXeN(SO_2F)_2$ is prepared from XeF_2 and $HN(SO_2F)_2$ in CF_2Cl_2 at 273 K. The structure of XeF_6 has been studied by X-ray diffraction and shown to contain associated $[XeF_5]^+$ and F^- ions in tetrameric and hexameric rings.[78a] ^{129}Xe n.m.r. studies on a series of Xe compounds[78b] also show that XeF_6 is present as Xe_4F_{24} when in solution in an inert solvent. A number of xenon compounds have been studied by ^{19}F n.m.r. spectroscopy, including the $[(FXe)_2SO_3F]^+$ cation, which is prepared by dissolving a salt containing $[Xe_2F_3]^+$ in HSO_3F.[79] The crystal structures of $[XeF_3]^-[SbF_6]^-$,[80a] $[Xe_2F_{11}]^+[AuF_6]^-$,[80b] $[Xe_2F_3]^+$ $[AsF_6]^-$, and $[XeF_5]^+[AsF_6]^-$[80c] have been reported. XeF_2 complexes with

[72] (a) M. Schmeisser and K. H. Tytko, *Z. anorg. Chem.*, 1974, **403**, 231; (b) H. C. Leung and A. Anderson, *Canad. J. Chem.*, 1974, **52**, 1081.

[73] (a) W. W. Wilson and F. Aubke, *Inorg. Chem.*, 1974, **13**, 326; (b) D. J. Merryman and D. J. Corbett, *ibid.*, p. 1258.

[74] (a) C. G. Davies, R. J. Gillespie, P. R. Ireland, and J. M. Sowa, *Canad. J. Chem.*, 1974, **52**, 2048; (b) P. L. Markila and J. Trotter, *ibid.*, p. 2197.

[75] R. Paetzold and K. Niendorf, *Z. anorg. Chem.*, 1974, **405**, 129.

[76] B. Frlec and J. H. Holloway, *J.C.S. Chem. Comm.*, 1974, 89; R. J. Gillespie and G. S. Schrobilgen, *ibid.*, p. 90.

[77] R. D. LeBlond and D. D. DesMarteau, *J.C.S. Chem. Comm.*, 1974, 555.

[78] (a) R. D. Burbank and G. R. Jones, *J. Amer. Chem. Soc.*, 1974, **96**, 43; (b) K. Seppelt and H. H. Rupp, *Z. anorg. Chem.*, 1974, **409**, 331, 338.

[79] R. J. Gillespie and G. J. Schrobilgen, *Inorg. Chem.*, 1974, **13**, 765, 1455, 1694, 2370.

[80] (a) P. Boldrini, R. J. Gillespie, P. R. Ireland, and G. J. Schrobilgen, *Inorg. Chem.*, 1974, **13**, 1690; (b) K. Leary, A. Zalkin, and N. Bartlett, *ibid.*, p. 775; (c) N. Bartlett, B. G. De Boer, F. J. Hollander, F. O. Sladky, D. H. Templeton, and A. Zalkin, *ibid.*, p. 780.

rare-earth fluorides have been synthesized[81a] and the standard enthalpies of formation of XeF_2,MF_5, $XeF_2,2MX_5$ (M = Sb, Ta, or Nb), and $2XeF_2,MF_5$ (M = Sb or Ta) have been estimated from their heats of alkaline hydrolysis.[81b]

$Li_4XeO_6,2H_2O$ has been synthesized and shown to decompose at 573—673 K to Li_2O, Xe, and O_2.[82]

[81] (a) V. I. Spitsyn, Yu. M. Kiselev, and L. I. Martynenko, *Zhur. neorg. Khim.*, 1974, **19**, 1152; (b) J. Burgess, B. Frlec, and J. H. Holloway, *J.C.S. Dalton*, 1974, 1740.
[82] N. N. Aleinikov, V. K. Isupov, and I. S. Kirin, *Izvest. Akad. Nauk S.S.S.R., Ser. khim.*, 1974, 278 (*Chem. Abs.*, 1974, **81**, 20 264).

11   The Transition Elements

Part I: Main Coverage

By D. W. A. SHARP

Department of Chemistry, University of Glasgow, Glasgow G12 8QQ

1 Introduction

As last year, this chapter is divided into two main sections. Properties of ligands and properties that result mainly from the ligands are covered in Part II. Binary compounds and complex oxides and halides are covered under the appropriate elements. In order to allow for cover of the chemistry of the elements, review articles are, in general, not listed.

2 Group III, Lanthanides, and Actinides

Scandium.—This element is very similar in its behaviour to the lanthanides, although its smaller size tends to restrict its co-ordination number to six except in network structures. $HSc(tropolonato)_4$ is a hydrogen-bonded dimer with D_{2d} eight-co-ordinate scandium [1a] $Cl_3Sc(THF)_3$ has a *mer*-octahedral environment about the scandium.[1b] Scandium is generally found in octahedral co-ordination in its chalcogenides; detailed studies of the compounds Sc_2S_3, Sc_3S_4, ScS, $ScS_{1.1-0.75}$, Sc_2O_2S, Sc_2Se_3, $ScSe$, and $SeTe$ have been described. Many of these are non-stoicheiometric and polytypic.[2]

The Lanthanides.—The values of the redox potentials $E^0[M^{3+}|M^{2+}]$ for the lanthanides have been estimated by means of a simple ionic model and compared with observation. These and related data form regular series across the lanthanides.[3] There is currently a great deal of interest in intermetallic compounds of the lanthanides because of the unusual magnetic and physical properties of many of the alloys. Diatomic molecular compounds of the type CeRh, CeIr, CePt, and LaRh and also ThRu are expected to have multiple bonds and can be identified in the vapour phase by mass spectrometry.[4] Amongst work described on the solid alloys may be mentioned phase-diagram and structural studies on

[1] (a)T. J. Anderson, M. A. Neuman, and G. A. Melson, *Inorg. Chem.*, 1974, **13**, 1884; (b) J. L. Atwood and K. D. Smith, *J.C.S. Dalton*, 1974, 921.

[2] V. Brozek, J. Flahaut, M. Guittard, M. Julien-Pouzol, and M.-P. Pardo, *Bull. Soc. chim. France*, 1974, 1740.

[3] D. A. Johnson, *J.C.S. Dalton*, 1974, 1671.

[4] K. A. Gingerich, *J.C.S. Faraday II*, 1974, 471; *Chem. Phys. Letters*, 1974, **25**, 523, and references therein.

cobalt-containing alloys[5] (*e.g.* the phases Sm_2Co_7, $SmCo_{5-x}$, $SmCo_5$, $SmCo_{5+x}$, and Sm_2Co_{17} are found in the Sm–Co system), nickel-containing alloys,[6] and the hexaurides, $R.E.Au_6$, which have structures comprising interpenetrating centred polyhedra.[7] Intermetallic compounds containing lanthanides and yttrium absorb hydrogen under moderate pressure and activate other metals for hydrogen uptake, and they are thus potentially important in hydrogen storage.[8] Ternary borides of the types YNi_4B, $Ce_3Co_{11}B_4$, and $YbCo_3B_2$ are receiving wide attention; the structures contain $CaCu_5$-type layers interleaved with boron-containing layers.[9a] The series of isostructural borides with the $ThMoB_4$ structure also contain boron layers.[9b] The new phase Ce_2Ge_3 has been identified in the Ce–Ge system.[10] Oxygen can replace nitrogen in the lattice of lanthanide nitrides over very wide ranges (*e.g.* to $CeN_{0.50}O_{0.50}$ at 1800 K);[11a] Ce_2N_2O is formed from Li_2CeN_2 (from Li_3N and Ce) and Li_2O or CeN and CeO_2.[11b] Many new lanthanide phosphides (of types $R.E.P_{\sim 1}$, $R.E.P_2$, and $R.E.P_5$) and non-stoicheiometric arsenides, R.E.As, have also been described.[12]

Whereas $K_3La(NH_2)_6$ has a distorted close-packed lattice of amide ions with lanthanum in approximately octahedral co-ordination, $KLa_2(NH_2)_7$ has a more irregular structure with eight-co-ordinate lanthanum.[13]

Infrared studies on matrix-isolated lanthanide and actinide oxides have shown linear structures for PrO_2 and UO_2 and non-linear structures for CeO_2, TbO_2, and ThO_2.[14] New structural theories have been developed for the defect fluorite lattices $R.E.O_x$ ($1.7 \leqslant x \leqslant 2.0$) in terms of octahedrally co-ordinated anion vacancies, a very different description from the crystallographic shear theories of other transition-metal oxides.[15] $Nd(OH)CO_3$, and probably other hydroxy-carbonates, contains layers of $(NdOH^{2+})_n$ ions held together by carbonate ions; the lanthanide atom is nine-co-ordinate.[16] The low-temperature orthorhombic compounds M_2S_3 (M = La—Gd) form an isostructural series containing seven- and eight-co-ordinate metal atoms[17a] whereas YbS_2 contains S_2 groups.[17b]

[5] Y. Khan, *Acta Cryst.*, 1974, **B30**, 861, 1533; *J. Less-Common Metals*, 1974, **34**, 191.
[6] J. M. Moreau, D. Paccard, and D. Gignoux, *Acta Cryst.*, 1974, **B30**, 2122; H. H. Van Mal, H. H. J. Buschow, and A. R. Miedema, *J. Less-Common Metals*, 1974, **35**, 65.
[7] J. M. Moreau and E. Parthé, *Acta Cryst.*, 1974, **B30**, 1743.
[8] T. Takeshita, W. E. Wallace, and R. S. Craig, *Inorg. Chem.*, 1974, **13**, 2282, 2283; D. H. W. Carstens and J. D. Farr, *J. Inorg. Nuclear Chem.*, 1974, **36**, 461.
[9] (*a*) P. Rogl, *Monatsh.*, 1973, **104**, 1623; K. Niihara, Y. Katayama, and S. Yajima, *Chem. Letters*, 1973, 613; N. S. Bilonizhko and Y. B. Kuz'ma, *Inorg. Materials*, 1974, **16**, 227; *Soviet Phys. Cryst.*, 1974, **18**, 447; (*b*) P. Rogl and H. Nowotny, *Monatsh.*, 1974, **105**, 1082.
[10] P. G. Rustamov, I. O. Nasibov, and M. M. Alieva, *Inorg. Materials*, 1974, **10**, 7.
[11] (*a*) R. C. Brown and N. J. Clark, *J. Inorg. Nuclear Chem.*, 1974, **36**, 1777, 2287, 2507; (*b*) M. G. Barker and I. C. Alexander, *J.C.S. Dalton*, 1974, 2166.
[12] S. Ono, K. Nomura, and H. Hayakawa, *J. Less-Common Metals*, 1974, **38**, 119; J. B. Taylor, L. D. Calvert, J. G. Despault, E. J. Gabe, and J. J. Murray, *ibid.*, p. 217.
[13] G. Hadenfeldt, B. Gieger, and H. Jacobs, *Z. anorg. Chem.*, 1974, **408**, 27; 1974, **403**, 319.
[14] S. D. Gabelnick, G. T. Reedy, and M. G. Chasanov, *J. Chem. Phys.*, 1974, **60**, 1167.
[15] R. L. Martin, *J.C.S. Dalton*, 1974, 1335.
[16] A. N. Christensen, *Acta Chem. Scand.*, 1973, **27**, 2973.
[17] (*a*) A. A. Eliseev, S. I. Uspenskaya, A. A. Fedorov, and V. A. Tolstova, *J. Struct. Chem.*, 1972, **13**, 66; (*b*) C. L. Teske, *Z. Naturforsch.*, 1974, **29b**, 16.

There has been considerable interest in the structures of ternary sulphides; NdYbS$_3$ has six-co-ordinate Yb and eight-co-ordinate Nd, with co-ordination polyhedra similar to those in the Y$_5$S$_7$ structure.[18a] The compounds R.E.$_4$MS$_7$ (M = Co or Ni) have a distorted K$_2$NiF$_4$ structure[18b] whilst La$_{32.66}$M$_{11}$S$_{60}$ (M = Mn or Fe) are unique in this series in possessing M—M bonds.[18c] La$_4$Ge$_3$S$_{12}$ is an orthothiogermanate containing GeS$_4$ tetrahedra.[18d] A study of LuSBr (and thus of isostructural ErSCl) shows planes of Lu$_4$S tetrahedra with octahedral Lu atoms.[19]

The two structure types known for the lanthanide trifluorides can each be stabilized outside the normal range by non-stoicheiometry involving the presence of M^{2+} ions.[20] There is currently great interest in complex fluorides of the lanthanide elements, and preparative and phase studies continue to illustrate the great range, comparable with that found for the actinides, of compounds that are possible: there is probably comparable complexity with other halides.[20] KHoBeF$_6$ is a holmium complex of BeF$_4$ groups; other lanthanides also form this type of compound, and Na$_5$BeTh$_{10}$F$_{45}$ contains BeF$_4$ groups and nine-co-ordinate thorium.[21] EuFCl is isostructural with PbFCl and can be obtained in solid solutions in R.E.OCl.[22] The complex molecule CsNdI$_4$, like some chlorides, can be detected mass-spectrometrically in the vapour phase.[23]

The Actinides.—The normal oxidation state of nobelium in aqueous solution is $+2$, and the ionic radius for No^{2+} is estimated at 1.1 Å.[24] Thin films of californium metal have been prepared by reduction of Cf$_2$O$_3$ with La; it has a melting point of 1170 K and exists in both cubic and hexagonal close-packed structures.[25] Four-co-ordinate ThCl[N(SiMe$_3$)$_2$]$_3$ is formed from ThCl$_4$ and LiN(SiMe$_3$)$_2$.[26] Matrix-isolated UO$_3$ has a T-shaped structure of C_{2v} symmetry.[27] There is currently a great interest in uranates [and the structurally related niobates and tungstates (q.v.)]. It has been shown that Na$^+$ can be

[18] (a) D. Carré and P. Laruelle, *Acta Cryst.*, 1974, **B30**, 952; (b) G. Collin and J. Flahaut, *J. Solid State Chem.*, 1974, **9**, 352; (c) G. Collin and P. Laruelle, *Acta Cryst.*, 1974, **B30**, 1134; (d) A. Mazurier and J. Etienne, *ibid.*, p. 759.

[19] G. Collin, C. Dagron, and F. Thevet, *Bull. Soc. chim. France*, 1974, 418.

[20] O. Greis and T. Petzel, *Z. anorg. Chem.*, 1974, **403**, 1; C. Hebecker, *Naturwiss.*, 1973, **60**, 518; D. Avignant and J.-C. Cousseins, *Compt. rend.*, 1974, **278**, C, 613; A. Védrine, R. Boutonnet, and J.-C. Cousseins, *ibid.*, 1973, **277**, C, 1129; R. C. Russo and H. M. Haendler, *J. Inorg. Nuclear Chem.*, 1974, **36**, 763; P. P. Fedorov, O. E. Izotova, V. B. Alexandrov, and B. P. Sobolev, *J. Solid State Chem.*, 1974, **9**, 368; J. Kutscher and A. Schneider, *Z. anorg. Chem.*, 1974, **408**, 135.

[21] Y. Le Fur, I. Tordjman, S. Aléanard, G. Bassi, and M. T. Roux, *Acta Cryst.*, 1974, **B30**, 2049; G. Brunton, *ibid.*, 1973, **B29**, 2976; cf. C. J. Barton, L. O. Kilpatrick, and H. Insley, *J. Inorg. Nuclear Chem.*, 1974, **36**, 1271.

[22] V. G. Lambrecht, jun., M. Robbins, and R. C. Sherwood, *J. Solid State Chem.*, 1974, **10**, 1.

[23] C. S. Liu and R. J. Zollweg, *J. Chem. Phys.*, 1974, **60**, 2384.

[24] R. J. Silva, W. J. McDowell, O. L. Keller, jun., and J. R. Tarrant, *Inorg. Chem.*, 1974, **13**, 2233.

[25] R. G. Haire and R. D. Baybarz, *J. Inorg. Nuclear Chem.*, 1974, **36**, 1295.

[26] D. C. Bradley, J. S. Ghotra, and F. A. Hart, *Inorg. Nuclear Chem. Letters*, 1974, **10**, 209.

[27] S. D. Gabelnick, G. T. Reedy, and M. G. Chasanov, *J. Chem. Phys.*, 1973, **59**, 6397.

incorporated into the vacant cation sites of α-UO_3 to give a sodium uranium bronze.[28] $K_2U_7O_{22}$ may be formulated as $K_2(UO_2)_2(UO_2)_5O_8$, and it contains layers of composition $(UO_2)_5O_8$ with other UO_2 and K atoms between the layers.[29] Series of perovskite-type uranates have been described. $M_2R.E._{0.67}^{3+}U^{5+}O_{5.5}$ phases are oxidized to $M_2R.E._{0.67}U^{6+}O_6$ ($M = Sr^{2+}$ or Ba^{2+}), and tungsten(VI) can replace uranium(VI).[30] $Li_2U_3O_{10}$ also has a framework-type structure with seven-co-ordinate uranium, and $UMo_{10}O_{32}$ and $U_3Mo_{20}O_{64}$ have hexagonal-bipyramidal uranium.[31] The oxysulphate $Cf_2O_2SO_4$ is formed by heating the tripositive sulphate in air at *ca.* 1000 K; heating the oxysulphate in hydrogen or in a vacuum gives the oxysulphide Cf_2O_2S;[32] the tellurides of plutonium are similar to those of the other actinides, and $NpTe_3$, $NpTe_{2-x}$, and η- and γ-Np_2Te_3 have been identified.[33]

Two new oxide fluorides of uranium have been reported. UOF_4 is formed by hydrolysis of UF_6 in an HF slurry and its thermal decomposition gives $U_2O_3F_6$. In contrast to tetrameric $MoOF_4$ and WOF_4, UOF_4 is involatile and has a polymeric structure with pentagonal-bipyramidal co-ordination about uranium, with oxygen equatorial.[34] Double pentagonal bipyramids with bridging fluorines and apical oxygens are present in the $[(UO_2)_2F_n]^{(n-4)-}$ (n = 9, 8, or 7) ions.[35] Octahedral tetrapositive actinide complexes containing PaF_6^{2-}, UF_6^{2-}, NpF_6^{2-}, and PuF_6^{2-} ions result from reactions in oxygen-free propylene carbonate.[36] β-NH_4UF_5 has UF_9 polyhedra with one terminal fluorine on each uranium atom.[37] Uranium(III) is stable in solid lattices, and $UZrF_7$ (isostructural with $SmZrF_7$) and UZr_2F_{11} react with oxygen only slowly at room temperature.[38] $ThCl_4$ has two forms, with thorium eight-co-ordinate in both structures.[39] UBr_4 contains equatorial edge-fused pentagonal bipyramids, and a similar arrangement is found in UO_2Cl_2,H_2O.[40] Cs_2UCl_6 reacts with antimony(III)

[28] C. Greaves, A. K. Cheetham, and B. E. F. Fender, *Inorg. Chem.*, 1973, **12**, 3003.
[29] L. M. Kovba, *J. Struct. Chem.*, 1972, **13**, 235.
[30] S. Kemmler-Sack, *Z. anorg. Chem.*, 1973, **402**, 232, 255; 1974, **403**, 149; H.-J. Schittenhelm and S. Kemmler-Sack, *ibid.*, 1974, **407**, 181.
[31] L. M. Kovba, *J. Struct. Chem.*, 1972, **13**, 428; V. N. Serezhkin, L. M. Kovba, and V. K. Trunov, *Soviet Phys. Cryst.*, 1974, **18**, 603; 1974, **19**, 231; *J. Struct. Chem.*, 1973, **14**, 689.
[32] R. D. Baybarz, J. A. Fahey, and R. G. Hare, *J. Inorg. Nuclear Chem.*, 1974, **36**, 2023.
[33] D. Damien, *J. Inorg. Nuclear Chem.*, 1974, **36**, 307.
[34] P. W. Wilson, *J. Inorg. Nuclear Chem.*, 1974, **36**, 303, 1783; J. C. Taylor and P. W. Wilson, *Acta Cryst.*, 1974, **B30**, 1701.
[35] Yu. N. Mikhailov, A. A. Udovenko, V. G. Kuznetsov, and R. L. Davidovich, *J. Struct. Chem.*, 1972, **13**, 694, 879; Yu. N. Mikhailov, A. A. Udovenko, V. G. Kuznetsov, and R. N. Shchelokov, *ibid.*, p. 695; Yu. N. Mikhailov, A. A. Udovenko, V. G. Kuznetsov, L. A. Butman, and K. A. Kokh, *ibid.*, 1973, **14**, 154; H. Brusset, N. Q. Dao, and S. Chourou, *Acta Cryst.*, 1974, **B30**, 768; N. Q. Dao and S. Chourou, *Compt. rend.*, 1974, **278**, *C*, 879.
[36] J. L. Ryan, J. M. Cleveland, and G. H. Bryan, *Inorg. Chem.*, 1974, **13**, 214; D. Brown, B. Whittaker, and N. Edelstein, *ibid.*, p. 1805.
[37] R. A. Penneman, R. R. Ryan, and A. Rosenzweig, *Acta Cryst.*, 1974, **B30**, 1966.
[38] G. Fonteneau and J. Lucas, *J. Inorg. Nuclear Chem.*, 1974, **36**, 1515.
[39] J. T. Mason, M. C. Jha, and P. Chiotti, *J. Less-Common Metals*, 1974, **34**, 1439.
[40] J. C. Taylor and P. W. Wilson, *J.C.S. Chem. Comm.*, 1974, 598; *Acta Cryst.*, 1974, **B30**, 169.

oxide to give Cs_2UOCl_4.[41] Octahedral uranium(v) complexes $[UCl_5L]^-$ are formed with many neutral unidentate ligands.[42]

3 Group IV

Titanium.—$(Me_2N)_2TiF_2$ is tetrameric, with both nitrogen and fluorine bridging to give octahedrally co-ordinated titanium.[43] $[(Me_3Si)_2N]_2TiCl_2$ has a chain structure with trigonal-bipyramidal titanium; the $(TiN)_2$ rings are planar, which is consistent with strong π-bonding within the rings.[44] A similar co-ordination arrangement about titanium is found in Tl_2TiO_3[45a] although, surprisingly in view of the normal tendency for titanium(IV) to be six-co-ordinate, it has approximately tetrahedral co-ordination in Rb_2TiO_3.[45b] The important $BaO-TiO_2$ system has now been extended to the stable compounds Ba_2TiO_4, $BaTiO_3$, $Ba_6Ti_{17}O_{40}$, $Ba_4Ti_{13}O_{30}$, $BaTi_4O_9$, and $Ba_2Ti_9O_{20}$.[46] Ti_8S_3 has a structure closely related to that of Ti_2S.[47] $Ti(SC_6F_5)_4$, $Hf(SC_6F_5)_4$, $Nb(SC_6F_5)_5$, and $TaCl_3(SC_6F_5)_2$ are formed from the metal chloride and pentafluorothio-phenol; each derivative is apparently monomeric.[48] Halogen-bridged polymeric anions, *e.g.* $Ti_2F_{11}^{3-}$, $Ti_2Br_9^-$, and $Zr_2Cl_9^-$, are readily formed by titanium and zirconium; many substituted derivatives of the TiF_6^{2-} ion can be detected in solution by ^{19}F n.m.r. spectroscopy.[49] Titanium oxide fluorides, *e.g.* $NiFeTiO_3F_3$ and $CoVTiO_4F_2$, with a rutile lattice have been obtained by direct synthesis from the constituent oxides and fluorides.[50]

Hafnium and Zirconium.—A nitride amide, $ZrN(NH_2)$, analogous to the nitride halides is prepared by ammonolysis of ZrNI with $NaNH_2$ in liquid ammonia or with ammonia at high temperature and pressure. On thermal decomposition it gives $ZrH_{0.6}N$ and blue Zr_xN.[51] Basic salts of hafnium and zirconium have been shown to contain infinite-chain cations of the type $[M(OH)_2]_n^{2n+}$, although it is significant that the hafnium in $Hf(OH)_2SO_4$ is eight-co-ordinate whereas the metal atoms have pentagonal-bipyramidal co-ordination in $Zr(OH)_2SO_4,H_2O$

[41] G. W. Watt, D. J. Baugh, jun., and K. F. Gadd, *Inorg. Nuclear Chem. Letters*, 1974, **10**, 987.
[42] J. G. H. du Preez, R. A. Edge, M. L. Gibson, H. E. Rohwer, and C. P. J. van Vuuren, *Inorg. Chim. Acta*, 1974, **10**, 27.
[43] W. S. Sheldrick, *J. Fluorine Chem.*, 1974, **4**, 415.
[44] N. W. Alcock, M. Pierce-Butler, and G. R. Willey, *J.C.S. Chem. Comm.*, 1974, 627.
[45] (a) A. Verbaere, M. Dion, and M. Tournoux, *J. Solid State Chem.*, 1974, **11**, 60; (b) W. Schartau and R. Hoppe, *Z. anorg. Chem.*, 1974, **408**, 60.
[46] T. Negas, R. S. Roth, H. S. Parker, and D. Minor, *J. Solid State Chem.*, 1974, **9**, 297.
[47] J. Powens and H. F. Franzen, *Acta Cryst.*, 1974, **B30**, 427.
[48] R. J. H. Clark and D. Kamineris, *Inorg. Chim. Acta*, 1974, **11**, L7.
[49] P. A. W. Dean, *Canad. J. Chem.*, 1973, **51**, 4024; P. A. W. Dean and B. J. Ferguson, *ibid.*, 1974, **52**, 667; J. I. Bullock, F. W. Barrett, and N. J. Taylor, *ibid.*, p. 2880; J. Sala-Pala and J. E. Guerchais, *Bull. Soc. chim. France*, 1973, 2913.
[50] R.-H. Odenthal, J. Grannec, J.-M. Dance, J. Portier, and P. Hagenmuller, *J. Solid State Chem.*, 1974, **9**, 120.
[51] H. Blunck and R. Juza, *Z. anorg. Chem.*, 1974, **406**, 145.

and $Hf_4(OH)_8(CrO_4)_4,H_2O$.[52a] It is probable that the series of hydrates formed by the zirconium oxide halides have similar structures.[52b] The structure of $HfTe_5$ (and of $ZrTe_5$) contains bicapped trigonal prisms linked together by zig-zag chains of Te atoms (in the related $ZrSe_3$ structure the linkage is direct to neighbouring prisms).[53] Pure crystalline zirconium(III) halides (Cl, Br, I) can be formed from ZrX_4 using AlX_3 melts. Reduction of hafnium(IV) is slow.[54]

4 Group V

Vanadium.—Vanadium compounds give metavanadates and then orthovanadates in sodium–potassium nitrite melts.[55] $VOCl$ reacts with ammonia to give the oxide amide $OVNH_2$, which gives V_2O_3 and $V(NH_2)_3$ at 430 K and V_2O_3 and VN at higher temperatures; at low temperatures $VOCl_2$ gives a penta-ammine complex with ammonia, but reduced products corresponding to ammonolysis of VOCl are also observed;[56a] $NbOCl_3$ reacts similarly.[56b] $(Me_3SiO)_3V=NSiMe_3$ is formed from VCl_3 and $(Me_3Si)_2NH$ and is a further example of a compound containing SiEV linkages.[57]

V(O)OH can be synthesized hydrothermally by reduction of $NaVO_3$ in hydrogen; it has the diaspore structure.[58] VO_2 reacts with liquid potassium to give the new compound KVO_2; other oxides give vanadates(V) although VO is observed.[59a] All oxides of niobium and tantalum give the pentapositive derivatives with liquid sodium.[59b] Most vanadates(V) contain tetrahedrally co-ordinated vanadium, although very distorted octahedral environments $(5 + 1)$ are also known, and regular octahedral co-ordination is observed in high-pressure phases.[60] Many vanadyl complexes contain octahedrally co-ordinated vanadium.[61] Vanadium pentafluoride is associated in SO_2ClF at low temperatures, a result which is consistent with the solid-state structure. It is a strong Lewis acid and gives, e.g., $[F_5TaFVF_5]^-$ with the TaF_6^- ion.[62] $VOF_3,2L$ and $VO_2F,2L$ complexes are readily prepared; $VO_2F,(bipyridyl)$ has two oxygen bridges in a dimeric molecule.[63] $VOBr_3$ is formed by the reaction between

52 (a) M. Hansson, *Acta Chem. Scand.*, 1973, **27**, 2455; 2614; M. Hansson and W. Mark, *ibid.*, p. 3467; (b) D. A. Powers and H. B. Gray, *Inorg. Chem.*, 1973, **12**, 2721.
53 S. Furuseth, L. Brattås, and A. Kjekshus, *Acta Chem. Scand.*, 1973, **27**, 2367.
54 E. M. Larsen, J. W. Moyer, F. Gil-Arnao, and M. J. Camp, *Inorg. Chem.*, 1974, **13**, 574.
55 S. S. Al-Omer and D. H. Kerridge, *Inorg. Chim. Acta*, 1973, **7**, 665.
56 (a) N. I. Vorob'ev, V. V. Pechkovskii, and L. V. Kobets, *Russ. J. Inorg. Chem.*, 1973, **18**, 1687: 1974, **19**, 1; (b) *ibid.*, p. 1581.
57 A.-F. Shihada, *Z. anorg. Chem.*, 1974, **408**, 9.
58 J. Muller and J. C. Joubert, *J. Solid State Chem.*, 1974, **11**, 79.
59 (a) M. G. Barker and A. J. Hooper, *J.C.S. Dalton*, 1973, 2614; M. G. Barker, A. J. Hooper, and R. M. Lintonbon, *ibid.*, p. 2618; (b) M. G. Barker, A. J. Hooper, and D. J. Wood, *J.C.S. Dalton*, 1974, 55.
60 B. D. Jordan and C. Calvo, *Canad. J. Chem.*, 1974, **52**, 2701; B. L. Chamberland and P. S. Danielson, *J. Solid State Chem.*, 1974, **10**, 249; M. Gondrand, A. Collomb, J. C. Joubert, and R. D. Shannon, *ibid.*, 1974, **11**, 1.
61 *E.g.* R. E. Drew, F. W. B. Einstein, and S. E. Gransden, *Canad. J. Chem.*, 1974, **52**, 2184; F. Théobald and J. Galy, *Acta Cryst.*, 1973, **B29**, 2732.
62 S. Brownstein and G. Latremouille, *Canad. J. Chem.*, 1974, **52**, 2236.
63 J. Sala-Pala and J. E. Guerchais, *J. Mol. Structure*, 1974, **20**, 169.

$VOCl_3$ and HBr at 273 K ; $VOBr_3$ decomposes thermally to give $VOBr_2$. Both oxide bromides give ranges of complexes.[64]

Niobium and Tantalum.—Cluster compounds of these elements are becoming characterized, although it is not clear as to whether M—M bonds are present in every case where bridging ligands are present. Examples of this type of compound are $[(Me_6C_6)_3Nb_3Cl_6]Cl$, $[\{(C_5H_5)Nb(formato)(OH)\}_3O]H$ (both with triangular Nb_3 groups and Nb—Nb distances of ca. 3.2 Å), and $[Me_4N]_3[(Nb_6Cl_{12})Cl_6]$ (with an Nb—Nb distance of 2.97 Å). Comparison of the structure of the latter compound with the structure of the $[(Nb_6Cl_{12})Cl_6]^-$ (n = 4 or 2) ions suggests that the redox electrons are accommodated in a bonding orbital associated with the metal atoms.[65]

A new carbide sulphide, $Ta_3C_2S_4$, is formed by interaction of tantalum and CS_2 at about 1200 K.[66a] The carbide sulphide Ta_2S_2C forms intercalation compounds,[66b] and it has been shown that metals and ammonia can be electrochemically intercalated from salt solutions into disulphides such as TaS_2.[66c] Some of these intercalation compounds are superconductors at low temperatures.[66d] Ta_3N_5 is isostructural with anosovite, Ti_3O_5, and has TaN_6 octahedra linked by edge and corner sharing;[67a] TaON has complete ordering of the anions in alternate layers similar to the ordering observed in oxyfluorides, e.g. YOF.[67b] $TaCl_4N_3$, which can be prepared from $TaCl_5$ and ClN_3 or NaN_3, is dimeric, with the two azides acting as bridges through one of the terminal nitrogen atoms.[68] Niobium alkoxides are similar to titanium alkoxides in that they readily undergo insertion reactions with phenyl isocyanate, e.g. $Nb(OMe)_5 \rightarrow (MeO)_4NbN(Ph)CO_2Me$.[69]

The oxides in the range NbO_2—Nb_2O_5 have lattices based on the ReO_3 structure. Wadsley-type defects have been observed by electron diffraction in $Nb_{22}O_{54}$, and similar defects are found in the Nb_2O_5–WO_3 system.[70] There is great interest in complex oxides of niobium and tantalum (and also of Group VI elements) because of the physical properties of the complexes. From a structural viewpoint it is to be noted that many stoicheiometries are possible for comparatively simple lattices. Thus perovskite lattices have been characterized for

[64] A. Anagnostopoulos, D. Nicholls, and M. E. Pettifer, *J.C.S. Dalton*, 1974, 569; D. Nicholls and K. R. Seddon, *ibid.*, 1973, 2747.
[65] M. R. Churchill and S. W.-Y. Chang, *J.C.S. Chem. Comm.*, 1974, 248; N. I. Kirillova, A. I. Gusev, A. A. Pasynskii, and Yu. T. Struchkov, *J. Struct. Chem.*, 1973, **14**, 1008 (cf. *J. Organometallic Chem.*, 1974, **74**, 91); F. W. Koknat and R. E. McCarley, *Inorg. Chem.*, 1974, **13**, 295.
[66] (a) M. Caillet, A. Galerie, and J. Besson, *J. Less-Common Metals*, 1974, **37**, 111; (b) R. Schöllhorn and A. Weiss, *Z. Naturforsch.*, 1973, **28b**, 716; (c) M. S. Whittingham, *J.C.S. Chem. Comm.*, 1974, 328; (d) G. V. Subba Rao, M. W. Shafer, S. Kawarazaki, and A. M. Toxen, *J. Solid State Chem.*, 1974, **9**, 323.
[67] (a) J. Strähle, *Z. anorg. Chem.*, 1973, **402**, 47; (b) D. Armytage and B. E. F. Fender, *Acta Cryst.*, 1974, **B30**, 809.
[68] J. Strähle, *Z. anorg. Chem.*, 1974, **405**, 139.
[69] R. C. Mehrotra, A. K. Rai, and R. Bohra, *J. Inorg. Nuclear Chem.*, 1974, **36**, 1887.
[70] S. Iijima, S. Kimura, and M. Goto, *Acta Cryst.*, 1974, **A30**, 251; S. Iijima and J. G. Allpress, *ibid.*, p. 29; cf. J. F. Maruclo, *J. Solid State Chem.*, 1974, **10**, 211.

compounds such as $Ca_2Nb_2O_7$, Ba_2NbVO_6, $Ba_4Nb_3LiO_{12}$, and Eu_2TaCrO_6[71] and pyrochlore-type lattices for $(H_3O)NbWO_6$, NH_4TaWO_6, and $Pb_{2-x}Nb_2$-$O_{7-\beta}$.[72] The new compound $TlNbB_2O_6$ contains octahedral NbO_6 and planar BO_3 units in a network structure.[73] In this Group the structural chemistry of complex oxide fluorides is often closely similar to that of the complex oxides. Thus the series MgF_2—Nb_2O_5 is very complex and has block-type structures based on the ReO_3 lattice, the series $NaR.E.TiNbO_6F$ has pyrochlore-type phases, and the series M_2NbO_5F (M = Ti, V, or Cr) has rutile lattices.[74] In many of the lattices mentioned above, titanium can replace niobium. The system $KF-Ta_2O_5$ is also complex, and $K_2Ta_4O_9F_4$ has a lattice structure of octahedra sharing corners, although $K_3TaO_2F_4$ contains discrete octahedra, and discrete octahedra are found for many other complex oxide fluorides.[75] NH_4NbOF_4 appears to contain oxygen-bridged chains.[76] K_3NbF_7 has been synthesized by reduction of molten $KF-NbF_5$ at 1120 K with niobium metal; it is isostructural with several similar niobium and tantalum compounds and contains pentagonal-bipyramidal niobium.[77] Complexes of TaF_5 and ligand exchange in derivatives of TaF_5 continue to be studied; anionic, neutral, and cationic species are formed.[78] Ligand exchange in complexes $TaCl_5$,L in non-co-ordinating solvents occurs by an associative mechanism.[79] Mixed halogeno-complexes $MF_nX_{6-n}^-$ (M = Nb^V or Ta^V; X = Cl or Br) are formed in solution in CH_3CN by allowing the appropriate pentahalides to react with HF.[80]

5 Group VI

Chromium.—The heavy-atom Group IV donors $SnCl_3^-$ and $[(Me_3Si)_2CH]_2Sn(L)$ form stable bonds to chromium (and Mo and W) in complexes such as $LCr(CO)_5$.[81] The oxidation of $N_2H_5^+$ by chromium(VI) at high acidity proceeds

[71] K. Scheunemann and H. Müller-Buschbaum, *J. Inorg. Nuclear Chem.*, 1974, **36**, 1965; J.-C. Bernier, C. Chauvel, and O. Kahn, *J. Solid State Chem.*, 1974, **11**, 265; M. Nanot, F. Queyroux, J.-C. Gilles, A. Carpy, and J. Galy, *ibid.*, p. 272; E. F. Jendrek, jun., A. D. Potoff, and L. Katz, *ibid.*, 1974, **9**, 375; B. M. Collins, A. J. Jacobson, and B. E. F. Fender, *ibid.*, 1974, **10**, 29; H. Parent, J.-C. Bernier, and P. Poix, *Compt. rend.*, 1974, **278**, *C*, 49.
[72] D. Groult, C. Michel, and B. Raveau, *J. Inorg. Nuclear Chem.*, 1974, **36**, 61; J. Bachelier, M. Hervieu, and E. Quemeneur, *Bull. Soc. chim. France*, 1973, 2593.
[73] M. Gasperin, *Acta Cryst.*, 1974, **B30**, 1181.
[74] J. L. Hutchison, F. J. Lincoln, and J. S. Anderson, *J. Solid State Chem.*, 1974, **10**, 312; *J.C.S. Dalton*, 1974, 115; J. Grannec, H. Baudry, J. Ravez, and J. Portier, *J. Solid State Chem.*, 1974, **10**, 66; J. Senegas and J. Galy, *Compt. rend.*, 1973, **277**, *C*, 1243.
[75] J. P. Chaminade, M. Vlasse, M. Pouchard, and P. Hagenmuller, *Bull. Soc. chim. France*, 1974, 179; G. Pausewang, *Z. Naturforsch.*, 1974, **29b**, 49.
[76] V. I. Pakhomov, R. L. Davidovich, T. A. Kaidalova, and T. F. Levcheshina, *Russ. J. Inorg. Chem.*, 1973, **18**, 654.
[77] L. O. Kilpatrick and L. M. Toth, *Inorg. Chem.*, 1974, **13**, 2242.
[78] Yu. A. Buslaev, Yu. V. Kokunov, V. D. Kopanev, and M. P. Gustyakova, *J. Inorg. Nuclear Chem.*, 1974, **36**, 1569; S. Brownstein and M. J. Farrall, *Canad. J. Chem.*, 1974, **52**, 1958.
[79] R. Good and A. E. Merbach, *J.C.S. Chem. Comm.*, 1974, 163.
[80] Yu. A. Buslaev and E. G. Ilyin, *J. Fluorine Chem.*, 1974, **4**, 271.
[81] T. Kruck and H. Breuer, *Chem. Ber.*, 1974, **107**, 263; J. D. Cotton, D. E. Goldberg, M. F. Lappert, and K. M. Thomas, *J.C.S. Chem. Comm.*, 1974, 893.

through an N-bonded chromate ester; the corresponding oxidation of NH_3OH^+ involves an O-bonded chromate, and chromium-(IV) and -(V).[82] For a first-row transition metal, chromium is exceptional in that it forms many distinct oxides; crystallographic shear phases have been established in the series Cr_nO_{2n-1} ($n = 4$ or 6).[83] $SrCr_2O_4$ and α-$CaCr_2O_4$ have a new structure type, with octahedral Cr^{3+} and trigonal-prismatic M^{2+}.[84] Chromates(VI) contain tetrahedral chromium, and lithium dichromate should be formulated as $Li_2[\mu$-O-Cr_2O_5-$(OH)](OH),H_2O$.[85] $BaLiCrF_6$ and a series of isostructural compounds with other tripositive ions have framework structures of BaF_{12} icosahedra, LiF_4 tetrahedra, and CrF_6 octahedra. Cobalt(II) can replace some lithium to give tetrahedral CoF_4^{2-} ions.[86]

Molybdenum and Tungsten.—There is still a great deal of interest in devising new methods for preparing compounds of this Group containing metal–metal bonds. The mixed acetate $[CrMo(OAc)_4]$ is formed from $Mo(CO)_6$ in HAc and $[Cr_2(OAc)_4(H_2O)_2]$.[87] A general method for the preparation of such species is to allow a halogenometallate to react with a metal carbonyl halide. In addition to previously known species, this method yields the $Mo_2Cl_9^{2-}$ anion.[88] Previous theoretical studies on species such as $[Mo_6Cl_8]^{4+}$ have suggested a closed-shell configuration, but magnetic susceptibility and c.d. studies suggest that this is not the case.[89] Structural studies on $Mo_2(NMe_2)_6$ and $Li_4[Mo_2Me_8],4Et_2O$ show that both compounds contain strong $Mo-Mo$ bonds.[90] Controversies over the nature of molybdenum(II) and molybdenum(III) species in aqueous solution continue, but the preparation of derivatives, *e.g.* $K_4Mo_2(SO_4)_4$, has enabled cleaner spectra to be obtained, and it might be expected that true identification of some of these species will occur soon.[91]

Dimethylamido-derivatives of the early transition metals generally undergo insertion reactions in all $M-N$ bonds with carbon dioxide, but $W(NMe_2)_6$ gives the triscarbamato-derivative in a reaction which is probably limited by the nucleophilicity of the NMe_2 ligands.[92] Molybdenum nitrido-complexes, *e.g.* Et_4NMoCl_5N, are conveniently prepared using trimethylsilyl azide.[93]

[82] G. P. Haight, jun., T. J. Huang and H. Platt, *J. Amer. Chem. Soc.*, 1974, **96**, 3137; R. A. Scott, G. P. Haight, jun., and J. N. Cooper, *ibid.*, p. 4136.
[83] M. A. Alario Franco, J. M. Thomas, and R. D. Shannon, *J. Solid State Chem.*, 1974, **9**, 261.
[84] H. Pausch and H. Müller-Buschbaum, *Z. anorg. Chem.*, 1974, **405**, 1, 113.
[85] I. D. Datt and R. P. Ozerov, *Soviet Phys. Cryst.*, 1974, **19**, 63.
[86] D. Babel, *Z. anorg. Chem.*, 1974, **406**, 23; W. Viebahn and D. Babel, *ibid.*, p. 38.
[87] C. D. Garner and R. G. Senior, *J.C.S. Chem. Comm.*, 1974, 580.
[88] W. H. Delphin and R. A. D. Wentworth, *J. Amer. Chem. Soc.*, 1974, **95**, 7920.
[89] B. Briat, J. C. Rivoal, O. Kahn, and S. Moreau, *Chem. Phys. Letters*, 1974, **26**, 604.
[90] M. Chisholm, F. A. Cotton, B. A. Frenz, and L. Shive, *J.C.S. Chem. Comm.*, 1974, 480; F. A. Cotton, J. M. Troup, T. R. Webb, D. H. Williamson, and G. Wilkinson, *J. Amer. Chem. Soc.*, 1974, **96**, 3824; *cf.* M. Chisholm and W. Reichert, *ibid.*, p. 1249.
[91] A. R. Bowen and H. Taube, *Inorg. Chem.*, 1974, **13**, 2245; M. Ardon and A. Pernick, *ibid.*, p. 2275.
[92] M. H. Chisholm and M. Extine, *J. Amer. Chem. Soc.*, 1974, **96**, 6214.
[93] J. Chatt and J. R. Dilworth, *J.C.S. Chem. Comm.*, 1974, 517.

Reduction of large crystals of WO_3 to $WO_{\sim 2.91}$ at high temperatures gives members of the series W_nO_{3n-2} with shear planes.[94] White α-molybdic acid contains isolated double chains of $[MoO_5(H_2O)]$ octahedra sharing edges; the structure is closely related to that of MoO_3, which can be considered as a condensation product of these chains.[95] Vaporization of mixtures of V_2O_5 and WO_3 gives a series of volatile vanadium and tungsten oxides (including new W_6O_{18}) together with mixed oxides VW_2O_8, VW_3O_{11}, VW_4O_{14}, $V_2W_2O_{10}$, $V_2W_3O_{14}$, V_3WO_{10}, $V_3W_2O_{13}$, and $V_4W_2O_{16}$.[96] Vapour-phase studies have also been made on Cs_2MoO_4, Cs_2WO_4, and Tl_2MoO_4, all of which contain approximately tetrahedral arrangements about tungsten or molybdenum.[97] The mixed metal oxides containing these elements continue to receive extensive attention. $AlWO_4$, which is synthesized from the oxides at high pressures, has alternate long and short $W—W$ distances in the crystal,[98] and W^{5+} has been confirmed by e.p.r. in the triclinic bronze $Na_{0.33}WO_3$.[99] The unusual magnetic properties of the inverse spinel Fe_2MoO_4 are interpreted in terms of high-spin Mo^{4+},[100] and a new structure type related to that of α-PbO_2 has been examined in Fe_2WO_6, which has a framework of chains of MO_6 octahedra sharing corners.[101]

The ready availability of X-ray crystallography has greatly stimulated work on polyanions of molybdenum and tungsten. $Na_2W_4O_{13}$ has four edge-sharing octahedra linked to further tetramic units by corner sharing;[102] this structure is very different from that of $K_2W_4O_{13}$, showing the great importance of the counter-ion on the structure adopted. The hexamolybdate ion in $[HN_3P_3$-$(NMe_2)_6]_2^+ [Mo_6O_{19}]^{2-}$ has a cage of six molybdenums about a central oxygen, with each molybdenum bridged to its neighbours and having one terminal oxygen.[103] Fluoride can replace oxygen in isopolytungstates to give a range of fluorine-containing polyanions.[104] E.s.r. studies might be expected to give definite information on the electron distribution in reduced heteropolyacid derivatives. However, studies on various derivatives give conflicting results, and the electronic situation is clearly complex.[105] In the heteropolyacid series, $(NH_4)_5[HMo_5P_2O_{23}],3H_2O$ has an anion consisting of a ring of five distorted MoO_6 octahedra sharing edges and apices; the ring is capped on one side by a

[94] M. Sundberg and R. J. D. Tilley, *J. Solid State Chem.*, 1974, **11**, 150.
[95] I. Böschen and B. Krebs, *Acta Cryst.*, 1974, **B30**, 1795.
[96] S. L. Bennett, S.-S. Lin, and P. W. Gilles, *J. Phys. Chem.*, 1974, **78**, 266.
[97] S. M. Tolmachev and N. G. Rambidi, *J. Struct. Chem.*, 1972, **13**, 1; V. V. Ugarov, Yu. S. Ezhov, and N. G. Rambidi, *ibid.*, 1973, **14**, 317.
[98] J.-P. Doumerc, M. Vlasse, G. Demazeau, and M. Pouchard, *Compt. rend.*, 1974, **279**, *C*, 201.
[99] H. F. Mollet and B. C. Gerstein, *J. Chem. Phys.*, 1974, **60**, 1440.
[100] J. Ghose, N. N. Greenwood, G. C. Hallam, and D. A. Read, *J. Solid State Chem.*, 1974, **11**, 239.
[101] J. Senegas and J. Galy, *J. Solid State Chem.*, 1974, **10**, 5.
[102] K. Viswanathan, *J.C.S. Dalton*, 1974, 2170.
[103] H. R. Allcock, E. C. Bissell, and E. T. Shawl, *Inorg. Chem.*, 1973, **12**, 2963.
[104] F. Chauveau and P. Souchay, *J. Inorg. Nuclear Chem.*, 1974, **36**, 1761.
[105] H. So, C. M. Flynn, jun., and M. T. Pope, *J. Inorg. Nuclear Chem.*, 1974, **36**, 329; R. A. Prados, P. T. Meiklejohn, and M. T. Pope, *J. Amer. Chem. Soc.*, 1974, **96**, 1261; M. Otake, Y. Komiyama, and T. Otaki, *J. Phys. Chem.*, 1973, **77**, 2896.

PO_4^{3-} group and on the other by an HPO_4^{2-} group.[106a] The anion in $Na_3H_6Mo_9PO_{34}(H_2O)_x$ has a central PO_4 tetrahedron surrounded by nine MoO_6 groups.[106b] The cerium in $Na_6H_2CeW_{10}O_{36}$,$30H_2O$ is in square anti-prismatic co-ordination from two $W_5O_{18}H^{5-}$ ions, which can be considered to be derived from $W_6O_{19}^{2-}$ anions by loss of one tungsten and its apical oxygen.[106c]

Tungsten hexafluoride forms both 1:1 and 1:2 complexes with dimethyl sulphide, although there is no information on the geometry of the resultant complexes.[107] Although the vapour phases of the oxide tetrafluorides $MoOF_4$ and WOF_4 contain a small amount of polymers, the main species present are square-pyramidal monomers, of C_{4v} symmetry, and the $WOCl_4$ molecule is apparently similar.[108] $(NH_4)_2MoO_3F_2$ has a polymeric oxygen-bridged anion with *cis*-fluorines; corresponding molybdenum(v) derivatives are also oxygen-bridged.[109] The reaction of HF with $WOCl_4$ gives the $[W_2O_2F_9]^-$ ion, which is fluorine-bridged, and the $[Mo_2O_2F_9]^-$ ion results from the action of acetyl-acetone on $MoOF_4$.[110] Complex dioxo-molybdenum(vi) bromide species have been confirmed in solution and in the solid state.[111]

6 Group VII

Manganese.—The structure of tris(tropolonato)manganese(III) demonstrates unequivocally the departures from octahedral symmetry due to the Jahn–Teller distortion expected for a d^4 high-spin species.[112] The first polymanganate has been characterized in the products of dehydration of permanganic acid as $(H_3O)_2[Mn^{IV}(Mn^{VII}O_4)_6]$,$11H_2O$.[113] $NaMn_7O_{12}$ has already been characterized as a perovskite, $(NaMn^{III}_3)(Mn^{III}_2Mn^{IV}_2)O_{12}$, and the series can be extended to include dipositive and tripositive cations in place of sodium by synthesis under high pressure and temperature.[114] $BaMnS_2$ has the $SrZnO_2$ structure with MnS_4 tetrahedra; it is an unusual example of a sulphide isostructural with an oxide.[115] The new fluoride Mn_2F_5 has been prepared by direct interaction of

[106] (a) J. Fischer, L. Ricard, and P. Toledano, *J.C.S. Dalton*, 1974, 941; B. Hedman, *Acta Chem. Scand.*, 1973, **27**, 3335; (b) R. Strandberg, *ibid.*, 1974, **A28**, 217; (c) J. Iball, J. N. Low, and T. J. R. Weakley *J.C.S. Dalton*, 1974, 2021.

[107] A. Steigel and S. Brownstein, *J. Amer. Chem. Soc.*, 1974, **96**, 6227.

[108] R. T. Paine and R. S. McDowell, *Inorg. Chem.*, 1974, **13**, 2366; L. E. Alexander, I. R. Beattie, A. Bukovsky, P. J Jones, C. J. Marsden, and G. J. Van Schalkwyk, *J.C.S. Dalton*, 1974, 81; W. Bues, W. Brockner, and F. Demiray, *Spectrochim. Acta*, 1974, **30A**, 579; K. Iijima and S. Shibata, *Bull. Chem. Soc. Japan*, 1974, **47**, 1393.

[109] R. Mattes and G. Müller, *Naturwiss.*, 1973, **60**, 550; R. Mattes and G. Lux, *Angew. Chem. Internat. Edn.*, 1974, **13**, 600.

[110] Yu. A. Buslaev, Yu. V. Kokunov. V. A. Bochkareva, and E. M. Shustorovich, *J. Struct. Chem.*, 1972, **13**, 491, 570.

[111] R. Kergoat, J.-M. Mauguen, and J. E. Guerchais, *Bull. Soc. chim. France*, 1974, 397.

[112] A. Avdeef, J. A. Costamagna, and J. P. Fackler, jun., *Inorg. Chem.*, 1974, **13**, 1854; cf. V. W. Day, B. R. Stults, E. L. Tasset, R. O. Day, and R. S. Marianelli, *J. Amer. Chem. Soc.*, 1974, **96**, 2496.

[113] B. Krebs and K.-D. Hasse, *Angew. Chem. Internat. Edn.*, 1974, **13**, 603.

[114] B. Bochu, J. Chenavas, J. C. Joubert, and M. Marezio, *J. Solid State Chem.*, 1974, **11**, 88.

[115] D. Schmitz and W. Bronger, *Z. anorg. Chem.*, 1973, **402**, 225.

MnF_2 and MnF_3 as a purple powder.[116] $CsMn_4Cl_9$ has Cs and Cl close-packed with manganese in octahedral holes.[117]

Technetium and Rhenium.—The reaction of tertiary phosphines with Re_3Cl_9 or $Re_2Cl_8^{2-}$ gives a new class of rhenium(II) complexes, $Re_2Cl_4L_4$, with an eclipsed rotational configuration and a very strong Re—Re bond; in $(NH_4)_2[Re_2Cl_6-$ (formato)$_2]$ there are two bridging formates but the Re—Re distance is still short.[118] The quadruple Re—Re bond in $Re_2Cl_8^{2-}$ undergoes ready photochemical cleavage to monomeric Re^{III} species.[118] In the vapour phase the ReORe angle of (symmetrical) Re_2O_7 appears to be large and near to 165°; the structure would thus be appreciably different from that of Cl_2O_7.[119] Violet TcO_4^{2-} and olive-green ReO_4^{2-} ions, analogous to manganates(VI), are produced by cathodic reduction of the heptapositive anions.[120]

7 The Iron Group

Iron.—Matrix-isolated iron atoms are conveniently generated by u.v. photolysis of $Fe(CO)_5$.[121] The intermetallic compound FeTi forms hydrides FeTiH and $FeTiH_2$;[122] the detailed structure of a compound, Cp(diphos)FeMgBr,3THF, containing a direct Fe—Mg bond, is now known; the bond length is 2.59 Å.[123] Although metal–silicon bond strengths are generally assumed to be substantially greater than the metal–carbon bond strengths, this is not so, and indeed the Fe—Si bond may be the weaker.[124] Fe_6Ge_5 (and Fe_6Ga_5) has a new structure type containing pentagonal pyramids with a heteroatom at the apices.[125]

Ferrate(VI), FeO_4^{2-}, is a very strong oxidizing agent; it appears to have potential for use as an oxidizer in organic reactions, and the oxidation products formed with polyhydric alcohols have been studied.[126] Most mixed metal oxides containing iron(III) are based on close-packed layers of anions but $K_6Fe_2O_6$, formed from $KO_{0.56}$ and 'FeO', contains discrete Fe_2O_6 anions with tetrahedral co-ordination about iron.[127] Mössbauer spectra enable the cation distribution in the iron–vanadium oxygen spinel system to be studied. For compounds $Fe_{1+x}V_{2-x}O_4$ ($0 \leqslant x \leqslant 0.35$) the compound is $Fe^{2+}[Fe^{3+}{}_xV^{3+}{}_{2-x}]O_4$;

[116] A. Tressaud and J.-M. Dance, *Compt. rend.*, 1974, **278**, C, 463.
[117] J. Goodyear and D. J. Kennedy, *Acta Cryst.*, 1973, **B29**, 2677.
[118] F. A. Cotton, B. A. Frenz, J. R. Ebner, and R. A. Walton, *J.C.S. Chem. Comm.*, 1974, 4; P. A. Koz'min, M. D. Surazhskaya, and T. B. Larina, *J. Struct. Chem.*, 1974, **15**, 56; G. L. Geoffroy, H. B. Gray, and G. S. Hammond, *J. Amer. Chem. Soc.*, 1974, **96**, 5565.
[119] V. S. Vinogradov, V. V. Ugarov, and N. G. Rambidi, *J. Struct. Chem.*, 1972, **13**, 661.
[120] K. Schwochau, L. Astheimer, J. Hauck, and H.-J. Schenk, *Angew. Chem. Internat. Edn.*, 1974, **13**, 346.
[121] M. Poliakoff and J. J. Turner, *J.C.S. Faraday II*, 1974, 93.
[122] J. J. Reilly and R. H. Wiswall, jun., *Inorg. Chem.*, 1974, **13**, 218.
[123] H. Felicin, P. J. Knowles, B. Meunier, A. Mitschler, L. Ricard, and R. Weiss, *J.C.S. Chem. Comm.*, 1974, 44.
[124] C. Windus, S. Sujishi, and W. P. Giering, *J. Amer. Chem. Soc.*, 1974, **96**, 1951.
[125] B. Malaman, M. J. Philippe, B. Roques, A. Courtois, and J. Protas, *Acta Cryst.*, 1974, **B30**, 2081.
[126] D. H. Williams and J. T. Riley, *Inorg. Chim. Acta*, 1974, **8**, 177.
[127] H. Rieck and R. Hoppe, *Z. anorg. Chem.*, 1974, **408**, 151.

for $1 \leqslant x \leqslant 2$ it is $Fe^{3+}[Fe^{2+}Fe^{3+}_{x-1}V^{3+}_{2-x}]O_4$.[128] Several ferrates(II), *e.g.* Na_4FeO_3, have been reported.[129] The Ba–Fe–S system contains an infinite number of ordered phases, $Ba_pFe_{2q}S_{2q}$, with superstructures based on the NH_4CuMoS_4 structure-type.[130] Iron(IV), an unusual oxidation state for this element, can be stabilized by dithiocarbamate ligands; the co-ordination arrangement is midway between octahedral and trigonal-prismatic.[131] $CsFeF_4$ has a new superstructure based on that of $TlAlF_4$ with FeF_6 octahedra;[132] studies of the Fe–O–F and Fe–S–F systems have shown phases FeO_xF_{2-x} (rutile) and $Fe_3O_{4-x}F_x$ (spinel) in the former system but no vacancies in the latter.[133] FeOCl, which has a layer lattice, forms intercalation compounds with pyridine.[134]

Ruthenium and Osmium.—Gaseous RuC_2, PtC_2, and IrC_2 have been observed mass-spectrometrically (MC have been previously observed).[135] The formation of silyl compounds of the Group VIII elements by oxidative addition of hydrosilanes to phosphine complexes generally leads to species of low co-ordination number, but it has now been shown that seven-co-ordinate $RuH_3(SiR_3)(PPh_3)_3$ results from addition of R_3SiH to $RuH_2(PPh_3)_4$.[136] Kinetic and mechanistic studies on ruthenium complexes provide information of general interest. A study of the spectroscopic quantum yields for the photoaquation of $Ru(NH_3)_5L^{2+}$ species shows that the photochemical initiation is due to low-lying, substitution-reactive, ligand-field-excited states – not charge-transfer states.[137] Heteronuclear complexes are frequently postulated as reaction intermediates but $[(NH_3)_5RuOV(H_2O)_n]^{4+}$ has been isolated as a stable dark-green species from a variety of reactions.[138] The ruthenium bronze $Na_{3-x}Ru_4O_9$ contains a single, double, and triple chain of RuO_6 octahedra sharing edges.[139] By using stereospecific reactions depending upon the *trans*-effect, it has proved possible to prepare many of the mixed-ligand hexahalogenometallates of osmium, $[OsCl_xBr_yI_{6-x-y}]^{2-}$.[140]

128 M. Abe, M. Kawachi, and S. Nomura, *J. Solid State Chem.*, 1974, **10**, 351.
129 H. Rieck and R. Hoppe, *Naturwiss*, 1974, **61**, 126; N. Tannières, O. Évrard, and J. Aubry, *Compt. rend.*, 1974, **278**, C, 247.
130 I. E. Grey, *J. Solid State Chem.*, 1974, **11**, 128.
131 R. L. Martin, N. M. Rohde, G. B. Robertson, and D. Taylor, *J. Amer. Chem. Soc.*, 1974, **96**, 3645.
132 D. Babel, F. Wall, and G. Heger, *Z. Naturforsch.*, 1974, **29b**, 139.
133 J. Claverie, L. Lozano, J. P. Odile, J. Portier, and P. Hagenmuller, *J. Fluorine Chem.*, 1974, **4**, 57.
134 F. Kanamaru, S. Yamanaka, M. Koizumi, and S. Nagai, *Chem. Letters*, 1974, 373.
135 K. A. Gingerich, *J.C.S. Chem. Comm.*, 1974, 199.
136 H. Kono and Y. Nagai, *Chem. Letters*, 1974, 931.
137 G. Malouf and P. C. Ford, *J. Amer. Chem. Soc.*, 1974, **96**, 601.
138 H. de Smedt, A. Persoons, and L. de Maeyer, *Inorg. Chem.*, 1974, **13**, 90.
139 J. Darriet, *Acta Cryst.*, 1974, **B30**, 1459.
140 W. Preetz, H. J. Walter, E. W. Fries, A. Scheffler, and H. Homborg, *Z. anorg. Chem.*, 1973, **402**, 169, 180; 1974, **406**, 92; 1974, **407**, 1.

8 The Cobalt Group

Cobalt.—The reduction of cobalt (II) or nickel(II) salts with sodium borohydride produces metal borides $(M_2B)_5H_3$ containing some hydrogen; in the presence of tertiary phosphines L, metal(I) halide and then borohydride species, MXL_3 (X = Cl or BH_4), are formed.[141] There is currently great interest in oxocobaltates. $NaCoO_2$ has the α-$NaFeO_2$ structure and $NaCoO_4$ is isostructural with K_2CoO_4; the rubidium and caesium salts have related structures. A series of perovskites $R.E.CoO_3$ can be prepared under a high oxygen pressure. Oxocobaltates(IV) M_2CoO_3 (M = K, Rb, or Cs) are also prepared by this method, and the caesium and rubidium salts have a $[CoO_3]$ chain with tetrahedrally co-ordinated cobalt.[142] The activation parameters for the reduction of $[Co(NH_3)_5X]^{2+}$ species by V^{2+} suggest that for the fluoride the reaction proceeds by an inner-sphere substitution-controlled mechanism (hard acid–hard base) whereas the chloride, bromide, and iodide react by outer-sphere mechanisms (softer bases).[143] A tetrahedral CoF_4^{2-} ion can be stabilized in a $BaLiCrF_6$ lattice.[86] The crystal structures of some mercaptoacetatobis(ethylenediamine)cobalt(III) complexes show clearly a lengthening of the Co—N bond *trans* to sulphur; this *trans* lengthening does not appear to occur in chromium(III) complexes.[144]

Rhodium and Iridium.—A detailed study of the interaction of four-co-ordinate rhodium(I) complexes with boron-containing Lewis acids confirms the presence of Rh $\rightarrow$ B linkages in some of the adducts [*e.g.* $RhCl(CO)(PPh_3)_2,BCl_3$], although other adducts have halogen or carbonyl bridges, and anion formation must also be considered.[145] Compounds containing M—Hg bonds formed by these elements are becoming increasing well characterized; the product of the reduction of *trans*-$[Rh(en)_2Cl_2]^+$ at mercury electrodes is the rhodium(I) species $[(en)_2Rh]_2Hg^{4+}$, and very many adducts have been reported from the interaction of mercury(II) halides and rhodium or iridium halides, hydride halides, or hydrides.[146] Ligand exchange in $Rh_2[P(OMe)_3]_8$ (which has a bicapped trigonal antiprismatic structure) is specifically between equatorial positions.[147] A new phase, $Na_4Ir_3O_8$, dissociating at 1313 K, has been reported in the Na_2O–IrO_2 system; the Na_2O–PtO_2 system contains a similar phase.[148]

[141] P. C. Maybury, R. W. Mitchell, and M. F. Hawthorne, *J.C.S. Chem. Comm.*, 1974, 534; D. G. Holah, A. N. Hughes, and B. C. Hui, *Inorg. Nuclear Chem. Letters*, 1974, **10**, 427.

[142] M. Jansen and R. Hoppe, *Z. anorg. Chem.*, 1974, **408**, 75, 97, 104; G. Demazeau, M. Pouchard, and P. Hagenmuller, *J. Solid State Chem.*, 1974, **9**, 202.

[143] M. R. Hyde, R. S. Taylor, and A. G. Sykes, *J.C.S. Dalton*, 1973, 2730.

[144] R. C. Elder, L. R. Florian, R. E. Lake, and A. M. Yacynych, *Inorg. Chem.*, 1973, **12**, 2690.

[145] D. D. Lehman and D. F. Shriver, *Inorg. Chem.*, 1974, **13**, 2203; P. Bird, J. F. Harrod, and K. A. Than, *J. Amer. Chem. Soc.*, 1974, **96**, 1222.

[146] J. Gulens and F. C. Anson, *Inorg. Chem.*, 1973, **12**, 2568; P. R. Brookes and B. L. Shaw, *J.C.S. Dalton*, 1974, 1702.

[147] R. Mathieu and J. F. Nixon, *J.C.S. Chem. Comm.*, 1974, 147.

[148] C. L. McDaniel, *J. Solid State Chem.*, 1974, **9**, 139.

Hexahydroxyiridates(IV) have been precipitated from solution.[149] A series of hexafluororhodates(IV), $M^{II}RhF_6$ (M = e.g. Ni, Cu, Hg, or Zn), have been prepared by direct fluorination of appropriate mixtures or compounds.[150] IrF_4 is formed from IrF_5 and iridium metal; on heating it disproportionates to IrF_3 and IrF_5.[151]

9 The Nickel Group

Nickel.—Compounds, e.g. $Ni(bipy)(SiX_3)_2$, containing Si—Ni bonds are found by treatment of silicon hydrides with alkyl-nickel derivatives.[152] N_2O reacts quite specifically with a nickel surface to give surface species Ni_2N_2, Ni_2O, and Ni_2N.[153] The complex fluorides $A_2BNi^{III}F_6$, like most other complexes of this stoicheiometry, have the elpasolite lattice, but the $NiF_6{}^{3-}$ octahedra are elongated because of Jahn–Teller distortion in the $t_{2g}^6 e_g^1$ low-spin configuration.[154]

Palladium and Platinum.—Dihydroplatinum(II) complexes *trans*-$[PtH_2L_2]$ are stable when L is a bulky tertiary phosphine; solvolysis of platinum(0) complexes of alkyl diphenylphosphinites gives hydride derivatives.[115] Palladium carbonyl halides have been given various formulations in the past but it now seems established that $Pd(CO)X$ (X = Cl or Br) are the products of the action of carbon monoxide on palladium(II) halides in aqueous HX solution; complexes $MPd(CO)X_2$ have also been prepared.[156] $Hg(SiMe_3)_2$ forms both Pt—$SiMe_3$ and Pt—$HgSiMe_3$ bonds on reaction with suitable Pt—Cl groups; diphenyl-mercury can form Pt—HgPh bonds.[157] Pd_6P, which, with Pd_8P, represents the most metal-rich phosphides of the transition metals, has a structure related to that of Re_3B, with Pd_6 trigonal prisms of which half contain a phosphorus atom.[158] Phosphine complexes of these elements continue to show unusual co-ordination features; $Pd(PPhBu^t{}_2)_2$ has an almost linear P—Pd—P skeleton, although there is some evidence for interaction of the metal with *ortho*-hydrogens of the phenyl group, and five-co-ordinate phosphite complexes of Pd and Pt have been reported for the first time. $Pd(PBu^t{}_3)_2$ is remarkably stable towards air but undergoes photochemical absorption of hydrogen to give hydrides.[159] Platinum and iridium siloxanes, e.g. $(Me_3P)_2Pt(OSiMe_3)_2$, are formed from

[149] M. B. Bardin and P. M. Ketrush. *Russ. J. Inorg. Chem.*, 1973, **18**, 693; R. F. Sarkozy and B. L. Chamberland, *J. Solid State Chem.*, 1974, **10**, 145.
[150] V. Wilhelm and R. Hoppe, *Z. anorg. Chem.*, 1974, **407**, 13.
[151] N. Bartlett and A. Tressaud, *Compt. rend.*, 1974, **278**, C, 1501.
[152] Y. Kiso, K. Tamao, and M. Kumada, *J. Organometallic Chem.*, 1974, **76**, 95.
[153] M. Barber and J. C. Vickerman, *Chem. Phys. Letters*, 1974, **26**, 277.
[154] E. Alter and R. Hoppe, *Z. anorg. Chem.*, 1974, **405**, 167; D. Reinen, C. Friebel, and V. Propach, *ibid.*, 1974, **408**, 187.
[155] B. L. Shaw and M. F. Uttley, *J.C.S. Chem. Comm.*, 1974, 918; P.-C. Kong and D. M. Roundhill, *J.C.S. Dalton*, 1974, 187.
[156] R. Colton, R. H. Farthing, and M. J. McCormick, *Austral. J. Chem.*, 1973, **26**, 2607.
[157] F. Glockling and R. J. I. Pollock, *J.C.S. Dalton*, 1974, 2259.
[158] Y. Andersson, V. Kaewchansilp, M. R. C. Soto, and S. Rundqvist, *Acta Chem. Scand.*, 1974, **A28**, 797.
[159] M. Matsumoto, H. Yoshioka, K. Nakatsu, T. Yoshida, and S. Otsuka, *J. Amer. Chem. Soc.*, 1974, **96**, 3324; P. Meakin and J. P. Jesson, *ibid.*, p. 5751.

phosphine metal halides and $NaOSiMe_3$; the corresponding palladium derivatives are not formed.[160] The oxometallate K_2PdO_2 is formed from K_2O and PdO; it has a chain-like structure with square-planar co-ordination about palladium.[161] Further studies have been reported on platinum bronzes, and details have been given of the preparations of $M_xPt_3O_4$ (M = Na or Cd) and bronze-like $CaPt_2O_4$. The latter has chains in two dimensions and there is extensive Pt—Pt bonding. A poorly defined oxide, Pt_5O_6, has also been reported to result from attempts to form a barium-containing bronze.[162] Diplatinum complexes containing SH bridges results from the action of H_2S on solutions of platinum(II) salts.[163] A series of double selenides, e.g. $Cs_2Pd_3Se_4$, similar to the corresponding sulphides, has been reported.[164] Heating the binary fluorides gives the perovskites $MPdF_3$ (M = K, Rb, or Tl).[165] A new mixed chloride bromide $Pt_6Cl_{12}Br_2$ is formed from $PtCl_2$ and Br_2; addition of bromine to platinum(II) complexes $PtCl_2L_2$ or mixing the appropriate platinum(IV) complexes results in halogen scrambling and a statistical distribution of $PtBr_xCl_{4-x}L_2$.[166] The salts K_2PtX_4 undergo thermal decomposition to give platinum(0) and K_2PtCl_6.[167]

10 Group IX

Copper.—There is still very considerable interest in cluster compounds of copper and in determining structures to throw some light on whether there is Cu—Cu bonding in the species. Amongst new groups described are a Cu_2 group in $[Cu(CF_3COO)_2(quinoline)]_2$; a triangular Cu_3 species in a basic 2-propylamino-2-methyl-3-butanoneoximato-derivative; Cu_4 groups in $Cu_4(ethylenethiourea)_9$-$(NO_3)_4,6H_2O$ and in $[Cu(CF_3OO)]_4,2C_6H_6$; and a Cu_8 group in $(Ph_4P)_4Cu_8$-$(1,2-dithiosquarato)_6$. It should be noted that cluster compounds occur in both copper(I) and copper(II) species,[168a] and that copper(I) carboxylates are dimeric in the vapour phase.[168b] Many polymeric halogenocuprate species have at least the possibility of Cu—Cu bonding. Reaction of borohydrides or tetrahydroaluminates with $LiCuMe_2$ or Cu^I gives complex hydrides of copper, probably

[160] H. Schmidbaur and J. Adlkofer, *Chem. Ber.*, 1974, **107**, 3680.
[161] H. Sabrowsky, W. Bronger, and D. Schmitz, *Z. Naturforsch.*, 1974, **29b**, 10.
[162] D. Cahen, J. A. Ibers, M. H. Mueller, and J. B. Wagner, jun., *Inorg. Chem.*, 1974, **13**, 110, 1377.
[163] Y. N. Kukushkin, V. V. Sibirskaya, S. D. Banzargashieva, and T. K. Mikha'chenko, *Russ. J. Inorg. Chem.*, 1973, **18**, 1385.
[164] J. Huster and W. Bronger, *Z. Naturforsch.*, 1974, **29b**, 594.
[165] E. Alter and R. Hoppe, *Z. anorg. Chem.*, 1974, **408**, 115.
[166] V. Bonora, M. Jawork, and M. F. Pilbrow, *J.C.S. Chem. Comm.*, 1974, 616; B. T. Heaton and K. J. Timmins, *ibid.*, 1973, 931.
[167] L. K. Shubochkin, M. A. Golubnichaya, E. F. Shubochkina, V. I. Gushchin, G. M. Larin, and V. A. Kolosov, *Russ. J. Inorg. Chem.*, 1973, **18**, 1735; 1974, **19**, 249.
[168] (a) J. A. Moreland and R. J. Doedens, *J.C.S. Chem. Comm.*, 1974, 28; A. L. Crumbliss, L. J. Gestaut, R. C. Rickard, and A. T. McPhail, *J.C.S. Chem. Comm.*, 1974, 545; P. F. Ross, R. K. Murmann, and E. O. Schlemper, *Acta Cryst.*, 1974, **B30**, 1120; P. F. Rodesiler and E. L. Amma, *ibid.*, p. 599; F. J. Hollander and D. Coucouvanis, *J. Amer. Chem. Soc.*, 1974, **96**, 5646; (b) T. Ogura and Q. Fernando, *Inorg. Chem.*, 1973, **12**, 2611.

MCuH$_2$.[169] Carbon monoxide is absorbed by Cu^I in BF$_3$–H$_2$O systems to give the [Cu(CO)$_4$]$^+$ ion.[170] Although cyanide ions normally reduce copper(II) to copper(I), in the presence of ligands such as 1,10-phenanthroline the Cu^{II} in stable, and apparently trigonal-bipyramidal complexes [CuL$_2$CN]X,nH$_2$O can be isolated.[171] CaCuSb, SrCuSb, and the corresponding bismuthides have a filled NiAs structure[172] whilst SrCuSn$_2$ and BaCuSn$_2$ have the BiOCl structure.[173] The copper(III) oxide SrLaCuO$_4$ has the K$_2$NiF$_4$ structure. The d^8 Cu^{3+} species has a low-spin state with linear O—Cu—O association.[174] Some of the Cu^{II} ions in copper-exchanged faujasite are in very exposed positions at the pore entrances to the supercage, and are effectively bonded to only one framework oxygen.[175] CuSiF$_6$,6H$_2$O has a very large unit cell but one quarter of the copper atoms are in regular octahedral co-ordination.[176] BaCu$_2$S$_2$ and BaCu$_2$Se$_2$ are isostructural and contain chains of CuX$_4$ tetrahedra linked by corner and edge-sharing.[177] A whole range of copper(II) salts of hexafluorometallates(III), M^{I}Cu^{II}M^{III}F$_6$ (M^{III} = e.g. Co or Rh) have been reported.[178] Cu(BF$_4$)(PPh$_3$)$_3$ is a copper(I) complex containing a weakly co-ordinated BF$_4^-$ group.[179] It is well known that the geometry of CuX$_4^{2-}$ groups can range from tetrahedral towards square planar, and the first authentic planar CuCl$_4^{2-}$ ion has been characterized in green bis-[N-methyl(2-phenylethyl)ammonium] tetrachlorocuprate(II), although the yellow modification has flattened tetrahedral co-ordination about copper. The Pr$^i_4$N salt contains both planar and distorted tetrahedral CuCl$_4^{2-}$ groups, and the dimeric Cu$_2$Cl$_6^{2-}$ ion in Ph$_4$AsCuCl$_4$ shows a distorted tetrahedral co-ordination.[180] Halogenocuprates of a higher degree of polymerization that have been reported include Cu(bipy)$_2$(CuCl$_2$)$_2$, which contains infinite chains of alternating linear CuCl$_2^-$ and tetrahedral CuCl$_4^{3-}$ groups together with some linear CuCl$_2^-$ ions,[181] and Cu$_3$Cl$_6$(H$_2$O)$_2$,-2(tetramethylenesulphone), which contains the trans-planar Cu$_3$Cl$_6$(H$_2$O)$_2$ complex, with bridging halogens.[182] It appears that iodocuprate(I) species may

[169] E. C. Ashby, T. F. Korenowski, and R. D. Schwarz, *J.C.S. Chem. Comm.*, 1974, 157; T. Yoshida and E.-I. Negishi, *ibid*., p. 762.
[170] Y. Matsushima, T. Koyano, T. Kitamura, and S. Wada, *Chem. Letters*, 1973, 433.
[171] M. Wicholas and T. Wolford, *Inorg. Chem.*, 1974, **13**, 316.
[172] B. Eisenmann, G. Cordier, and E. Schäfer, *Z. Naturforsch.*, 1974, **29b**, 457.
[173] N. May and H. Schäfer, *Z. Naturforsch.*, 1974, **29b**, 20.
[174] J. B. Goodenough, G. Demazeau, M. Pouchard, and P. Hagenmuller, *J. Solid State Chem.*, 1973, **8**, 325.
[175] I. E. Maxwell and J. J. de Boer, *J.C.S. Chem. Comm.*, 1974, 814.
[176] S. Ray, A. Zalkin, and D. H. Templeton, *Acta Cryst.*, 1973, **B29**, 2748.
[177] J. E. Iglesias, K. E. Pachali, and H. Steinfink, *J. Solid State Chem.*, 1974, **9**, 6.
[178] R. Hoppe and R. Jesse, *Z. anorg. Chem.*, 1973, **402**, 29.
[179] A. P. Gaughan, jun., Z. Dori, and J. A. Ibers, *Inorg. Chem.*, 1974, **13**, 1657.
[180] R. L. Harlow, W. J. Wells, G. W. Watt, and S. H. Simonsen, *Inorg. Chem.*, 1974, **13**, 2106; D. N. Anderson and R. D. Willett, *Inorg. Chim. Acta*, 1974, **8**, 167; R. D. Willett and C. Chow, *Acta Cryst.*, 1974, **B30**, 207.
[181] J. Kaiser, G. Brauer, F. A. Schröder, I. F. Taylor, and S. E. Rasmussen, *J.C.S. Dalton*, 1974, 1490.
[182] D. D. Swank and R. D. Willett, *Inorg. Chim. Acta*, 1974, **8**, 143.

be more highly polymerized than has previously been supposed, and species such as $Cu_4I_6^{2-}$ may be present.[183]

Silver.—$(CF_3CO_2Ag)_2C_6H_6$ has several most unusual features in its structure. Carboxylate groups and benzene bridge the two dissimilar silver atoms (one of these has planar three-co-ordination) and the silver–silver distance (2.893 Å) is short, although this probably arises from the nature of the bridges rather than from the presence of an Ag—Ag bond.[184] The compound (naphthalene)-$(AgClO_4)_4$,$4H_2O$ has a more normal structure, with weak naphthalene—Ag— $OClO_3$ bonding and the layers (aromatic compound)–$AgClO_4$–H_2O–$AgClO_4$– (aromatic compound). Macrocyclic tetra-aza-ligands stabilize the higher oxidation states of silver; Ag^{II} species are obtained by disproportionation of Ag^I derivatives and the Ag^{III} species can be obtained by oxidation.[185] The reduction of Ag^{II} by V^{IV} proceeds by way of an $[Ag \cdot VO]^{4+}$ intermediate although, in contrast to the corresponding reduction of ruthenium derivatives (see p. 261) these intermediates have not been isolated.[186] The preparation and structures of ternary and quarternary silver-containing sulphides (many of them ores) continue to be of interest. New double sulphides MAg_3S_2 (M = Rb or Cs) and $K_2Ag_4S_3$ have been reported; the effect of pressure on $AgAlS_2$ is to give a new phase with close-packed sulphur atoms, octahedral Al, and tetrahedral Ag. Aramoyite, $Ag(Sb,Bi)S_2$, has a distorted NaCl structure, whilst freislebenite, $PbAgSbS_3$, and samsonite, $(SbS_3)_2Ag_4Mn^{IV}$, both contain SbS_3 pyramids and triangularly co-ordinated AgS_3 groups.[187] Although $BaAgF_4$ contains planar AgF_4^{2-} groups, other fluoroargentates(II) contain octahedrally co-ordinated silver.[188]

Gold.—Gold cluster compounds continue to be described, and this year has seen structure determinations for $[Au_6\{P(p\text{-tolyl})_3\}_6][BPh_4]_2$, which contains a distorted Au_6 octahedron, and $[CpFe(\pi\text{-}C_5H_4)Au_2(PPh_3)_2][BF_4]$, containing an Au—Au—Fe chain.[189] It has become apparent that some polymeric gold derivatives do not contain metal clusters but instead contain many-membered rings containing several gold atoms.[190] $(Ph_3P)_2AuCl$ is the first compound for which the co-ordination about the Au^I has been found to be essentially planar.[191] The

[183] G. A. Bowmaker, L. D. Brockless, C. D. Earp, and R. Whiting, *Austral. J. Chem.*, 1973, **26**, 2593.
[184] G. W. Hunt, T. C. Lee, and E. L. Amma, *Inorg. Nuclear Chem. Letters*, 1974, **10**, 909; E. A. H. Griffith and E. L. Amma, *J. Amer. Chem. Soc.*, 1974, **96**, 743.
[185] E. K. Barefield and M. T. Mocella, *Inorg. Chem.*, 1973, **12**, 2829; D. Karweik, N. Winograd, D. G. Davis, and K. M. Kadish, *J. Amer. Chem. Soc.*, 1974, **96**, 591.
[186] E. Baumgartner and D. S. Hornig, *J. Inorg. Nuclear Chem.*, 1974, **36**, 196.
[187] J. Eyck, C. Burschka, and W. Bronger, *Naturwiss.*, 1973, **60**, 518; K.-J. Range, G. Engert, and A. Weiss, *Z. Naturforsch.*, 1974, **29b**, 186; D. J. E. Mullen and W. Nowacki, *Z. Krist.*, 1974, **139**, 54; T. Ito and W. Nowacki, *ibid.*, p. 85; A. Edenharter and W. Nowacki, *ibid.*, 1974, **140**, 87.
[188] R. H. Odenthal, D. Paus, and R. Hoppe, *Z. anorg. Chem.*, 1974, **407**, 144, 151.
[189] P. Bellon, M. Manassero, and M. Sansoni, *J.C.S. Dalton*, 1973, 2423; V. G. Andrianov, Yu. T. Struchkov, and E. R. Rossinskaya, *J. Struct. Chem.*, 1974, **15**, 65.
[190] G. Banditelli, G. Minghetti, and F. Bonati, *Inorg. Chem.*, 1974, **13**, 1600; *J.C.S. Chem. Comm.*, 1974, 88.
[191] N. C. Baenziger, K. M. Dittemore, and J. R. Doyle, *Inorg. Chem.*, 1974, **13**, 805.

structure of $AuTe_2X$ (X = Cl or I) shows corrugated two-dimensional $AuTe_2$ nets interconnected by halogen atoms.[192] Mass-spectroscopic studies of O_2AuF_6 provide fairly convincing evidence for the existence of AuF_5, probably as a polymer, in the vapour phase.[193] AuCl is a polymeric chain molecule with linear co-ordination about gold.[194] The oxide chloride AuOCl is formed from Au_2O_3 and concentrated hydrochloric acid; it decomposes thermally to Cl_2, O_2, and gold.[195]

11 Group X

Zinc and Cadmium.—Zinc hydride, ZnH_2, is formed free from anionic species from the 1 : 1 reactions between NaH and $ZnCl_2$, and LiH and $ZnBr_2$, and from the 2 : 1 reaction between NaH and ZnI_2; complex hydrides tend to be formed from other ratios of reactants or from other reactants, and are formed particularly readily from anionic zinc-containing organometallic compounds and aluminium hydrides.[196] The Ba—Cd system has been little studied previously, and the new compounds Ba_2Cd, $BaCd_5$, and BaCd are described.[197] Zinc and cadmium derivatives containing Ge—M or Sn—M bonds are readily formed by metathetical reactions, e.g. Ph_3GeH and Et_2Zn give $EtZnGePh_3$ and ethane.[198] $ZnAs_2$ contains $ZnAs_4$ units and $AsAs_2Zn_2$ units and, contrary to previous assumptions, does not contain Zn—Zn bonds.[199] The double chalcogenides Ba_2CdS_3, Ba_2CdSe_3, and $BaCdS_2$ have tetrahedral or trigonal-bipyramidal co-ordination about cadmium, and are similar to sulphides of other metals.[177] Fused zinc chloride has been the subject of extensive studies to determine reactivity and species resulting from dissolving oxo-anions.[200] Cadmium species in melts do not always follow the corresponding zinc systems; $CdCl_4^{2-}$ is the major species in molten $KAlCl_4$.[201] Solid $CdCl_2,2NaCl,3H_2O$ has a structure similar to that of $CdCl_2$, with sheets of chloride atoms and cadmium in octahedral holes; the sheets are held together by sodium ions and water molecules.[202]

Mercury.—Photoelectron spectroscopy provides direct evidence for the involvement of the 5d electrons in Hg^{II} compounds in covalent bonding, with significant

[192] H. M. Haendler, D. Mootz, A. Rabenau, and G. Rosenstein, *J. Solid State Chem.*, 1974, **10**, 175.
[193] A. J. Edwards, W. E. Falconer, J. E. Griffiths, W. A. Sunder, and M. J. Vasile, *J.C.S. Dalton*, 1974, 1129.
[194] J. Strähle and K.-P. Lörcher, *Z. Naturforsch.*, 1974, **29b**, 266; E. W. Janssen, J. C. W. Folmer, and G. A. Wieges, *J. Less-Common Metals*, 1974, **38**, 71.
[195] E. Schwarzmann, E. Schulze, and J. Mohn, *Z. Naturforsch.*, 1974, **29b**, 561.
[196] E. C. Ashby and J. J. Watkins, *Inorg. Chem.*, 1973, **12**, 2493; 1974, **13**, 2350.
[197] R. T. Dirstine, *J. Less-Common Metals*, 1975, **39**, 181.
[198] V. T. Bychkov, N. S. Vyazankin, and G. A. Razuvaev, *J. Gen. Chem. (U.S.S.R.)*, 1973, **43**, 792; G. S. Kalinina, O. A. Kruglaya, B. I. Petrov, E. A. Shchupak, and N. S. Vyazankin, *ibid.*, p. 2215.
[199] M. E. Fleet, *Acta Cryst.*, 1974, **B30**, 122.
[200] D. H. Kerridge and I. A. Sturton, *Inorg. Chim. Acta*, 1973, **7**, 701; 1974, **8**, 27, 31, 37.
[201] J. H. R. Clarke and P. J. Hartley, *J. Phys. Chem.*, 1974, **78**, 595.
[202] R. Boistelle, G. Pèpe, B. Simon, and A. Leclaire, *Acta Cryst.*, 1974, **B30**, 2200.

$d\pi \rightarrow \pi^*$ back donation in the Hg—CN bond.[203] In the presence of traces of arsenic, mercury crystallizes with a hexagonal lattice.[204] A novel mercury cluster compound $Hg_{2.86}AsF_6$, is formed in the reaction between mercury and AsF_5 in SO_2; it contains metallically bonded infinite-chain cations.[205] The oxidation of Hg_2^{2+} is now well understood: one-electron oxidants give $Hg^I + Hg^{II}$, with subsequent fast oxidation of Hg^I, whereas two-electron oxidants give $Hg^O + Hg^{II}$ initially.[206] A whole range of new phases has been identified in the Sr–Hg system.[207] Compounds containing Sn–Hg linkages are formed from Bu^t_2Hg and R_2SnH; they are unstable to air but useful in preparing Sn or Hg linkages to other elements.[208] Further examples continue to be found of heteronuclear polymeric cations containing mercury. $[O(HgI)_2HgOH]^+$ ions are found in the perchlorate, and the compounds $Hg_2Ni_2F_6X$ (X = O or S) contain $[Hg_2E]^{2+}$ units.[209] $Hg_3O_2Cl_2$ contains three-co-ordinate and two-co-ordinate mercury, and the O and Cl atoms are approximately close-packed.[210] There is currently considerable controversy as to the nature of HgX_2 species in aromatic hydrocarbons, with the possibilities being that linear monomers or associated species are present.[211] The isolation of solid complexes (and incidentally the knowledge that chloride complexes are less stable than $HgBr_2$ or HgI_2 complexes) may permit structural determinations on these complexes.

[203] P. Burroughs, S. Evans, A. Hamnett, A. F. Orchard, and N. V. Richardson, *J.C.S. Chem. Comm.*, 1974, 921.
[204] M. Pušelj and Z. Ban, *J. Less-Common Metals*, 1974, **37**, 213.
[205] I. D. Brown, B. D. Cutforth, C. G. Davies, R. J. Gillespie, P. R. Ireland, and J. E. Vekris, *Canad. J. Chem.*, 1974, **52**, 79.
[206] R. Davies, B. Kipling, and A. G. Sykes, *J. Amer. Chem. Soc.*, 1973, **95**, 7250.
[207] G. Bruzzone and F. Merlo, *J. Less-Common Metals*, 1974, **35**, 153.
[208] U. Blaukat and W. P. Neumann, *J. Organometallic Chem.*, 1973, **63**, 27.
[209] K. Köhler, G. Thiele, and D. Breitinger, *Angew. Chem. Internat. Edn.*, 1974, **13**, 545; F. Champion, D. Bernard, J. Pannetier, and J. Lucas, *Compt. rend.*, 1974, **278**, C, 1185.
[210] K. Aurivillius and C. Stålhandske, *Acta Cryst.*, 1974, **B30**, 1907.
[211] I. M. Vezzosi, G. Peyronel, and A. F. Zanoli, *Inorg. Chim. Acta*, 1974, **8**, 229; I. Eliezer and G. Algavish, *ibid.*, 1974, **9**, 257; C. L. Cheng, R. K. Pierens, D. V. Radford, and G. L. D. Ritchie, *J. Chem. Phys.*, 1973, **59**, 5209.

11 The Transition Elements

Part II: Ligands and Complexes

By S. M. NELSON

Department of Chemistry, Queen's University of Belfast, Belfast BT9 5AG

1 Dinitrogen and Related Ligands

An article by Sellman[1] summarizes much of the recent work on the properties of dinitrogen complexes and discusses the outlook for catalytic reduction of co-ordinated dinitrogen under mild conditions. In the dinitrogen complex of permethyltitanocene $(\eta^5\text{-}C_5Me_5)_2Ti(N_2)$ there is evidence for two types of bonding for the dinitrogen to the metal, *monohapto* (end-on) and *dihapto* (π), in rapid equilibrium in solution in heptane above 208 K.[2] The related complex $[(\eta^5\text{-}C_5Me_5)_2Zr]_2(N_2)_3$, thought to contain both terminally bound and bridging N_2, yields a stoicheiometric quantity of hydrazine on treatment with an excess of HCl at 193 K in toluene and subsequent warming to room temperature.[3] X-Ray analysis[4] of $[W(N_2H_2)Cl(diphos)_2]BPh_4$ has demonstrated the essential linearity of $W-N(1)-N(2)$. This precludes protonation at $N(1)$, in its formation from $W(N_2)_2(diphos)_2$, and establishes the ligand as the *monohapto* hydrazido(2 −) form $W=N\cdots NH_2$. The protonation and reduction of terminally co-ordinated dinitrogen in trans-$[M(N_2)_2(diphos)_2]$ (M = Mo or W) or cis-$[W(N_2)_2(PPhMe_2)_4]$ by HX to diazene and hydrazido-complex salts is the first chemical evidence that N_2 is as likely to be reduced at a mono- as at a bi-metal site in nitrogenase.[5] Nitrogenase model systems composed of molybdate and thiol ligands such as L-(+)-cysteine reduce N_2 to NH_3 slowly in the presence of $NaBH_4$. In the presence of substrate amounts of ATP, the reduction is significantly stimulated but leads to accumulation of di-imide, which disproportionates but is not itself reduced as such.[6] It has been claimed[7a] that the reaction of HCl and trans-$[Mo(N_2)_2(dpe)_2]$ (dpe = 1,2-diphenylphosphinoethane) with the cluster anion $[Fe_4S_4(SEt)_4]^{2-}$ gives NH_3 in low yield. However, in a re-investigation[7b] in

[1] D. Sellman, *Angew. Chem. Interna*. *Edn.*, 1974, **13**, 639.
[2] J. E. Bercaw, E. Rosenberg, and J. D. Roberts, *J. Amer. Chem. Soc.*, 1974, **96**, 612; J. E. Bercaw, *ibid.*, p. 4087.
[3] J. M. Manriquez and J. E. Bercaw *J. Amer. Chem. Soc.*, 1974, **96**, 6230.
[4] G. A. Heath, R. Mason, and K. M. Thomas, *J. Amer. Chem. Soc.*, 1974, **96**, 259.
[5] J. Chatt, G. A. Heath, and R. L. Richards, *J.C.S. Dalton*, 1974, 2074.
[6] G. N. Schrauzer, G. W. Kiefer, K. Tano, and P. A. Doemeny, *J. Amer. Chem. Soc.*, 1974, **96**, 641; see also J. Chatt and J. R. Dilworth, *J.C.S. Chem. Comm.*, 1974, 517.
[7] (a) E. E. van Tamelen, J. A. Gladysz, and C. R. Brulet, *J. Amer. Chem. Soc.*, 1974, **96**, 3020; (b) J. Chatt, C. M. Elson, and R. L. Richards, *J.C.S. Chem. Comm.*, 1974, 189.

another laboratory no NH_3 was found, the main products being N_2 (in quantitative yield), H_2S, EtSH, H_2, iron chlorides, and $[MoCl_2(dpe)_2]^+$. The complexes mer-$[Re(S_2CNR_2)(N_2)(PMe_2Ph)_3]$ (R = Me or Et) have been prepared from trans-$[ReCl(N_2)(PMe_2Ph)_4]$. With excess of HCl or HBr in THF the dihydro-complexes $[ReH_2(S_2CNR_2)X(PMe_2Ph)_3]X$ are produced.[8] Two bridging dinitrogen ligands occur in the trinuclear complex $[ReCl(N_2)(PMe_2Ph)_4]_2$-$TiCl_4$, indicating the Lewis-base properties of the co-ordinated N_2.[9] A linear Mo—N—N—Re bridge occurs in the bimetallic complex $[(MeO)Cl_4MoN_2$-$ReCl(PMe_2Ph)_4]$.[10] The complex $[Re(N_2)Cl(LL)_2]$ could be prepared where LL was $Ph_2P(CH_2)_2AsPh_2$ but not $Ph_2As(CH_2)_2AsPh_2$.[11] The metal is protonated on addition of HBF_4 to μ-dinitrogen-bis[1,2-bis(dimethylphosphino)-ethane]hydrido-[η-(1,3,5-trimethylbenzene)]molybdenum, the N_2 remaining co-ordinated.[12] Oxidative addition to N_2-containing complexes usually leads to loss of N_2, but the reaction of $[IrCl(N_2)(PPh_3)_2]$ with methyl trifluoromethane-sulphonate gives $[Ir\{OS(O)_2CF_3\}(CH_3)Cl(N_2)(PPh_3)_2]$.[13] Treatment of $[Ru(en)_3]^{3+}$ (en = ethylenediamine) with NO in alkaline solution results in production of the N_2 complex $[Ru(en)_2(N_2)(OH_2)]BF_4$, a reaction involving cleavage of a C—N bond of co-ordinated en.[14] A dinitrogen complex of an Os^{II}–porphyrin complex has been reported.[15] Dinitrogen has a greater affinity for the $RuH_2(PPh_3)_3$ moiety than has 2-pentene.[16] Pale yellow $Ni(PPh_3)_4$ dissociates under argon in organic solvents to purple $Ni(PPh_3)_3$; exposure to N_2 produces a colour change, and solutions exhibit an i.r. absorption at ~ 2070 cm^{-1}. No complexes could be isolated.[17] The complex $RhCl(PCy_3)_2$ (Cy = cyclohexyl) spontaneously adds N_2 to produce trans-$RhCl(N_2)(PCy_3)_2$.[18]

The substituted hydrazide group is asymmetrically bonded to two (π-C_5H_5)Mo(NO)I units; Mo(1) is bound to the terminal N atom and Mo(2) to both N atoms.[19] I.r. and n.m.r. studies[20] have indicated that 1,3-diaryltriazenes react with hydrido complexes or phosphine complexes of the platinum metals in low oxidation states to afford products containing unidentately co-ordinated or chelated $[ArNNNAr]^-$. The latter bonding mode occurs in $[Co(PhNNNPh)_2]$.[21] Transition-metal complexes of aryldiazonium groups can exist in three modes

8 J. Chatt, R. H. Crabtree, J. R. Dilworth, and R. L. Richards, *J.C.S. Dalton*, 1974, 2358.
9 R. Robson, *Inorg. Chem.*, 1974, **13**, 475.
10 M. Mercer, *J.C.S. Dalton*, 1974, 1637.
11 D. J. Darensburg and D. Madrid, *Inorg. Chem.*, 1974, **13**, 1532.
12 M. L. H. Green and W. E. Silverthorn, *J.C.S. Dalton*, 1974, 2164.
13 D. M. Blake, *J.C.S. Chem. Comm.*, 1974, 815.
14 S. Pell and J. N. Armor, *J.C.S. Chem. Comm.*, 1974, 259.
15 J. W. Buchler and P. D. Smith, *Angew. Chem. Internat. Edn.*, 1974, **13**, 745.
16 F. Pennella, *J. Organometallic Chem.*, 1974, **65**, C17.
17 C. A. Tolman, D. H. Gerlach, J. P. Jesson, and R. A. Schunn, *J. Organometallic Chem.*, 1974, **65**, C23.
18 H. L. M. van Gaal, F. G. Moers, and J. J. Steggerda, *J. Organometallic Chem.*, 1974, **65**, C43.
19 W. G. Kita, J. A. McCleverty, B. E. Mann, D. Seddon, G. A. Sim, and D. I. Woodhouse, *J.C.S. Chem. Comm.*, 1974, 132.
20 K. R. Laing, S. D. Robinson, and M. F. Ulley, *J.C.S. Dalton*, 1974, 1205.
21 M. Corbett and B. F. Hoskins, *Austral. J. Chem.*, 1974, **27**, 667.

$$R-\overset{+}{N}\equiv N-M \overset{H^+}{\longrightarrow} \text{No reaction}$$

(1)

$$\underset{R}{\overset{\cdot\cdot}{N}}=\overset{-}{N}=M \overset{H^+}{\longrightarrow} \underset{R}{\overset{H}{\underset{\diagdown}{\overset{+}{N}}}}=\overset{+}{N}=M$$

(2)

$$\underset{R-\overset{\cdot\cdot}{N}}{\overset{N-M}{\diagup}} \overset{H^+}{\longrightarrow} \underset{R-N}{\overset{H}{\underset{\diagdown}{\overset{N-M}{\diagup}}}}$$

(3)

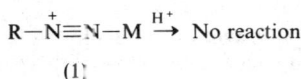

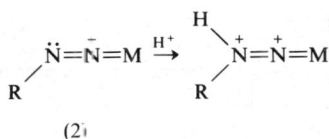

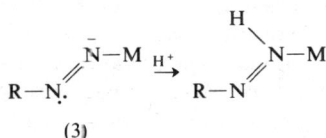

of co-ordination, for which the predominant VB structures (1)—(3) may be written, and for which different behaviour towards H^+ may be expected. An X-ray structural analysis[22] of $[Pt(PEt_3)_2(HNNC_6H_4F)Cl]ClO_4$, an intermediate in a proposed model system for nitrogen fixation, and formed by protonation of the aryldiazonium complex of type (3), has confirmed the site of protonation as the metal-bonded nitrogen $N(1)$; the $Pt-N(1)-N(2)$ bond angle is $125.3(6)°$. In contrast, treatment of mer-$[ReCl_3(PMe_2Ph)_3]$ with phenylhydrazine affords $ReCl_2(HN_3)(N_2Ph)(PMe_2Ph)_2$, which with aqueous HBr gives $[ReCl_2(NH_3)-(NNHPh)(PMe_2Ph)_2]Br$. In this reaction the protonation occurs at $N(2)$, as shown by the linearity of $Re-N(1)-N(2)$ as well as by difference electron-density synthesis.[23] The aryldiazo-cations $[Fe(N_2Ph)(CO)_2L_2]^+$ (L = tertiary phosphine) are inert to substitution of CO by excess of L.[24] This contrasts with the displacement of CO by L in corresponding nitrosyl complexes and is attributed to a smaller π-acceptor capacity on the part of PhN_2^+ than of the isoelectronic NO^+. Some aryldiazo-complexes and aryldi-imine complexes of Ru[25] and Pt[26] have been described. Interaction of $(\eta-C_5H_5)MH_2$ (M = Mo or W) with azo- and diaza-complexes proceeds via an incipient formation of hydridohydrazino-complexes $(\eta-C_5H_5)_2MH(\sigma-NRNHR)$ to π-complexes, e.g. $(\eta-C_5H_5)_2M(RN=NR)$.[27] The 1 : 1 adduct of CuCl and 2,3-diazabicyclo[2,2,1]hept-2-ene is a linear polymer with the Cu atoms bridged by both nitrogen and chlorine atoms.[28] 5-Substituted tetrazoles are ambidentate and may co-ordinate to Pd^{II} via $N(1)$ or $N(2)$.[29] The

[22] S. D. Ittel and J. A. Ibers, *J. Amer. Chem. Soc.*, 1974, **96**, 4804.

[23] R. Mason, K. M. Thomas, J. A. Zubieta, P. G. Douglas, A. R. Galbraith, and B. L. Shaw, *J. Amer. Chem. Soc.*, 1974, **96**, 260.

[24] W. E. Carroll, M. E. Deane, F. A. Deeney, and F. J. Lalor, *J.C.S. Dalton*, 1974, 1430, 1837.

[25] S. Cerini, F. Porta, and M. Pizzotti, *Inorg. Nuclear Chem. Letters*, 1974, **10**, 983.

[26] A. W. B. Garner and M. J. Mays, *J. Organometallic Chem.*, 1974, **67**, 173.

[27] A. Nakamura, M. Aokate, and S. Otsuka, *J. Amer. Chem. Soc.*, 1974, **96**, 3456.

[28] G. S. Chandler, C. L. Raston, G. W. Walker, and A. H. White, *J.C.S. Dalton*, 1974, 1797.

[29] D. A. Redfield, J. H. Nelson, R. A. Henry, D. W. Moore, and H. B. Jonassen, *J. Amer. Chem. Soc.*, 1974, **96**, 6298.

2-thiazolylazo-2-naphthalato-anion is terdentate in its complexes with Pd^{II} and Ni^{II}, using the phenolic oxygen and the azo and thiazole nitrogens.[30]

2 Nitrosyls

Interest in nitrosyl complexes has been stimulated in recent years by the recognition that NO may co-ordinate to transition metals in either a linear or bent manner. Valence Bond and MO approaches have been employed to account for the different bonding behaviour of NO in different situations. The most widely held current interpretation associates a linear or near-linear $M—N—O$ linkage with the NO^+ ligand configuration and a severely bent $M—N—O$ arrangement with NO^-. Nitric oxide is thus analogous to the aryldiazonium moiety in this respect. Enemark and Feltham[31] have given a MO description of the effect of stereochemistry (five-co-ordinate or six-co-ordinate) of the complex on the nature and geometry of the $M—NO$ bond. A rather different VB approach has also been suggested.[32] A MO treatment of trigonal-bipyrimidal complexes has led to the conclusion that the greater the σ- and π-donating capacity of associated ligands in the equatorial plane the more likely is the NO ligand to adopt a bent conformation.[33] The cation $[IrH(NO)(PPh_3)_3]^+$ has a distorted trigonal-bipyramidal structure with NO, H, and one PPh_3 group in the trigonal plane;[34] here, a linear $M—NO$ occurs in accord with the formulation of the nitrosyl group as NO^+. The nearly linear $M—NO$ configuration is also present in octahedral $RuCl_2(NO)(PMePh_2)_2$ (176.4°)[35] and in the distorted tetrahedral complex $Ru(NO)_2(PPh_3)_2$ (170—178°).[36] The data for the latter compound support its formulation as a d^{10} NO^+ complex of Ru^{-II}. In the distorted tetrahedral complex $CoI(NO)_2(LL')$, where L = the mono-oxide of 1,2-diphenylphosphinoethane, both nitrosyl groups are disordered between two positions, all corresponding to distinctly bent $M—N—O$ linkages.[37] It may be that many of the so-called '20 electron' dinitrosyls may be better formulated as '18-electron' N,N-bonded cis-dinitrogen dioxide complexes in which there is a significant $N\cdots N$ interaction.[38,39] Thus, $IrBr(NO)_2L_2$ (L = phosphine), for example, would become $IrBr(N_2O_2)L_2$. Several cobaloxime nitrosyls react with O_2 to afford corresponding nitrato-(O-bonded) complexes and, in smaller yield, nitro-complexes.[40] The new complexes $Re(NO)(CO)_2(PPh_3)_2$ and

[30] M. Kurahashi, Bull. Chem. Soc. Japan, 1974, 47, 2045.
[31] J. H. Enemark and R. D. Feltham, J. Amer. Chem. Soc., 1974, 96, 5002, 5004.
[32] R. D. Harcourt and J. A. Bowden, J. Inorg. Nuclear Chem., 1974, 36, 1115.
[33] R. Hoffman, M. M. L. Chen, M. Elian, A. Rossi, and D. M. P. Mingos, Inorg. Chem., 1974, 13, 2666.
[34] C. R. Clark, J. M. Waters, and K. R. Whittle, Inorg. Chem., 1974, 13, 1628.
[35] A. J. Schultz, R. L. Henry, J. Reed, and R. Eisenberg, Inorg. Chem., 1974, 13, 732.
[36] A. P. Gaughan, B. J. Corden, R. Eisenberg, and J. A. Ibers, Inorg. Chem., 1974, 13, 786.
[37] J. S. Field, P. J. Wheatley, and S. Bhaduri, J.C.S. Dalton, 1974, 74.
[38] B. L. Haymore and J. A. Ibers, J. Amer. Chem. Soc., 1974, 96, 3325.
[39] S. Bhaduri, B. F. G. Johnson, C. J. Savoy, J. A. Segal, and R. H. Walter, J.C.S. Chem. Comm., 1974, 809.
[40] W. C. Taylor and L. G. Marzilli, Inorg. Chem., 1974, 13, 1008.

$ReH(NO)_2(PPh_3)_2$ react with halogens to form new halogenonitrosyl complexes.[41] The five-co-ordinate complex $ReCl_2(NO)(PPh_3)_2$ is paramagnetic. It adds on other ligands L to form six-co-ordinate diamagnetic complexes.[42] The nitrosyl in these compounds appears inert to both electrophilic and nucleophilic attack. Thionitrosyl complexes $Mo(NS)(S_2CNR_2)_3$ have been reported.[43] These are produced in high yield from the nitrido-complexes $MoN(S_2CNR_2)_3$ on reaction with sulphur; on treatment with tributylphosphine the sulphur is removed and the nitrido-compound regenerated.

3 Nitrogen Ligands

New Rh^I complexes of the three-membered cyclic ligand aziridine (ethyleneimine) have been reported.[44] The hexakis(hydroxylamine)nickel(II) cation has an octahedral NiN_6 co-ordination polyhedron.[45] Amides usually bond to metal ions via the oxygen atom but i.r. spectra suggest N-co-ordination in MCl_2L_4 ($M = Co^{II}$, Ni^{II}, or Cu^{II}).[46] A number of papers by Preti et al. on the complexing behaviour of ambidentate ligands such as thiomorpholin-3-one, thiomorpholine-3-thione, and thiazolidine-2-thione and -2-selenone have appeared.[47] With the first-row transition-metal ions, nitrogen appears to be the preferred donor atom, as expected, but exceptionally the Group VI donor atom is employed. The metal atom has a distorted pentagonal-bipyramidal stereochemistry in two Ta^V complexes of the NN-dialkylacetamidinato-anion,[48] and in the methylimido-complex $trans$-$[Re(NMe)(MeNH_2)_4Cl]^{2+}$ the Re—N bond is considered to have double-bond character.[49] A variety of stereochemistries has been reported for the transition-metal complexes of 2-(N-morpholinylmethyl)-3-quinuclidinone and related ligands.[50,51] The nitrile-nitrogen lone pair is used in the bonding of various aryl nitriles,[52] aliphatic dinitriles,[53] cyanopyridines,[52] and 2-cyano-8-hydroxyquinoline[54] to several transition-metal ions. In the case of the cyanopyridines, evidence for ambidentate behaviour was observed in some cases. Co-ordinated nitrile is susceptible to hydrolysis to amido-compounds.

[41] G. La Monica, M. Freni, and S. Cenini, *J. Organometallic Chem.*, 1974, **71**, 57.
[42] R. W. Adams, J. Chatt, N. E. Hooper, and G. J. Leigh, *J.C.S. Dalton*, 1974, 1075.
[43] J. Chatt and J. R. Dilworth, *J.C.S. Chem. Comm.*, 1974, 508.
[44] M. R. Hoffman and J. O. Edwards, *Inorg. Nuclear Chem. Letters*, 1974, **10**, 837.
[45] L. M. Englehardt, P. W. G. Newman, C. L. Raston, and A. H. White, *Austral. J. Chem.*, 1974, **27**, 503.
[46] M. G. Barvinok and L. V. Mashkov, *Russ. J. Inorg. Chem.*, 1974, **19**, 310.
[47] C. Fregni, C. Preti, G. Tosi, and G. Verani, *J. Inorg. Nuclear Chem.*, 1974, **36**, 3695; C. Preti and G. Tosi, *Canad. J. Chem.*, 1974, **52**, 2845.
[48] M. G. B. Drew and J. D. Wilkins, *J.C.S. Dalton*, 1974, 1579, 1973.
[49] R. S. Shandles, R. K. Murmann, and E. O. Schlemper, *Inorg. Chem.*, 1974, **13**, 1373.
[50] R. C. Dickinson and G. J. Long, *Inorg. Chem.*, 1974, **13**, 262; *J. Inorg. Nuclear Chem.*, 1974, **36**, 1235.
[51] G. J. Long and E. O. Schlemper, *Inorg. Chem.*, 1974, **13**, 279.
[52] R. J. Balahura, *Canad. J. Chem.*, 1974, **52**, 1762.
[53] D. L. Greene, P. G. Sears, and J. Kripkendorf, *Inorg. Nuclear Chem. Letters*, 1974, **10**, 894.
[54] C. R. Clark and R. W. Hay, *J.C.S. Dalton*, 1974, 2148.

There have been many attempts to prepare simple substitution products of $Fe(CO)_5$ and $Fe_2(CO)_9$ with nitrogen donor ligands. These have always led to redox products containing carbonylate anions. The pyridine and pyrazine tetracarbonylirons have now been prepared in high yield by refluxing the amine and $Fe_2(CO)_9$ in dry THF under CO flush at room temperature.[55] The intermediate is believed to be $Fe(CO)_4(THF)$. An X-ray study of these complexes has shown that the nitrogen occupies an axial position of a trigonal bipyramid. ^{13}C n.m.r. spectra of the pyridine complex indicate that it is fluxional in solution down to 183 K. Complex formation between metal ions and alk-2-enyl-substituted pyridines has been studied. Where the metal is Ag^+, chelation is observed when the alkene function is two or more atoms removed from the nitrogen.[56a] 2-Allyl and 2-(1-methylallyl) groups may be uni- or bi-dentate in their complexes with Pt^{II}, depending on the preparative conditions.[56b] A particularly interesting series of bidentate ligands are those represented in (4), in which the pyridine nitrogen and the terminal olefinic group span *trans*-positions in the square-planar complexes with $PtCl_2$; the large rings may have 10, 11, 13, or 17 member atoms.[57]

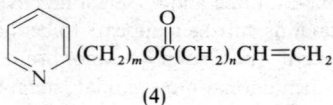

$$\text{pyridine}\quad (CH_2)_mOC(CH_2)_nCH{=}CH_2$$

(4)

Hitherto neutral imidazole has been regarded as bonding exclusively through the tertiary(imine) nitrogen. New studies have now shown that it possesses ambidentate character and may bond alternatively *via* the amine nitrogen or *via* C(2), the intervening carbon atom. Adducts of imidazole and N-methylimidazole with $Ni(acac)_2$, $Co(acac)_2$ (acac = acetylacetonate), and several Co^{II} Schiff bases undergo rapid intermolecular exchange on the n.m.r. time-scale in solution in the presence of excess ligand.[58] From an analysis of the n.m.r. spectra of the bases and a study of the temperature dependence of the n.m.r. shifts it was concluded that two types of adduct are present, one bonded *via* N(1) and the other *via* N(3). Similar behaviour was noted for pyrazole. Good yields of Ru^{II} and Ru^{III} complexes having imidazole bonded *via* N(3) (the imine nitrogen) were obtained by the reaction of the base with $[Ru(NH_3)_5(H_2O)]^{2+}$ followed by air oxidation.[59] Below pH ~ 2 the Ru^{II} complexes revert to the aquopentammine, the aquation leading to low conversion to derivatives of $Ru(NH_3)_4$ in which an imidazole ligand is bound through C(2). 4-Methyl imidazole and benzimidazole behave similarly. An analogous C-bound complex was the principal reaction product of $[Ru(NH_3)_5(H_2O)]^{2+}$ with 4,5-dimethyl-

[55]   F. A. Cotton and J. M. Troup, *J. Amer. Chem. Soc.*, 1974, **96**, 1233, 3438.
[56]   (a) M. Israeli, D. K. Laing, and L. D. Pettit, *J.C.S. Dalton*, 1974, 2194; (b) B. T. Heaton and D. J. A. McCaffrey, *J. Organometallic Chem.*, 1974, **70**, 455.
[57]   J. C. Chottard, E. Mulliez, J. P. Girault, and D. Mansuy, *J.C.S. Chem. Comm.*, 1974, 780; see also Ref. 125.
[58]   B. S. Tovrog and R. S. Drago, *J. Amer. Chem. Soc.*, 1974, **96**, 2743.
[59]   R. J. Swindberg, R. F. Bryan, I. F. Taylor, and H. Taube, *J. Amer. Chem. Soc.*, 1974, **96**, 381.

imidazole. An X-ray study of one derivative having C-bonded imidazole has been carried out.[59] A novel type of nine-membered Au_3N_6 ring structure involving bridging pyrazolyl units has been suggested[60] for the reaction product of $AuCl(SMe_2)$ with 1-substituted pyrazoles in methanolic KOH. The first flavin complexes of molybdenum have been reported. These have the general formula $MOCl_3(Hfl)$, where Hfl = a monocation which is chelated *via* the primary binding site, N(5) and C=O(4).[61] $[Ru(NH_5)_5(H_2O)]^{2+}$ shows a unique selectivity for heterocyclic nitrogen, and this fact has been used to synthesize a number of purine complexes with Ru bound to N(7).[62] Both N(1) and N(7) donor atoms are involved in co-ordination of adenosine and purine riboside to two different Pt atoms in solutions of [Pt(dien)Cl]Cl in D_2O when the reactants are present in 1 : 1 ratio.[63] In cytidine, N(3) is the binding site, whereas uridine does not interact with Pt under the same conditions.

2,2'-Bipyridyl and 1,10-phenanthroline continue to be much-studied ligands. New, or re-investigated, complexes include those of In^{III}, Fe^{II}, Rh^I, and Pt^{II}.[64] In *cis*-PtCl(phen)(PEt_3)_2 the ratio of the axial Pt—N to the equatorial Pt—N distances is unusually large (1.33), indicating virtual unidentate co-ordination of 1,10-phenanthroline.[65] The nickel atom has a formal oxidation state of 1.5 in the dinuclear cation $[Ni_2(naph)_4X_2]^+$ of 1,8-naphthyridine (X = halide, NCS, or NO_3).[66] The co-ordination polyhedron consists of distorted bicapped prisms in which each Ni atom is bound to four N atoms of four bridging naph molecules. The Ni···Ni distance is 2.41 Å and the magnetic moment of the dimeric unit is 4.2 BM, corresponding to three unpaired electrons. Bisdimethylbis-(1-pyrazolyl)-gallatonickel(II) contains two six-membered Ge(N—N)Ni rings in the boat conformation, with a planar arrangement of the four nitrogen atoms.[67] Further studies on the geometry and spin-state of Fe^{II} complexes of 2-(2-pyridyl)imidazole, dipyrromethenes, and terdentate ligands based on 2-substituted 1,10-phenanthrolines have been reported.[68] Oxidation of aqueous Ni^{2+} and 2,6-diacetyl-pyridine dioxime yields the Ni^{IV} complex; the Ni—N bonds are shorter by 0.17 Å than in corresponding Ni^I complexes.[69] Bis-(N-picolinoyl)-3-amino-1-propoxidoaquo copper(II) dihydrate is dimeric and contains a four-membered

[60] F. Bonati, G. Minghetti, and G. Banditelli, *J.C.S. Chem. Comm.*, 1974, 88.
[61] J. Selbin, J. Sherrill, and C. H. Bigger, *J.C.S. Dalton*, 1974, 2544.
[62] M. J. Clark and H. Taube, *J. Amer. Chem. Soc.*, 1974, **96**, 5413; H. I. Heitner and S. J. Lippard, *Inorg. Chem.*, 1974, **13**, 815.
[63] P.-C. Kong and T. Theophanides, *Inorg. Chem.*, 1974, **13**, 1167, 1981; G. R. Clark and J. D. Orbell, *J.C.S. Chem. Comm.*, 1974, 139.
[64] J. G. Contreras, F. W. Einstein, and D. G. Tuck, *Canad. J. Chem.*, 1974, **52**, 3793; W. M. Reiff and G. J. Long, *Inorg. Chem.*, 1974, **13**, 2150; G. Mestroni, A. Cannas, and G. Zassinovich, *J. Organometallic Chem.*, 1974, **65**, 119; R. S. Osborn and D. Rogers, *J.C.S. Dalton*, 1974, 1002.
[65] G. W. Bushnell, K. R. Dixon, and M. A. Khan, *Canad. J. Chem.*, 1974, **52**, 1367.
[66] L. Sacconi, C. Mealli, and D. Gatteschi, *Inorg. Chem.*, 1974, **13**, 1985.
[67] D. F. Rendle, A. Storr, and J. Trotter, *J.C.S. Chem. Comm.*, 1974, 406.
[68] Y. Sasaki and T. Shigematsu, *Bull. Chem. Soc. Japan*, 1974, **47**, 109; Y. Marakami, Y. Matsuda, K. Sakata, and K. Harada, *ibid.*, p. 458; H. A. Goodwin and D. W. Mather, *Austral. J. Chem.*, 1974, **27**, 965, 2121; *Inorg. Nuclear Chem. Letters*, 1974, **10**, 99.
[69] G. Sprout and G. D. Stucky, *Inorg. Chem.*, 1973, **12**, 2898.

chelate ring.[70] Complexes of terdentate polyamines having five,six-membered condensed chelate rings have higher heats of formation than corresponding complexes having five,five- or six,six-membered chelate rings.[71] Complexes of the terdentate 1,3-bis(pyridylimino)isoindolines have been prepared by the template method, starting from phthalonitrile, an arylamine, and nickel(II) acetate.[72] I.r. and n.m.r. exchange studies in D_2O have indicated that the alcoholic hydroxy-group is unco-ordinated in the bis-complexes $[Co^{III}L_2]^-$ of the potentially quadridentate N-hydroxyethyliminodiacetic acid. Moreover, the hydroxy-group can be acetylated without dissociation of the complex.[73] However, the ligand is deprotonated and quadridentate in Co(en)L. A somewhat similar situation obtains in the Cu^{II} complex of 2,6-bis(N-2'-aminoethylaminomethyl)-p-cresol. This phenolate anion is quinquedentate, the metal having an N_4O donor set.[74]

Amongst the linear quinquedentate nitrogen-donor ligands reported are 1,8-bis-(2'-pyridyl)-3,6-diazaoctane (in a distorted square-planar Cu^{II} complex)[75] and NN'-di-(3-aminopropyl)piperazine (in a square-pyramidal Ni^{II} complex).[76] The conformations of linear tetramines are dependent on chelate ring size. Both cis- and $trans$-forms of $[Rh(tet)Cl_2]Cl$ have been isolated for 2,3,2-tet but only $trans$-forms for 3,2,3-tet and 3,4,3-tet.[77] The ^{129}I Mössbauer spectra[78] of the two I atoms in the Ni^{II} and Zn^{II} complexes of the tripod ligand Me_6tren seem contrary to their earlier formulation as trigonal-bipyrimidal salts, $[M(Me_6tren)I]I$.

A large number of complexes of Schiff-base ligands have been reported, with emphasis on synthesis and structure. Usually the Schiff base complexes as a chelating anion. An interesting exception is the product of reaction in 90% yield of two moles of N-methylsalicylaldimine with bis(cyclo-octa-1,5-diene) nickel(o) in toluene at room temperature.[79] The two molecules of base disproportionate into the anion and a protonated species. The latter is bound to the Ni^O via a $C=N$ π-bond, and the metal is trigonal planar if this bond is taken as one co-ordinating site. Pd(salen) has been suggested as a suitable model for hydrogenase.[80] $Na[Co^I(salen)]$ is a reversible CO_2 carrier.[81] The Schiff base

[70] J. A. Bertrand, E. Fujita, and P. G. Elder, *Inorg. Chem.*, 1974, **13**, 2067.
[71] R. Barbucci, L. Fabrizzi, and P. Paoletti, *J.C.S. Dalton*, 1974, 2403.
[72] W. O. Siegl, *Inorg. Nuclear Chem. Letters*, 1974, **10**, 825.
[73] P. Horrigan, R. A. Canelli, J. R. Kashmann, C. A. Hoffman, R. Bauer, D. R. Boston, and J. C. Bailar, *Inorg. Chem.*, 1974, **13**, 1108.
[74] I. E. Dickson and R. Robson, *Inorg. Chem.*, 1974, **13**, 1301.
[75] D. A. Wright and J. D. Quinn, *Acta Cryst.*, 1974, **B20**, 2132.
[76] J. G. Gibson and E. D. McKenzie, *J.C.S. Dalton*, 1974, 989.
[77] R. W. Halliday and R. H. Court, *Canad. J. Chem.*, 1974, **52**, 3469.
[78] M. J. Potasek, P. G. Debrunner, W. H. Morrison, and D. N. Hendrickson, *J.C.S. Chem. Comm.*, 1974, 174.
[79] M. Matsumoto, K. Nakatsu, K. Jani, A. Nakamura, and S. Otsuka, *J. Amer. Chem. Soc.*, 1974, **96**, 6778.
[80] G. Henrici-Olivé and S. Olivé, *Angew. Chem. Internat. Edn.*, 1974, **13**, 549.
[81] C. Floriani and G. Fachinetti, *J.C.S. Dalton*, 1974, 615.

derived from 2,6-diacetylpyridine and picolinoylhydrazine uses eight of the total of nine potential donor atoms in co-ordination to both metal atoms in a dinuclear Cu^{II} chelate.[82] A series of seven-co-ordinate complexes of first-row transition-metal ions with the bis(semicarbazone) of 2,6-diacetylpyridine has been described. Single-crystal X-ray analyses of the Ni^{II} and Cu^{II} derivatives have shown that the donor atoms of the quinquedentate ligand occupy the equatorial positions of an approximate pentagonal bipyramid, the axial positions being occupied by H_2O molecules.[83]

Much of the interest in amino-acids as ligands rests on the identification of binding sites and in the metal-assisted hydrolysis of amino-acids, peptides, and related species. Potentiometric methods and visible spectra have provided evidence for the nature of the complexed species formed in aqueous solutions of Co^{II}, Ni^{II}, Cu^{II}, and Zn^{II} with 3-[(carboxymethyl)thiol]-L-alanine and 3-[(2-aminoethyl)thiol]-L-alanine. A variety of structures were detected, their nature being critically dependent on pH. In some, chelation is *via* the amino- and carboxylate groups of the amino-acid residue; in others, the sulphur atom is involved in co-ordination.[84] The equilibrium constants for protonation and metal ion (Cu^{II} and Zn^{II}) complex formation of L,L- and D,L-diastereoisomers of the dipeptides Ala-Ala, Ala-Phe, Leu-Leu, and Leu-Tyr have been measured. The higher pK_a values of NH_2 and lower pK_a values of $CO_2{}^-$ in the 'mixed' (*i.e.* D,L) isomers are explained on the assumption that the dipeptides have a predominantly β conformation regardless of pH.[84] Formation constants are higher for the D,L- than for the L,L-isomers. The hydrolysis of glycine methyl ester is catalysed by $Cu(dpa)^{2+}$ [dpa = bis(2-pyridylmethyl)amine].[85] An interesting intramolecular chelation reaction involves condensation between a carbonyl group of a pyruvate ion, co-ordinated to Co^{III} *via* the carboxylate group, and a co-ordinated NH_3 ligand on the same metal.[86] This is presumed to occur by deprotonation of the NH_3 group to give a co-ordinated amide ion, which then attacks the carbonyl centre to give, after elimination of H_2O, a chelated imine. This type of reaction has been applied to the stereospecific synthesis of one isomer of a quinquedentate amine complex of Co^{III} where the synthesis of the organic ligand was conducted on the metal ion. A novel terdentate amino-acid complex was obtained from the reaction of the above chelated imine with acetylacetonate ion.[86] DL-α-Alanine displaces Cl^- from Zeise's salt in H_2O at 25 °C by a stepwise mechanism to give $[PtCl(C_2H_4)(Ala)]$; in the intermediate the amino-acid is linked to Pt by the N atom only.[87] The multidentate nature of many biologically important amino-acids coupled with the bridging capacity

[82] A. Mangia, C. Pelizzi, and G. Pelizzi, *Acta Cryst.*, 1974, **B20**, 2146.
[83] D. Webster and G. J. Palenik, *J. Amer. Chem. Soc.*, 1974, **96**, 7565.
[84] R. Nakon, E. M. Beadle, and R. J. Angelici, *J. Amer. Chem. Soc.*, 1974, **96**, 719; R. Nakon and R. J. Angelici, *ibid.*, p. 4178; G. Brookes and L. D. Pettit, *J.C.S. Chem. Comm.*, 1974, 813.
[85] R. Nakon, P. R. Rechani, and R. J. Angelici, *J. Amer. Chem. Soc.*, 1974, **96**, 2117.
[86] B. T. Golding, J. M. Harrowfield, and A. M. Sargeson, *J. Amer. Chem. Soc.*, 1974, **96**, 2634, 3003; B. T. Golding, J. M. Harrowfield, G. B. Robertson, A. M. Sargeson, and B. O. Whimp, *ibid.*, p. 3692.
[87] G. Carturan, P. Uguagliati, and U. Belluco, *Inorg. Chem.*, 1974, **13**, 542.

of carboxylate oxygen is responsible for the di- and tri-nuclear nature of many complexes with transition metals.[88] One such complex, $[Fe(Ala)_2(H_2O)]_3(ClO_4)_7$, has been suggested as an analogue for ferritin.[84]

4 Macrocyclic Ligands

There has been a great deal of activity during 1974 on the synthesis and properties of complexes of macrocyclic ligands, the porphyrins and related quadridentate groups having received the most attention. Under this heading, the more significant advances in the synthetic and structural fields are reviewed; oxygen-carrying and related properties are considered separately below. Recent aspects of metalloporphyrin chemistry have been reviewed.[89] Single-crystal X-ray structure determinations have been carried out on new and previously known porphyrin or substituted porphyrin complexes of Mn,[90a,90b] Fe,[90a] Co,[90c—f] Ni,[90g—i] and Zn.[90j] These have four-, five-, or six-co-ordinate geometries depend-in on the metal, its oxidation state, and the nature of the axial ligands. Where the axial ligands atoms are nitrogen, Co—N(ax) is almost always significantly longer than Co—N(eq). An Os^{II} complex of octaethylporphyrin has been prepared by treatment of H_2oep with OsO_4 at 473 K,[91a] and a new Rh^{III} complex $Rh(tpp)(CO_2Et)$ of tetraphenylporphyrin by treatment of $RhCl(tpp)CO$ with NaOEt.[91b] Surprisingly, the dication $[H_4oep]^{2+}$ is planar[91c] whereas in the diacid of tpp, for example, the pyrrole rings are inclined at 33°. It seems likely that steric crowding of the inner hydrogen atoms is relieved by out-of-plane displacements. In contrast to all other Cu^{II} porphyrins studied to date, chloro-N-methyltetraphenylporphyrincopper(II) undergoes dissociation readily on dilution in DMF, a result, presumably, of the distortion of the porphyrin nucleus from planarity arising from interaction of the N-methyl group with the N-hydrogen atoms.[92] There is strong evidence from X-ray studies of haemoproteins for a molecular interaction between the porphyrin π-system and certain aromatic side-chains of the protein. The function of these interactions in myoglobin and haemoglobin is thought to be to position the non-covalently held protoporphyrin

[88] E. M. Holt, S. L. Holt, W. F. Tucker, R. O. Asplund, and K. J. Watson, *J. Amer. Chem. Soc.*, 1974, **96**, 2621; T. Yasui, H. Kawaguchi, Z. Kanda, and T. Ama, *Bull. Chem. Soc. Japan*, 1974, **47**, 2393.

[89] D. Ostfeld and N. Tsutsui, *Accounts Chem. Res.*, 1974, **7**, 52; J.-H. Fuhrhop, *Angew. Chem. Internat. Edn.*, 1974, **13**, 321.

[90] (a) P. L. Picculo, G. Rupprecht, and W. R. Scheidt, *J. Amer. Chem. Soc.*, 1974, **96**, 5293; (b) B. R. Stults, E. L. Tassett, R. O. Day, and R. S. Marianelli, *ibid.*, p. 2650; (c) R. G. Little, J. W. Lauher, and J. A. Ibers, *ibid.*, pp. 4440, 4447; (d) P. N. Dwyer, P. Madura, and W. R. Scheidt, *ibid.*, p. 4815: (e) W. R. Scheidt, *ibid.*, p. 84; (f) J. A. Kaduk and W. R. Scheidt, *Inorg. Chem.*, 1974, **13**, 1875; (g) P. N. Dwyer, J. W. Buchler, and W. R. Scheidt, *J. Amer. Chem. Soc.*, 1974, **96**, 2789; (h) D. L. Cullen and E. F. Meyer, *ibid.*, p. 2095; (i) B. Chevrier and R. Weiss, *J.C.S. Chem. Comm.*, 1974, 884; (j) L. D. Spaulding, P. G. Elder, J. A. Bertrand, and R. H. Fellon, *J. Amer. Chem. Soc.*, 1974, **96**, 982.

[91] (a) J. W. Buchler and K. Rohhoch, *J. Organometallic Chem.*, 1974, **65**, 223; (b) I. A. Cohen and B. C. Chow, *Inorg. Chem.*, 1974, **13**, 488; (c) E. Cetinkaya, A. W. Johnson, M. F. Lappert, G. M. McLaughlin, and K. W. Muir, *J.C.S. Dalton*, 1974, 1236.

[92] D. K. Lavallee and A. E. Gebala, *Inorg. Chem.*, 1974, **13**, 2004.

in the haem cavity, whereas, in cytochromes, they are related to control of the redox process. Two recent studies[93] have provided evidence that these donor–acceptor interactions at a point remote from the metal can significantly alter the properties of the metal. These results have important implications for a fuller understanding of co-operative effects in the binding of oxygen. In one study it has been shown that the electron donor 1,10-phenanthroline interacts in a 1 : 1 stoicheiometry with the low-spin bis(imidazole)deuterioporphyrin-IX dimethyl ester iron(III) chloride at a point remote from the metal. This results in an enhancement of the binding of the metal with imidazole. Thus, equilibrium (1) is displaced to the right in solution. In the second study the effect of addition

$$Fe(porph)Cl + 2im \rightleftharpoons Fe(porph)(im)_2 + Cl^- \qquad (1)$$

$$\text{(high-spin)} \qquad\qquad \text{(low-spin)}$$

of the π-electron acceptor trinitrobenzene on equilibrium (1) was followed by ^1H n.m.r. techniques. (In this case porph = tetraphenylporphyrin and im = N-methylimidazole). Here the effect of the π-acceptor was to favour the five-co-ordinate, high-spin species. The first crystallographic studies of a metal-substituted haem protein have been reported.[94a] $Mn^{III}Hb$ is isomorphous with ferrihaemoglobin, $Fe^{III}Hb$, and undergoes similar changes in quaternary structure on reduction to $Mn^{II}Hb$. Thus, $Mn^{II}Hb$ retains the major structural and functional[94b] properties of $Fe^{II}Hb$. Oxidation or ligation of $Fe^{II}Hb$ induces a motion of the metal atom relative to the mean porphyrin plane as well as a spin change. However, no spin change occurs on co-operative oxidation of $Mn^{II}Hb$ (nor of $Co^{II}Hb$).[94b] The kinetics (first order) of the reduction of ferricytochrome by $[Fe(edta)]^{2-}$ are compatible with outer-sphere attack of $[Fe(edta)]^{2-}$ at an exposed haem edge, though the possibility of adjacent attack through the haem pocket is not ruled out.[95] A novel photoreduction in which octaethyl-α-hydroxy-porphinatozinc(II) is transformed with loss of one CO into a product containing a biliverdin residue has been reported;[96] metalloporphyrins are oxidized *in vitro* by H_2O_2 or O_2 to hydroxy-derivatives, which have also been detected as primary products in the biological degradation of haem to biliverdin. The complexes $Fe(porph)L_2$ (L = *e.g.* 2-methylimidazole, piperidine, pyridine, or imidazole) reversibly bind CO in toluene with displacement of one molecule of L.[97a] Equilibrium constants for the displacement decrease in the order of bases as written. A dissociative mechanism is involved. A five-co-ordinate intermediate is also suggested for the photo-induced insertion of O_2 into the Co—C bond of alkylcobaloxime complexes Co(cobalox)R(base); the base re-co-ordinates

[93] (a) E. H. Abbott and P. A. Rafson, *J. Amer. Chem. Soc.*, 1974, **96**, 7378; (b) G. N. La Mar, J. D. Satterlee, and R. V. Snyder, *ibid.*, p. 7137.
[94] (a) K. Moffat, R. S. Lee, and B. M. Hoffman, *J. Amer. Chem. Soc.*, 1974, **96**, 5259; (b) C. Bull, R. G. Fisher, and B. M. Hoffman, *Biochem. Biophys. Res. Comm.*, 1974, **59**, 140.
[95] H. L. Hodges, R. A. Holwerda, and H. B. Gray, *J. Amer. Chem. Soc.*, 1974, **96**, 3132.
[96] S. Besecke and J.-H. Fuhrhop, *Angew. Chem. Internat. Edn.*, 1974, **13**, 150.
[97] (a) D. V. Stynes and B. R. James, *J. Amer. Chem. Soc.*, 1974, **96**, 2733; (b) G. Giannotti, C. Fontaine, and B. Septe, *J. Organometallic Chem.*, 1974, **71**, 107.

immediately after the insertion step.[97b] The poor co-ordinating tendency of oxygen-donors for axial ligation in methylcobaloxime *vis-a-vis* sulphur donors is due to the entropy term, not the enthalpy term; thus, interpretations based on Class a/Class b ligand preferences are unreliable.[98]

There has been a significant increase in the number of synthetic macrocyclic complexes reported during the year. A useful summary of the published synthetic work in this field up to the end of 1972 has appeared.[99] The greatly enhanced stability of macrocyclic complexes *vis-a-vis* that of comparable chelates of open-chain ligands of the same denticity has been recognized for some time. Margerum and Busch, respectively, have termed this the 'macrocyclic effect' and 'multiple juxtapositional fixedness'. New equilibrium and calorimetric results[100] for the formation of Ni^{II} complexes of cyclam and 2,3,2-tet reveal that the higher stability of the former rests entirely in the enthalpy term (the difference in $-\Delta H$ is $\sim 60\ kJ\ mol^{-1}$). This striking effect has its origin not in stronger metal–ligand bonds in the macrocyclic complex but in a smaller degree of hydration of the free ligand. The free macrocycle cannot accommodate as many hydrogen-bonded water molecules with its nitrogen donor atoms because of steric hindrance. A difference of two in the number of solvating water molecules would account for the observed difference in ΔH. Clearly, ligand-solvation effects, often disregarded in the past, are of prime importance in governing complex stability not only of macrocyclic complexes. The effect of ring size on metal–ligand distances and on D_q for a series of fully saturated cyclic tetra-aza-ligands has been noted.[101] The strain energy for the free ligands is minimized in the fourteen-membered ring. Metal–nitrogen distances increase with the increase in ring size at the rate of 0.10 to 0.15 Å per ring member added. The fourteen-membered ring is the best fit for Co^{3+} and the fifteen-membered ring for high-spin Ni^{2+}. The analysis shows that mechanical constraints of molecular origin can profoundly influence the strength of the M—N bonds. This is reflected in the abnormally high Dq_{xy} values exhibited by rings smaller than the 'best-fit' and the markedly low Dq_{xy} values found for over-sized rings. One of the more interesting aspects of metal–macrocycle chemistry is the influence of the stereochemical preferences of a particular metal on ligand conformation and, in turn, the capacity of the metal ion to accommodate to the geometrical constraints imposed by the macrocycle. By far the greater number of synthetic macrocycles are quadridentate, with nitrogen donor atoms. Usually the ligating atoms lie on a plane, or nearly so, with the metal atom positioned either in the same plane or slightly above it. Thus, the most commonly encountered polyhedra are square planar, square pyrimidal, and octahedral, and distorted forms of these. This is the expected consequence of macrocycles having full or substantial degrees of unsaturation. With fully or mainly saturated macrocycles, ring folding may

[98] R. L. Cartwright, R. S. Drago, J. A. Nusz, and M. S. Nazari, *Inorg. Chem.*, 1973, **12**, 2809.

[99] J. J. Christensen, D. J. Eatough, and R. M. Izatt, *Chem. Rev.*, 1974, **74**, 351.

[100] F. P. Hinz and D. W. Margerum, *J. Amer. Chem. Soc.*, 1974, **96**, 4993.

[101] L. Y. Martin, L. J. De Hayes, L. J. Zompa, and D. H. Busch, *J. Amer. Chem. Soc.*, 1974, **96**, 4046.

occur. A recent example is the high-spin, five-co-ordinate Fe^{II} complex $[FeCl(C_{16}H_{32}N_4)]I$, where $C_{16}H_{32}N_4$ is the fourteen-membered hexamethyl-1,4,8,11-tetra-azacyclotetradeca-4,11-diene.[102a] The macrocycle is in a folded conformation with the Cl occupying the apical position. This geometry is adopted because the steric constraints of the ligand do not permit the metal ion to fit into the plane of the four N atoms, yet the ring is not flexible enough to permit sufficient folding to accommodate a sixth ligand *cis* to Cl. The more flexible, fully saturated, sulphur macrocycle 1,4,8,11-tetrathiacyclododecane (and substituted derivatives) is capable of greater folding, and its complexes with Co^{III} of the type $[Co(S_4)X_2]^+$ are either *cis* or *trans* depending on the nature of X.[102b] These are the first Co^{II} complexes of a macrocycle having sulphur donor atoms exclusively. The ligand-field strength of this macrocycle is comparable to that of analogues having secondary amino-donor atoms. Complexes of a number of macrocycles containing oxygen,[103a—c] sulphur,[103a,103d] or phosphorus[103e] donor atoms as well as nitrogen have been described. Some of these are quinquedentate and extend the number of pentagonal-bipyrimidal complexes having a planar array of macrocyclic donor atoms from the previously described Fe^{III} derivatives to Mn^{II}, Zn^{II}, and Cd^{II}.[103a,103d] An important property of both naturally occurring and synthetic macrocycles having varying degrees of unsaturation is their ability to stabilize both high and low oxidation states of the complexed metal. Such species are almost certainly implicated in the metabolic function of the naturally occurring porphyrins. The redox behaviour of several transition-metal porphyrin analogues[104] and of an extensive series of Ni^{II} macrocyclic complexes varying in the nature and degree of unsaturation, ring size, and charge[105] has been elucidated in detail.

5 Dioxygen, Peroxide, Superoxide, and Related Complexes

Some of the most important developments in co-ordination chemistry in 1974 have concerned the interaction of dioxygen with transition-metal complexes.[106] Significant new results have followed the discovery in the previous year of the reversible O_2-transport properties of porphyrin-type and synthetic iron(II) macrocycles. The factors affecting O_2-adduct formation are complex and include protection of the bound O_2 molecule, with respect to bimolecular oxidation, by steric hindrance, the neighbouring-group effect of covalently attached base, the

[102] (a) V. L. Goedkin, J. Molin-Case, and G. G. Christoph, *Inorg. Chem.*, 1973, **12**, 2894; (b) K. Travis and D. H. Busch, *ibid.*, p. 2591.
[103] (a) N. W. Alcock, D. C. C. Liles, M. McPartlin, and P. A. Tasker, *J.C.S. Chem. Comm.*, 1974, 727; (b) F. Vogtle and E. Weber, *Angew. Chem. Internat. Edn.*, 1974, **13**, 149; (c) R. B. King and P. R. Heckley, *J. Amer. Chem. Soc.*, 1974, **96**, 3118; (d) L. F. Lindoy and D. H. Busch, *Inorg. Chem.*, 1974, **13**, 2494; (e) J. Riker-Nappier, and D. W. Meek, *J.C.S. Chem. Comm.*, 1974, 442.
[104] N. Takvoryan, K. Farmery, V. Katovic, F. V. Lovecchio, E. S. Gore, L. B. Anderson, and D. H. Busch, *J. Amer. Chem. Soc.*, 1974, **96**, 731.
[105] F. V. Lovecchio, E. S. Gore, and D. H. Busch, *J. Amer. Chem. Soc.*, 1974, **96**, 3109.
[106] G. Henrici-Olivé and S. Olivé, *Angew. Chem. Internat. Edn.*, 1974, **13**, 29.

nature and orientation of the co-ordinated base, and solvent effects.[107] Several models for the nature of the iron–dioxygen bond in haemoglobin and myoglobin have been proposed at different times. The first X-ray study of a reversibly formed Fe^{II}–O_2 complex has now been carried out. This complex is one of the 'picket-fence' derivatives and contains an axially co-ordinated imidazole.[108a] The O_2 molecule is co-ordinated in an end-on, angular fashion and is therefore in accord with the Pauling model. The FeOO angle is 136°, and the Fe—O and O—O distances are 1.75 and 1.25 Å, respectively. A sharp intense peak at 1385 cm^{-1} in the low-temperature i.r. spectrum of the same compound has been assigned to the O—O stretching vibration of the co-ordinated O_2.[108b] Comparison of this value with those for other O_2-containing compounds suggests that the O_2 co-ordinated to iron has substantial double-bond character; in fact it is only ~ 100 cm^{-1} below $\nu(O_2)$ of uncomplexed singlet O_2. Thus, both the X-ray and spectroscopic results are in accord with the formulation of a low-spin Fe^{II}–singlet O_2 interaction. New e.s.r. and n.m.r. evidence suggests that in O_2 adducts of cobalt a similar formulation may apply; such complexes have hitherto been more commonly viewed as superoxide complexes of Co^{III}.[108c]

A low-spin, five-co-ordinate Co^{II} complex of a 'picket-fence' porphyrin with axially bound 1-methylimidazole reversibly forms O_2 adducts both in the solid state and in solution in toluene.[109] Comparison of the extent of oxygenation of this complex with CoHb and other Co^{II} complexes reinforces the view that restrictions to solvation of the unoxygenated and/or the oxygenated complexes (e.g. by the 'picket-fence' or the globin cavity) may be critically important and that the environment of haemoproteins may approximate more to that of a solid than a solution. The e.s.r. spectra of the mononuclear O_2-carrying complex Co(bzacen)(py)(O_2) are consistent with the presence of magnetically equivalent oxygen atoms in solution. Previous crystallographic studies have established a bent CoOO bond (126°) for this complex in the solid state. It is suggested that the observed equivalence in solution is the result of a rapid flipping of the O—O group between two bent positions, or possibly, of a π-bonded structure.[110a] A high-spin Co^{II} complex $[Co(pfp)_2(H_2O)_2]^{2-}$ (pfp = perfluoropinacolato dianion) reacts reversibly with O_2 to form the ion $[Co(pfp)(Hpfp)(O_2)]^-$, isolable as the NR_4^+ salt; a π-bonding interaction is suggested for this compound.[110b] In $[NEt_4]_3[Co(CN)_5(O_2)]$,5H_2O the Co—O—O linkage is nearly linear (175°).[110b] Structural data for (1-methylimidazole)(octaethylporphinato)cobalt(II), a model

[107] (a) J. P. Collman, R. R. Gagne, and C. A. Reed, J. Amer. Chem. Soc., 1974, 96, 2631; (b) J. Almog, J. E. Baldwin, R. L. Dyer, J. Huff, and C. J. Wilderson, ibid., p. 5600; (c) G. C. Wagner and R. J. Kassner, ibid., p. 5593; (d) D. L. Anderson, C. J. Weschler, F. Basolo, ibid., p. 5599; (e) W. S. Brinigar and C. K. Chang, ibid., p. 5597; (f) W. S. Brinigar, C. K. Chang, J. Geibel, and T. G. Traylor, ibid., p. 5599.

[108] (a) J. P. Collman, R. R. Gagne, C. A. Reed, W. T. Robinson, and G. A. Rodley, Proc. Nat. Acad. Sci. U.S.A., 1974, 71, 1326; (b) J. P. Collman, R. R. Gagne, H. B. Gray, and J. W. Hare, J. Amer. Chem. Soc., 1974, 96, 6522; (c) B. S. Torrog and R. S. Drago, ibid., p. 6765.

[109] J. P. Collman, R. R. Gagne, J. Kouba, and H. Ljusberg-Wahren, J. Amer. Chem. Soc., 1974, 96, 6802.

[110] (a) E. Melamud, L. Silver, and Z. Dori, J. Amer. Chem. Soc., 1974, 96, 4690; (b) C. J. Willis, J.C.S. Chem. Comm., 1974, 117; (c) L. D. Brown and K. N. Raymond, J.C.S. Chem. Comm., 1974, 470.

for CoHb, suggest that the out-of-plane displacement of the metal atom is small compared with that in FeHb but comparable with the situation in methaemoglobin.[111] The oxidation of dihydroxybenzene and other organic substrates by O_2 activated by co-ordination to Co^{II} and its possible relevance to the mode of action of cytochrome oxidase has been noted.[112] Polypyrazolylborate ligands have been suggested as useful synthetic models of the histidine chelating sites of some metalloproteins, including the O_2-carrier haemocyanin, tyrosinase, and other proteins where copper atoms occur in pairs. The basis for this view is the dimeric nature of $[Cu(HBpz)_3]_2$ in solution and its reaction with CO and with O_2 to form adducts.[113a] Reversible uptake of CO and O_2 has been reported for a polyoxime complex of Cu.[113b] Co-ordinately unsaturated Cr^{2+} ions held in zeolite cavities reversibly bind O_2.[114]

6 Oxygen Ligands

New anhydrous hexanitrato-complexes of Ir^{IV} and Pt^{IV} have been prepared; e.s.r. spectra of the former reveal an unexpectedly high degree of $Ir-NO_3$ π-bonding.[115a] The reaction of $Re(CO)_5(NO_3)$ with $[AsPh_4]NO_3$ yielded $[AsPh_4][Re(CO)_4(NO_3)_2]$, in which the nitrato-groups are unidentate and *trans*.[115b] The metal has a distorted pentagonal-bipyramidal stereochemistry in $MeTa^VCl_2[ON(Me)NO]_2$ but is six-co-ordinate in $Ta^VCl_3(OCH_2CH_2OMe)_2$, in which one methoxyethoxide group is bidentate and one unidentate.[116] Seven-co-ordination occurs in the pyridine adduct of dioxobis(tropolonato)uranium(IV)[117] while several tetrakis(tropolonato)scandium(III) derivatives are eight-co-ordinate.[118] The four-co-ordinate alkoxo titanium(IV) compounds $(C_5H_5)_nTi(OCH_2CH_2CH=CH_2)_{4-n}$ ($n = 1$ or 2) have been synthesized; in none of the derivatives prepared is the terminal olefinic group used in co-ordination.[119] L-3,4-Dihydroxyphenylalanine, used in the treatment of Parkinson's disease, is ambidentate towards Cu^{2+} in aqueous solution. Above pH 7 it uses the catecholate (O, O) group and below pH 7 the amino-acid (N, O) end of the molecules for co-ordination.[120] The crystal structure of pentakis-(ethyleneurea)copper(II) perchlorate has been determined; the co-ordination

[111] R. G. Little and J. A. Ibers, *J. Amer. Chem. Soc.*, 1974, **96**, 4452.

[112] E. W. Abel, J. M. Pratt, B. Whelan, and P. J. Wilkinson, *J. Amer. Chem. Soc.*, 1974, **96**, 7120.

[113] (a) C. S. Arcus, J. L. Wilkinson, C. Mealli, T. J. Marks, and J. A. Ibers, *J. Amer. Chem. Soc.*, 1974, **96**, 7565; (b) S. J. Kim and T. Takizawa, *J.C.S. Chem. Comm.*, 1974, 356.

[114] R. Kellerman, P. J. Hutta, and K. Klier, *J. Amer. Chem. Soc.*, 1974, **96**, 5946.

[115] (a) B. Harrison, N. Logan, and J. B. Raynor, *J.C.S. Chem. Comm.*, 1974, 202; (b) C. C. Addison, R. Davis, and N. Logan, *J.C.S. Dalton*, 1974, 1073.

[116] J. D. Wilkins and M. G. B. Drew, *J. Organometallic Chem.*, 1974, **69**, 111; *Inorg. Nuclear Chem. Letters*, 1974, **10**, 5-9.

[117] S. Degetto, M. G. Bombieri, E. Forsellini, L. Baracco, and R. Graziani, *J.C.S. Dalton*, 1974, 1933.

[118] D. J. Olszanski, T. J. Anderson, M. A. Neuman, and G. A. Melson, *Inorg. Nuclear Chem. Letters*, 1974, **10**, 137; A. R. Davis and F. W. B. Einstein, *Inorg. Chem.*, 1974, **13**, 1880.

[119] R. J. H. Clark and M. A. Coles, *J.C.S. Dalton*, 1974, 1462.

[120] R. J. Majeote and L. M. Trefonas, *Inorg. Chem.*, 1974, **13**, 1062.

polyhedron is a square pyramid with the metal bonded to the oxygen atoms.[120]
Pyrazine mono-N-oxide has a bridging function in its polymeric 1 : 1 complexes
with Ag^I but uses the nitrogen atom only in its 2 : 1 complexes,[121a] and the
mono-N-oxide of 1,10-phenanthroline is bidentate towards Cr^{III}, Fe^{III}, and Ni^{II}.
As expected, it exerts a much weaker field than 1,10-phenanthroline itself.[121b]
Co-ordination compounds of a variety of phosphoryl-containing ligands have
been studied during the year.[122] A new class of 'spin-labelled' ligand has been
defined as one having two functional groups, one of which provides a Lewis-base
site suitable for co-ordination to metal ions while the other comprises a site
bearing an unpaired electron; spin-density distributions, via e.s.r. measurements,
in t-butyl nitroxide have provided information on the relative electron-accepting
abilities of different metal ions in various ligand environments.[123]

7 Phosphine, Arsine, and Stibine Ligands

Some hundreds of papers concerning complexes of ligands of this class have been
published during the year. Many of these are considered in the Organometallic
and other Sections of this volume. Here, we restrict attention to those new ligands
having a novel structure or co-ordinating function and to complexes having
novel properties. Reactions of co-ordinated phosphines, arsines, $etc.$ have been

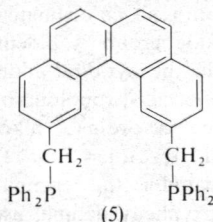

$$\text{CH}_2 \qquad \text{CH}_2$$
$$\text{Ph}_2\text{P} \qquad \text{PPh}_2$$
(5)

reviewed.[124] A bidentate diphosphine (5), which spans the $trans$-positions of
some Ni^{III}, Pd^{II}, and Pt^{II} complexes, has been synthesized.[125a] The rigid polynuclear aromatic moiety serves to position the donor atoms in a manner suitable
for spanning $trans$-sites. Significantly, the ligand does not appear to form complexes with Co^{II}, which might be expected to have a tetrahedral structure. However, rigidity in the ligand is not a necessary requirement for '$trans$'-co-ordination

[121] (a) P. J. Huffman and J. E. House, $J. Inorg. Nuclear Chem.$, 1974, 36, 2618; (b) A. N.
Speca, L. L. Pytlewski, and N. M. Karayannis, $J. Inorg. Nuclear Chem.$, 1974, 36,
1227; A. N. Speca, N. M. Karayannis, and L. L. Pytlewski, $Inorg. Chim. Acta$, 1974,
9, 87.

[122] H. L. Gillman, $Inorg. Chem.$, 1974, 13, 1921; D. V. Naik, G. J. Palenik, S. Jacobson,
and A. J. Carty, $J. Amer. Chem. Soc.$, 1974, 96, 2286; B. D. Catsikis and M. L. Good,
$J. Inorg. Nuclear Chem.$, 1974, 36, 1039; M. G. Newton, H. D. Caughman, and R. C.
Taylor, $J.C.S. Dalton$, 1974, 1031; D. A. Crouch, S. D. Robinson, and J. N. Wingfield,
$ibid.$, p. 1309.

[123] C. T. Cazianis and D. R. Eaton, $Canad. J. Chem.$, 1974, 52, 2454.

[124] C. E. Kraihanzel, $J. Organometallic Chem.$, 1974, 73, 137.

[125] (a) N. J. DeStefano, D. K. Johnson, and L. M. Venanzi, $Angew. Chem. Internat. Edn.$,
1974, 13, 133; (b) B. L. Shaw, Abstracts of XVIth Conference on Co-ordination
Chemistry, Dublin, 1974.

of diphosphines. Shaw[125b] has used $Bu^t_2P(CH_2)_nPBu^t_2$ ($n = 9$, 10, or 12) to bond to Ir with a mutual *trans*-arrangement of donor atoms in *e.g.* [IrCl(CO)-(diphos)]. Bulky tertiary phosphines or arsines frequently give complexes in which the metal is in an unusual environment or oxidation state. They also promote hydride and metal–metal bond formation. The longer than usual Pt—P (2.371 Å) and Pt—I (2.612 Å) bonds in *trans*-di-iodobis(tricyclohexylphosphine) platinum(II) and the ready elimination of PCy_3 on heating to give the iodide-bridged complex $[PtI(PCy_3)_2]_2$, are ascribed to 'over-crowding'.[126] An interesting new P-donor ligand is the cyclic compound triphenylcyclotriphosphane $[C_6H_5P]_3$. The reactions of this ligand with $Fe(CO)_4(THF)$ and $Ni(CO)_4$ gave $Fe(CO)_4L$ and $Ni(CO)_3L$, respectively; neither complex is particularly stable. [31]P n.m.r. spectra have shown that the three phosphorus atoms are equivalent; a *trihapto* interaction is therefore indicated.[127] Complexes of tris(aziridino)phosphines, which are related to some anti-cancer drugs, of Pd^{II} and Pt^{II} and Group VI carbonyls have been described.[128] The mono-quaternized diphosphines $Ph_2P(CH_2)_n$-$\overset{+}{P}Ph_2Me$ ($n = 2$ or 3) form zwitterionic carbonyl halides with tungsten.[129] Fluorocarbon biphosphines $[(F_3C)_2P]_2E$ (E = O, S, or Se) react with $Fe_2(CO)_2$-(NO), to yield $Fe(CO_4)(diphos)$ and $Fe(CO)(NO)_2(diphos)$ wherein the diphosphine is unidentate (*via* P) in most cases, though in the reaction with the selenide there is evidence for a product containing a five-membered Fe_2P_2Se ring.[130a] Phosphinohydrazines, *e.g.* $Ph_2PNHNMe_2$, have a number of potential co-ordinating modes, *viz.* unidentate *via* any one of three donor atoms or bidentate *via* the P and terminal N atoms. Reaction with Group VI carbonyls in methylcyclohexane at 100 °C gave $M(CO)_5L$, for which i.r. spectra suggest P-co-ordination.[130b] U.v. irradiation of THF solutions of $M(CO)_6$ with $Me_2AsSAs(S)Me_2$ gave $M(CO)_5L$ (M = Cr or W, L = $Me_2AsSAsMe_2$) and $MM'(CO)_{10}L$ (M = M' = Cr, L = $Me_2AsAsMe_2$; M = Cr, M' = W or M = M' = W, L = $Me_2AsSAsMe_2$). The dinuclear compounds with L = $Me_2AsSAsMe_2$ are thermally unstable and lose sulphur.[130c] [31]P and [1]H n.m.r. data have been obtained for Group VI metal carbonyl complexes of the new unsymmetrical bidentate ligand $Ph_2PCH_2OPPh_2$ and related compounds. Unusually large, and unexplained, downfield phosphorus chemical shifts were observed for all chelates with five-membered rings.[131] Equilibrium studies of the chelate ring-opening reaction (2) [M = Cr, Mo, or W; PN = $Ph_2P(CH_2)_nNR_2$] support a

$$M(CO)_4(PN) + CO \rightleftharpoons M(CO)_5P—N \qquad (2)$$

[126] N. W. Alcock and P. G. Leviston *J.C.S. Dalton*, 1974, 1834.
[127] M. Baudler and M. Bock, *Angew. Chem. Internat. Edn.*, 1974, **13**, 147.
[128] R. B. King and O. von Stetten, *Inorg. Chem.*, 1974, **13**, 2449.
[129] R. L. Keiter and D. Marcovitch, *Inorg. Nuclear Chem. Letters*, 1974, **10**, 1099.
[130] (a) R. C. Dobbie and M. J. Hopkinson, *J.C.S. Dalton*, 1974, 1290; (b) G. E. Graves and L. W. Houk, *J. Inorg. Nuclear Chem.*, 1974, **36**, 232; (c) E. W. Ainscough, A. M. Brodie, and G. Leng-Ward, *J.C.S. Dalton*, 1974, 2437.
[131] S. O. Grim, W. L. Briggs, R. C. Barth, C. A. Tolman, and J. P. Jesson, *Inorg. Chem.*, 1974, **13**, 1095.

dissociative mechanism.[132a] The reaction is more complete for seven-membered than for six-membered chelates. The reaction of fac-M(CO)$_3$(PNP) with CO gave cis-M(CO)$_4$(PNP), in which the nitrogen atom is uncoordinated {PNP = [Ph$_2$P(CH$_2$)$_2$]$_2$NEt}. Displacement reactions of CO in Mn(CO)$_5$X (X = Cl, Br, or I) by the potentially quadridentate hexaphenyl-1,4,7,10-tetraphosphadecane give products exemplifying mono-metallic bi- and ter-dentate, and bimetallic ter- and quadri-dentate co-ordination.[132b] Activation parameters for the reverse reaction of equation (2) have been measured for R$_2$P(CH$_2$)$_n$PR$_2$ (R = Ph or Me).[132c] The more rapid chelation for Mo than for Cr is due to the ΔH^+ term. An interesting series of Rh^I complexes of a series of β-phosphino-ethers Ph$_2$P-(CH$_2$CH$_2$O)$_n$CH$_2$CH$_2$PPh$_2$(n = 1, 2, or 3) have been prepared.[133] X-Ray analyses of the complexes RhClL have been carried out for n = 1 and n = 3. In both complexes there is a P$_2$OCl donor set. In the former case (n = 1) the donor oxygen is the oxygen of the ligand whereas in the latter (n = 3) it belongs to a H$_2$O molecule, which in turn is H-bonded to two (unco-ordinated) oxygen atoms of the ligand. Complexes of new open-chain tetratertiary arsines[134a] and of several alkenyl phosphines and arsines[134b] have been studied. Zerovalent NiL$_4$ phosphine complexes dissociate into NiL$_3$ + L in C$_6$H$_6$ to an extent governed mainly by steric effects; in contrast, dissociation rates are determined mainly by electronic influences.[135] New tertiary stibines have been synthesized[136a] and various new stibine complexes described.[136b] In general, it seems that stibines form less stable complexes with transition metals than do phosphines and arsines, and that they exercise somewhat weaker ligand fields.

8 Sulphur, Selenium, and Tellurium Ligands

Interest in sulphur-containing ligands has become increasingly evident during the past few years, stimulated in part by efforts to synthesize useful analogues for sulphur-containing proteins and related molecules of natural occurrence. One of the more impressive achievements in this area has been the investigation[137] of the structure, properties, and reactions of the tetranuclear cluster anions

[132] (a) W. J. Knebel and R. J. Angelici, *Inorg. Chem.*, 1974, **13**, 627; (b) I. S. Butler and N. J. Coville, *J. Organometallic Chem.*, 1974, **66**, 111; (c) J. A. Connor and G. A. Hudson, *ibid.*, 1974, **73**, 351.

[133] N. W. Alcock, J. M. Brown, and J. C. Jeffrey, *J.C.S. Chem. Comm.*, 1974, 829.

[134] S. T. Chow and C. A. McAuliffe, *J. Organometallic Chem.*, 1974, **77**, 401; B. Bosnich, W. G. Jackson, and S. T. O. Lo, *Inorg. Chem.*, 1974, **13**, 2598; (b) M. A. Bennett, R. N. Thomas, and I. B. Tompkins, *J. Amer. Chem. Soc.*, 1974, **96**, 61; C. A. McAuliffe and D. G. Watson, *J.C.S. Dalton*, 1974, 1531; M. K. Cooper, R. S. Nyholm, P. W. Carreck, and M. McPartlin, *J.C.S. Chem. Comm.*, 1974, 343.

[135] C. A. Tolman, W. C. Seidel, and L. W. Gosser, *J. Amer. Chem. Soc.*, 1974, **96**, 53.

[136] (a) E. Shewchuk and S. B. Wild, *J. Organometallic Chem.*, 1974, **71**, C1; L. Volpnoi, B. Zarli, and G. de Paoli, *Gazzetta*, 1974, **104**, 897; (b) R. F. Bryan and W. C. Schmidt, *J.C.S. Dalton*, 1974, 2337; W. Levason and C. A. McAuliffe, *Inorg. Chim. Acta*, 1974, **11**, 33; *Inorg. Chem.*, 1974, **13**, 2765; C. D. Hem, M. Dartiguenave, and Y. D. Dartiguenave, *Inorg. Nuclear Chem. Letters*, 1974, **10**, 1039; T. W. Beall and L. W. Houk, *Inorg. Chem.*, 1974, **13**, 2549.

[137] See a series of papers by R. H. Holm and H. B. Gray and collaborators in the 1974 issues of *J. Amer. Chem. Soc.*, *e.g.* L. Que, J. R. Anglin, M. A. Bobrik, A. Davison, and R. H. Holm, *J. Amer. Chem. Soc.*, 1974, **96**, 6042.

$[Fe_4S_4(SR)_4]^{2-}$. These serve as models for the active centres found in bacterial proteins, including those of the 'high-potential' (HP) and ferredoxin types. These compounds undergo facile exchange of the bound SR^1 groups with added thiol R^2SH. Such exchange reactions provide a valuable synthetic route to peptide complexes which approximate more closely to the natural systems. This has been demonstrated by the remarkable similarity in ^1H n.m.r. and electronic spectra, redox potentials, and magnetic properties of the synthetic systems with those of oxidized ferredoxin and reduced HP proteins. ^1H n.m.r., Mössbauer, and X-ray photoelectron spectra all point to equivalence of the Fe atoms of the cluster. The properties of the Fe_4S_4 core are therefore consistent with a purely localized type of bonding rather than with the existence of trapped valence states.

The triangular Fe_3 cluster in $Fe_3(CO)_8(C_4H_8S)_2$ contains bridging thiophen molecules on two edges and two very unsymmetrical CO bridges on the other.[138] Bridging sulphide occurs also in several Pt^{II} complexes.[139] Halide sulphides and selenides of W and Mo have similarly been prepared by the reaction of the metal halide with Sb_2S_3 and Sb_2Se_3.[140] Pd^{II} complexes of o-$C_6H_4(SMe)EPh_2$ (E = P or As) and related terdentate ligands undergo S-demethylation in hot DMF.[141] Pyridine-2-thiol provides a model for the S- and N(1) donor atoms in the thiolated nucleoside 6-thioguanosine (6). It exists as a tautomer and is

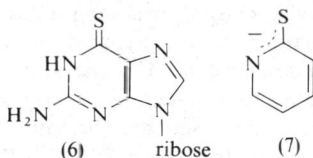

(6) ribose (7)

possibly best considered as a thione. It has three potential co-ordinating modes, *via* the N atom, or the S atom, or *via* both, as in (7). Complexes with several, mostly dipositive, transition-metal ions were prepared, and with the exception of $PdL_2(LH)$ they all contain unidentate ligand; the appearance of an N—H stretching vibration in the i.r. in most of the complexes indicates, in conjunction with other evidence, that the donor atom is sulphur.[142] The donor properties of dipositive tellurium are seen in a number of Cu^I halide complexes of diaryl tellurides.[143]

The preparation, structure, and properties of several metal complexes of monothio- and dithio-β-diketones have been reported.[144] Unlike the acetyl-acetonates of Co^{II} and Ni^{II}, the corresponding thio-derivatives are monomeric

[138] F. A. Cotton and J. M. Troup, *J. Amer. Chem. Soc.*, 1974, **96**, 5070.
[139] P. L. Goggin, R. J. Goodfellow, and F. J. S. Reed, *J.C.S. Dalton*, 1974, 576; Yu. N. Kukushkin, V. I. Vshivlsev, and V. S. N. Shvedova, *Russ. J. Inorg. Chem.*, 1974, **19**, 558.
[140] D. Britnell, G. W. A. Fowles, and D. A. Rice, *J.C.S. Dalton*, 1974, 2191.
[141] T. N. Lockyer, *Austral. J. Chem.*, 1974, **27**, 259.
[142] I. P. Evans and G. Wilkinson, *J.C.S. Dalton*, 1974, 946.
[143] W. R. McWhinnie and V. Rattanaphani, *Inorg. Chim. Acta*, 1974, **9**, 153.

(square planar).[144a] *Cis-* and *trans-*isomers are possible for both bis-planar and tris-octahedral complexes of monothio-β-diketones; the planar bis-complex of monothioacetylacetone with Ni^{II} has the *cis*-configuration.[144b] In the dinuclear zerovalent tungsten complex $(diphos)_2(CO)WC\equiv SW(CO)_5$ the thiocarbonyl ligand has an end-to-end bridging function.[145] Stable trigonal planar cationic complexes of Cu^I with tertiary phosphine and arsine sulphides and selenides have been prepared, and an X-ray structure determination of one of them, $[Cu(SPMe_3)_3]ClO_4$, has been reported.[146] The SO_2 complex $Pt(SO_2)(PPh_3)_3$ is essentially trigonal-pyramidal, with the SO_2 molecule weakly co-ordinated in the apical position *via* the sulphur atom.[147] The tetrahedral co-ordination in the bisethylenethiourea complex of Zn^{II} thiosulphate is made up of three sulphur atoms and one oxygen atom, the $S_2O_3{}^{2-}$ group bridging the metal ions binding through sulphur and oxygen atoms.[148] Nb^V is in a distorted pentagonal-bipyramidal polyhedron in trischlorobis(*N*-methylthioacetamido)niobium(v)[149a] and Pd^{II} is five-co-ordinate in $[Pd(ntu)_4Cl]Cl$ (ntu = naphthylthiourea).[149b] The tetrathio- and thio-oxo- complex ions of Mo and W, $M(O_nS_{4-n})^{2-}$, can themselves act as bidentate ligands to other transition-metal ions.[150]

A large number of papers on metal complexes of 1,2-dithio- and 1,1-dithio-chelated anions (dithiolens, dithiocarbamates, and dithiocarboxylates) have appeared. An important property of ligands of this class is their capacity for stabilizing metal ions in a wide range of oxidation states. A series of Re^I, Re^{II}, Re^{IV}, and Re^V dithiocarbamates have been prepared.[151] Mn^{III} dithiocarbamates are both readily oxidized and reduced to Mn^{IV} and Mn^{II} complexes.[152] Several Fe^{IV} compounds of this class have been reported,[153] as has a singlet–triplet spin equilibrium in a neutral mixed bisdithiocarbamato-1,2-dithioleniron complex.[154] Seven co-ordination occurs in some Ti^{III}, Ti^{IV}, Re^{III}, and Mo^{VI} dithiocarbamates;[155] tetrakis(dithiocarboxylato)-complexes of V^{IV} and Mo^{IV} are

[144] (a) R. Beckett and B. F. Hoskins, *J.C.S. Dalton*, 1974, 622; (b) O. Siiman, D. D. Titus, C. D. Cowman, J. Fresco, and H. B. Gray, *J. Amer. Chem. Soc.*, 1974, **96**, 2353; (c) M. Das and S. E. Livingstone, *Austral. J. Chem.*, 1974, **27**, 53, 2109, 2115.

[145] B. Dombek and R. J. Angelici, *J. Amer. Chem. Soc.*, 1974, **96**, 7568.

[146] J. A. Tiethof, A. T. Hetey, and D. W. Meek, *Inorg. Chem.*, 1974, **13**, 2505.

[147] J. P. Linsky and C. G. Pierpont, *Inorg. Chem.*, 1973, **12**, 2959.

[148] S. Baggio, R. F. Baggio, and P. K. de Perazzo, *Acta. Cryst.*, 1974, **B20**, 2166.

[149] (a) M. G. B. Drew and J. D. Wilkins, *J.C.S. Dalton*, 1974, 198; (b) M. M. Khan, *J. Inorg. Nuclear Chem.*, 1974, **36**, 299.

[150] A. Müller, H.-H. Heinsen, and G. Vandrish, *Inorg. Chem.*, 1974, **13**, 1001; E. Königer-Ahlhorn and A. Müller, *Angew. Chem. Internat. Edn.*, 1974, **13**, 672.

[151] J. F. Rowbottom and G. Wilkinson, *J.C.S. Dalton*, 1974, 685.

[152] A. R. Hendrickson, R. L. Martin, and N. M. Rohde, *Inorg. Chem.*, 1974, **13**, 1933; L. R. Gahan and M. J. O'Connor, *J.C.S. Chem. Comm.*, 1974, 68; K. L. Brown, R. M. Golding, P. C. Healy, K. J. Jessoh, and W. C. Tennant, *Austral. J. Chem.*, 1974, **27**, 2075.

[153] R. L. Martin, N. M. Rohde, G. B. Robertson, and D. Taylor, *J. Amer. Chem. Soc.*, 1974, **96**, 3647; F. J. Hollander, R. Pedelty, and D. Coucouvanis, *J. Amer. Chem. Soc.*, 1974, **96**, 4032.

[154] L. H. Pignolet, G. S. Patterson, J. F. Weiher, and R. H. Holm, *Inorg. Chem.*, 1974, **13**, 1263.

[155] A. N. Bhat, R. C. Fay, D. F. Lewis, A. F. Lindmark, and S. H. Strauss, *Inorg. Chem.*, 1974, **13**, 886; D. F. Lewis and R. C. Fay, *J. Amer. Chem. Soc.*, 1974, **96**, 3843; S. R. Fletcher and A. C. Skapski, *J.C.S. Dalton*, 1974, 486.

eight-co-ordinate.[156] Some dimeric Mo^V dithiocarbamates have been suggested[157] as models for the active sites in nitrogenase enzymes. Iron chelates of thioselenocarbamates[158] and unidentately co-ordinated monothio- and monoselenocarbamate complexes[159] have been reported.

9 Halide and Pseudohalide Complexes

Further examples of linkage isomerism in thiocyanate complexes have been reported, and the controversy[163] over the relative importance of steric *versus* electronic factors in the control of the bonding mode of NCS^- continues. Evidence for the existence of an S-bonded $Fe^{III}(SCN)$ species in conc. $HClO_4$ solutions has been adduced.[61] $Rh(NCO)(PPh_3)_3$ and $Rh(OCN)(PPh_3)_3$ are the first reported linkage isomers of NCO^- in the solid state.[162] Two types of bridging co-ordination have previously been suggested for the NCO^- ion, an end-to-end $M-NCO-M$ bridge and a single-atom (N) bridge $M-N(CO)-M$, also found for the azide N_3^- ion (*cf.* p. 255).[163a] An X-ray structural analysis has demonstrated the occurrence of the former type of bridge in $[Ni_2(tren)_2-(NCO)_2](BPh_4)_2$.[163b] The latter type of bridge is thought to occur in $Cu(NCO)_2-(aniline)_2$ [163c] and in $Cu(N_3)_2(bipy)$.[163d]

Although Cu^{II} is reduced by CN^- in aqueous solution, it has been shown that CN^- can function as a donor to Cu^{II} in the presence of ligands such as 1,10-phenanthroline, as in the complexes $[Cu(phen)_2CN]Y,nH_2O$ (X = *e.g.* halide, NO_3^-, or ClO_4^-).[164] Previously prepared $K_5Ti(CN)_8$ has been reformulated as the seven-co-ordinate complex $K_4[Ti(CN)_7],KCN$.[165] A linear $Cu-CN-Cu$ bridge occurs in a number of dinuclear Cu^{II} (N_4)-macrocyclic complexes[166a] and in some dicobaloximes,[166b] while the complex $[(NH_3)_5CoCNCo(CN)_5],H_2O$ contains both terminally bound and bridging CN.[166c] This latter compound is a linkage isomer of the previously reported complex $[(NH_3)_5CoNCCo(CN)_5]$, H_2O.

[156] M. Bonamico, G. Dessey, V. Fares, and L. Scaramuzza, *J.C.S. Dalton*, 1974, 1258; O. Piovesana and L. Sestile, *Inorg. Chem.*, 1974, **13**, 2745.
[157] W. E. Newton, J. L. Corbin. D. C. Bravard, J. E. Searles, and J. W. McDonald, *Inorg. Chem.*, 1974, **13**, 1100.
[158] S. Nakajuma and T. Tanaka, *Bull. Chem. Soc. Japan*, 1974, **47**, 763.
[159] F. W. Pijpers, A. H. Dix, and J. G. M. van der Linden, *Inorg. Chim. Acta*, 1974, **11**, 41.
[160] See, *e.g.*, J. L. Burmeister, R. L. Hassel, K. A. Johnson, and J. C. Lim, *Inorg. Chim. Acta*, 1974, **9**, 23; J. E. Huhey and S. O. Grim, *Inorg. Nuclear Chem. Letters*, 1974, **10**, 913.
[161] H. N. Po, W.-K. Wong, and K. D. Chien, *J. Inorg. Nuclear Chem.*, 1974, **36**, 3872.
[162] S. J. Anderson, A. H. Norbury, and J. Songstad, *J.C.S. Chem. Comm.*, 1974, 37.
[163] (*a*) J. Nelson and S. M. Nelson, *J. Chem. Soc. (A)*, 1969, 1597; (*b*) D. M. Duggan and D. N. Hendrickson, *Inorg. Chem.*, 1974, **13**, 2056; (*c*) J. Kohout, M. Quastlerova-Hvastijova, and J. Gazo. *Inorg. Chim. Acta*, 1974, **8**, 241; (*d*) G. W. Bushnell and M. A. Khan, *Canad. J. Chem.*, 1974, **52**, 3125.
[164] M. Wicholas and T. Wolford, *Inorg. Chem.*, 1974, **13**, 316.
[165] D. Nicholls, T. A. Ryan, and K. R. Seddon, *J.C.S. Chem. Comm.*, 1974, 635.
[166] (*a*) D. M. Duggan, R. G. Jungst, K. R. Mann, G. D. Stucky, and D. N. Hendrickson, *J. Amer. Chem. Soc.*, 1974, **96**, 3443; (*b*) A. L. Crumbliss, P. L. Gans, and P. M. Gross, *Inorg. Nuclear Chem. Letters*, 1974, **10**, 485; (*c*) F. R. Fronczek and W. P. Schaefer, *Inorg. Chem.*, 1974, **13**, 727.

12 Transition-metal Carbonyl, Organometallic, and Related Complexes

By R. J. CROSS

Department of Chemistry, The University, Glasgow G12 8QQ

1 Introduction

The layout and material covered in this section is similar to that of recent years. With the firm establishment of the Specialist Periodical Reports on Organometallic Chemistry, any need here for a comprehensive survey has passed. Increasing pressure for space brought about by an expanding chemical literature has, in any case, already initiated a trend towards more selective reports. This means that the choice of material is necessarily subjective, and inevitably, much good work is excluded.

2 Metal Carbonyls

Photolysis and metal-atom condensation reactions are two methods of proven value for the production of unstable metal carbonyls or reactive metal-carbonyl fragments. $Fe(CO)_3$ has now been produced by the prolonged u.v. irradiation of $Fe(CO)_5$ isolated in matrices of CH_4, Xe, or Kr.[1] I.r. examination along with isotopic labelling (^{13}CO) indicates a pyramidal C_{3v} structure for the new species, with bond angles at iron near 108°. This geometry was unexpected and matrix interactions could be responsible, though it seems that it can be accounted for by ground-state bonding effects. Further photolysis leads to the formation of $Fe(CO)_2$ or $Fe(CO)$, but these were not examined in detail. The effect of sunlight on heptane solutions of $Ru_3(CO)_{12}$ in the presence of CO is to produce monomeric $Ru(CO)_5$, a somewhat surprising result in view of the high strength of $Ru-Ru$ bonds compared to $Ru-CO$.[2] Irradiation of $Ru_3(CO)_{12}$ in the presence of ethylene or Ph_3P leads, respectively, to $Ru(CO)_4C_2H_4$ and $Ru(CO)_3(Ph_3P)_2$. Vibronic excitation cannot be ruled out, but electronic $\sigma \rightarrow \sigma^*$ transitions are more likely to be responsible for the bond cleavage reactions.

The dinuclear complexes $Rh_2(CO)_8$ and $Ir_2(CO)_8$ have been sought after for many years, and reactions such as $Rh_4(CO)_{12}$ with CO under pressure have been tried, possibly with success, but the problems of identifying the products have

[1] M. Poliakoff, *J.C.S. Dalton*, 1974, 210.
[2] B. F. G. Johnson, J. Lewis, and M. V. Twigg, *J. Organometallic Chem.*, 1974, **67**, C75.

been frustrating. Condensation of Rh or Ir atoms onto a matrix of pure CO at 15 K have produced $M(CO)_4$ and $M_2(CO)_8$, all identified by their i.r. spectra.[3] High metal-atom densities favour production of the dinuclear species, and warming the lattices to 45—50 K promoted the dimerization of $M(CO)_4$. By comparison with the spectra of $Co_2(CO)_8$, the new dinuclear species seem bridge-bonded. $Ir_2(CO)_8$ dimerizes to $Ir_4(CO)_{12}$ more readily than its rhodium analogue.

The oxidation of $Cr(CO)_6$ at a platinum electrode in a $0.2M$-Bu_4NBF_4 solution in MeCN leads to a novel 17-electron species, $Cr(CO)_6{}^+$. The only other known 17-electron metal carbonyl is the isoelectronic $V(CO)_6$. Two-electron oxidations for $Cr(CO)_6$ and other metal carbonyls, all leading to unstable ions, were also reported.[4]

Chemical studies on technetium compounds usually reveal properties intermediate between those of manganese and rhenium. The i.r. spectrum of $Tc_2(CO)_{10}$ studied this year, however, indicates that both the axial and equatorial C—O stretching force constants are higher than those of the Mn and Re analogues.[5] Other parameters adopt the expected intermediate values. The trinuclear hydride $Mn_3H_3(CO)_{12}$, reported last year, reacts with alcoholic base to form the anion $Mn_3(CO)_{14}{}^-$. Isolated as red-orange air-sensitive crystals with the cation Ph_4As^+, X-ray analysis[6] reveals a collinear array of manganese atoms, with a staggered conformation of equatorial carbonyls. Like the isoelectronic $FeMn_2(CO)_{14}$, steric considerations probably dictate the structure adopted.

The proposition that bridging carbonyls are obtained only between metals formally joined by a metal–metal bond has been made in a letter which suggests that the only exception to this, $[CpRh(CO)_2]_2$, appears not to exist and may in reality be $Cp_2Rh_2(CO)_3$[7] ($Cp = h^5$-C_5H_5). A bonding theory which claims that the M—C and M—M orbitals are inseparable moieties is cited in support. The intermolecular redistribution of CO, X, and PR_3 between $M(CO)X(PR_3)_2$ compounds (M = Rh or Ir; X = halogen) has been reported as proceeding *via* double CO or X bridges.[8] Similar routes for CO migration between metals were discussed last year. In this case, at least, the presence of accompanying metal–metal bonds does not seem necessary.

One of the compounds isolated from the reaction of $Fe_2(CO)_9$ and Ph_2PCH_2-PPh_2 has been characterized crystallographically as (1).[9] It has a rather long Fe—Fe bond (270.9 pm) and a symmetrically bridging CO. ^{13}C N.m.r. shows that all the carbonyls are rapidly scrambled at 298 K, but the scrambling mechanism is not yet established.

[3] L. A. Hanlan and G. A. Ozin, *J. Amer. Chem. Soc.*, 1974, **96**, 6324.
[4] C. J. Pickett and D. Pletcher, *J.C.S. Chem. Comm.*, 1974, 660.
[5] G. Bor and G. Sbrignadello, *J.C.S. Dalton*, 1974, 440.
[6] R. Bau, S. W. Kirtley, T. N. Sorrell, and S. Winarko, *J. Amer. Chem. Soc.*, 1974, **96**, 988.
[7] F. A. Cotton and D. L. Hunter, *Inorg. Chem.*, 1974, **13**, 2044.
[8] P. E. Garrou and G. E. Hartwell, *J.C.S. Chem. Comm.*, 1974, 381.
[9] F. A. Cotton and J. M. Troup, *J. Amer. Chem. Soc.*, 1974, **96**, 4422.

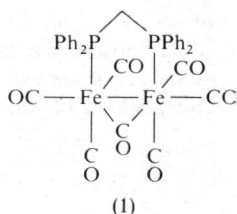

(1)

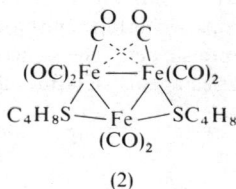

(2)

The X-ray crystal structure of $Fe_3(CO)_8(SC_4H_8)_2$ (2) reveals a pair of very unsymmetrical ('borderline') bridging carbonyls.[10] The degree of asymmetry of pairs of bridging groups varies in a series of compounds from the completely symmetric $Fe_3(CO)_9(PPhMe_2)_3$, through the slightly asymmetric $Fe_3(CO)_9$-$(PPh_3)_3$ and the markedly asymmetric $Fe_3(CO)_{12}$, to the present compound (2). Interestingly, the greater degrees of asymmetry are accompanied by longer $Fe-Fe$ bonds [from 254 pm in $Fe_3(CO)_9(PPhMe_2)_3$ to 264 pm in (2)]. The variations found in these four compounds have been suggested as models for intermediate structures passed through during CO exchange *via* double-bridge formation.

Bipyridyl reacts with $Fe_2(CO)_9$ to produce (3), with one only of the two bridging carbonyls very asymmetric.[11] An i.r. band at 1850 cm^{-1} means that this structure

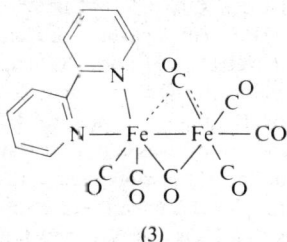

(3)

persists in solution. This structure is rationalized in terms of bipy being a good σ-electron donor, but a poor π-acceptor compared to CO. Substitution by bipy therefore generates a need for greater π-acidity amongst the remaining ligands. The semi-bridging carbonyl provides a route for relieving electron density from the electron-rich metal atom by a $d \rightarrow \pi^*$ overlap onto a ligand associated with a different metal. A symmetrically bridging CO is a less efficient π-acid than a terminal one.

The complexes $Cp_2Fe_2(CO)(CNMe)_3$ and $Cp_2Fe_2(CO)_2(CNMe)_2$ exist in several isomeric forms due to the bent configuration of bridging CNMe ligands.[12] One isomer of the latter compound was isolated and examined crystallographically (4). Isomerization *via* N-inversion is fast on the n.m.r. time-scale at 298 K, but

[10] F. A. Cotton and J. M. Troup, *J. Amer. Chem. Soc.*, 1974, **96**, 5070.
[11] F. A. Cotton and J. M. Troup, *J. Amer. Chem. Soc.*, 1974, **96**, 1233.
[12] (*a*) R. D. Adams and F. A. Cotton, *Inorg. Chem.*, 1974, **13**, 249; (*b*) F. A. Cotton and B. A. Frenz, *ibid.*, p. 253; (*c*) R. D. Adams, F. A. Cotton, and J. M. Troup, *ibid.*, p. 257.

separate sharp signals are discerned for the isomers at 173 K. Exchange of terminal groups *via* bridge positions also involves inversion, but takes place about 10^5 times slower. The structure of (4) suggests that there will be a considerable barrier against CNBu^t adopting a bridge position because of steric

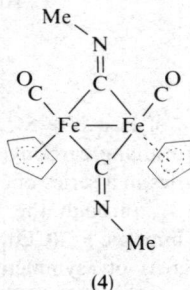

(4)

strain. In fact the only isomer of Fe$_2$Cp$_2$(CO)$_3$(CNBu^t) isolated pure was a *cis* complex with a terminal isocyanide. This terminal ligand flips from iron to iron, presumably through a bridged intermediate, but this process is slower than *cis–trans* isomerism, which involves bridge–terminal migrations of carbonyls.

The structures and reactivities of trinuclear metal carbonyl clusters continue to excite interest. Several Co$_3$(CO)$_9$CCR^1=CR^{2}R^3 have been prepared, and some of these react with HPF$_6$ *via* β-carbon proton addition to form stable carbonium ions Co$_3$(CO)$_9$CCHR$^+$. ^{1}H and ^{13}C n.m.r. spectrometry provides further evidence of charge delocalization into the CCo$_3$ system.[13] The action of ethylene on Os$_3$(CO)$_{12}$ or Ru$_3$(CO)$_{12}$ leads to H$_2$M$_3$(CO)$_9$(C=CH$_2$) (the structure of the osmium compound was reported last year) and H$_2$Ru$_3$(CO)$_9$-(CH=CH), for which structure (5) is suggested[14] (the position of the metal-bridging protons in these complexes is uncertain). Protonation of the tri-osmium cluster by CF$_3$CO$_2$H leads to [H$_3$Os$_3$(CO)$_9$(C=CH$_2$)]$^+$, which appears to have structure (6). This is related to the Co$_3$(CO)$_9$CCHR$^+$ species and might represent

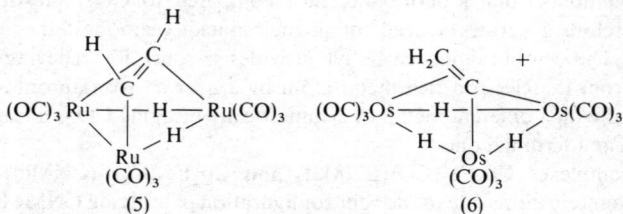

(5)   (6)

an alternative formulation. The neutral compound H$_3$Os$_3$(CO)$_9$(C—CH$_3$), made by passing H$_2$ through solutions of H$_2$Os$_3$(CO)$_9$(C=CH$_2$), has been

[13] (a) D. Seyferth, C. S. Eschbach, G. H. Williams, P. L. K. Hung, and Y. M. Cheng, *J. Organometallic Chem.*, 1974, **78**, C13: (b) D. Seyferth, G. H. Williams, and D. D. Traficante, *J. Amer. Chem. Soc.*, 1974, **96**, 604.

[14] A. J. Deeming, S. Hasso, M. Underhill, A. J. Canty, B. F. G. Johnson, W. G. Jackson, J. Lewis, and T. W. Matheson, *J.C.S. Chem. Comm.*, 1974, 807.

examined by X-ray powder photography and nematic-phase ^{1}H n.m.r. and assigned a symmetrical structure similar to that of the ruthenium analogue reported last year.[15] The Os—H (bridge) separations are 182 pm, with angle OsHOs 103°.

The high reactivity of these metal clusters seems quite general. $Os_3(CO)_{12}$ reacts with Me_3P or Et_3P to form $Os_3(CO)_{11}PR_3$ or $Os_3(CO)_{10}(PR_3)_2$, and heating these in nonane produces $H_2Os_3(CO)_9(R_2PCX)$ and $H_2Os_3(CO)_8$-$(R_3P)(R_2PCX)$.[16] These have skeletons (7), clearly related to the structures

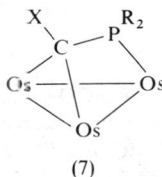

(7)

discussed in the last paragraph Ligands such as 1,4-*trans*-diphenylbuta-1,3-diene, hexadiene, or isoprene react with $Ru_3(CO)_{12}$ to give $HRu_3(CO)_9(L - H)$, as well as known Ru and Ru_2 derivatives.[17]

The reduction of $PtCl_6^{2-}$ in the presence of CO by alkali metals or alkaline methanol leads to a remarkable series of anions $[Pt_3(CO)_6]_n^{2-}$ ($n = 2, 3, 4,$ or 5).[18] Isolated as coloured, crystalline salts of Ph_4P^+ or Ph_4As^+, crystal structure determinations show them to be built up from eclipsed $Pt_3(CO)_3(\mu\text{-}CO)_3$ units (8), with only slight leaning or twisting. Metal-atom frameworks for 'dimer', 'trimer', and 'pentamer' are shown in (9)—(11). Pt—Pt separations in a given

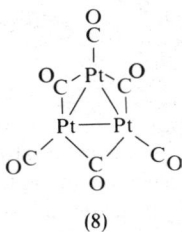

(8) (9) (10) (11)

triangle are all near 266 pm, whereas individual triangles are separated by Pt—Pt distances near 310 pm. The analogous nickel complex has similar $Ni_3(CO)_6^{2-}$ triangles, but there is anti-prismatic packing in $[Ni_3(CO)_3$-

[15] J. P. Yesinowski and D. Bailey, *J. Organometallic Chem.*, 1974, **65**, C27.
[16] A. J. Deeming and M. Underhill, *J.C.S. Dalton*, 1973, 2727.
[17] O. Gambino, M. Valle, S. Aime, and G. A. Vaglio, *Inorg. Chim. Acta*, 1974, **8**, 71.
[18] J. C. Calabrese, L. F. Dahl, P. Chini, G. Longoni, and S. Martinengo, *J. Amer. Chem. Soc.*, 1974, **96**, 2614.

$(\mu\text{-CO})_3]_2{}^{2-}$ (12).[19] Here, too, the inter-triangle Ni—Ni distances (277 pm) are greater than those in the triangles (238 pm).

The oxidation of $K_2[Rh_6(CO)_{15}C]$ in water and under CO leads to two new carbido-carbonyl clusters, $Rh_8(CO)_{19}C$ (13) and $[H_3O][Rh_{15}(CO)_{28}C_2]$, characterized crystallographically;[20] the carbido carbon atom of (13) is located in the Rh_6 prism. Treatment of (13) with MeCN forms the known $[Rh_6(CO)_{15}C]^{2-}$,

(12)   (13)

along with $[Rh(CO)_2(MeCN)_2]^+$. $Co(CO)_4{}^-$ reacts with $Co_3(CO)_9CCl$ to form the dianion $[Co_6(CO)_{15}C]^{2-}$, the first carbido-carbonyl cluster of cobalt.[20] The ^{13}C n.m.r. spectrum of $[Rh_{12}(CO)_{30}]^{2-}$ agrees with the reported crystallographic structure, $[Rh_{12}(CO)_{20}(\mu_2\text{-CO})_2(\mu_3\text{-CO})_8]^{2-}$, showing that the structure persists in solution.[21] This is the first reported ^{13}C resonance for a triply-bridging carbonyl. The crystal structure of $(Me_4N)_2[Co_4Ni_2(CO)_{14}]$ reveals a near-octahedral metal cluster with triply-bridging CO's over each face and one terminal CO per metal atom.[22]

Crystals of $Ir_2Co_2(CO)_{12}$ proved to be twinned and disordered, making X-ray analysis difficult.[23] Nevertheless, refinement suggested structure (14),

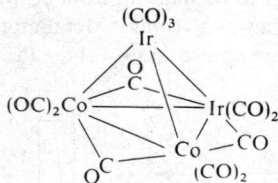

(14)

with three bridging carbonyls. The disorder amongst the metal atoms was not complete, with the heavier metal showing a preference for the apical position. ^{13}C N.m.r. examination of the related $RhCo_3(CO)_{12}$ indicates that site-exchange of CO takes place *via* two mechanisms, one operating at 230 K, and the other becoming rapid only above 243 K.[24] The complexes $Ir_4(CO)_8L_4$, prepared by the action of L (various P and As donors) on $Ir_4(CO)_{12}$, also all contain bridge carbonyls.[25] Scrambling of the carbonyls above ambient temperature is revealed

[19] J. C. Calabrese, L. F. Dahl, A. Cavalieri, P. Chini, G. Longoni, and S. Martinengo, *J. Amer. Chem. Soc.*, 1974, **96**, 2616.

[20] V. G. Albano, P. Chini, S. Martinengo, M. Sansoni, and D. Strumbolo, *J.C.S. Chem. Comm.*, 1974, 299.

[21] P. Chini, S. Martinengo, D. J. A. McCaffrey, and B. T. Heaton, *J.C.S. Chem. Comm.*, 1974, 310.

[22] V. G. Albano, G. Ciani, and P. Chini, *J.C.S. Dalton*, 1974, 432.

[23] V. G. Albano, G. Ciani, and S. Martinengo, *J. Organometallic Chem.*, 1974, **78**, 265.

[24] B. F. G. Johnson, J. Lewis, and T. W. Matheson, *J.C.S. Chem. Comm.*, 1974, 441.

[25] P. E. Cattermole, K. G. Orrell, and A. G. Osborne, *J.C.S. Dalton*, 1974, 328.

here by variable-temperature n.m.r. studies. In acid solution, non-rigid H and then H_2 adducts were formed.

Heterocyclic compounds containing azo links, *e.g.* (15), react with photochemically produced $Cr(CO)_5(THF)$ to displace the THF and bond chromium *via* one nitrogen atom only.[26] Related N_1-bonded derivatives $(h^6\text{-}C_6H_6)Cr(CO)_2L$ have also been made. The same ligands can act as bridges, however, and some

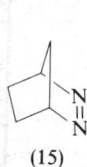

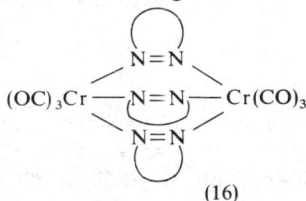

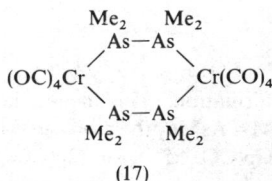

(15)   (16)   (17)

very symmetrical complexes (16) have resulted. Cacodylic acid, $Me_2As(O)OH$, reacts with $Cr(CO)_6$ to form an unexpected complex (17) with two As_2Me_4 bridges.[27] Treatment of $[CpMo(NO)X_2]_2$ with Me_2NNH_2 affords (18), for which crystal structure has been determined for $X = I$.[28] This very unusual

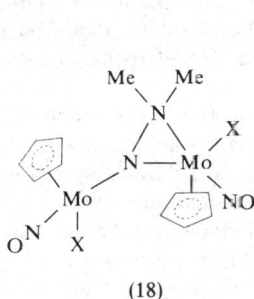

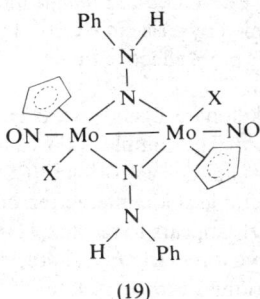

(18)   (19)

structure contains a bridge hydrazide ligand, attached to one Mo by one N only and to the other by both N atoms. The Mo-N-N-Mo skeleton is essentially planar. Phenylhydrazine affords a product containing two hydrazide bridges when allowed to react with $[CpMo(NO)X_2]_2$. The n.m.r. spectra of the product favour (19) as the most likely structure. $SbCl_3$ forms a $1:2$ adduct with $CpFe(CO)_2Cl$. The adduct contains chlorine atoms in unusual positions bridging Fe and Sb (20).[29]

In general, complexes with terminal di-organophosphido groups, $M-PR_2$, and their arsenic analogues, were thought to be obtainable only when R is sufficiently electron-withdrawing (*e.g.* CF_3, C_6F_5, or Cl) to prevent bridge

[26] (a) M. Herberhold and W. Golla, *Chem. Ber.*, 1974, **107**, 3199; (b) M. Herberhold, W. Golla, and K. Leonhard, *ibid.*, p. 3209; (c) M. Herberhold, K. Leonhard, and C. G. Kreiter, *ibid.*, p. 3222.

[27] F. A. Cotton and T. R. Webb, *Inorg. Chim. Acta*, 1974, **10**, 127.

[28] W. G. Kita, J. A. McCleverty, B. E. Mann, D. Seddon, G. A. Sim, and D. I. Woodhouse, *J.C.S. Chem. Comm.*, 1974, 132.

[29] F. W. B. Einstein and A. C. MacGregor, *J.C.S. Dalton*, 1974, 778.

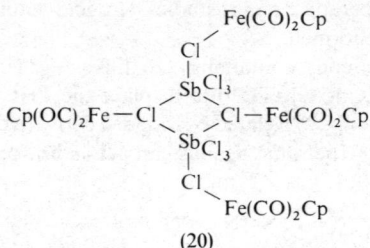

(20)

formation. This makes the report this year of a series of complexes $Cp(CO)_n$-M—$AsMe_2$ the more surprising[30] (M = Cr, Mo, W, or Fe). Prepared by treating $Cp(CO)_nMNa$ or $Cp(CO)_nMSiMe_3$ with Me_2AsCl at 298 K in a non-polar solvent, the complexes are remarkably stable, even in solution. Organic halides react with them to form quaternary arsonium salts, and heat causes decomposition to Me_4As_2 and $[Cp(CO)_nM]_2$. Molecular structures of $Cp(OC)_2FeP(CF_3)_2$ and its oxide $Cp(OC)_2FeP(O)(CF_3)_2$ have been determined.[31] On oxidation at phosphorus, the Fe—P bond length decreases from 226.5 to 219.1 pm, whilst the mean Fe—CO bond length increases from 176.8 pm to 178.0 pm, the latter accompanied by a rise in $\nu(CO)$. The explanation for this is that π-bonding from Fe to P^V is more effective than to P^{III}, and occurs at the expense of back donation from Fe to CO.

A β-diketonate ligand has been detected in its *trans* isomeric form for the first time.[32] The complex (21), examined crystallographically, was a unidentate thiodiketonate derivative of $W(CO)_5$. The usual *cis* configuration would lead to considerable steric interaction between ligand and carbonyls.

$Fe_2(CO)_9$ appears to react in THF in a different manner than in other solvents. Perhaps most remarkably, simple amine substitution products like $Fe(CO)_4(py)$ (py = pyridine) can be isolated.[33] It had been suspected that they might not be available due to disproportionation. In fact they are quite stable and the nitrogen donor adopts an axial position both in the solid and in solution.

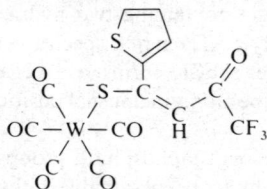

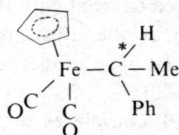

(21)   (22)

[30] W. Malisch and M. Kuhn, *Angew. Chem. Internat. Edn.*, 1974, **13**, 84.
[31] M. J. Barrow, G. A. Sim, R. C. Dobbie, and P. R. Mason, *J. Organometallic Chem.*, 1974, **69**, C4.
[32] M. McPartlin, G. B. Robertson, G. H. Barnett, and M. K. Cooper, *J.C.S. Chem. Comm.*, 1974, 305.
[33] F. A. Cotton and J. M. Troup, *J. Amer. Chem. Soc.*, 1974, **96**, 3438.

In complex (22), where C* is an asymmetric centre, the two carbonyls should be diastereotopic. This diastereotopy has been detected by ^{13}C n.m.r. spectrometry.[34] Two chemically shifted ^{13}C (carbonyl) resonances were detected, when only one is apparent in *e.g.* $Cp(OC)_2FeCH_2Ph$.

The photochemical decarbonylation of $(+)-(R)-CpFe(CO)(PPh_3)COEt$ to $(-)-(S)-CpFe(CO)(PPh_3)Et$ proceeds with 43% optical purity.[35] The simplest interpretation is that the ethyl group migrates into the co-ordination site left by the eliminated CO. Decarbonylation of $ArCOCl$ or $ArCH_2COCl$ by Wilkinson's catalyst, $(Ph_3P)_3RhCl$, possibly proceeds *via* similar migrations.[36] Intermediates $Rh(PPh_3)_2Cl_2(COR)$ and $Rh(PPh_3)_2Cl_2(CO)R$ were isolated. In this case, free co-ordination sites for R-migration were already present. Decarbonylation of $(-)-(S)-\alpha-CF_3PhCHCOCl$ gives a racemic product, and racemization probably takes place at the acyl → alkyl step (the asymmetric centre is carbon in this case, as opposed to the metal atom for the photochemical reaction).

Insertion of CO into metal–carbon bonds is enhanced by the effects of amines, phosphines, alcohols, or thiocyanate in S-donor *ortho*-metallated iron complexes.[37] Isolation of the intermediates establishes Scheme 1.

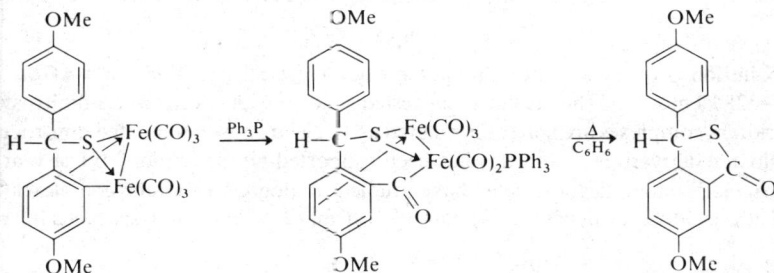

Scheme 1

Oxygen bonding of co-ordinated carbonyls to form bridges to main-group or other transition elements is well established and might be a very common phenomenon despite the lateness of its discovery. The i.r. spectra of $LiMn(CO)_5$ or $NaMn(CO)_5$ in ether, for example, indicate the presence of ion-paired species of C_{3v} symmetry with the alkali metal bonded to a carbonyl oxygen.[38] In THF, ion-paired $NaMn(CO)_5$ is in equilibrium with separate ions. The interaction of $NaMn(CO)_5$ with Mg^{2+} produces species whose i.r. spectra are consistent with $Mg-O-C$ interactions. ^7Li N.m.r. spectra of mixtures of $(MeLi)_4$ and $LiMn(CO)_5$ or $LiCo(CO)_4$ show the presence of these pure species only, so

[34] T. Yu. Orlova, P. V. Petrovskii, V. N. Setkina, and D. N. Kursanova, *J. Organometallic Chem.*, 1974, **67**, C23.
[35] A. Davison and N. Martinez, *J. Organometallic Chem.*, 1974, **74**, C17.
[36] (a) J. K. Stille and M. T. Regan, *J. Amer. Chem. Soc.*, 1974, **96**, 1508; (b) J. K. Stille and R. W. Fries, *ibid.*, p. 1514.
[37] H. Alper, W. G. Root, and A. S. K. Chan, *J. Organometallic Chem.*, 1974, **71**, C14.
[38] C. D. Pribula and T. L. Brown, *J. Organometallic Chem.*, 1974, **71**, 415.

metal carbonyl fragments do not invade the tetrameric structure of methyl-lithium. The reactions of various chlorosilanes with $Co_2(CO)_8$ or $NaCo(CO)_4$ in ethers have been studied,[39] and these appear to feature electrophilic attack on oxygen also, producing, for example, $Cl_3Si—O—CCo_3(CO)_9$.

Methyl-lithium reacts with $Mn(CO)_5(COPh)$ to yield $Li[cis-(OC)_4Mn(COPh)-(COMe)]$, isolated as the Me_4N^+ salt.[40] Attack might have been expected not at CO, but at the COPh group, as this resembles a carbene. Heating the product to 343 K in THF produced acetophenone, and the action of bromine was to form $PhC(O)C(O)Me$. The reaction of the ylide $(Ph_3P)_2C$ with $Mn(CO)_5Br$ leads to Ph_3PO and $cis-(OC)_4\overset{-}{M}nBr(C\equiv\overset{+}{C}PPh_3)$.[41] This is the first Wittig-type reaction on a metal carbonyl. Normally an ylide carbene adduct forms, which does not eliminate triphenylphosphine oxide.

The ability of $Ir(PPh_3)_2(NO)_2^+$ to oxidize CO and Ph_3P was reported last year, and this year sees nitric oxide added to the list (equation 1), although the catalytic reaction of CO and NO (equation 2) at the iridium dinitrosyl complex remains more noteworthy because of its potential application to environmental

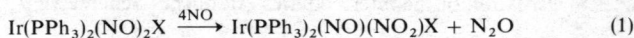

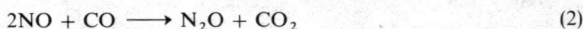

$$Ir(PPh_3)_2(NO)_2X \xrightarrow{4NO} Ir(PPh_3)_2(NO)(NO_2)X + N_2O \qquad (1)$$

$$2NO + CO \longrightarrow N_2O + CO_2 \qquad (2)$$

pollution problems.[42] For this latter reaction, ΔH°_{298} is -382 and ΔG°_{298} is $-328\,kJ\,mol^{-1}$. The authors suggested that the 20-electron cis-dinitrosyl-iridium complexes are better formulated as 18-electron $N—N$ bonded dinitrogen dioxide derivatives,[42] and this has been supported by the authors of last year's original communication, who have studied analogous rhodium complexes.[43] They produce evidence in support of Scheme 2 as the catalysis mechanism.

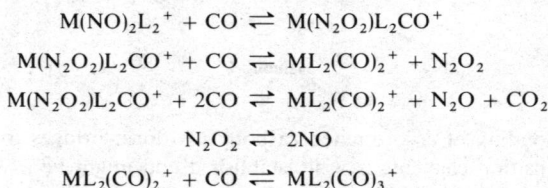

$$M(NO)_2L_2^+ + CO \rightleftharpoons M(N_2O_2)L_2CO^+$$

$$M(N_2O_2)L_2CO^+ + CO \rightleftharpoons ML_2(CO)_2^+ + N_2O_2$$

$$M(N_2O_2)L_2CO^+ + 2CO \rightleftharpoons ML_2(CO)_2^+ + N_2O + CO_2$$

$$N_2O_2 \rightleftharpoons 2NO$$

$$ML_2(CO)_2^+ + CO \rightleftharpoons ML_2(CO)_3^+$$

M = Rh or Ir. L is tertiary phosphine.

Scheme 2

Interestingly, they also observe that the coupling of two NO molecules to a co-ordinated N_2O_2 group was probably discovered first in 1972, with the formation of (23).

[39] B. K. Nicholson, B. H. Robinson, and J. Simpson, *J. Organometallic Chem.*, 1974, **66**, C3.
[40] C. P. Casey and C. A. Bunnell, *J.C.S. Chem. Comm.*, 1974, 733.
[41] W. C. Kaska, D. K. Mitchell, R. F. Reichelderfer, and W. D. Korte, *J. Amer. Chem. Soc.*, 1974, **96**, 2847.
[42] B. L. Haymore and J. A. Ibers, *J. Amer. Chem. Soc.*, 1974, **96**, 3325.
[43] S. Bhaduri, B. F. G. Johnson, C. J. Savory, J. A. Segal, and R. H. Walter, *J.C.S. Chem. Comm.*, 1974, 809.

$$Ph_3P \diagdown \atop Ph_3P \diagup Pt \diagdown_{N\diagup}^{N=O} \diagdown_{O}$$

(23)

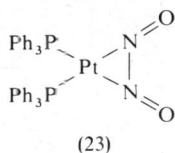

The complexes *trans*-$(Ph_3P)_2M(CO)OH$ (M = Rh or Ir) react in crystalline form with CO_2 under normal conditions.[44] The 1 : 1 adducts are formed, and their i.r. spectra show that the symmetry of the CO_2 is reduced. The carbon dioxide is released on pumping to regenerate the original complex. Carbon dioxide also reacts with $[(Ph_3P)Rh(CO)_2]_2 \cdot C_6H_6$ in the presence of Ph_3P in solution, the product being $(Ph_3P)_3Rh_2(CO)_2(CO_2)_2, C_6H_6$.[45] With Wilkinson's catalyst in solution with $HSi(OEt)_3$, CO_2 carbonylates the metal to form $(Ph_3P)_2Rh(CO)Cl$ quantitatively.[46] Some ruthenium complexes behave similarly. These are the first reported examples of carbonylation by CO_2 under mild conditions. The interaction of CO_2 with transition-metal complexes could prove to be an exciting area of chemistry in the near future.

Salts of thionitrosyl, NS^+, were reported in 1971, and a thionitrosyl complex has been isolated this year.[47] $MoN(S_2CNR_2)_3$ [R_2 = Me_2, Et_2, or $(CH_2)_5$] reacts with S_8 or propylene sulphide in refluxing MeCN to produce 90% yields of $Mo(NS)(S_2CNR_2)_3$. The same complex is formed in 50% yield by the action of Me_3SiN_3 on $MoO_2(S_2CNMe_2)_2$ in the presence of tetramethylthiuram disulphide. N.m.r. spectrometry suggests a pentagonal-bipyramidal structure with apical NS. The mass spectra of these monomeric orange-red crystalline solids show parent ions, and loss of NS. $v(NS)$ is *ca.* 1100 cm^{-1}, and Bu_3P destroys the complexes to regenerate the molybdenum nitride.

The crystal structure of $[Pt_2Cl_6(NO)_2]^{2-}$ shows the unusual unsymmetrically bridged anions (24).[48] The NO groups bend towards each other, suggesting

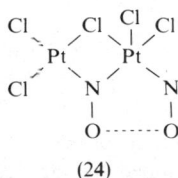

$$\begin{array}{c} Cl \diagdown \atop Cl \diagup Pt \diagup^{Cl} \diagdown_{N}^{Cl} \diagdown_{N}^{Cl} \diagup Pt \\ \quad O \cdots\cdots O \end{array}$$

(24)

opposite charges. *X*-ray crystallographic analysis of $Cp_2Cr_2(NO)_4$, on the other hand, reveals it to be the first symmetrical doubly-bridged nitrosyl complex

[44] B. R. Flynn and L. Vaska, *J.C.S. Chem. Comm.*, 1974, 703.
[45] I. S. Kolomnikov, T. S. Belopotapova, T. V. Lysyak, and M. E. Vol'pin, *J. Organometallic Chem.*, 1974, **67**, C25.
[46] P. Svoboda, T. S. Belopotapova, and J. Hetflejš, *J. Organometallic Chem.*, 1974, **65**, C37.
[47] J. Chatt and J. R. Dilworth, *J.C.S. Chem. Comm.*, 1974, 508.
[48] J. M. Epstein, A. H. White, S. B. Wild, and A. C. Willis, *J.C.S. Dalton*, 1974, 436.

(25)   (26)

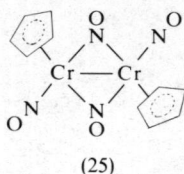

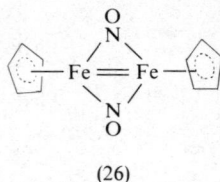

studied.[49] Although *trans* in the crystalline state (25), n.m.r. and i.r. spectroscopic examination indicate fluxional behaviour in solution. CpFeNO (26) is also dimeric and has symmetrical (NO)$_2$ bridges.[50] It has a short Fe—Fe bond (232.6 pm compared to the normal Fe—Fe separation of *ca.* 260 pm) indicative of double bonding.

A full report on the molecular structure of (CpMn)$_3(\mu_2$-NO)$_3(\mu_3$-NO) has appeared.[51] The molecules (27), of C_{3v} symmetry, show longer bonds to triply bridging nitrosyl (124.7 pm) than to μ_2-NO (121.3 pm). Several new complexes

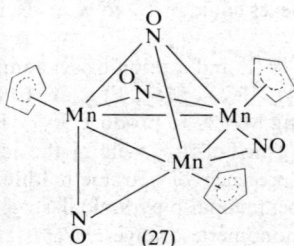

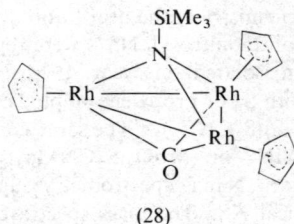

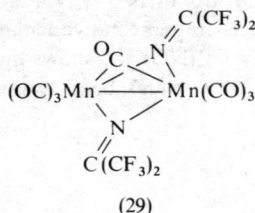

(27)   (28)

(29)

of trimethylsilylnitrene have been reported.[52] Formed by decomposition of Me$_3$SiN$_3$, this ligand can be trapped by various transition-metal complexes. The structure (28) is typical; Me$_3$SiN acts as a formal four-electron donor. The complex (29) is the first containing a bridging, one-electron methyleneamido-group. The molecule is analogous to Fe$_2$(CO)$_9$.[53]

[49] J. L. Calderón, S. Fontana, E. Frauendorfer, and V. W. Day, *J. Organometallic Chem.*, 1974, **64**, C10.
[50] J. L. Calderón, S. Fontana, E. Frauendorfer, V. W. Day, and S. D. A. Iske, *J. Organometallic Chem.*, 1974, **64**, C16.
[51] R. C. Elder, *Inorg. Chem.*, 1974, **13**, 1037.
[52] E. W. Abel, T. Blackmore, and R. J. Whitley, *Inorg. Nuclear Chem. Letters*, 1974, **10**, 941.
[53] E. W. Abel, C. A. Burton, M. R. Churchill, and K.-K. G. Lin, *J.C.S. Chem. Comm.*, 1974, 917.

The first thiocarbonyl complexes of chromium, molybdenum, and tungsten, $M(CO)_5CS$ and $M(CO)_4(PPh_3)CS$, have been isolated.[54] Made in low yield *via* the reaction of dianions, such as $Cr(CO)_5^{2-}$, and Cl_2CS, they are stable to air and moisture. Interestingly, the Mo derivatives are less stable than those of Cr and W. The ligand CS resembles PF_3 in being a good electron acceptor in these compounds. Like the known manganese thiocarbonyls, CS appears more firmly bonded than CO, as ligand substitution reactions eliminate the latter. $W(CO)_5CS$ reacts at room temperature with Me_2NH to produce $W(CO)_5$-$[C(SH)NMe_2]$, so the co-ordinated group is quite reactive.

3 Hydride Complexes

One of the most celebrated and oft-quoted examples of oxidative addition of a C—H bond to a transition metal, the tautomerism between $(Me_2PC_2H_4PMe_2)_2Ru$ and $(Me_2PC_2H_4PMe_2)Ru(H)(Me_2PC_2H_4PMeCH_2)$, has been shown to be incorrect in one detail. The C—H addition is not intramolecular, but intermolecular, leading to the dimeric structure (30).[55]

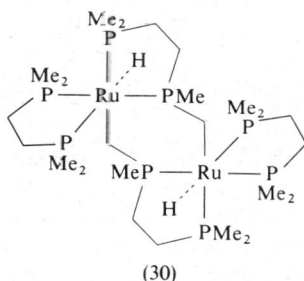

(30)

Complex hydrides $MCuH_2$ are useful reducing agents, converting not only sp^3C-halides but sp^2C-halides into the corresponding hydrides, a difficult reaction and one not achievable using one of the parent reagents KBu^s_3BH.[56,57]

The preparation of platinum dihydrides by the action of $AlEt_3$ and bulky tertiary phosphines on $Pt(acac)_2$ was reported last year, but this appears to have been in error, as $(Cx_3P)_2PtH_2$ has now been prepared from $(Cx_3P)_2PtCl_2$ and $NaBH_4$ and has different properties (Cx = cyclohexyl).[58] Curiously, the Pt—H frequency (1710—1750 cm^{-1}) in *trans*-L_2PtH_2 is close to that assigned for $(Ph_3P)_2PtH_2$, reported 11 years ago[59] but subsequently shown to be $(Ph_3P)_2$-$PtCO_3$.

The controversy over the cause of signal broadening of one n.m.r. hydride signal of *trans*-$(Et_3P)_2PtH(SCN)$ is finally closed, ^{14}N decoupling confirming

[54] B. D. Dombek and R. J. Angelici. *J. Amer. Chem. Soc.*, 1973, **95**, 7516.
[55] F. A. Cotton, B. A. Frenz, and D. L. Hunter, *J.C.S. Chem. Comm.*, 1974, 755.
[56] E. C. Ashby, T. F. Korenowski, and R. D. Schwartz, *J.C.S. Chem. Comm.*, 1974, 157.
[57] T. Yoshida and E.-I. Negishi, *J.C.S. Chem. Comm.*, 1974, 762.
[58] B. L. Shaw and M. F. Uttley, *J.C.S. Chem. Comm.*, 1974, 918.
[59] L. Malatesta and R. Ugo, *J. Chem. Soc.*, 1963, 2080.

last year's support of quadrupolar coupling to ^{14}N in the N-bonded isomer as the source.[60]

Irradiation of Cp_2WH_2 at 366 nm in mesitylene produces $Cp_2W[CH_2(3,5-Me_2C_6H_3)]_2$.[61] Similar reactions forming Cp_2WHR by insertion of W into $C-H$ are known, but there is no precedent for the present disubstitution process. A plausible mechanism involves the reversible transfer of H or R between tungsten and a cyclopentadienyl ring, giving $CpWR(C_5H_6)$, which would react further to form $CpWHR_2(C_5H_6)$ by a second insertion into $C-H$. Elimination of H_2 finally forms the product. The reaction of sodium trichloroacetate in diglyme with Cp_2WH_2 produces $Cp_2WH(CCl_2H)$.[62] The analogous reaction with NaO_2CCFCl_2, however, forms only $Cp_2W(O_2CCFCl_2)_2$.

White phosphorus reacts in toluene with Cp_2MoH_2 to form red, crystalline $Cp_2Mo(P_2H_2)$.[63] Spectroscopic investigation of this oxygen-sensitive material suggests structure (31), but it is not yet known if the P_2H_2 moiety is planar, cis,

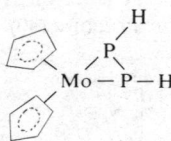

(31)

or trans. The complex is probably related to the known $Cp_2Mo(C_2H_4)$, and provides a model for a di-imine derivative, since P_2H_2, N_2H_2, and C_2H_4 are isoelectronic. Cp_2MoH_2 reacts with dimethyl fumarate or maleate by specific cis addition of hydrogen to the olefinic bond, then elimination to racemic or meso products.[64] The rate-determining step is elimination to form reactive Cp_2Mo, and this takes place with retention of configuration at the σ-bonded α-carbon atom. Kinetic studies suggest that the 1 : 1 adduct $Cp_2MoH_2(olefin)$ is formed prior to cis insertion.

The formation of chiral products from achiral reactants with an optically active catalyst is known in several cases. An example this year is the formation of chiral alkoxysilanes from the asymmetric hydroxylation of ketones at a chiral rhodium catalyst.[65] Asymmetry at the catalyst is achieved using the chiral diphosphine 2,3-O-isopropylidene-2,3-dihydroxy-1,4-bisdiphenylphosphinobutane (diop). An insoluble rhodium(I) catalyst prepared from a resin-bonded form of diop has also been described.[66] This heterogeneous catalyst (32) is

[60] B. E. Mann, B. L. Shaw, and A. J. Stringer, J. Organometallic Chem., 1974, 73, 129.
[61] K. Elmitt, M. L. H. Green, R. A. Forder, I. Jefferson, and K. Prout, J.C.S. Chem. Comm., 1974, 747.
[62] K. S. Chen, J. Kleinberg, and J. A. Landgrebe, Inorg. Chem., 1973, 62, 2826.
[63] J. C. Green, M. L. H. Green, and G. E. Morris, J.C.S. Chem. Comm., 1974, 212.
[64] A. Nakamura and S. Otsuka, J. Amer. Chem. Soc., 1973, 95, 7262.
[65] R. J. P. Corriu and J. J. E. Moreau, J. Organometallic Chem., 1974, 64, C51.
[66] W. Dumont, J.-C. Poulin, T.-P. Dang, and H. B. Kagan, J. Amer. Chem. Soc., 1973, 95, 8295.

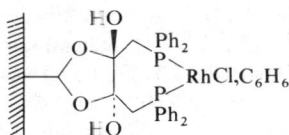

(32)

less efficient than the homogeneous version for hydrogenation of olefins, but is as efficient a hydrosilylation catalyst, producing alkoxysilanes from ketones in 58 % optical purity. The use of $(Ph_3P)_3RuCl_2$ as a catalyst for the addition of Et_3SiH to $R^1R^2C=O$, forming $R^1R^2HC-O-SiEt_3$, has been compared to Wilkinson's catalyst.[67] The rhodium catalyst is more efficient.

The activation of Grignard reagents by Cp_2TiCl_2 in catalytic amounts forms potentially valuable reducing agents.[68] Stereospecific reduction of alkoxy-chloro-, and -fluoro-silanes has been achieved by their use. The reducing properties of Pr^iMgBr so activated appear to rival those of $LiAlH_4$.

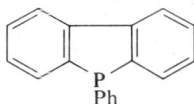

(33)

The preparation of L_4RhH, where L is the dibenzophosphole derivative (33), has been reported. The compound is a very effective catalyst for the selective hydrogenation of terminal olefins.[69] The stoicheiometric hydrogenation of olefins by $(Ph_3P)_3RuHCl$ in the absence of molecular hydrogen produces the *o*-metallated product $[(Ph_3P)ClRu(o-C_6H_4PPh_2)]_2$.[70] The action of H_2 or HCl on this dimer forms the catalytically important bis-phosphine complex $(Ph_3P)_2RuXCl$ (X = H or Cl).

A thorough examination of the general behaviour and reaction kinetics of $(Ph_3P)_3RhCl$ has been reported.[71] The catalytically active compound does not dissociate to $(Ph_3P)_2RhCl$ to any spectroscopically detectable extent in solution, but it is in equilibrium with the chloride-bridged dimer $[(Ph_3P)_2RhCl]_2$. Despite this, the presence of the monomeric species $(Ph_3P)_2RhCl$ must be postulated in order to account for the kinetics of the reaction between H_2 and $Rh(PPh_3)_3Cl$, and the reaction between H_2 and cyclohexene catalysed by $[(Ph_3P)_2RhCl]_2$. Hydrogen transfer from 1,4-dioxan to olefins catalysed by $(Ph_3P)_3RhCl$ has also been examined.[72] The products of the reaction are dioxen and paraffins,

[67] C. Eaborn, K. Odell, and A. Pidcock, *J. Organometallic Chem.*, 1973, **63**, 93.
[68] R. J. P. Corriu and B. Meunier, *J. Organometallic Chem.*, 1974, **65**, 187.
[69] D. E. Budd, D. G. Holah, A. N. Hughes, and B. C. Hui, *Canad. J. Chem.*, 1974, **52**, 775.
[70] B. R. James, L. D. Markham, and D. K. W. Wang, *J.C.S. Chem. Comm.*, 1974, 439.
[71] C. A. Tolman, P. Z. Meakin, D. L. Lindner, and J. P. Jesson, *J. Amer. Chem. Soc.*, 1974, **96**, 2762.
[72] T. Nishiguchi and K. Fukuzumi, *J. Amer. Chem. Soc.*, 1974, **96**, 1893.

along with $(Ph_3P)_2RhCl(C_4H_8O_2)$. A large kinetic isotope effect was noted when $C_4D_8O_2$ was used ($R^H/R^D = 3.1$), indicating that oxidative addition of C—H (or C—D) to form $(Ph_3P)_2RhClH(C_4H_7O_2)$ was the rate-determining step.

Hydrido-complexes of the Group VIII metals Fe, Co, Ru, Rh, and Pd promote C—O bond cleavage in vinyl or allyl carboxylates, liberating ethylene or propylene and forming metal carboxylates.[73] Scheme 3 depicts the suggested reaction route for the reaction between $(Ph_3P)_4RuH_2$ and vinyl acetate.

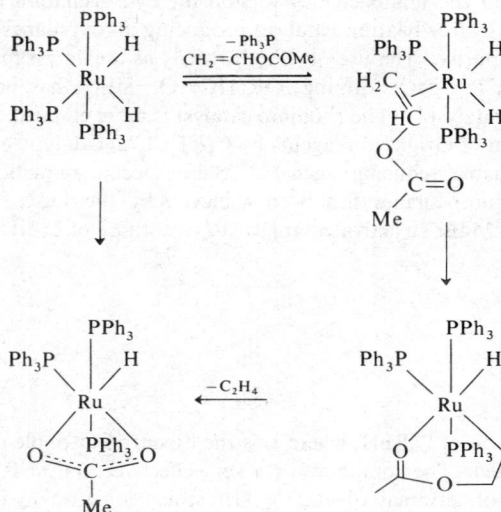

Scheme 3

The fluxional cations $(Et_3P)_4MH^+$ have been examined by n.m.r. spectrometry[74] (M = Ni, Pd, or Pt). A dissociative phosphine exchange involving $(Et_3P)_3MH^+$ and free Et_3P operates, but intramolecular rearrangements are also significant. $TaH(CO)_2(Me_2PC_2H_4PMe_2)_2$ has a distorted capped octahedron with H as the capping atom. The geometry of the diphosphine molecules is disordered, but they have a preferred conformation resembling other chelate molecules of this type.[75] Variable-temperature n.m.r. spectra between 263 and 373 K show that this molecule is fluxional also.

The complexes $Os(CO)(NO)L_2Cl$ (L = PPh_3 or PCx_3) react with $AgPF_6$ in the presence of H_2 to form $[OsH_2(CO)(NO)L_2]PF_6$.[76] The ^1H n.m.r. spectra of these reveal non-rigid molecules. The action of CO promotes the elimination of H_2 to form $[Os(CO)_2(NO)L_2]PF_6$. Ph_3P causes a similar elimination to

[73] S. Komiya and A. Yamamoto, *J.C.S. Chem. Comm.*, 1974, 523.
[74] P. Meakin, R. A. Schunn, and J. P. Jesson, *J. Amer. Chem. Soc.*, 1974, **96**, 277.
[75] P. Meakin, L. J. Guggenberger, F. N. Tebbe, and J. P. Jesson, *Inorg. Chem.*, 1974, **13**, 1025.
[76] B. F. G. Johnson and J. A. Segal, *J.C.S. Dalton*, 1974, 981.

form [Os(CO)(NO)(PPh$_3$)$_3$]PF$_6$ when L = Ph$_3$P. PCx$_3$ reacts when L = PCx$_3$, however, to eliminate H$^+$ and produce neutral OsH(CO)(NO)(PCx$_3$)$_3$.

The reduction of CpCoNO to form cluster Co$_4$Cp$_4$H$_4$ was reported last year. To test the generality of this method as a means of producing cyclopentadienyl-metal clusters, the action of LiAlH$_4$–AlCl$_3$ in THF on [CpFe(NO)]$_2$ and CpNiNO has been tried.[77] The iron compound produced only ferrocene, but the nickel nitrosyl formed both nickelocene and the cluster (34), Cp$_4$Ni$_4$H$_3$. Hydrogen atoms (not shown) are located over three of the four tetrahedral faces. This remarkable deep-violet crystalline compound contains three unpaired electrons per molecule and is air-sensitive in solution.

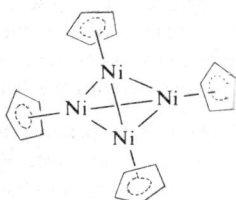 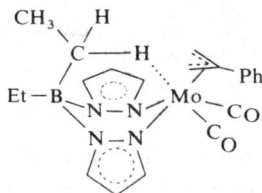

(34)   (35)

The *X*-ray crystal structure of (diethyl-1-pyrazolylborato)(*h*^{3}-2-phenylallyl)-dicarbonylmolybdenum (35) has been determined.[78] The molybdenum atom would be formally in a 16-electron environment but for a very strong CH···Mo interaction (215 pm) from a CH$_2$ group of an ethyl. N.m.r. spectra from 198 to 383 K of this and the unsubstituted allyl derivative show that the structures in solution are the same as in the solid, and that two types of fluxional motion take place.[79] Below room temperature, exchange of the α-CH's is observed, with an activation energy of *ca.* 58 kJ mol^{-1}. Above room temperature, a second motion involving flipping of the NNBNNMo boat becomes rapid, and this effectively leads to exchange of the two ethyl groups. Activation energy for this process is *ca.* 77 kJ mol^{-1}, and may approximate to the strength of the CH···Mo interaction. Finally, the strength and importance of CH–metal interactions of this type is established from the molecular structure of (Et$_2$Bpz$_2$)Mo(CO)$_2$C$_7$H$_7$.[80] An even stronger CH···Mo interaction is found (*ca.* 193 pm), and the C$_7$H$_7$ is *h*^{3}-bonded. Thus although the 16-electron moiety could interact with the *h*^{3}-C$_7$H$_7$ ring to produce an *h*^{5}-C$_7$H$_7$ structure (examples are known), it interacts instead with a saturated C—H electron-pair, despite the fact that in doing so it strains the molecule to the extent of buckling the pyrazolylborate ligand!

[77] J. Müller, H. Dorner, G. Huttner, and H. Lorenz, *Angew. Chem. Internat. Edn.*, 1973, **12**, 1005.
[78] F. A. Cotton, T. LaCour. and A. G. Stanislowski, *J. Amer. Chem. Soc.*, 1974, **96**, 754.
[79] F. A. Cotton and A. G. Stanislowski, *J. Amer. Chem. Soc.*, 1974, **96**, 5074.
[80] F. A. Cotton and V. W. Day, *J.C.S. Chem. Comm.*, 1974, 415.

4 Organometallic Compounds

One-carbon Ligands.—*Carbyne Complexes.* The complex $Br(OC)_4W\equiv CPh$ has been obtained by a new route. The carbene complex $(OC)_5WC(OLi)Ph$ reacts with Ph_3PBr_2, eliminating CO, LiBr, and Ph_3PO, as well as other species including $W(CO)_6$ and $[(OC)_5WBr]^-$, to form the carbyne derivative.[81] The established BX_3 reaction, meanwhile, has been used to prepare a new acetylene carbyne complex. $(OC)_5WC(OEt)(C\equiv CPh)$ reacts in pentane at 228 K to produce $X(OC)_4W\equiv C-C\equiv CPh$, eliminating CO and X_2BOEt.[82] Dimethylamine adds to the acetylene link to form $X(OC)_4W\equiv C-CH=CPh(NMe_2)$. Another resonance form of this carbyne is the carbene complex $X(OC)_4\overline{W}=C=CH-CPh-$ $(=\overset{+}{N}Me_2)$. Molecular structures have been reported for $I(OC)_4W\equiv CPh$, $I(OC)_4Cr\equiv CMe$, and $Br(OC)_3(Ph_3P)Cr\equiv CMe$.[83] They include the shortest $W-C$ and $Cr-C$ bonds known to date. Interestingly, where the chromium complexes are linear at the carbyne carbon, the tungsten derivative is bent at 162°.

Carbene Complexes. The cation $Rh(CNBu^t)_4{}^+$ reacts with primary amines to form diaminocarbene complexes, but $RhR^1X(CNBu^t)_4{}^+$ reacts with R^2NH_2 to form chelated dicarbene complexes.[84] Oxidative addition of R^1X to the diaminocarbene derivatives causes chelate formation, and Scheme 4 summarizes the

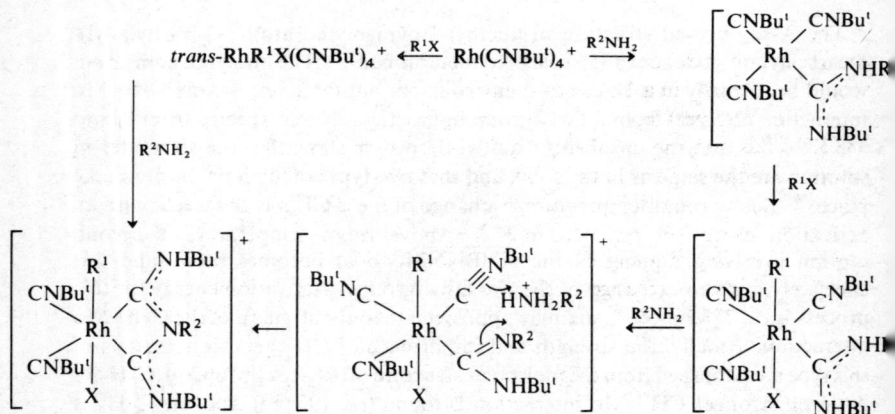

Scheme 4

processes involved. The possibility that the carbenoid ligand may be constrained and directed towards an isocyanide ligand in the octahedral Rh^{III} complexes may in part account for the chelate formation.

[81] H. Fischer and E. O. Fischer, *J. Organometallic Chem.*, 1974, **69**, C1.
[82] E. O. Fischer, H. J. Kalder, and F. H. Köhler, *J. Organometallic Chem.*, 1974, **81**, C23.
[83] G. Huttner, H. Lorenz, and W. Gartzke, *Angew. Chem. Internat. Edn.*, 1974, **13**, 609.
[84] P. R. Branson, R. A. Cable, M. Green, and M. K. Lloyd, *J.C.S. Chem. Comm.*, 1974, 364.

The ^{13}C n.m.r. spectra and X-ray crystal structures of [Rh(CPh$_2$)Cl(py)]$_2$CO (36), [Rh(CPh$_2$)Cp]$_2$CO, and [Rh(CPh$_2$)Cp]$_2$ show that diphenylcarbene bridges the metal atoms in each case.[85] The germoxycarbene complex of manganese, Me$_2$GeMn(CO)$_4$C(O)Me, reported last year, is now known to be in equilibrium in solution with its dimer.[86] The structure of the rhenium analogue is shown by (37).

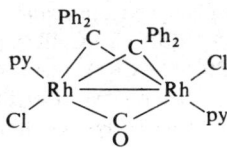

(36)

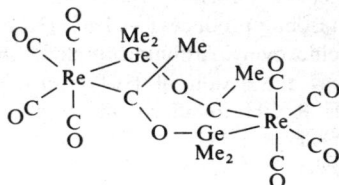

(37)

The acidity of protons attached to the α-carbon atom of carbene complexes is well known, and the anion (OC)$_5$CrC(OMe)CH$_2^-$ has now been isolated as its (Ph$_3$P)$_2$N$^+$ salt.[87] Spectroscopic examination is consistent with a contribution from the vinyl-chromium anion [OC)$_5$$\bar{C}$rC(OMe)=CH$_2$ to its structure. The X-ray crystal structure of [trans-{(p-MeC$_6$H$_4$NH)$_2$C}$_2$AuI$_2$]ClO$_4$,Et$_2$O indicates that the Au—C bonds are essentially σ-bonds, with C—N showing considerable multiple-bond character.[88] The mean Au—C bond length of 208 pm suggests that the carbene ligands exert a considerable *trans*-influence upon each other.

The ^1H n.m.r. spectra of various palladium and platinum carbene complexes allow configurational assignments to be made.[89] In general the lowest energy configuration reflects the least steric interaction. Restricted rotation about the metal–carbon bonds also seems best assigned to steric interactions. Similar methods have been used for the complex RhL$_2$(CO)Cl [L = $\overline{CN(Et)C_2H_4N}$(Et)].[90] Activation parameters for rotation about a metal–carbon bond include a low activation energy ($\leqslant 42$ kJ mol^{-1}) and negative activation entropy. The *cis* → *trans* isomerism of the carbene complexes (OC)$_4$Cr(PR$_3$)C(OMe)Me occurs by a first-order mechanism.[91] Excess CO or R$_3$P have essentially no effect, thus no bond breaking of CO or R$_3$P seems involved. An intramolecular rearrangement, possibly through a prismatic form, is proposed.

The carbene complex (OC)$_5$CrCPh(OMe) reacts with the carbene-transfer agents PhHgCX$_2$Y (X and Y are halogens) to produce Ph(MeO)C=CX$_2$.[92]

[85] T. Yamamoto, A. R. Garber, J. R. Wilkinson, C. B. Boss, W. E. Streib, and L. J. Todd, *J.C.S. Chem. Comm.*, 1974, 354
[86] M. J. Webb, M. J. Bennett, L. Y. Y. Chan, and W. A. G. Graham, *J. Amer. Chem. Soc.*, 1974, **96**, 5931.
[87] C. P. Casey and R. L. Anderson, *J. Amer. Chem. Soc.*, 1974, **96**, 1230.
[88] L. Manojlović-Muir, *J. Organometallic Chem.*, 1974, **73**, C45.
[89] B. Crociani and R. L. Richards, *J.C.S. Dalton*, 1974, 693.
[90] M. J. Doyle and M. F. Lappert, *J.C.S. Chem. Comm.*, 1974, 679.
[91] H. Fischer, E. O. Fischer, and H. Werner, *J. Organometallic Chem.*, 1974, **73**, 331.
[92] A. de Renzi and E. O. Fischer, *Inorg. Chim. Acta*, 1974, **8**, 185.

Benzophenone is found amongst the organic products, and this is formed from the interaction of the by-product PhHgY and $(OC)_5CrCPh(OMe)$. The preparation, resolution, and use in asymmetric syntheses of diastereomeric (+)- and (−)-CpFe(CO)PPh$_3$(h^1-CH$_2$OC$_{10}$H$_{19}$) have been described.[93] This is the first example of a characterized optically-active transition-metal organometallic carbene-transfer agent. In their reactions with *trans*-1-phenylpropene, the (+)-isomer yields (−)-*trans*-(1R,2R)-1-methyl-2-phenylcyclopropane, whereas the (−)-isomer produces (+)-*trans*-(1S,2S)-1-methyl-2-phenylcyclopropane.

The chlorovinylplatinum complexes *trans*-(PMe$_2$Ph)$_2$PtCl(CCl=CHR^1), produced by the action of HCl on *trans*-(PMe$_2$Ph)$_2$PtCl(C≡CR^1), react with alcohols R^{2}OH to afford alkoxycarbene complexes, {*trans*-(PMe$_2$Ph)$_2$PtCl-[C(OR^2)CH$_2$R^1]}$^+$Cl$^-$. Vinyl halides tend to be quite unreactive, and alcoholysis is unexpected. A suggested mechanism[94] is given in Scheme 5.

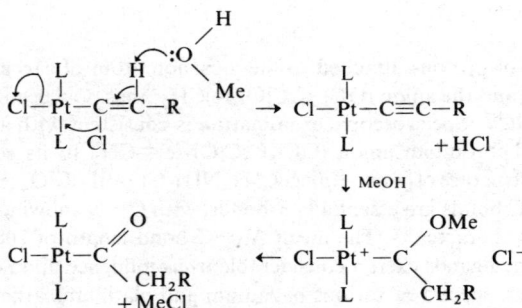

Scheme 5

Alkyl, Aryl, and Related Complexes. The success of arylcadmiums and arylmercurials in the formation of arylgold(III) complexes naturally leads to the expectation that diphenylzinc will also be a valuable reagent for such syntheses. In fact, Ph$_2$Zn reacts with AuCl$_3$ or Au(CO)Cl to form the orange-red complex (38).[95] Ph$_3$ZnAu is thermally stable, and the proposed structure is supported by molecular weight determinations and its ^{1}H and ^{13}C n.m.r. spectra. The reaction of silver acetate with diazomethane at 268 K in aprotic solvents produces disilverdiazomethane, Ag$_2$CN$_2$.[96] Silver salts of stronger acids require base to react fully, but the method seems quite general. This red, solid material is the first of its type. It is highly explosive and shock-sensitive when pure, but can be stored in the dark. Potassium cyanide liberates CH$_2$N$_2$.

Methyl, ethyl, or n-propyl derivatives of copper with bipyridyl or PCx$_3$ ligands have been prepared from the reaction between Cu(acac)$_2$ and R$_2$AlOEt in ether

[93] A. Davison, W. C. Krusell, and R. C. Michaelson, *J. Organometallic Chem.*, 1974, **72**, C7.
[94] R. A. Bell and M. H. Chisholm, *J.C.S. Chem. Comm.*, 1974, 818.
[95] P. W. J. de Graaf, J. Boersma, and G. J. M. van der Kerk, *J. Organometallic Chem.*, 1974, **78**, C19.
[96] E. T. Blues, D. Bryce-Smith, J. G. Irwin, and I. W. Lawston, *J.C.S. Chem. Comm.*, 1974, 466.

$$\text{(38)}$$

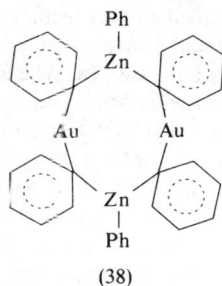

(38)

at low temperatures in the presence of ligands.[97] Without added ligand, MeCu was formed, but this is thermally unstable and decomposes explosively. Bipy has little effect on stabilizing the copper alkyls, but tricyclohexylphosphine imparts considerable stability. Metal complexes of phosphorus ylides reported last year were remarkable for their stability. Analogous sulphonium derivatives are also possible, and a palladium derivative (39) has been isolated.[98]

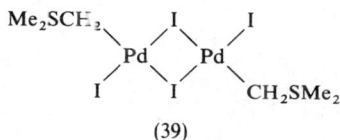

(39)

The past few years have seen the isolation of organometallic derivatives of Groups IV—VII, the existence of which would have been unsuspected previously. This year, red-purple crystals of Me_4ReO have been isolated from the action of methyl-lithium on $ReCl_4O$. The methylrhenium(VI) derivative is air-sensitive, but thermally stable.[99] Perhaps even more remarkably, methyl-lithium reacts with $Mo_2(OAc)_4$ in ethers to produce $Li_4[Mo_2Me_8]$,4ether.[100] These complexes are very pyrophoric, but are stable at 298 K. They react at 195 K with HOAc (OAc = acetate) to regenerate starting complex. The symmetry is essentially D_{4h} with eclipsed $MoMe_4$ units. This and the short (214 pm) Mo—Mo separation indicate a quadruple metal–metal bond.

In Group V, several niobium and tantalum oxohalides of general formula $MeMOX_2$,2L have been prepared from the action of MeMgI on $MOCl_3$ (with L = Ph_3P, *etc.*), or from the action of oxo-ligands such as Ph_3PO on $MeMX_4$.[101] The monomethylhalide derivatives $MeMX_4$ can conveniently be prepared from MX_5 and Me_2Hg or Me_4Sn. Pentamethyltantalum[102] has been added to the

[97] T. Ikariya and A. Yamamoto, *J. Organometallic Chem.*, 1974, **72**, 145.

[98] P. Bravo, G. Fronza, C. Ticozzi, and G. Gaudiano, *J. Organometallic Chem.*, 1974, **74**, 143.

[99] K. Mertis, J. F. Gibson, and G. Wilkinson, *J.C.S. Chem. Comm.*, 1974, 93.

[100] F. A. Cotton, J. M. Troup, T. R. Webb, D. H. Williamson, and G. Wilkinson, *J. Amer. Chem. Soc.*, 1974, **96**, 3824.

[101] C. Santini-Scampucci and J. G. Riess, *J.C.S. Dalton*, 1974, 1433; 1973, 2436.

[102] R. R. Schrock and P. Meakin, *J. Amer. Chem. Soc.*, 1974, **96**, 5288.

list of simple metal methyls, which already includes Me_4Ti, Me_4Zr, Me_4Cr, and Me_6W. Prepared from MeLi and $TaMe_3Cl_2$, it is a yellow oil, rather unstable, and decomposing to *inter alia* CH_4 and Ta. The addition of $Me_2PC_2H_4PMe_2$ (dmpe) affords more stable (dmpe)$TaMe_5$, and (dmpe)$NbMe_5$ can be isolated similarly. Dimethylzinc reacts with $VO(OPr^i)_3$ to form $(MeZnOPr^i)_n$ and the new methylvanadium derivative $MeVO(OPr^i)_2$.[103] From Group IV, tetra-adamantyltitanium has been formed in 18% yield from $TiCl_4$ and $C_{10}H_{15}Cl$ by a Wurtz–Fittig reaction.[104] It is remarkably stable compared with Me_4Ti.

DDT reacts with the cobaloxime bis(dimethylglyoximato)pyridinecobalt(I) to produce a vinylcobalt derivative (40), characterized crystallographically.

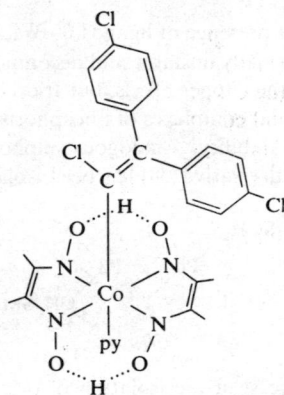

(40)

HCl elimination from the S_N2 reaction product involving Co^I and C-2 of DDT is a likely source of this unexpected product.[105] A similar environmental process could be responsible for the eventual degradation of DDT in the biosphere.

The quaternization of triphenylphosphine by aryl halides is catalysed by $(Ph_3P)_3Ni$.[106] The mechanism appears to involve oxidative addition of ArX to form $(Ph_3P)_2NiArX$, then reaction of co-ordinated aryl with Ph_3P. The exchange of halide substituents of aryl groups catalysed by copper(I) salts is also believed to proceed *via* oxidative addition to an organocopper(III) derivative.[107]

The oxidative addition of various halides RX to $(Ph_3P)_3Pt$, producing *trans*-$(Ph_3P)_2PtRX$, may proceed by a non-chain free-radical process.[108] Using Bu^tNO as a spin-trap, $Bu^rRNO\cdot$ was detected from these reactions by e.s.r. The presence of Bu^tNO did not appear to affect the reactions in other ways.

[103] K.-H. Thiele, B. Adler, H. Grahlert, and A. Lachowicz, *Z. anorg. Chem.*, 1974, **403**, 279.
[104] R. M. G. Roberts, *J. Organometallic Chem.*, 1973, **63**, 159.
[105] R. H. Prince, G. M. Sheldrick, D. A. Stotter, and R. Taylor, *J.C.S. Chem. Comm.*, 1974, 854.
[106] L. Cassar and M. Foà, *J. Organometallic Chem.*, 1974, **74**, 75.
[107] T. Cohen, J. Wood, and A. G. Dietz, *Tetrahedron Letters*, 1974, 3555.
[108] M. F. Lappert and P. W. Lednor, *J.C.S. Chem. Comm.*, 1973, 948.

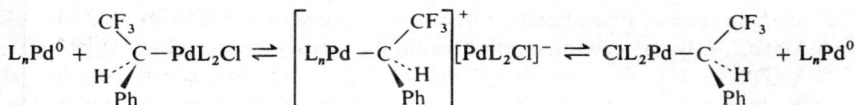

$$L_nPd^0 + \overset{\displaystyle CF_3}{\underset{\displaystyle Ph}{\overset{|}{\underset{|}{C}}}\!\!-}PdL_2Cl \;\rightleftharpoons\; \left[L_nPd - C\overset{\displaystyle CF_3}{\underset{\displaystyle Ph}{\overset{}{\diagdown}}}\!\!_{'H} \right]^+ [PdL_2Cl]^- \;\rightleftharpoons\; ClL_2Pd - C\overset{\displaystyle CF_3}{\underset{\displaystyle Ph}{\overset{}{\diagdown}}}\!\!_{'H} + L_nPd^0$$

Scheme 6

The use of optically active halides in oxidative additions has often provided valuable mechanistic information. The interpretation when racemization is observed is often questionable, however, and an intermolecular nucleophilic displacement (Scheme 6) has been forwarded as a possible explanation in one such case.[109] Elegant work using rapid CO insertion into the palladium–carbon bonds (known to proceed with 100% retention of configuration) to prevent such racemization processes demonstrated that oxidative addition of PhMeCHBr to $(Ph_3P)_nPd$ took place with inversion of configuration, implying an S_N2 type reaction with nucleophilic Pd^0. The mechanism of addition of benzyl halides to $(Ph_3P)_3Pd$ seems to be similar, but, somewhat alarmingly, when the reaction was carried out in the presence of Bu^tNO, $Bu^t(PhCH_2)NO\cdot$ was found! While a free-radical addition cannot be altogether ruled out, the formation of radicals by decomposition of $(Ph_3P)_2Pd(CH_2Ph)X$ in the presence of the Bu^tNO spin-trap is possible, and such experiments require careful interpretation.

Oxidative addition of *trans*-bromostyrene to $(Ph_3P)_3Pt^0$ produces *trans*-$(Ph_3P)_2PtBr(trans$-styryl).[110] A possible intermediate in this stereospecific oxidative addition is $(Ph_3P)_2Pt(\pi$-olefin).

o-Metallation of azobenzene is a well-known process, and may take place by either electrophilic or nucleophilic mechanisms. Fluorocarbons are prone to such nucleophile attack mechanisms, and $Cp(Ph_3P)_2RuMe$ has been found to *o*-metallate decafluoroazobenzene, producing (41).[111] Loss of fluoride is a new

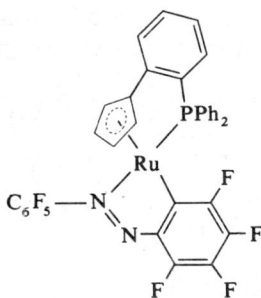

(41)

[109] (a) K. S. Y. Lau, R. W. Fries, and J. K. Stille, *J. Amer. Chem. Soc.*, 1974, **96**, 4983; (b) P. K. Wong, K. S. Y. Lau, and J. K. Stille, *ibid.*, p. 5957.

[110] J. Rajaram, R. G. Pearson, and J. A. Ibers, *J. Amer. Chem. Soc.*, 1974, **96**, 2103.

[111] M. I. Bruce, R. C. F. Gardner, B. L. Goodall, F. G. A. Stone, R. J. Doedens, and J. A. Moreland, *J.C.S. Chem. Comm.*, 1974, 185.

route to metal–carbon bonds. Oxidative additions of trans-[IrCl(CO){PMe$_2$(o-MeOC$_6$H$_4$)}$_2$] proceed much more readily than those of trans-[IrCl(CO){PMe$_2$(p-MeOC$_6$H$_4$)}$_2$] or trans-[IrCl(CO)(PMe$_2$Ph)$_2$]. An interaction between a methoxy oxygen and iridium in the o-substituted phosphine, increasing the nucleophilicity of Ir, is deemed responsible.[112]

Several reactions of hydrated chromium(III) organics, and organocobalamine derivatives, have been studied kinetically in water. The stoicheiometry of the reaction between aqueous Cr^{2+} and CrCH$_2$I^{2+} is

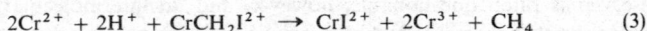

$$2Cr^{2+} + 2H^+ + CrCH_2I^{2+} \rightarrow CrI^{2+} + 2Cr^{3+} + CH_4 \qquad (3)$$

Kinetic data on this and the reaction between aqueous Cr^{2+} and CrCHI$_2$$^{2+}$ support the involvement of dinuclear carbon-bridged intermediates, and the mechanisms proposed are given in Scheme 7.[113] The exchange of the organic

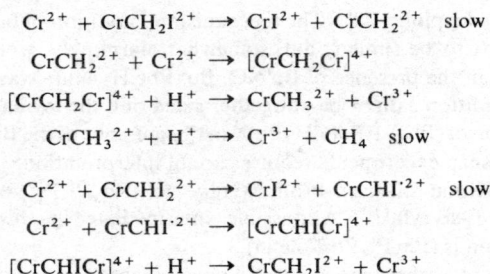

$$Cr^{2+} + CrCH_2I^{2+} \rightarrow CrI^{2+} + CrCH_2^{2+} \quad slow$$
$$CrCH_2^{2+} + Cr^{2+} \rightarrow [CrCH_2Cr]^{4+}$$
$$[CrCH_2Cr]^{4+} + H^+ \rightarrow CrCH_3^{2+} + Cr^{3+}$$
$$CrCH_3^{2+} + H^+ \rightarrow Cr^{3+} + CH_4 \quad slow$$

$$Cr^{2+} + CrCHI_2^{2+} \rightarrow CrI^{2+} + CrCHI^{·2+} \quad slow$$
$$Cr^{2+} + CrCHI^{·2+} \rightarrow [CrCHICr]^{4+}$$
$$[CrCHICr]^{4+} + H^+ \rightarrow CrCH_2I^{2+} + Cr^{3+}$$

Scheme 7

group in (42) to labelled ^{51}Cr^{2+} in the redox reaction takes place by a direct exchange, as kinetic data are inconsistent with homolysis then recombination as the mechanism.[114] This creates difficulties as it clashes with the accepted decomposition mechanism of organochromium complexes as being via homolysis of Cr—C.

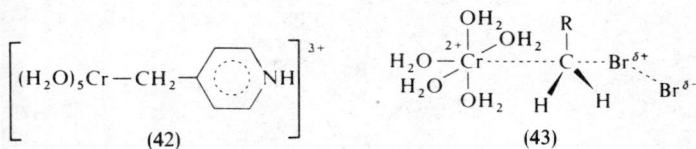

$$\left[(H_2O)_5Cr-CH_2-\bigcirc-NH \right]^{3+}$$

(42)

$$\begin{array}{c} OH_2 \quad R \\ H_2O-\overset{2+}{\underset{|}{Cr}}\cdots\cdots\overset{|}{C}\cdots Br^{\delta+} \\ H_2O \quad | \quad /\backslash \quad Br^{\delta-} \\ OH_2 \quad H \quad H \end{array}$$

(43)

The cleavage of [(H$_2$O)$_5$Cr-R]$^{2+}$ by bromine to produce Cr$_{aq}$$^{3+}$, Br$^-$, and RBr has been examined kinetically.[115] An S_E2 mechanism with an 'open transition state' (43) is proposed, as the (kinetically inert) species (H$_2$O)$_5$CrBr^{2+} is

[112] E. M. Miller and B. L. Shaw, J.C.S. Dalton, 1974, 480.
[113] (a) R. S. Nohr and L. O. Spreer, Inorg. Chem., 1974, 13, 1239; (b) R. S. Nohr and L. O. Spreer, J. Amer. Chem. Soc., 1974, 96, 2618.
[114] J. H. Espenson and J. P. Leslie, J. Amer. Chem. Soc., 1974, 96, 1954.
[115] J. H. Espenson and D. A. Williams, J. Amer. Chem. Soc., 1974, 96, 1008.

not a product. The varying rates for different alkyls can correspond to inversion at carbon. The same mechanism is suggested by the second-order kinetics[116] for halogen cleavage of p-$ZC_6H_4CH_2$—$Cr(OH_2)_5^{2+}$. The second-order rate law and lack of pH dependence [between 0 and 2.3) support direct homolytic attack of Cr^{2+} at carbon in the Co—C cleavage reaction of methyl (or ethyl) cobalamin.[117] In the reaction between (Schiff-base)cobalt organic derivatives and cobalt(II), a direct nucleophilic attack of Co^{II} on the Co^{III}-bonded carbon seems likely.[118] The inversion of carbon during the interaction of Hg^{2+} with cobalamine derivatives has been followed by deuterium-decoupled ^1H n.m.r. spectrometry. Intermediate (44) is proposed, and this type of process is obviously

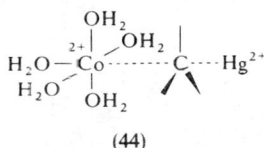

(44)

predominant in organocobalt reactions of this type.[119] The transfer of methyl from methylcobalamin to palladium (of $PdCl_4^{2-}$) displays two kinetically distinguishable steps.[120] The first is the co-ordination of Pd^{II} to the nitrogen of 5,6-dimethylbenzimidazole, and the second, slower, step is the methyl-group transfer.

Palladium(II) compounds are able to cleave Si—R bonds in various alkyl and aryl silanes and in linear and cyclic siloxanes and silazanes at 333—393 K.[121] The Group VIII metals and compounds $PdCl_2$, H_2PtCl_6, $H_2PtCl_6 + SnCl_2$, $(Ph_3P)_3Pt$, $(Ph_3P)_4Pt$, Pt/C, and Pd catalyse the cleavage of Si—CH_3 bonds in hexamethyldisiloxane at 373—473 K to give linear trisiloxane as the main product. The reactions involve transfer of a methyl or phenyl from silicon to palladium, and can be used for methylating or phenylating olefins. Thus silicon compounds might be able to compete with the more reactive mercury or tin derivatives in this useful transfer process.

erythro-2,3-Dimethylpentanoyl(pentacarbonyl)manganese (45), a mixture of this and the *threo*-isomer (46), and 4-methylhexanoyl(pentacarbonyl)manganese

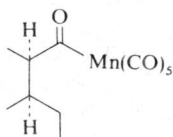   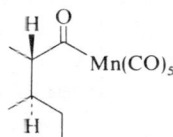   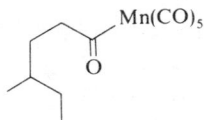

(45)   (46)   (47)

[116] J. C. Chang and J. H. Espenson, *J.C.S. Chem. Comm.*, 1974, 233.
[117] J. H. Espenson and T. D. Sellers, *J. Amer. Chem. Soc.*, 1974, **96**, 94.
[118] V. van den Bergen and B. O. West, *J. Organometallic Chem.*, 1974, **64**, 125.
[119] H. L. Fritz, J. H. Espenson, D. A. Williams, and G. A. Molander, *J. Amer. Chem. Soc.*, 1974, **96**, 2378.
[120] W. M. Scovell, *J. Amer. Chem. Soc.*, 1974, **96**, 3451.
[121] I. S. Akhrem, N. M. Chistovalova, E. I. Mysov, and M. E. Vol'pin, *J. Organometallic Chem.*, 1974, **72**, 163.

(47), all decompose thermally to produce the same mixture: namely 4—11%
3-methylpent-1-ene, 56—62% *trans*-3-methylpent-2-ene, and 31—32% *cis*-3-
methylpent-2-ene.[122] Alkene isomerization does not occur under these conditions, so it appears that these $RMn(CO)_5$ species and the corresponding
(olefin)$HMn(CO)_4$ interconvert faster than the latter loses alkene.

The insertion of SO_2 into the $Fe—CH_3$ bond of optically active (48) is stereospecific, probably with retention of configuration.[123] The reaction of (48) with
I_2, HI, or HgI_2 appears to proceed *via* initial attack at Fe, followed by CH_3I,
CH_4, or MeHgI elimination, rather than by direct electrophilic attack at CH_3.

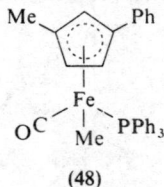

(48)

The convenient syntheses of *threo*- and *erythro*-3,3-dimethyl[1,2-$^2H_2$]butan-1-
ol allow the formation of cyclopentadienyl iron dicarbonyl derivatives with
inversion of configuration at carbon (Scheme 8).[124] Cleavage of the Fe—C

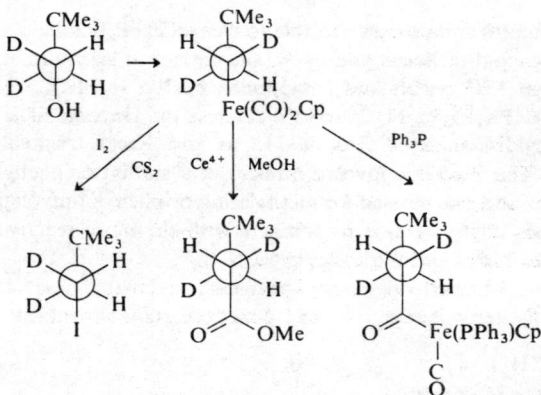

Scheme 8

bond by Br_2 or I_2 proceeds with inversion at carbon, whereas reaction with
Ph_3P, Bu^tNC, O_2, Ce^{4+} or other oxidizing agents produces derivatives of
4,4-dimethyl[2,3-$^2H_2$]pentanoic acid with retention of configuration at carbon.

[122] C. P. Casey, C. R. Cyr, and J. A. Grant, *Inorg. Chem.*, 1974, **13**, 910.
[123] T. G. Attig and A. Wojcicki, *J. Amer. Chem. Soc.*, 1974, **96**, 262.
[124] P. L. Bock, D. J. Boschetto, J. R. Rasmussen, J. P. Demers, and G. M. Whitesides,
    J. Amer. Chem. Soc., 1974, **96**, 2814.

SO_2 insertion also leads to inversion. Using *threo*-3,3-dimethyl[1,2-$^2H_2$]butyl brosylate or triflate, *erythro*-derivatives have also been prepared of phenyl-selenide, pyridinecobaloxime, and cyclopentadienyltricarbonylmolybdenum,[125] making these complexes available also for cleavage reaction study. Similar studies on *threo*-PhCHDCHDFe(CO)$_2$Cp indicate that Fe—C cleavage by Br_2, I_2, or $HgCl_2$ proceed with retention of configuration at carbon,[126] in apparent conflict with the above results.

In H–D exchange reactions catalysed by Pt^{II}, cyclobutane shows a large multiple exchange factor, and this is attributed to the interaction of Pt^{II} with a C—C bond.[127] Tetraphenylcyclobutane similarly forms *trans*-stilbene in the presence of K_2PtCl_4 (Scheme 9).

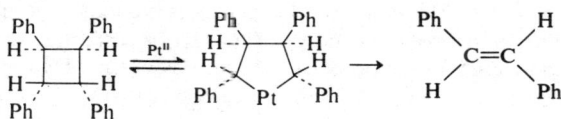

Scheme 9

β-Hydrogen eliminations are well documented for alkyl–metal compounds, but α-hydrogen transfer is not so well known. This year sees some new examples of this latter phenomenon. A remarkable reaction sequence starting from the treatment of $Cp_2W(C_2H_4)Me^+$ with PMe_2Ph and ending at the equilibrium $Cp_2W(PMe_2Ph)Me^+ \rightleftharpoons Cp_2W(CH_2PMe_2Ph)H^+$ has been described.[128] Scheme 10 outlines the proposed intermediates, a key step in which is the equilibrium

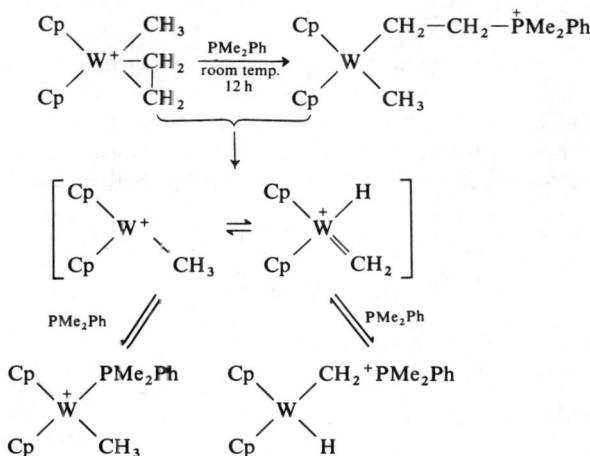

Scheme 10

[125] P. L. Bock and G. M. Whitesides, *J. Amer. Chem. Soc.*, 1974, **96**, 2827.
[126] D. Slack and M. C. Baird, *J.C.S. Chem. Comm.*, 1974, 701.
[127] I. J. Harvie and F. J. McQuillin. *J.C.S. Chem. Comm.*, 1974, 806.
[128] (a) N. J. Cooper and M. L. H. Green, *J.C.S. Chem. Comm.*, 1974, 208; (b) N. J. Cooper and M. L. H. Green, *ibid.*, 1974, 761.

between W—CH$_3$ and HW=CH$_2$. Mass-spectrometric analysis of poly-deuteriated methanes formed in the reaction of D$_2$ with (Ph$_3$P)$_3$CoCH$_3$ or (Ph$_3$P)$_3$RhCH$_3$ in toluene forms the basis of another claim for a reversible α-hydrogen abstraction.[129] The equilibrium M—CH$_3$ ⇌ H—M=CH$_2$ appears to operate, leading to H—D exchange. The reaction between Fe(acac)$_3$, Me$_2$AlOEt, and Ph$_2$PC$_2$H$_4$PPh$_2$ (diphos) leads to the new iron compound (diphos)$_2$FeMe$_2$.[130] This, on thermolysis and treatment with CH$_2$Cl$_2$, liberates ethylene, and carbenoid intermediates are suspected.

The reaction of 1,4-dilithiobutane with Cp$_2$TiCl$_2$ leads to the metallacycle Cp$_2$TiCH$_2$CH$_2$CH$_2$CH$_2$ (A).[131] Carbon monoxide at 195 K forms the insertion product Cp$_2$TiCH$_2$CH$_2$CH$_2$CH$_2$CO, which liberates cyclopentanone in toluene at room temperature. Thermolysis of (A) produces n-butane, but also ethylene, and an equilibrium between (A) and Cp$_2$Ti(C$_2$H$_4$)$_2$ is suspected, as (Cp$_2$Ti)$_2$N$_2$ and C$_2$H$_4$ lead to the formation of (A). Photolysis of Cp$_2$MMe$_2$ (M = Ti, Zr, or Hf) produces methyl radicals and the metallocenes Cp$_2$M.[132] These metallo-cenes do not appear to be the hydrogen-bridged variety already described for these stoicheiometries. Their e.s.r. spectra in THF solution show them to be diamagnetic. In the presence of PhC≡CPh, they react to form (49).

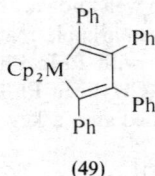

(49)

An improved method of C—C formation, devised in order to convert octan-1-ol into decane, involves making an organocopper reagent from R^{1}MgX and Li$_2$CuCl$_4$, then treating the reagent with R^{2}tosylate. Reductive elimination of R^{1}R^2 from Cu probably operates.[133] The palladium derivative (50), conveniently made from PdCl$_2$ and (51; X = N or CH), reacts with organolithium or Grignard

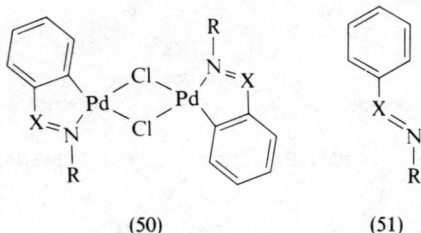

(50)   (51)

[129] L. S. Pu and A. Yamamoto, *J.C.S. Chem. Comm.*, 1974, 9.
[130] T. Ikariya and A. Yamamoto, *J.C.S. Chem. Comm.*, 1974, 720.
[131] J. X. McDermott and G. M. Whitesides, *J. Amer. Chem. Soc.*, 1974, **96**, 947.
[132] H. Alt and M. D. Rausch, *J. Amer. Chem. Soc.*, 1974, **96**, 5936.
[133] G. Fouquet and M. Schlosser, *Angew. Chem. Internat. Edn.*, 1974, **13**, 82.

reagents to form o-$C_6H_4R^1XNR$.[134] Reductive elimination from Pd is likely. If R^1 is prone to β-elimination, (51) may also be formed in the reaction. A recently described route[135] to azobenzene–palladium derivatives with specific and varied substituents increases the synthetic value of this procedure.

Oxidative addition of aryl halides to $(Et_3P)_4M$ (M = Ni, Pd, or Pt) leads to *trans*-$(Et_3P)_2MXAr$,[136] and treatment of these with organolithium reagents forms $(Et_3P)_2MR(Ar)$. Thermolysis of the nickel or palladium complexes in solution then affords R—Ar by reductive elimination. Reductive elimination as one step in the cross-coupling of Grignard reagents and organic halides catalysed by L_2NiX_2 is well established. When L_2 is the chiral diphosphine diop, the coupling reaction (4) proceeds in yields from 20—90%, and the product is optically active (*ca.* 10—20%).[137]

$$HR^1R^2CMgX + R^3X \rightarrow HR^1R^2R^3C \qquad (4)$$

The formation of $(Bu_3P)AgBu^n$ from BuLi and (Bu_3PAgI) at 195 K has been described. Between 223 and 273 K this complex decomposes to liberate octane. Free radicals are not involved (though they are in the photochemical decomposition of this compound) and a dinuclear process is implicated.[138] The thermolysis of $(Ph_3P)_2Ir(CO)(octyl)$ produces mainly octene and $(Ph_3P)_2Ir(CO)H$. Some octane is formed, however, though the presence of excess Ph_3P suppresses this process.[139] A dinuclear elimination process is involved, and deuterium labelling confirms this. Intermediate (52) is postulated. The thermolysis of $(Ph_3P)_2NiArX$ leads to Ar_2, Ph_2, and Ar—Ph.[140] In this case, both dinuclear eliminations and migrations from P to Ni appear to be involved.

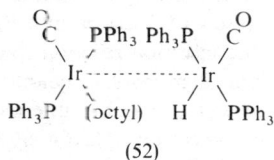

(52)

Carbon monoxide reacts with Cp_2VR *via* insertion to yield $Cp_2V(CO)COR$ when R is methyl or benzyl. When R is phenyl, however, there is migration to a C_5 ring to produce (53). This rapidly loses CO and hydrogen to afford paramagnetic (54).[141] A common reaction of nickelocene with nucleophiles is the loss of a cyclopentadienyl, and conversion into a $(h^1$-$C_5H_5)$ group prior to cleavage is

[134] S.-I. Murahashi, Y. Tanba, M. Yamamura, and I. Moritani, *Tetrahedron Letters*, 1974, 3749.
[135] R. J. Cross and N. H. Tennent, *J. Organometallic Chem.*, 1974, **72**, 21.
[136] G. W. Parshall, *J. Amer. Chem. Soc.*, 1974, **96**, 2360.
[137] Y. Kiso, K. Tamao, N. Miyake, K Yamamoto, and M. Kumada, *Tetrahedron Letters*, 1974, 3.
[138] G. M. Whitesides, D. E. Bergbreiter, and P. E. Kendall, *J. Amer. Chem. Soc.*, 1974, **96**, 2807.
[139] J. Schwartz and J. B. Cannon, *J. Amer. Chem. Soc.*, 1974, **96**, 2276.
[140] A. Nakamura and S. Otsuka, *Tetrahedron Letters*, 1974, 463.
[141] G. Fachinetti and C. Floriani, *J.C.S. Chem. Comm.*, 1974, 516.

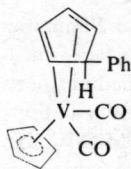   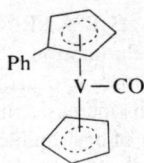

(53)   (54)

believed to occur. In 1971 the reaction of Cp_2Ni and $Me_2C=C=O$ appeared to produce an isolable crystalline σ-cyclopentadienyl of nickel, providing direct support for this type of cleavage mechanism. Alas, two crystal structure determinations this year reveal the complex to be (55),[142] and no $(\sigma\text{-}C_5H_5)Ni$ derivative has yet been isolated.

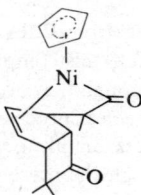

(55)

A new route to carbonyl phosphine derivatives of organoplatinum is the oxidative addition of MeI to $Pt_4(CO)_5(PR_3)_4$. Treatment of the $(R_3P)(OC)PtMeI$ with other phosphines can cause CO insertion into Pt—C, CO elimination, or both.[143] A kinetic study of the insertion reaction to produce $L^1L^2Pt(COMe)I$ reveals that the rate-determining step involves neither the incoming ligand and L^2 nor the solvent. This means a possible intermediate is the 14-electron $LPt(COMe)I$, rather than the 18-electron $L^1L^2(OC)PtMeI$.

The interaction of $CpCr(NO)_2R$ (R = CH_3 or CH_2Ph) with various unsaturated groups has been examined.[144] No insertion into Cr–alkyl was found with CO or COS, but SO_2 inserts to form the S-sulphinates. Tetracyanoethylene inserts into Cr—Me to produce $CpCr(NO)_2C(CN)_2C(CN)_2Me$, but with the

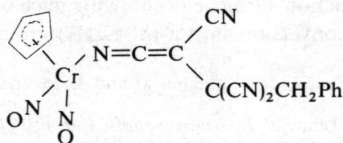

(56)

[142] (a) M. R. Churchill, B. G. DeBoer, and J. J. Hackbarth, *Inorg. Chem.*, 1974, **13**, 2098; (b) D. A. Young, *J. Organometallic Chem.*, 1974, **70**, 95.
[143] (a) C. J. Wilson, M. Green, and R. J. Mawby, *J.C.S. Dalton*, 1974, 421; (b) C. J. Wilson, M. Green, and R. J. Mawby, *ibid.*, 1974, 1293.
[144] J. A. Hanna and A. Wojcicki, *Inorg. Chim. Acta*, 1974, **9**, 55.

benzyl complex, insertion to form $CpCr(NO)_2C(CN)_2C(CN)_2CH_2Ph$ is accompanied by the production of (56). Insertion reactions of several methylniobium(v) and methyltantalum(v) compounds have been examined.[145] RNCO and RNCS insert to give $M—N(R)C(=O)Me$ and $M—N(R)C(=S)Me$ derivatives, respectively. Nitric oxide reacts with $Me_2MCl_3^-$ or Me_3MCl_2 to yield $MCl_3[ON(Me)-NO]_2$ and $MeMCl_2[ON(Me)NO]_2$.

Platinum(iv) aromatic compounds are not as inert as was once thought. The action of Cl_2 on $PhPt(PEt_3)_2Cl_3$ (C) in the presence of $AlCl_3$ leads to substitution of the phenyl ring to give a p-chlorophenyl complex, rather than cleavage of $Pt—C$.[146] Heating (C) in polar solvents forms the phosphonium complex $[PEt_3Ph]^+[PtCl_3PEt_3]^-$. $(Ph_3P)_2PtX$(vinyl), formed by the oxidative addition of vinyl halides to $(Ph_3P)_4Pt$, reacts with Cl_2 to give a Pt^{IV}–vinyl, again without $Pt—C$ cleavage.[147]

The He^I photoelectron spectra of MR_4 (M = Ti, Zr, Hf, Ge, or Sn; R = Me_3CCH_2 or Me_3SiCH_2) have been reported and the highest occupied MO (at 8—9 e.v.) has been assigned to σ-$M—C$.[148] The observed stability trends are not related to ground-state electronic effects. Nickel core-electron binding energies have been compared with calculated charges in *trans*-$(Et_3P)_2NiXY$ (alkyls, aryls, ethynyls, halides, *etc.*).[149] As expected, the π-bonding component of nickel–aryl bonding was found to be unimportant. Comparison of the ^{13}C shieldings and $^1J(^{13}C–^{195}Pt)$ of $[trans$-$(Me_3As)_2LPtPh^+]PF_6^-$ with those of corresponding methyl derivatives leads to the conclusion that electron density in the phenyl ring remains constant, and again σ rather than π bonding dominates the interactions in $Pt—Ph$ systems.[150]

Several complexes $CpCoI(PR_3)R_F$ and $[CpCo(CO)(PPh_3)C_2F_5]ClO_4$ have been characterized ($R_F = CF_3$, C_2F_5, or $C_3F_7^i$).[151] Their ^1H and ^{19}F n.m.r. spectra are indicative of hindered rotations about both $Co—C$ and $P—$aryl. The i.r. spectrum of the carbonyl derivative suggests the presence of diastereomers. Both Co and P are chiral centres.

$Li_3CrMe_6,3$dioxan contains octahedrally co-ordinated chromium, with $Cr—C$ *ca.* 230 pm.[152] A weak interaction between the methyls and lithium distorts three edges of the octahedron; $Li—CH_3$ is 217 pm. The structure can be represented as $(Cr[Me_2—Li—2(dioxan)_{\frac{1}{2}}]_3)_n$.

Although thermolysis studies on organo-transition-metal complexes are now quite common and allow definite reaction patterns to be recognized, oxidation of organometallics is still a little-understood area. The autoxidations of R_4Ti, R_4Zr, R_6Mo_2, and R_6W_2 have been examined this year. They proceed by rapid

[145] (a) J. D. Wilkins, *J. Organometallic Chem.*, 1974, **65**, 383; (b) J. D. Wilkins, *ibid.*, 1974, **67**, 269; (c) J. D. Wilkins and M. G. B. Drew, *ibid.*, 1974, **69**, 111.
[146] D. R. Coulson, *J.C.S. Dalton*, 1973, 2459.
[147] B. F. G. Johnson, J. Lewis, J. D. Jones, and K. A. Taylor, *J.C.S. Dalton*, 1974, 34.
[148] M. F. Lappert, J. B. Pedley, and G. Sharp, *J. Organometallic Chem.*, 1974, **66**, 271.
[149] D. R. Fahey and B. A. Baldwin, *J. Organometallic Chem.*, 1974, **70**, C11.
[150] H. C. Clark and J. E. H. Ward, *J. Amer. Chem. Soc.*, 1974, **96**, 1741.
[151] R. J. Burns, P. B. Bulkowski, S. C. V. Stevens, and M. C. Baird, *J.C.S. Dalton*, 1974, 415.
[152] J. Krausse and G. Marx, *J. Organometallic Chem.*, 1974, **65**, 215.

radical displacements at the metal centres. Unstable organometallic peroxides were detected, but even at 200 K they reacted rapidly with the alkyls to form alkoxy-derivatives.[153]

Two-carbon Ligands.—A method has been developed for treating metal atoms with compounds in solution in inert solvents at temperatures up to 273 K.[154] It has been used to prepare $(cod)_2Fe$, which appears to be paramagnetic and is possibly pseudotetrahedral in structure. Whilst the reactive nature of $(cod)_2Fe$ makes it difficult to characterize it structurally, it means that it is a useful starting material for preparing other derivatives by ligand displacement.

The platinum(0) complex $(Ph_3P)_2Pt(C_2H_4)$ reacts with $\Delta^{1,4}$-bicyclo[2,2,0]-hexane (57) at 253 K to form (58), an organometallic 'propellane'. This makes

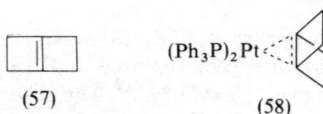

(57)   (58)

a convenient storage for (57), which usually has to be generated impure immediately prior to reaction.[155] Ketens usually react with metal carbonyls by either oxygen or carbonyl loss, leading to polynuclear metal derivatives. The first metal complex containing a structurally intact keten has now been isolated.[156] $(h^5\text{-}C_5H_4R)Mn(CO)_3$ reacts photochemically in THF to form $(h^5\text{-}C_5H_4R)\text{-}Mn(CO)_2THF$ (R = Me or H). This then reacts with diphenylketen to produce (59), formulated as a π-olefin from its spectroscopic properties and chemical reactions.

Dialkylaminomethyl(trialkyl)tin complexes react with $Mn(CO)_5Br$ by Scheme 11 to form (dialkylaminomethylene)manganese tetracarbonyls. The C=N group in the aziridinylmethylene complex (60) is π-bonded.[157]

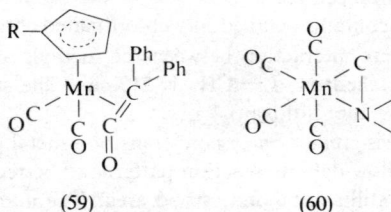

(59)   (60)

The addition, insertion, and oligomerization reactions of hexafluorobut-2-yne at transition-metal complexes have proved a very rich field in recent years. This year sees some interesting additions to the already impressive list. C_4F_6 adds

[153] P. B. Brindley and J. C. Hodgson, *J. Organometallic Chem.*, 1974, **65**, 57.
[154] R. Mackenzie and P. L. Timms, *J.C.S. Chem. Comm.*, 1974, 650.
[155] M. E. Jason, J. A. McGinnety, and K. B. Wiberg, *J. Amer. Chem. Soc.*, 1974, **96**, 6531.
[156] W. A. Herrmann, *Angew. Chem. Internat. Edn.*, 1974, **13**, 335.
[157] E. W. Abel, R. J. Rowley, R. Mason, and K. M. Thomas, *J.C.S. Chem. Comm.*, 1974, 72.

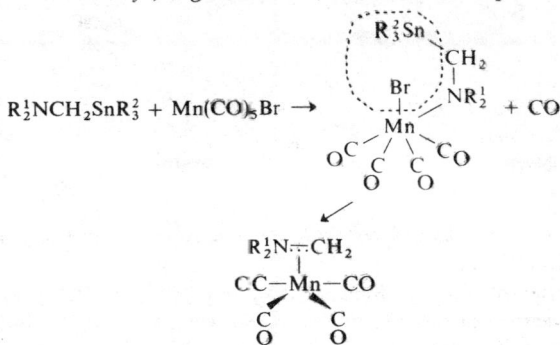

Scheme 11

in Diels–Alder fashion to co-ordinated cod of $[(cod)RhCl]_2$.[158] Treatment with Na(acac) produces (61), examined by X-rays. In contrast, hexafluorobut-2-yne reacts with Ir(acac)(cod) to form an iridiacyclopentene ring system (62).

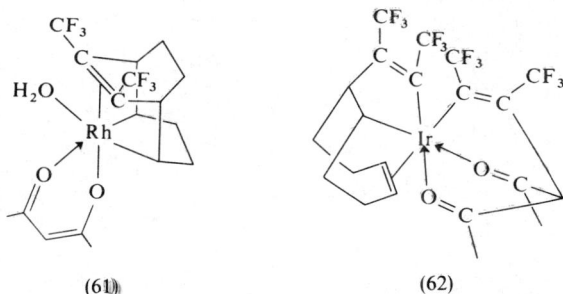

(61)   (62)

The oxidation of cod by $Pb(OAc)_4$ in the presence of $PdCl_2$ in acetic acid proceeds *via* *trans*-addition to $(cod)PdCl_2$ to produce (63), then *cis*-addition to (64). Pd is finally displaced by OAc^- in an S_N2 fashion with inversion of configuration to yield (65).[159] The rearrangement of polyolefins catalysed by transition-metal carbonyls is a wide and often confusing area. The compound (66) rearranges to (67) under the influence of *e.g.* $Mo(CO)_6$.[160] A possible inter-

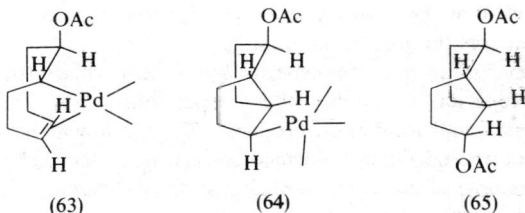

(63)   (64)   (65)

[158] A. C. Jarvis, R. D. W. Kemmitt, B. Y. Kimura, D. R. Russell, and P. A. Tucker, *J.C.S. Chem. Comm.*, 1974, 797.
[159] P. M. Henry, M. Davies, G. Ferguson, S. Phillips, and R. Restivo, *J.C.S. Chem. Comm.*, 1974, 112.
[160] L. A. Paquette, J. M. Photis, J. Fayos, and J. Clardy, *J. Amer. Chem. Soc.*, 1974, **96**, 1217.

(66)   (67)   (68)

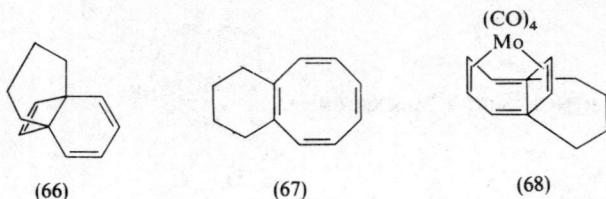

mediate is (68), which was isolated from the reaction mixture and characterized crystallographically.

The interaction of (o-styryl)diphenylphosphine (sp) with RhCl₃ was reported last year, and interesting dimerization reactions were involved. With $Ru_3(CO)_{12}$, sp forms mononuclear complexes $Ru(CO)_3(sp)$ and $Ru(CO)_2(sp)_2$.[161] Refluxing the latter in nonane leads to (69) and (70), and dehydrogenation of (70) forms (71). Obviously dimerization of co-ordinated vinyl-phenyl derivatives is quite general. The complex (cod)Ni(duroquinone) is protonated in HSO_3F to form (72).[162]

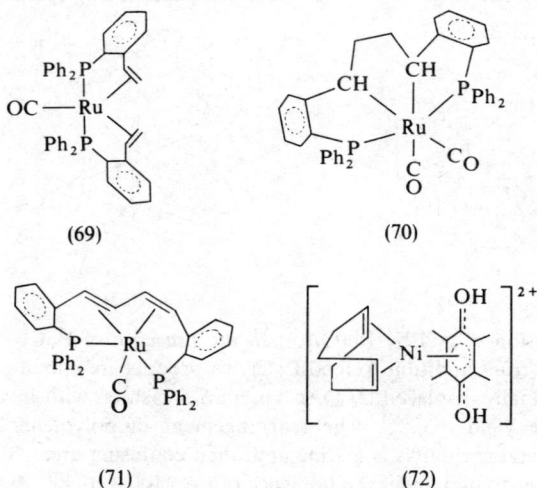

(69)   (70)

(71)   (72)

The chemical shift of the hydroxyl protons suggests extensive delocalization of the positive charge through back-bonding.

The photodimerization of norbornene (nb) by copper(I) triflate (trifluoromethyl-sulphonate) is superior to that by other copper salts.[163] A 2 : 1 nb : CuOTf complex is formed, and irradiation produces (73). Cyclopentene, cyclohexene, and cycloheptene (though not cyclo-octene) can also be dimerized by this method, and the first example of mixed photodimerization of different olefins by metal salts is reported. Norbornadiene (nbd) and benzonorbornadiene are dimerized

[161] M. A. Bennett, R. N. Johnson, and I. B. Tomkins, *J. Amer. Chem. Soc.*, 1974, **96**, 61.
[162] M. Brookhart and G. J. Young, *J.C.S. Chem. Comm.*, 1974, 205.
[163] (a) R. G. Salomon and J. K. Kochi, *J. Amer. Chem. Soc.*, 1974, **96**, 1137; (b) R. G. Salomon, K. Folting, W. E. Streib and J. K. Kochi, *ibid.*, p. 1145.

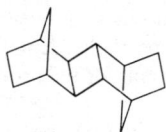 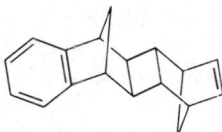 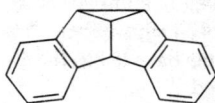

(73)　　　　　　　　(74)　　　　　　　　(75)

by $Fe(CO)_2(NO)_2$. When mixtures of the two are used, the crossed dimer (74) is also produced.[164]

Dibenzosemibullvalene (75) reacts with $Fe_2(CO)_9$ to form an $Fe(CO)_4$ complex. containing a novel 'ferretan' ring system (76).[165] The reaction of *in situ* generated tetrafluorobenzyne with $Fe_3(CO)_{12}$ was reported in 1970 to produce a complex

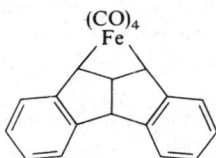 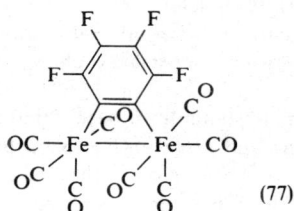

(76)　　　　　　　　　　　　(77)

$(C_6F_4)Fe_2(CO)_8$. An *X*-ray crystal structure determination this year confirms structure (77).[166] Co-ordination about each iron is near octahedral, and the C_6F_4 ring, Fe_2, and four CO's are nearly co-planar.

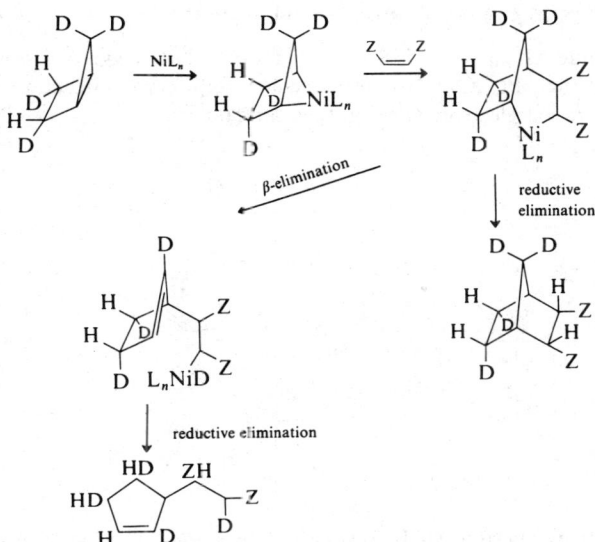

Scheme 12

[164] L. Lombardo, D. Wege, and S. P Wilkinson, *Austral. J. Chem.*, 1974, **27**, 143.
[165] J. L. Flippen, *Inorg. Chem.*, 1974, **13**, 1054.
[166] M. J. Bennett, W. A. G. Graham, R. P. Stewart, and R. M. Tuggle, *Inorg. Chem.*, 1973, **12**, 2944.

Nickel(0) complexes catalyse the addition or cycloaddition of olefins to bicyclo[2,1,0]pentane. Analyses of the products from partially deuteriated reactants have led to the establishment of Scheme 12 to account for the processes involved.[167]

The feeling that the π-acceptor part of metal–olefin bonding has been over-emphasized in the past is strengthened this year. The SCF X α-scattered wave method has been used to calculate the electronic structure of Zeise's anion, $PtCl_3(C_2H_4)^-$. The results predict optical transitions which agree well with experimental spectra, but present a more complicated picture than the simple Chatt–Dewar model.[168] The σ-component is much more important than the π back-bond. N.m.r., u.v., and calorimetry experiments on the displacement of olefins from $[PdCl_2(olefin)]_2$ by pyridine also support the greater importance of the σ-donor part of the metal–olefin bonds.[169] Comparison with other metals produces a series of increasing π-character of $Ag^I < Pd^{II} < Pt^{II} \sim Rh^I$ for such complexes.

Formation constants have been determined for complex formation between several olefins and tris-tri-*o*-tolylphosphitenickel(0).[170] Assuming an Ni—P bond energy of $126\,kJ\,mol^{-1}$, the nickel–olefin bonds vary in strength from $113\,kJ\,mol^{-1}$ to $176\,kJ\,mol^{-1}$. Enthalpies of ligand displacements, including cod, from Rh^I and Pd^{II} have been measured. Displacement energies from Rh follow the sequence $(PhO)_3P \gg cod \sim Ph_3P > py$, and from Pd, $Ph_3P > (PhO)_3P > py \gg cod$.[171] Calorimetric studies of reaction (5) produce the

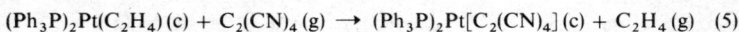

$$(Ph_3P)_2Pt(C_2H_4)\,(c) + C_2(CN)_4\,(g) \;\rightarrow\; (Ph_3P)_2Pt[C_2(CN)_4]\,(c) + C_2H_4\,(g) \quad (5)$$

enthalpy value $\Delta H_{298} = -155.8 \pm 8.0\,kJ\,mol^{-1}$. This suggests a weaker $Pt—C_2H_4$ bond than $Pt—C_2(CN)_4$, in agreement with crystal structure data, but contrary to recent electron-emission spectral data.[172]

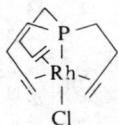

(78)

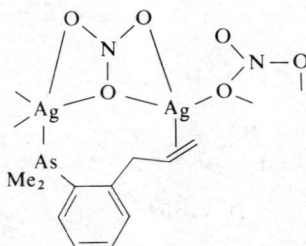

(79)

[167] R. Noyori, Y. Kumagai, and H. Takaya, *J. Amer. Chem. Soc.*, 1974, **96**, 634.
[168] N. Rösch, R. P. Messmer, and K. H. Johnson, *J. Amer. Chem. Soc.*, 1974, **96**, 3855.
[169] W. Partenheimer and B. Durham, *J. Amer. Chem. Soc.*, 1974, **96**, 3800.
[170] C. A. Tolman, *J. Amer. Chem. Soc.*, 1974, **96**, 2781.
[171] W. Partenheimer and E. F. Hoy, *Inorg. Chem.*, 1973, **12**, 2805.
[172] A. Evans, C. T. Mortimer, and R. J. Puddephatt, *J. Organometallic Chem.*, 1974, **72**, 295.

The crystal and molecular structure of $ClRhP(CH_2CH_2CH=CH_2)_3$ (78) reveals a trigonal-bipyramidal co-ordination.[173] The Rh—P bond is shorter than usual, presumably contracted to allow more favourable Rh–olefin bonding. $Me_2As(o$-allyl-Ph) is commonly a chelating ligand, but in the complex LAg_2-$(NO_3)_2$, arsenic and olefin bond to different metal atoms (79).[174] The geometry of crystalline (nbd)Ir(PMe_2Ph)_2SnCl_3 (*trans*-apical PMe_2Ph groups) has been elucidated by X-rays, and is believed to be the ground-state orientation for the fluxional processes observed in solution.[175] $CuC_2C_6F_5$ reacts with $ClRe(CO)_3$-$(PPh_3)_2$ to afford $ReCu(C_2C_6F_5)_2(CO)_3(PPh_3)_2$.[176] The molecular structure of this is shown by (80). No Cu—Re bond is suspected in this zwitterionic

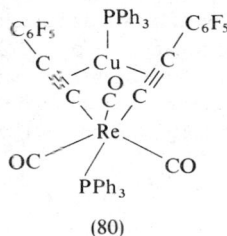

(80)

complex, which can be described as ion pairs $[Ph_3PCu]^+[Re(C_2C_6F_5)_2(CO)_3$-$(PPh_3)]^-$, additionally held together by Cu–acetylenic bonding. The molecular structure of $(C_{10}H_8)(AgClO_4)_4,4H_2O$ seems best described as clathrated naphthalene in hydrated $AgClO_4$, rather than as a complex of silver.[177]

N.m.r. studies have provided olefin rotation energies in *cis*-PtCl_2L(olefin) and (acac)PtCl(olefin).[178] $\Delta G_{Tc}^{\neq}$ depends on both the bulk and the electro-negativity of the substituents, and on olefin symmetry. The ground-state olefin orientation depends mainly on steric factors. N.m.r. studies of a different type have been performed on *trans*-PtCl_2(py)(C_2H_4).[179] The olefin resonance was recorded in nematic solution in poly-γ-benzyl-L-glutamate and CH_2Cl_2. The spectra were analysed despite non-rigidity problems, and the results are in good agreement with solid-state determinations of the structure.

Bis-trifluoromethyldiazomethane, $(CF_3)_2CN_2$, reacts with Cp_2Ni to form a 2 : 1 adduct, and with $Co(CO)_3(allyl)$ to form a 1 : 1 adduct. The structure of the latter complex appears to be (81).[180] Treatment of this by Ph_3P displaces CO. Di-cyanodiazomethane was reported in 1972 to react with $(Bu^tNC)_4Ni$ to form $[(Bu^tNC)_3NiC(CN)_2]_2$, now reformulated as (82), and formed *via* the intermediate $(Bu^tNC)_3Ni[N_2C(CN)_2]$.[181]

[173] M. O. Visscher, J. C. Huffman, and W. E. Streib, *Inorg. Chem.*, 1974, **13**, 792.
[174] M. K. Cooper, R. S. Nyholm, P. W. Carreck, and M. McPartlin, *J.C.S. Chem. Comm.*, 1974, 343.
[175] M. R. Churchill and K.-K. G. Lin. *J. Amer. Chem. Soc.*, 1974, **96**, 76.
[176] O. M. A. Salah, M. I. Bruce, and A. D. Redhouse, *J.C.S. Chem. Comm.*, 1974, 855.
[177] E. A. H. Griffith and E. L. Amma, *J. Amer. Chem. Soc.*, 1974, **96**, 743.
[178] J. Ashley-Smith, Z. Douek, B. F. G. Johnson, and J. Lewis, *J.C.S. Dalton*, 1974, 128.
[179] D. R. McMillin and R. S. Drago, *Inorg. Chem.*, 1974, **13**, 546.
[180] J. Clemens, M. Green, and F. G. A. Stone, *J.C.S. Dalton*, 1974, 93.
[181] D. J. Yarrow, J. A. Ibers, Y. Tatsuno, and S. Otsuka, *J. Amer. Chem. Soc.*, 1973, **95**, 8590.

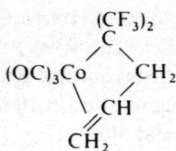

(81)

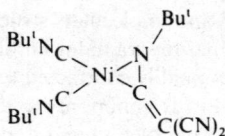

(82)

Bis-(2-pyridinato)acetylene complexes of platinum(0) have been prepared.[182] Their interaction with $CoCl_2$ leads to the unusual metallobicycle (83). Diphenylacetylene displaces CO from $Cp_2Ti(CO)_2$ to yield $Cp_2Ti(CO)(C_2Ph_2)$.[183] This compound is indefinitely stable in the solid phase under N_2 at 273 K. In solution, however, it slowly disproportionates to $Cp_2Ti(CO)_2$ and (49). It is an excellent mild-condition, homogeneous hydrogenation catalyst. A full report on the triosmium-benzyne clusters (communication 1972) has appeared,[184] emphasizing the inherent stability of the $C_6H_4Os_3$ frame (84), which is found in many derivatives.

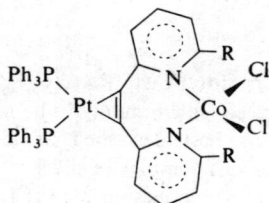

(83)

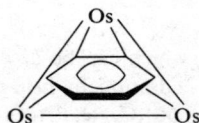

(84)

Hexafluorobut-2-yne reacts with $CpM(CO)_3Cl$ (M = Mo or W) to produce complexes (85).[185] These are remarkable in that they are 16-electron derivatives, uncommon in this part of the periodic table. Treatment of (85) with CpTl leads to (86). Unsaturated aldehydes, ketones, esters, and nitriles add acetylenes in

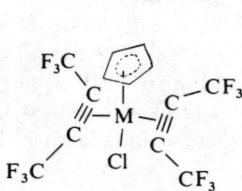

(85)

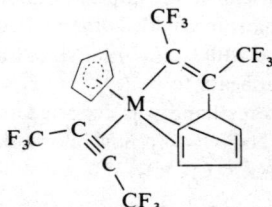

(86)

[182] G. R. Newkome and G. L. McClure, *J. Amer. Chem. Soc.*, 1974, **96**, 617.
[183] G. Fachinetti and C. Floriani, *J.C.S. Chem. Comm.*, 1974, 66.
[184] A. J. Deeming, R. A. Kimber, and M. Underhill, *J.C.S. Dalton*, 1973, 2589.
[185] J. L. Davidson, M. Green, D. W. A. Sharp, F. G. A. Stone, and A. J. Welch, *J.C.S. Chem. Comm.*, 1974, 706.

the form of their $Co_2(CO)_6$ complexes, to form the corresponding conjugated dienes.[186] For example $Co_2(CO)_6(PhC_2H)$ and $MeCH=CHCO_2Et$ produce $PhCH=CH-CMe=CHCO_2Et$. The reaction of $(NC)_2C=C(Cl)Mo(CO)_3Cp$ with diphenylacetylene in refluxing benzene results in the cyclization of dicyanovinylidene with two Ph_2C_2 molecules to afford (87).[187]

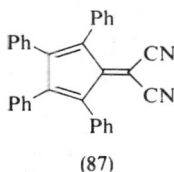

(87)

It is frequently desirable to detach organic ligands, L, from iron complexes $LFe(CO)_n$. A common method of doing so is to treat the complex with oxidizing agents such as Ce^{4+}, but an obvious limitation is the tendency of L itself to undergo oxidation. A variety of organic ligands have been detached from the complexes using amine oxides.[188] The reactions proceed according to equation (6) in aprotic solvents. The method may prove to be of great value in the elimination of interesting structures formed by rearrangements or additions brought about at the iron atoms.

$$LFe(CO)_n + R_3NO \rightarrow L + R_3N + CO_2 + \text{'iron compounds'} \qquad (6)$$

Three-carbon Ligands.—Allene reacts with complex (88) [from butadiene and nickel(0)] to produce the bisallylnickel derivative (89). Carbon monoxide or alkyl isocyanides cause ring closure to (90) and (91), but in addition, isocyanide forms (92). Hydrolysis and hydrogenation of this produces ($\pm$)-muscone, (93), in 40% overall yield.[189] The cycloaddition of allene molecules is achieved using various phosphine–nickel(0) catalysts.[190] Selectivity between the trimer, tetramer, and pentamer products depends on the nature of the phosphine. Intermediates with linear oligomeric structures were detected spectroscopically or isolated by tertiary-phosphine stabilization, and overall Scheme 13 was proposed.

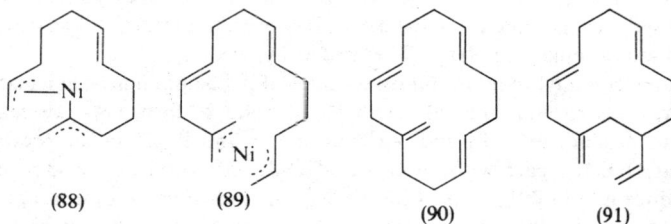

(88)   (89)   (90)   (91)

[186] I. U. Khand and P. L. Pauson, *J.C.S. Chem. Comm.*, 1974, 379.
[187] R. B. King and M. S. Saran, *J.C.S. Chem. Comm.*, 1974, 851.
[188] Y. Shvo and E. Hazum, *J.C.S. Chem. Comm.*, 1974, 336.
[189] R. Baker, R. C. Cookson, and J. R. Vinson, *J.C.S. Chem. Comm.*, 1974, 515.
[190] S. Otsuka, K. Tani, and T. Yamagata, *J.C.S. Dalton*, 1973, 2491.

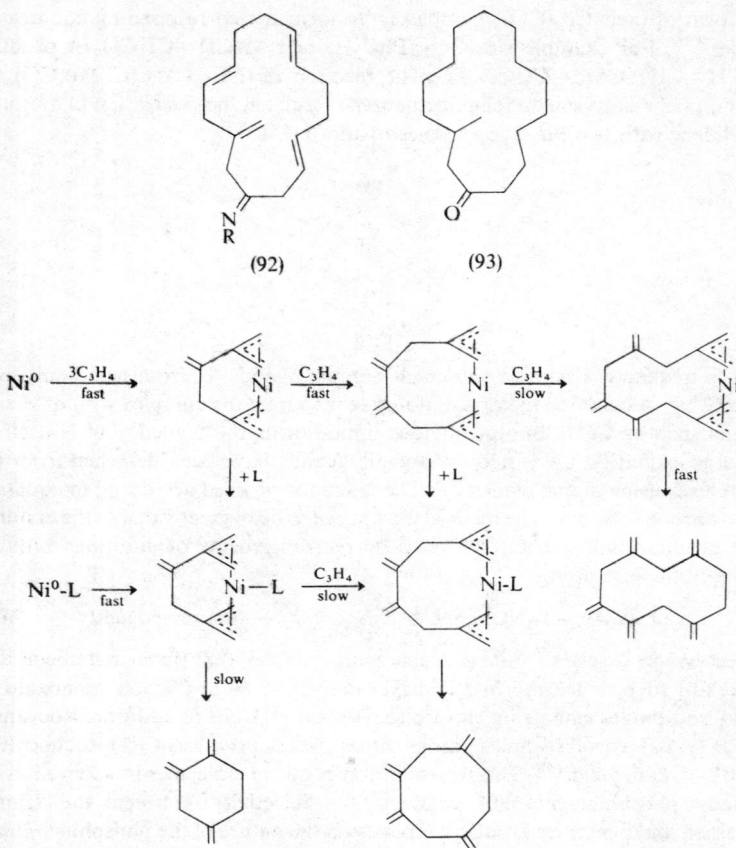

(92)   (93)

Scheme 13

The action of allylamine or 2-methylallylamine with cationic platinum hydrides has proved a facile route to π-allyl derivatives of platinum(II).[191] The complexes formed are cationic, and NH_3 is liberated as by-product.

A series of π-allyl monothio-β-diketonates of palladium has been isolated.[192] Butadiene inserts into the Pd–(methyl)allyl bond to form (94), whereas nbd reacts to produce (95). Ligands such as diphos and Ph_3P, in the presence of $NaHC(CO_2Et)_2$, react with *syn-syn*-bis-(1,3-dimethylallyl)dichlorodipalladium to produce (96) in 20% optical purity.[193] The induction of optical activity is uncommon under these conditions, and the potential of π-allyls is high. Allyl

[191] H. Kurosawa and R. Okawara, *J. Organometallic Chem.*, 1974, **81**, C31.
[192] J. A. Sadownick and S. J. Lippard, *Inorg. Chem.*, 1973, **12**, 2659.
[193] B. M. Trost and T. J. Dietsche, *J. Amer. Chem. Soc.*, 1973, **95**, 8200.

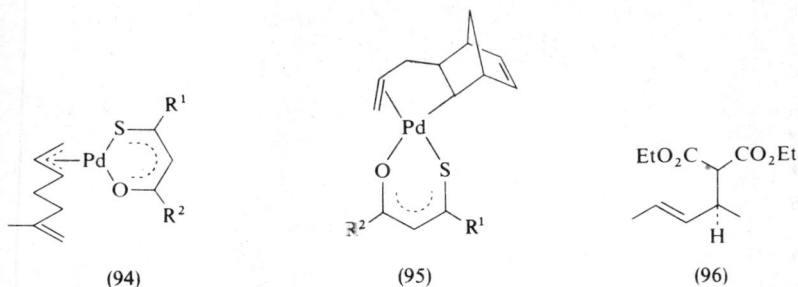

(94)   (95)   (96)

halides react with allylthioalkylcopper(I) by S_N2 displacement of Br. Alkylation is exclusively γ to sulphur (Scheme 14).[194]

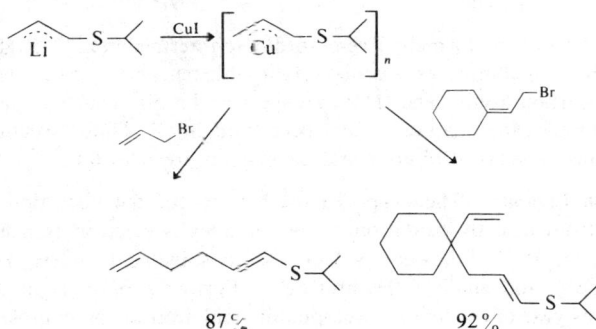

Scheme 14

The report of the reactions between (cod)PdCl$_2$ and Et$_3$N in 1972 suggested that the product, (C$_8$H$_{11}$PdCl)$_2$, was a π-allyl. *X*-Ray crystallography now shows this to be an example of a σ-allyl (97) in the solid form at least.[195] The crystal structure of (1-keto-2,3,4-triphenylcyclobutenyl)cobalt tricarbonyl (98) is reported: the first π-ketocyclobutenyl ligand.[196] An h^3-allylic structure was found by *X*-ray analysis for the Ph$_3$P adduct of the product of trimerization of Bu^{t}C≡CH on PdCl$_2$ (99).[197]

An extensive series of allyl and 2-methylallyl derivatives of rhodium and iridium, [(all)ML$_2$$^+$]BF$_4$$^-$, with neutral ligands R$_3P, R_3$As, py, MeCN, *etc.*, has been examined in solution by ^{1}H variable-temperature n.m.r. Right–left exchanges in the cationic derivatives take place by dissociation of L, then Berry

[194] K. Oshima, H. Yamamoto, and H. Nozaki, *J. Amer. Chem. Soc.*, 1973, **95**, 7926.
[195] F. Dahan, C. Agami, J. Levisalles, and F. Rose-Munch, *J.C.S. Chem. Comm.*, 1974, 505.
[196] J. Potenza, R. Johnson, D. Mastropaolo, and A. Efraty, *J. Organometallic Chem.*, 1974, **64**, C13.
[197] P. M. Bailey, B. E. Mann, A. Segnitz, K. L. Kaiser, and P. M. Maitlis, *J.C.S. Chem. Comm.*, 1974, 567.

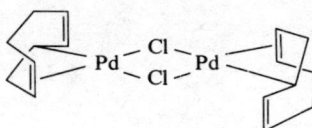

(97)

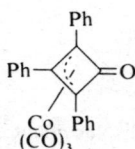

(98)

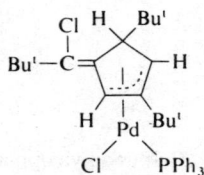

(99)

pseudorotation of the formally five-co-ordinate intermediate.[198] Right–left exchange in neutral complexes is concomitant with *syn–anti* exchange *via* rupture of a metal–carbon bond. The ^{1}H n.m.r. spectra of π-allyls such as [(allyl)PdI]$_2$ are more complex than had previously been thought.[199] Thus the centre hydrogen resonances consist of 15 lines, and the systems are AA′BB′C.

Four-carbon Ligands.—The proposed mechanism for the liberation of cyclobutadiene from iron by oxidation of the complex is electron transfer to give $(h^4\text{-}C_4H_4)Fe(CO)_3{}^{n+}$, followed by CO loss to solvated $(h^4\text{-}C_4H_4)Fe^{n+}$, then $(h^2\text{-}C_4H_4)Fe^{n+}$, and finally C_4H_4 and Fe^{n+}. Further support is gained by the isolation this year of the first h^2-cyclobutadienoid–iron cation complex (100)[200] prepared by the action of Ph_3C^+ on (101).

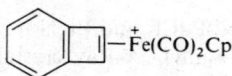

(100)

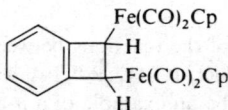

(101)

The syntheses of $(h^4\text{-}C_4H_3CO_2H)Fe(CO)_3$ and $(h^4\text{-}C_4H_3CH_2CO_2H)Fe(CO)_3$ have been reported.[201] pK_a measurements suggest that the $(h^4\text{-}C_4H_3)Fe(CO)_3$ group is electron-withdrawing by induction, but electron-releasing by resonance.

Nucleophilic attack at co-ordinated dienyls is predicted by electron-density considerations to occur at position 1 or 3, but some evidence suggests that attack may preferentially take place at position 2 to generate a h^4-1,3,4,5 bonding mode. An example has now come to light of attack at position 3, followed by isomerism

[198] M. Green and G. J. Parker, *J.C.S. Dalton*, 1974, 333.
[199] B. E. Mann, R. Pietropaolo, and B. L. Shaw, *J.C.S. Dalton*, 1973, 2390.
[200] A. Sanders, C. V. Magatti, and W. P. Giering, *J. Amer. Chem. Soc.*, 1974, **96**, 1610.
[201] D. Stierle, E. R. Biehl, and P. C. Reeves, *J. Organometallic Chem.*, 1974, **72**, 221.

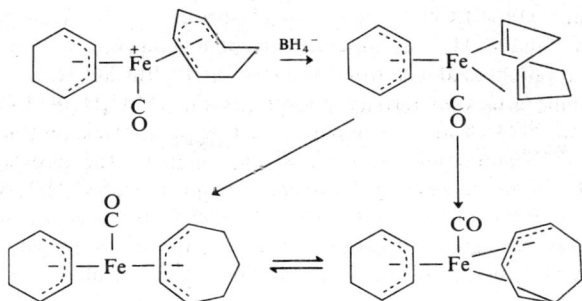

Scheme 15

to 1 and 2 substituted positions (Scheme 15).[202] Obviously great care is necessary in the interpretation of these reactions from product isolation alone.

A convenient synthesis of barbaralone (102), a compound previously made by quite laborious processes, has been described. Treatment of the readily available cyclo-octatetraene complex $(C_8H_8)Fe(CO)_3$ with $AlCl_3$ produces (103), a complex

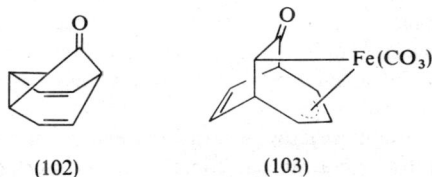

(102)   (103)

previously isolated from the action of iron carbonyls on barbaralone itself. Carbon monoxide (393 K, 100 atm) converts (103) into (102) in 90% yield.[203,204]

The reaction between atomic molybdenum or tungsten and butadiene produces monomeric air-stable complexes $M(C_4H_6)_3$.[205] These crystalline white (W) or yellow (Mo) complexes are unusual in that carbonyl ligands are usually required to give stable complexes of these elements. Co-condensation of butadiene and metal atoms of Ti, V, Cr, Mn, Fe, Co, or Ni, however, is reported to lead to oligomerization of the butadiene, though cocatalysts such as $[AlEt_2Cl]_2$ are necessary in some cases.[206] Both linear polymers and cyclic derivatives such as cod and cyclododecatriene (cdt) are formed, some metals favouring cyclization and others linear oligomers. $(Ph_3P)_3CoCl$ catalyses the dimerization of butadiene to cod and 4-vinyl-cyclohex-1-ene at 333 K. The residue contains butadiene

[202] B. F. G. Johnson, J. Lewis, T. W. Matheson, I. E. Ryder, and M. V. Twigg, *J.C.S. Chem. Comm.*, 1974, 269.
[203] V. Heil, B. F. G. Johnson, J. Lewis, and D. J. Thompson, *J.C.S. Chem. Comm.*, 1974, 270.
[204] A. H.-J. Wang, I. C. Paul, and R. Aumann, *J. Organometallic Chem.*, 1974, **69**, 301.
[205] P. S. Skell, E. M. Van Dam, and M. P. Silvon, *J. Amer. Chem. Soc.*, 1974, **96**, 626.
[206] V. M. Akhmedov, M. T. Anthony, M. L. H. Green, and D. Young, *J.C.S. Chem. Comm.*, 1974, 777.

polymers and $(Ph_3P)_2CoCl_2$, but no Co^I species.[207] No reaction between $(Ph_3P)_3CoCl$ and C_4H_6 was observed at room temperature over one week, though the rhodium analogue reacts to form $(Ph_3P)_2RhCl(C_4H_6)$.

Nucleophilic attack of tertiary phosphines on $[(h^4\text{-}C_4H_4)Fe(CO)_2NO]PF_6$ in acetone at 298 K leads to complexes (104) by *exo*-attack on the ring. U.v. photolysis or bromination reverse this and eliminate the phosphine. With triphenylphosphine, however, subsequent reaction to $[(h^4\text{-}C_4H_4)Fe(CO)(NO)PPh_3]PF_6$ is observed.[208] Electrophilic Friedel–Crafts acetylation of (*trans-trans*-hexa-2,4-diene)tricarbonyliron (105) is predicted to occur *via endo* attack. The molecular structure of the product (106) proves unambiguously that this is so.[209]

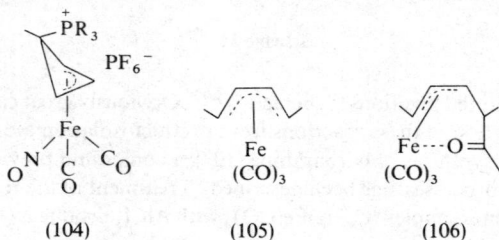

(104)   (105)   (106)

Five-carbon Ligands.—Although various metallocene cations are known, the first anion was reported only this year. Electrochemical reduction of cobaltocene leads to Cp_2Co^-. The solution is stable in monoglyme, though very air-sensitive.[210]

Diazoalkanes usually react with transition-metal complexes to form a diazoalkane complex, to form a carbene complex, or to insert a carbene species into M–halogen, M–H, or M–C. The cyclic diazo-compounds (107) (R = H or Ph) react with $[L_2RhX]_2$ $[L_2 = cod, (CO)_2, or (C_2H_4)_2; X = Cl or Br]$ to form halogen-substituted cyclopentadienyls (108).[211] These novel complexes are air-stable.

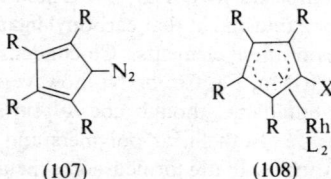

(107)   (108)

Cyclopentadienyls of the lanthanide elements continue to attract interest, and in general the picture of mainly ionic bonding between the C_5 rings and

[207] M. A. Cairns and J. F. Nixon, *J. Organometallic Chem.*, 1974, **64**, C19.
[208] A. Efraty, J. Potenza, S. S. Sandhu, jun., R. Johnson, M. Mastropaolo, R. Bystrek, D. Z. Denney, and R. H. Herber, *J. Organometallic Chem.*, 1974, **70**, C24.
[209] E. O. Greaves, G. R. Knox, P. L. Pauson, S. Toma, G. A. Sim, and D. I. Woodhouse, *J.C.S. Chem. Comm.*, 1974, 257.
[210] W. E. Geiger, *J. Amer. Chem. Soc.*, 1974, **96**, 2632.
[211] V. W. Day, B. R. Stults, K. J. Reimer, and A. Shaver, *J. Amer. Chem. Soc.*, 1974, **96**, 1227.

metal atoms is confirmed. The molecular structure of tetrameric $(C_5H_4Me)_3Nd$ is in accord with this view, though one of the (h^5-bonded) rings is also h^1-bonded to a different metal atom.[212] Mainly ionic bonding between metal and both cyclic ligands is suggested by the i.r. spectra of $(h^8$-$C_8H_8)Ln(h^5$-$C_5H_5)$ (Ln = Y, Nd, Sm, Ho, or Er).[213] Complexes $LHoCp(C_8H_8)$ were also isolated (L = py, NH_3, THF, or $C_6H_{11}NC$).

The debate over the existence and structure of titanocene derivatives continues (see also ref. 132). Pentamethyltitanocene, $(C_5Me_5)_2Ti$, has been prepared from the dihydride. In solution it is in equilibrium with $(h^5$-$C_5Me_5)(h^6$-C_5Me_4-$CH_2)TiH$, and in the presence of D_2, complete H–D exchange of all the protons readily takes place.[214] Various dinitrogen adducts of $(C_5Me_5)_2Ti$ have been identified. Meanwhile unambiguous spectroscopic proof has been presented that the dimeric green 'titanocene' obtained from the reduction of Cp_2TiCl_2 is a bridging fulvene complex (109).[215]

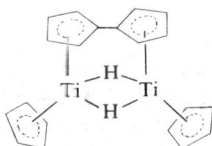

(109)

Ruthenocene reacts with bromine or iodine to form $[Cp_2RuX]^+X_3^-$. The crystal structure of $[Cp_2RuI]I$, shows that the cyclopentadienyl rings of the cation are still eclipsed (as in ruthenocene itself) despite being tilted back 16° to accommodate the iodine atom.[216] ^1H N.m.r. spectra of $Cp_2Cr_2(CO)_6$ show it to exist as a solvent- and temperature-dependent mixture of *anti* and *gauche* rotamers, with $\Delta G_{298}^{\neq}$ near 50 kJ mol^{-1}. Above 263 K, reversible line-broadening occurs, and this is presumed to be due to population of a paramagnetic state.[217] Cr—Cr is 328.1 pm, even longer than in the molybdenum analogue, and there is evidence for internal steric strain.

For some years there has been controversy over the location of the non-bonding electrons in d^1 or d^2 complexes Cp_2ML_2. An orbital lying between L_2 was originally proposed and has found support, but a non-bonding orbital perpendicular to the ML_2 plane has also been suggested. E.p.r. measurements on single crystals of Cp_2TiS_5 doped by Cp_2VS_5 have now yielded the first quantitative results on the shape and size of the non-bonding orbital, and these, together with non-parameterized Fenske–Hall MO calculations, support the original hypothesis of the orbital lying between L_2.[218] The temperature

[212] J. H. Burns, W. H. Baldwin, and F. H. Fink, *Inorg. Chem.*, 1974, **13**, 1916.
[213] J. Jamerson, A. P. Masino, and J. Takats, *J. Organometallic Chem.*, 1974, **65**, C33.
[214] J. E. Bercaw, *J. Amer. Chem. Soc.*, 1974, **96**, 5087.
[215] A. Davison and S. S. Wreford, *J. Amer. Chem. Soc.*, 1974, **96**, 3017.
[216] Y. S. Sohn, A. W. Schlueter, D. N. Hendrickson, and H. B. Gray, *Inorg. Chem.*, 1974, **13**, 301.
[217] R. D. Adams, D. E. Collins, and F. A. Cotton, *J. Amer. Chem. Soc.*, 1974, **96**, 749.
[218] J. L. Petersen and L. F. Dahl, *J. Amer. Chem. Soc.*, 1974, **96**, 2248.

dependence of the magnetic susceptibility of $Cp_2Ti(\mu\text{-}X)_2TiCp_2$ shows that the d^1–d^1 interactions of these titanium(III) complexes increase in the order $F < Cl \sim I < Br$.[219] The anomalous position of I is not understood.

The clathrate host thiourea has been observed to accommodate ferrocene molecules into its lattice.[220] Mixtures of ferrocene and nickelocene lead to a mixed-guest clathrate (Ni/Fe *ca.* 0.4), but neither nickelocene itself nor ruthenocene could be induced to enter. Ferrocene distorts the lattice, and since nickelocene molecules are larger, this might explain the difference.

Cp_2Co reacts with tertiary phosphites to form a paramagnetic linear trinuclear complex (110).[221] Two cobalt(I) atoms and a cobalt(III) atom are present.

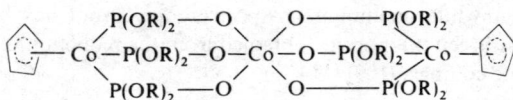

(110)

Methyl iodide adds oxidatively to $CpRh(PPh_3)_2$ to form $[CpRh(PPh_3)_2Me]I$. This contrasts with the behaviour of Pr^iI, which reacts to give a ring-substituted product $(C_5H_4Pr^i)Rh(PPh_3)I_2$. Carbon disulphide leads to a number of products including $CpRh(PPh_3)CS_2$, $CpRh(PPh_3)(CS_3)$, and $CpRh(PPh_3)S_5$.[222]

Nickelocene protonates in HF to $[CpNiC_5H_6]^+$, and BF_3 then affords $CpNi^+BF_4^-$, a 1 : 1 electrolyte in $MeNO_2$.[223] This cationic cyclopentadienyl-nickel species can add to nickelocene to produce 'triple-decker' sandwich $Cp_3Ni_2^+$. The X-ray crystal structure of $[Cp_3Ni_2]BF_4$ shows shorter Cp—Ni terminal separations than in the Cp—Ni bridge.[224] This is in accord with the compound's behaviour towards Lewis bases.

The diolefin complex $[CpNi(nbd)]BF_4$ is attacked by methoxide to form $CpNi(C_7H_8OMe)$, and $[CpNi(cod)]BF_4$ forms $CpNiC_8H_{12}OMe$.[225] Loss of methanol from this latter complex affords the cyclopentadienyl-allyl derivative (111). A 1 : 1 adduct is formed between $[CpNi(Bu_3P)_2]N_3$ and thiocyanates RNCS.[226] The thiocyanate CN link adds 1,3 to the azide. The product is an equilibrium between ionic and N-bonded ligand (Scheme 16). The silacyclopentadiene complex (112) has been described.[227] Its mass spectrum shows a strong ion at $M - 15$, and it is speculated that this might represent the silacyclopentadienyl derivative (113).

[219] R. S. P. Coutts, R. L. Martin, and P. C. Wailes, *Austral. J. Chem.*, 1973, **26**, 2101.
[220] R. Clement, R. Claude, and C. Mazieres, *J.C.S. Chem. Comm.*, 1974, 654.
[221] V. Harder, E. Dubler, and H. Werner, *J. Organometallic Chem.*, 1974, **71**, 427.
[222] Y. Wakatsuki and H. Yamazaki, *J. Organometallic Chem.*, 1974, **64**, 393.
[223] T. L. Court and H. Werner, *J. Organometallic Chem.*, 1974, **65**, 245.
[224] E. Dubler, M. Textor, H.-R. Oswald, and A. Salzer, *Angew. Chem. Internat. Edn.*, 1974, **13**, 135.
[225] G. Parker, A. Salzer, and H. Werner, *J. Organometallic Chem.*, 1974, **67**, 131.
[226] F. Sato, M. Etoh, and M. Sato, *J. Organometallic Chem.*, 1974, **70**, 101.
[227] H. Sakurai and J. Hayashi, *J. Organometallic Chem.*, 1973, **63**, C10.

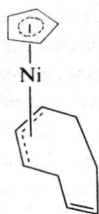

(111)

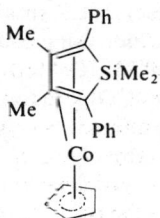

(112)

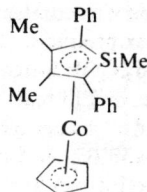

(113)

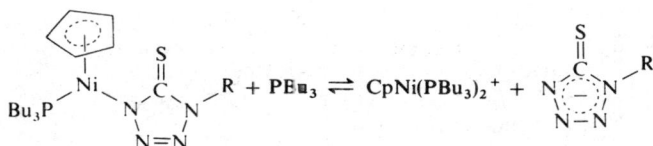

Scheme 16

The investigation of reactions between unsaturated cyclic hydrocarbons and ruthenium carbonyl derivatives continues to be a fruitful field in terms of novel reactions and structures.[228,229] Thus *trans-trans-trans*-cdt reacts by ring contractions with $Ru(CO)_4(GeMe_3)_2$ to form tetrahydropentalene derivatives (114)

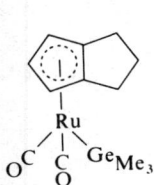

(114)

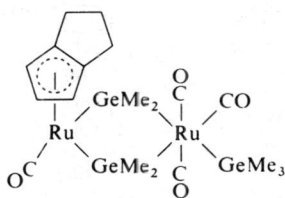

(115)

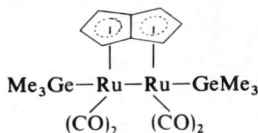

(116)

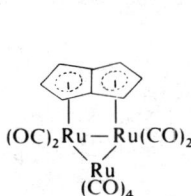

(117)

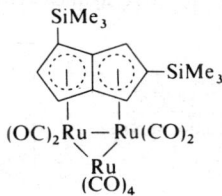

(118)

[228] (a) S. A. R. Knox, R. P. Phillips, and F. G. A. Stone, *J.C.S. Dalton*, 1974, 658; (b) J. A. K. Howard, S. A. R. Knox, V. Riera, F. G. A. Stone, and P. Woodward, *J.C.S. Chem. Comm.*, 1974, 452.

[229] J. A. K. Howard, S. A. R. Knox, F. G. A. Stone, A. C. Szary, and P. Woodward, *J.C.S. Chem. Comm.*, 1974, 788.

and (115), and the pentalene complex (116). A striking number of related reactions afford similar products. These include using cod instead of cdt, and dinuclear $(Me_3Ge)_2Ru_2(CO)_8$ instead of $Ru(CO)_4(GeMe_3)_2$. Cyclo-octatetraenes C_8H_7R (R = H, Me, or Ph) react with $Ru_3(CO)_{12}$ to form pentalene complexes of type (117). Similar trimethylsilyl-substituted derivatives, *e.g.* (118), form from $Ru_3(CO)_{12}$ and bis- and tris-(trimethylsilyl)-cyclo-octatrienes. These trinuclear pentalene derivatives are shown by variable-temperature n.m.r. to be fluxional, and a pendulum type of motion of the $Ru(CO)_4$ group (Scheme 17) is proposed to account for this.

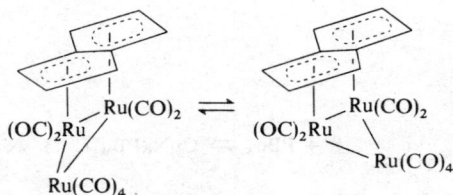

Scheme 17

Ligands with Six or more Carbon Atoms.—Hexa-*hapto* bonded phenyl rings from Ph_3P and BPh_4^- are known for a few metals, and new examples have been reported this year. The action of excess fluoroboric acid in methanol on $RuH(CO_2Me)(PPh_3)_3$ is known to give $[RuH(PPh_3)_3]BF_4$, an h^6-$C_6H_5PPh_2$ complex (119).[230] $[(EtO)_3P]_2CoBPh_4$[231] and $CpRuBPh_4$[232] (120) also have

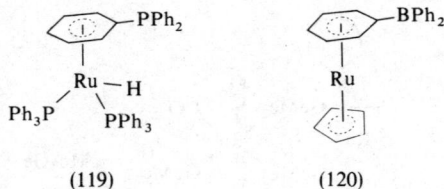

(119)   (120)

π-bonded arenes. Modification of an arene already bonded to the metal has led to a complex where two Ph_3P ligands are π-bonded (Scheme 18) to chromium.[233]

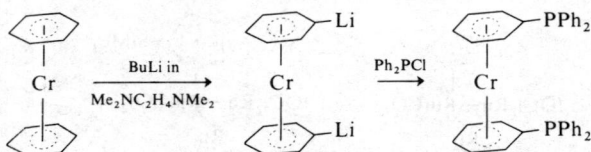

Scheme 18

[230] J. C. McConway, A. C. Skapski, L. Phillips, R. J. Young, and G. Wilkinson, *J.C.S. Chem. Comm.*, 1974, 327.
[231] L. W. Gosser and G. W. Parshall, *Inorg. Chem.*, 1974, **13**, 1947.
[232] G. J. Kruger, A. L. du Preez, and R. J. Haines, *J.C.S. Dalton*, 1974, 1302.
[233] Ch. Elschenbroich and F. Stohler, *J. Organometallic Chem.*, 1974, **67**, C51.

$$\pi\text{-}C_6Me_6$$

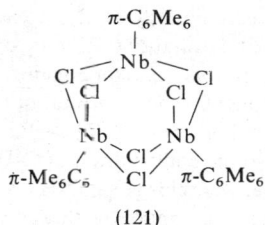

(121)

$[(C_6Me_6)_3Nb_3Cl_6]Cl$ (121) has molecules of *ca.* D_{3h} symmetry with h^6-C_6Me_6 ligands (see p. 255). The C_6Me_6 ligands may be slightly folded, but the crystals did not allow sufficient structure refinement to confirm this.[234]

The arene complex $[(h^6\text{-}C_6Me_6)_2Fe]^{2+}(PF_6)_2^-$ acts as an electron acceptor with a large number of aromatic compounds, including benzene, naphthalene, anthracene, benzo[*b*]thiophen, indole, aniline, pyridine, furan, and ferrocene.[235] The 1 : 1 charge-transfer complexes form as crystals from acetone solutions of the reactants. Redissolving destroys the complexes. The low solubility of the adducts in acetone has enabled separation of related aromatics by precipitating the charge-transfer complex of one of them.

Catalytic polymerization of acetylene derivatives frequently leads only to low molecular weight materials, probably due to the favourable trimerization to aromatic compounds. Arene derivatives $(Ar)M(CO)_3$ have now been found to catalyse alkynes to linear polymers (M = Cr, Mo, or W). The highest molecular weights for phenylacetylene polymerization (*ca.* 12 000) were obtained when toluene was used as the arene. A ladder-like intermediate (122) was isolated,

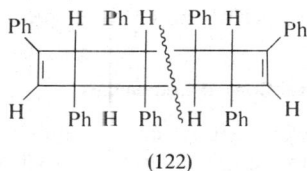

(122)

which converts catalytically into poly(phenylacetylene), $(-CPh=CH-)_n$.[236] $(Arene)Mo(CO)_3$ is also a good Friedel–Crafts catalyst.[237] Its reactions in this capacity are ionic, not free radical, and it is easier to use than $AlCl_3$ due to facile handling, storing, and removal. It has advantages in mildness also.

Metallation of $(h^5\text{-}C_5H_5)M(h^7\text{-}C_7H_7)$ by BuLi in ether becomes less easy along the sequence M = Ti > V > Cr.[238] Furthermore, metallation of the C_5 ring is found for the V and Cr derivatives, whereas the titanium compound is

[234] M. R. Churchill and S. W.-Y. Chang, *J.C.S. Chem. Comm.*, 1974, 248.
[235] D. M. Braitsch, *J.C.S. Chem. Comm.*, 1974, 460.
[236] M. F. Farona, P. A. Lofgren, and P. S. Woon, *J.C.S. Chem. Comm.*, 1974, 246.
[237] J. F. White and M. F. Farona, *J. Organometallic Chem.*, 1973, **63**, 329.
[238] (*a*) C. J. Groenenboom, H. J. de Liefde Meijer, and F. Jellinek, *J. Organometallic Chem.*, 1974, **69**, 235; (*b*) C. J. Groenenboom, H. J. de Liefde Meijer, and F. Jellinek, *Rec. Trav. chim.*, 1974, **93**, 6; (*c*) C. J. Groenenboom and F. Jellinek, *J. Organometallic Chem.*, 1974, **80**, 229.

metallated at either, but predominantly at C_7. Reasons for these changes are found from quantitative MO calculations. The e_2 orbital of $C_5H_5^-$ is of high energy and is unimportant in bonding, consequently the charge on this ring is not very dependent on the metal. The e_2 orbital of $C_7H_7^-$, on the other hand, is close in energy to the metal $3d$ orbitals. These $3d$ orbitals increase in energy in the order Cr < V < Ti, so the contribution of C_7H_7 to the δMO increases in the same order. Thus the negative charge on C_7H_7 is greater for Ti. The shorter (219 pm) $Ti-C_{(7)}$ bond lengths compared with $Ti-C_{(5)}$ (232 pm) bear this out, and ^{13}C n.m.r. spectra agree with this reversal of the more negative ligand of $CpM(C_7H_7)$ in moving from Ti to Cr. ^1H and ^{13}C n.m.r., and ^{55}Mn n.q.r. analyses of $(h^6\text{-}Me_nC_6H_{6-n})Mn(CO)_3^+$ find greater ring-to-metal charge migration in these compounds than in the analogous Cr^0 derivatives. Also Cp is a better electron donor than C_6H_6.[239]

^1H and ^{13}C variable-temperature n.m.r. spectra of $(C_7H_7)(C_7H_9)Ru_3(CO)_6$, a molecule whose crystal structure was reported last year, show three types of fluxional motion. The C_7H_7 ring is non-rigid even down to 173 K (the lowest temperature reported).[240] The $h^5\text{-}C_7H_9$ ring stops motion below room temperature, however. Fast exchange of carbonyl groups was also found at higher temperatures. The fluxional behaviour of $(h^6\text{-}C_8H_8)M(CO)_3$ (M = Cr, Mo, or W) is believed to proceed through symmetrical intermediates $(h^8\text{-}C_8H_8)M(CO)_3$ with random redistribution to hexa-*hapto* complexes.[241] ^{13}C N.m.r. measurements rule out 1–2, 1–4, and 1–5 shifts, and 1–3 is excluded logically.

A second organometallic protactinium derivative has been reported.[242] Prepared by the action of $K_2C_8H_8$ on $PaCl_4$ in THF, $Pa(h^8\text{-}C_8H_8)_2$ was isolated as a gold-yellow sublimate on a micromolar scale. X-Ray powder patterns show it to be isostructural with the Th^{IV} and U^{IV} analogues.

5 Metal–Metal Bonded Compounds

Attempts to prepare silylene, germylene, or stannylene derivatives of various transition metals by chloride ion removal from $Cl_3M^1-M^2(CO)_xCp_y$ failed. Treatment by $AgBF_4$, $AgPF_6$, or $AgSbF_6$ liberated BF_3, PF_5, or SbF_5, respectively, and replaced Cl^- by F^- on the complexes.[243] Complexes best described as stannylene derivatives have been isolated, however. Bulky $(Me_3Si)_2CH$ groups (R) led to R_2Sn, which acts as a ligand to many metals.[244] The crystal structure of $(R_2Sn)Cr(CO)_5$ shows planar co-ordination at tin. R_2Sn is also able to insert into M–H, M–alkyl, M–Cl, or M–M bonds. Examples are given in Scheme 19. The crystal structure of $cis\text{-}Cp_2Fe_2(CO)_2(\mu\text{-}CO)(\mu\text{-}GeMe_2)$ reveals both bridges

[239] T. B. Brill and A. J. Kotlar, *Inorg. Chem.*, 1974, **13**, 470.

[240] T. H. Whitesides and R. A. Budnik, *J.C.S. Chem. Comm.*, 1974, 302.

[241] F. A. Cotton, D. L. Hunter, and P. Lahuerta, *J. Amer. Chem. Soc.*, 1974, **96**, 4723.

[242] D. F. Starks, T. C. Parsons, A. Streitwieser, jun., and N. Edelstein, *Inorg. Chem.*, 1974, **13**, 1307.

[243] T. J. Marks and A. M. Seyam, *Inorg. Chem.*, 1974, **13**, 1624.

[244] J. D. Cotton, P. J. Davison, D. E. Goldberg, M. F. Lappert, and K. M. Thomas, *J.C.S. Chem. Comm.*, 1974, 893.

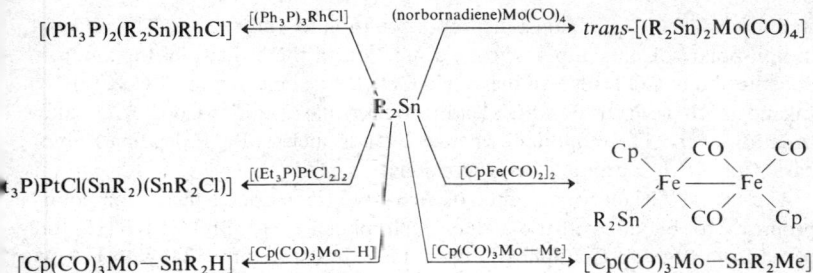

$[(Ph_3P)_2(R_2Sn)RhCl]$ $\xleftarrow{[(Ph_3P)_3RhCl]}$ (norbornadiene)Mo(CO)$_4$ $\longrightarrow$ *trans*-$[(R_2Sn)_2Mo(CO)_4]$

R_2Sn

$_3$P)PtCl(SnR$_2$)(SnR$_2$Cl)] $\xleftarrow{[(Et_3P)PtCl_2]_2}$ $[CpFe(CO)_2]_2$

Cp, CO CO
Fe —— Fe
R$_2$Sn CO Cp

$[Cp(CO)_3Mo-SnR_2H]$ $\xleftarrow{[Cp(CO)_3Mo-H]}$ $[Cp(CO)_3Mo-Me]$ $\longrightarrow$ $[Cp(CO)_3Mo-SnR_2Me]$

Scheme 19

to be symmetrical.[245] In solution, an 8 : 1 mixture of *cis* and *trans* isomers is present, and these interconvert rapidly on the n.m.r. time-scale at 363—433 K. This seems to occur *via* opening of both bridges, rotation, and re-establishment of the bridges.

The iron–antimony bonds of a number of cations $[X_nSb\{Fe(CO)_2Cp\}_{4-n}]^+$ have been examined by ^{57}Fe and ^{121}Sb Mössbauer spectra. They are isoelectronic with neutral tin analogues. The ^{121}Sb isomer shift is between that of Sb^{III} and Sb^{V}, so oxidation-level assignment is not very meaningful. Fe—Sb π-bonding appears more important than Sn—Fe π-bonding.[246] Contrary to general belief, evidence which suggests that Fe—Si and Fe—C bond strengths are quite similar and that Fe—C *may* be stronger has been presented. A number of complexes $(ClCH_2)Me_{2-n}Cl_nSi-Fe(CO)_2Cp$ ($n = 0$, 1, or 2) have been isolated and characterized.[247] At 373 K (in the presence of AlCl$_3$ when $n = 2$) a rearrangement to $(Me_{2-n})Cl_{n+1}SiCH_2-Fe(CO)_2Cp$ takes place. The equilibrium constant is > 200.

The crystal structures of gallium(III) and indium(III) complexes $Mn_2(CO)_8\{\mu$-$M-Mn(CO)_5\}_2$ have been described (123).[248] Several preparative routes to

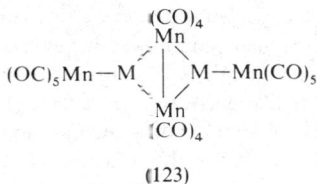

(CO)$_4$
Mn
$(OC)_5Mn-M$ $M-Mn(CO)_5$
Mn
(CO)$_4$

(123)

yellow, crystalline, air-sensitive TlCo(CO)$_4$ have been described[249] as well as to other Tl^I metal carbonyl derivatives. The crystal structure of TlCo(CO)$_4$ shows it to be essentially ionic Tl$^+$Co(CO)$_4^-$ in the solid, though ion-pairs with some covalency probably exist in solution in low-dielectric solvents. The

[245] R. D. Adams, M. D. Brice, and F. A. Cotton, *Inorg. Chem.*, 1974, **13**, 1080.
[246] W. R. Cullen, D. J. Patmore, J. R. Sams, and J. C. Scott, *Inorg. Chem.*, 1974, **13**, 649.
[247] C. Windus, S. Sujishi, and W. P. Giering, *J. Amer. Chem. Soc.*, 1974, **96**, 1951.
[248] H. Prent and H.-J. Haupt, *Chem. Ber.*, 1974, **107**, 2860.
[249] (a) D. P. Schussler, W. R. Robinson, and W. F. Edgell, *Inorg. Chem.*, 1974, **13**, 153; (b) J. M. Burlitch and T. W. Theyson, *J.C.S. Dalton*, 1974, 828.

stability of TlCo(CO)$_4$ against disproportionation to Tl metal and Tl[Co(CO)$_4$]$_3$ in non-polar solvents appears to be a function of the basicity of the anion.[250] Substitution of CO leads to either TlCo(CO)$_3$L or Tl metal and Tl[Co(CO)$_3$L]$_3$. Ligands such as P(OPh)$_3$ which lead to weakly basic anions Co(CO)$_3$L$^-$ allow isolation of the Tl^I compound, whereas ligands such as Ph$_3$P, leading to more basic Co(CO)$_3$L$^-$, promote disproportionation.

A class of Grignard reagents of first-row transition-element compounds promises to be of synthetic value. Cp(diphos)Fe—MgBr(THF)$_2$,THF, prepared from Mg and Cp(diphos)FeBr in THF, has structure (124) with Fe—Mg 259 pm.[251] [Cp$_2$MoHMgCx(μ-Br)$_2$Mg(OEt$_2$)]$_2$ is prepared from Cp$_2$MoH$_2$ and

(124)   (125)

(126)

(127)

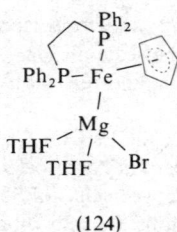

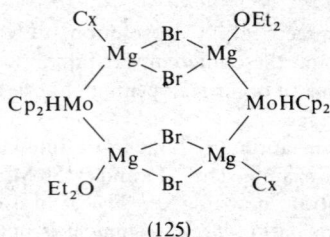

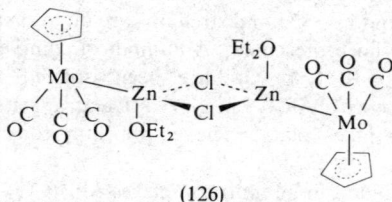

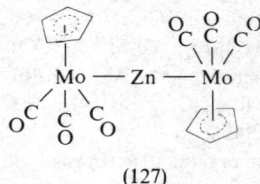

CxMgBr in ether and has a cyclic structure (125).[252] The molecular structures of two molybdenum–zinc bonded complexes have been described (126) and (127).[253] The Mo—Zn bond length is 263 pm in the chlorine-bridged dimer, and 254 pm in the linear Mo$_2$Zn derivative. Structure (128) reveals a distorted trigonal-bipyramidal geometry about cadmium, and bent (136°) Mn—Cd—Mn bonds;[254] Mn—Cd is 271 pm. When the terdentate ether is replaced by bipy or phen, a distorted tetrahedral arrangement results at cadmium.

The crystal structure of [Cp$_2$Fe$_2$Rh(CO)$_4$(PPh$_2$)$_2$]PF$_6$ contains an open-chain cation (129) instead of the closed triangular arrangement that would be

[250] S. E. Pedersen, W. R. Robinson, and D. P. Schussler, *J.C.S. Chem. Comm.*, 1974, 805.
[251] H. Felkin, P. J. Knowles, B. Meunier, A. Mitschler, L. Ricard, and R. Weiss, *J.C.S. Chem. Comm.*, 1974, 44.
[252] M. L. H. Green, G. A. Moser, I. Packer, F. Petit, R. A. Forder, and K. Prout, *J.C.S. Chem. Comm.*, 1974, 839.
[253] J. St. Denis, W. Butler, M. D. Glick, and J. P. Oliver, *J. Amer. Chem. Soc.*, 1974, **96** 5427.
[254] W. Clegg and P. J. Wheatley, *J.C.S. Dalton*, 1974, 424, 511.

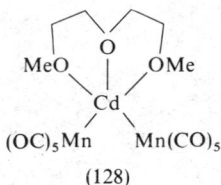

(128)

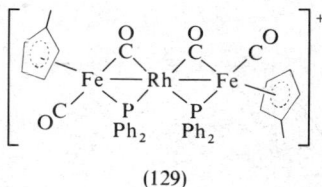

(129)

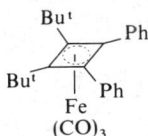

(130)

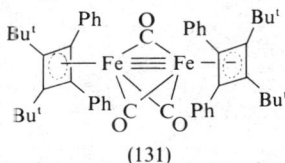

(131)

predicted from the 18-electron rule.[255] The high basicity of the bridge diphenyl-phosphido-groups may account for this, though 16-electron configurations are common at rhodium. Irradiation of (130) produces dinuclear (131), characterized crystallographically.[256] The 18-electron rule is served if an Fe≡Fe triple bond is postulated, and the short bond length [217.7(3) pm] is compatible with this Normal Fe—Fe single bonds are *ca*. 252 pm, and an Fe=Fe double bond has been determined at 221 pm (see also (26)[50]].

Arylcopper halide clusters $Cu_6Ar_4X_2$, prepared from ArCu and CuX, have been converted into mixed organocopper clusters $Cu_6Ar_4R_2$ by treatment with RLi (R is ethynyl).[257] I.r. spectra indicate σ-bonded R with no π-bonding, and a partial structure determination reveals an octahedron of copper atoms with bridging organic groups, as depicted by (132). Thermolysis of these green or

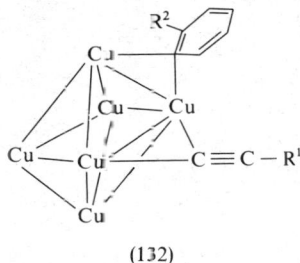

(132)

brown compounds at 373 K produces Ar—R, but no R—R, Ar—Ar, R—H, or Ar—H. This selective reductive elimination can be understood from the geometrical arrangement of the ligands.

The reaction between copper acetylide $[CuC_2Ph]_n$ and *trans*-$(Ph_3P)_2Ir(CO)Cl$ also produces a cluster compound $Cu_4Ir_2(PPh_3)_2(C_2Ph)_8$ with iridium atoms

[255] R. Mason and J. A. Zubieta, *J. Organometallic Chem.*, 1974, **66**, 279.
[256] S.-I. Murahashi, T. Mizoguchi, T. Hosokawa, I. Moritani, Y. Kai, M. Kohara, N. Yasuoka, and N. Kasai, *J.C.S. Chem. Comm.*, 1974, 563.
[257] G. van Koten and J. G. Noltes, *J.C.S. Chem. Comm.*, 1974, 575.

PPh₃ ... Ir —— C ≡ C — Ph / Cu—Cu / Cu—Cu / Ir / PPh₃

(133)

F₅C₆—C≡C ... PPh₃ ... C≡C—C₆F₅ / Rh / C≡C ... C≡C—C₆F / F₅C₆—C Ag—C—Ag C—C₆F / Ph₃P   C   PPh₃ / C₆F₅

(134)

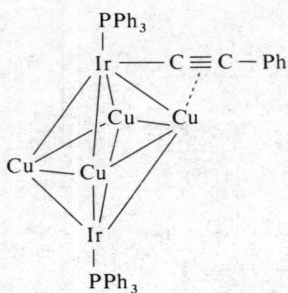

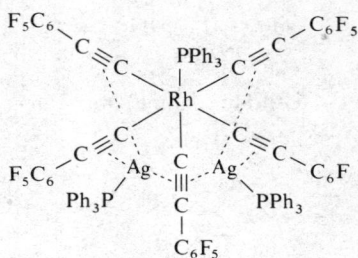

at opposite ends of an octahedron.[258] Each Ir atom is σ-bonded to four phenyl-acetylenes, and these acetylenes are π-bonded to the equatorial copper atoms. The geometry is shown in (133) where, for clarity, only one PhC₂ group is entered. By contrast, $(Ph_3P)_3RhCl$ reacts with $AgC_2C_6F_5$ to yield $RhAg_2(C_2C_6F_5)_5$-$(PPh_3)_3$.[259] The octahedrally co-ordinated rhodium atom is σ-bonded to the acetylenes, which in turn are π-bonded to the silver atoms (134). No metal–metal bonds are present, however, and thus the formulation as a Rh^{III}–Ag^{I} complex is more accurate than Rh^V–Ag^0.

[258] M. R. Churchill and S. A. Bezman, *Inorg. Chem.*, 1974, **13**, 1418.
[259] O. M. A. Salah, M. I. Bruce, M. R. Churchill, and B. G. DeBoer, *J.C.S. Chem. Comm.*, 1974, 688.

Erratum

Volume 70 A, 1973

Page 415: Reference 543 should read "T. Ottersen, G. Warner, and K. Seff, *J.C.S. Chem. Comm.*, 1973, 876" (*i.e.* not *J. Inorg. Nuclear Chem.*, 1973, **35**, 876).
A fuller account of this work has been published subsequently: *idem., Inorg. Chem.*, 1974, **13**, 1904.

Author Index